What's Wrong With My Mouse?

What's Wrong With My Mouse?

Behavioral Phenotyping of Transgenic and Knockout Mice

Second Edition

Jacqueline N. Crawley, Ph.D.

WILEY-INTERSCIENCE

A John Wiley & Sons, Inc., Publication

Published by John Wiley & Sons, Inc., Hoboken, New Jersey
Published simultaneously in Canada.

For general information on our other products and services or for technical support, please contact our Customer Care Department within the United States at (800) 762-2974, outside the United States at (317) 572-3993 or fax (317) 572-4002.

Wiley also publishes its books in a variety of electronic formats. Some content that appears in print may not be available in electronic formats. For more information about Wiley products, visit our web site at www.wiley.com.

Library of Congress Cataloging-in-Publication Data:

Crawley, Jacqueline N.
  What's wrong with my mouse? : behavioral phenotyping of transgenic and knockout mice / Jacqueline N. Crawley. —2nd ed.
       p. ; cm.
  Includes bibliographical references.
  ISBN: 978-0-471-47192-9 (cloth)
   1. Neurogenetics—Animal models. 2. Transgenic mice—Behavior. I. Title.
  [DNLM: 1. Mice, Transgenic. 2. Behavior, Animal. 3. Disease Models, Animal.
 4. Genetics, Behavioral—methods. 5. Mice, Knockout. 6. Phenotype. QY 60.R6 C911w 2006]
  QP356.22.C73 2006
  573.8′480724—dc22   2006033565

Printed in the United States of America

10 9 8 7 6 5 4 3 2 1

To Andy and Barry,
For the privilege of sharing your genes

Contents

Preface

Five years is not a long time in science. Imagine my amazement at the stunning number of new publications describing elegant behavioral phenotyping of transgenic and knockout mice. When the first edition of *What's Wrong With My Mouse?* was completed in 1999, approximately 500 papers had been published on behavioral phenotypes of mice with targeted gene mutations. Since then the literature has quintupled, with at least 2000 new papers between 2000 and 2005. *Science* magazine's *Breakthroughs of the Year* 2005* highlights genes linked to brain disorders, emphasizing that "connecting the dots between genetics and abnormal behavior has been anything but child's play." Functional consequences of dysfunctional genes, allelic variations, susceptibility genes, protective genetic backgrounds, and environmental modulation are increasingly revealed by careful analyses of the behaviors of mutant mice.

Writing the first edition of *What's Wrong With My Mouse? Behavioral Phenotyping of Transgenic and Knockout Mice* was a challenge in summarizing the large number of well-validated behavioral tests for mice. The challenge for the second edition was to read the exponentially expanding new literature on targeted gene mutations and then select just a few good examples to insert into existing chapters for each behavioral domain. Two new chapters are added. Chapter 12, "Neurodevelopment and Neurodegeneration," illustrates the application of behavioral phenotyping methods to some of the best mouse models of human brain diseases. Important genetic models of neurodegenerative diseases, including Alzheimer's, Parkinson's, Huntington's, and amyotrophic lateral sclerosis, employ robust behavioral phenotypes in mutant mice to test hypotheses about the causes of the neurodegenerative process, and to evaluate the efficacy of potential treatments. Mutant mouse models of neurodevelopmental diseases, including mental retardation and autism, provide intriguing new leads about genes regulating the exquisite biological processes controlling brain development. Chapter 13, "Putting It All Together," outlines strategies for comprehensive behavioral phenotyping. Recommendations are offered for designing the sequence of behavioral tests that

Science **310**: 1880–1885, December 23, 2005.

best addresses your hypothesis about the targeted gene, while incorporating the critical controls to rule out artifactual interpretations. Remarkably, some forefront techniques mentioned in the first edition's Chapter 12, "The Next Generation" are now established methodologics, including DNA microarrays and viral vector delivery of genes into the brain. Revised Chapter 14, "The Next Generation," contains discussion of new bioinformatics tools, including interactive websites that compile comprehensive phenotypes of each mutant mouse, database mining to discover single nucleotide polymorphisms across inbred strains of mice, and the role of micro RNAs in gene silencing.

The greatest reward for writing a scientific book is to see it used in advancing biomedical research. Thank you for the many communications I received about your successful uses of *What's Wrong With My Mouse?* in your ongoing experiments. It is beautiful to see the scientific community effectively applying information from the first edition—batteries of accepted behavioral tests to characterize a new mutation, backcrossing into a single genetic background, proper controls for procedural abilities, multiple corroborative tasks within a behavioral domain—thereby enhancing the rigor of experimental findings. Fascinating questions received from readers inspired many of the new inclusions in the second edition. Truly it was those emails out of the blue from unknown students and renowned molecular geneticists, plus the positive reinforcement of seeing *What's Wrong With My Mouse?* in labs and offices that I visited while giving lectures over the past five years, that convinced this behavioral neuroscientist to engage in the drudgery of a book update. Further I am grateful to Luna Han, the Life and Medical Sciences book editor at John Wiley & Sons who talked me into the second edition, Thomas Moore, Wiley's Senior Editor who shepherded the project to completion, Editorial Assistant Ian Collins, Senior Production Editor II Danielle Lacourciere, and Art Director Dean Gonzalez, for their essential guidance and much-needed practical and moral support. I hope this book continues to offer ideas that will aid neuroscientists and geneticists in your exciting future discoveries, to elaborate our understanding of how genes shape behavior, and to employ the behavioral phenotypes of transgenic and knockout mice in developing cures for human genetic diseases.

JACQUELINE N. CRAWLEY, PH.D.
CHEVY CHASE, MARYLAND, USA
March 2006

Preface to the First Edition

Targeted mutation of genes expressed in the nervous system is an exciting new research field that is forging a remarkable amalgam of molecular genetics and behavioral neuroscience. My laboratory in Bethesda has been the fortunate recipient of visits from many molecular geneticists over the past five years, who come to ask, "What's wrong with my mouse? Can you tell us what behaviors are abnormal in our null mutants? And how *do* you measure behavior, anyway?"

We have had some remarkable opportunities to collaborate with outstanding molecular geneticists in the National Institutes of Health Intramural Research Program and throughout the world on investigations of the behavioral effects of mutations in genes expressed in the mouse brain. Each of these collaborations has been a learning experience, increasing our understanding of the optimal experimental design for analyzing behavioral phenotypes of mutant mice. What are the best tests to address each specific hypothesis? Which methods work best for mice? Which rat tasks can be adapted for mice? What are the correct controls? What are the hidden pitfalls, lurking artifacts, false positives, and false negatives? Which statistical tests are most sensitive for detection of the genotype effect? What is the minimum number of animals necessary for each genotype, gender, and age? Our laboratory and many others are gradually working out the best methods for behavioral phenotyping of transgenic and knockout mice.

In the same conversations, molecular geneticists frequently asked me to recommend a book they could consult to learn more about behavioral tests for mice. Apparently the scientific book publishers are receiving similar queries. Ann Boyle and Robert Harington at John Wiley & Sons, convinced of a real need for such a book, sweet-talked me into filling the void. *What's Wrong With My Mouse?* is written for these pioneering molecular geneticists, and for the talented students who will be the next driving force in moving the field forward.

On a personal level, I would like to express deep appreciation to all of my behavioral neuroscientist colleagues around the world for their outstanding work, past, present, and future. Your contributions to the excellence and abundance of mouse behavioral tests provide the foundation for the rapidly expanding scientific discoveries forthcoming from behavioral phenotyping studies of transgenic and knockout mice. This book is a testament to your accomplishments.

JACQUELINE N. CRAWLEY, PH.D.
CHEVY CHASE, MARYLAND
July, 1999

Acknowledgments

The author expresses sincere gratitude to the many colleagues who generously contributed photographs, drawings, and graphics from their research programs:

Carol Barnes, University of Arizona (*Figure 6.20A*)

Caroline and Robert Blanchard, University of Hawaii (*Figure 9.6*)

Igor Branchi and Enrico Alleva, Istituto Superiore di Sanita, Rome (*Figures 8.8A, 12.1, 12.2*)

Richard Brown, Dalhousie University (*Table 3.3*)

Katrina Cuasay, Joanna Hill, and Jacqueline Crawley, National Institute of Mental Health, Bethesda, MD (*Figure 5.1A*).

Robert Drugan, University of New Hampshire (*Opening to Chapter 10*)

Steven Dunnett, Rebecca Carter, and Jenny Morton and Steve Dunnett, Cardiff University and Cambridge University (*Figures 4.4, 4.8, 4.12, 5.3A*)

Gerry Dawson, Merck Neuroscience Research Centre (*Figure 6.32*)

William Falls, University of Vermont (*Figure 10.3*)

Deborah Finn and John Crabbe, Oregon Health Sciences University, Portland, OR (*Figure 11.2*)

Nick Grahame and Chris Cunningham, Oregon Health Sciences University (*Figure 11.3*).

Howard Eichenbaum and Aras Petrulis (*Figure 6.22*)

Andrew Holmes, National Institute on Alcoholism and Alcohol Abuse, Rockville, MD (*Figures 6.31, 10.9A,B,C, 12.7*)

Dina Lipkind, Yoav Benjamini, and Ilan Golani, Tel Aviv University (*Figure 4.2*)

Timothy Moran, Johns Hopkins University (*Figure 7.9*)

Sheryl Moy, University of North Carolina (*Figures 12.4C,D*)

Randy Nelson, Johns Hopkins University (*Figure 9.4*)

Patrick Nolan, Mammalian Genetics Unit, Medical Research Council, Harwell, UK (*Figure 14.7*)

Richard Palmiter, University of Washington (*Openings to Chapters 2 and 7*)

Richard Paylor, Baylor College of Medicine (*Figures 2.7, 6.6, 10.15, Table 3.4*)

Petko Petkov, The Jackson Laboratory, Bar Habor, ME (*Figure 2.6*).

Marina Picciotto and Venetia Zachariou, Yale University and University of Crete (*Figure 11.11*).

Emilie Rissman, University of Virginia (*Figure 8.2A*)

John Robinson, State University of New York at Stony Brook (*Figure 6.29A*)

Selen Tolu, National Institute of Mental Health (*Figures 12.4A,B*).

Harry Shair, New York State Psychiatric Institute (*Figures 8.8B, 12.3*)

Uri Shalev and Ina Weiner, Tel Aviv University (*Figure 10.13*)

George Uhl, National Institute on Drug Abuse (*Figure 11.5*)

Douglas Wahlsten, University of Windsor (*Figure 13.1*)

Jeanne Wehner, Institute of Behavioral Genetics, University of Colorado (*Figures 2.4, 14.5, Table 2.1*)

Scott Wersinger, University of Buffalo (*Figures 9.7*)

Craige Wrenn, Drake University (*Figures 5.1B, 6.24B, 6.24C*)

Jeana Yates, Northwestern University (*Table 14.2*)

Scott Young, National Institute of Mental Health (*Figure 8.6B*)

The author's admiration and appreciation is extended to Jonathan Klein, the outstanding artist who contributed his creative illustrations:

Openings to Chapters 3, 5, 8, 9, and 14

Gratitude goes to John Ward, the impressive photographer whose patience captured the images of mice performing behavioral tasks:

Figures 4.1A, 4.1B, 4.5, 4.6A, 5.2, 6.3A, 6.3B, 6.11B, 6.14, 6.18, 6.26A,B, 7.2, and 10.7A,B, Opening to Chapter 4

Special thanks to Andy Wolfe for his creative design:

Opening to Chapter 11

The author is grateful to the many excellent companies that manufacture behavioral test equipment or products used in generating and testing transgenic and knockout mice, for their generosity in contributing the following photographs and schematic diagrams:

Accuscan (*Figure 7.8*)

Affymetrix, Inc. (*Figure 14.15*)

American Association of Laboratory Animal Science (*Figure 13.2*)

Clontech (*Figure 14.2*)

Columbus Instruments (*Figures 3.3, 4.7, 4.10, 7.8*)

Hamilton-Kinder (*Figure 3.4*)

Lafayette Instrument Company, Inc. (*Figure 4.9*)

Med Associates, Inc. (*Figure 6.26 concept, 10.11*)

Mini-Mitter Company (*Figure 4.11A*)

New Behavior (*Figure 3.5*)

San Diego Instruments (*Figure 5.3A, 6.3E, 6.11A*)

Stoelting Company (*Figures 4.3A, 5.5, 5.6, 5.7*)

Taconic Farms, Inc (*Figure 2.1, 2.3*)

Permission for the use of previously published figures was kindly granted by the following journals:

Academic Press (*Figures 3.2, 8.2B, Table 9.1, Table 9.2*)

American Physiological Society (*Figure 7.5*)

American Psychological Association (*Figures 6.16, 6.17, 6.21, 6.23, 6.29B, 6.32B*)

American Society for Clinical Investigation (*Figure 14.14*)

American Society for Pharmacology and Experimental Therapeutics (*Figure 11.8*)

Annals of the New York Academy of Sciences (*Figure 10.1*)

Blackwell (*Figures 2.8; 12.1*)

Cell Press (*Figures 3.3A,B, 4.11B, 6.1, 6,2, 7.4, 8.3A, 8.4, 9.2, 9.5B, 10.17, Tables 5.2, 10.4*)

Cold Spring Harbor Laboratory Press (*Figure 2.6, Table 13.1*).

Elsevier Science (*Figures 2.10, 5.6B, 6.3D, 6.6, 6.10, 6.22, 6.25, 6.28, 7.9, 8.1B, 8.5, 9.7, 10.10, 10.12, 10.14, 11.6, Tables 5.1, 7.2, 10.1, 10.2, 10.3, 10.4, 11.1, 12.5A,B, 12.6, 12.9, Frontpiece to Chapter 13*)

John Wiley & Sons, Inc. (*Figures 2.2, 2.4, 4.4, 4.8, 6.3C, 6.19, 6.24, 6.27, 6.32A,B, 7.11, 8.7, 9.3A,B, 9.6, 10.3A,B, 10.4, 10.8A, 10.9B, 12.2, 12.3, Tables 2.1, 2.4, 12.1, 12.10*)

Lippincott Williams & Wilkins (*Figures 2.10 by Montkowski Brain Research, 5.4, Table 10.4*)

Nature (*Openings to Chapters 2 and 7, Figures 4.1B, 11.7, 12.8. Table 14.2*)

Nature Genetics (*Figure 2.5, 2.9, 4.6B, 4.4, 6.12A, 11.4, 14.4, 14.5, 14.9, Table 14.1*)

Nature Medicine (*Table 7.2*)

Oxford University Press (*Figures 7.1, 7.6, 7.7, 7.10, 14.8*)

Plenum Publishing Company (*Figures 8.3B, 9.5A, 10.7*).

Proceedings of the National Academy of Sciences USA (*Figures 6.4, 6.5, 6.7, 6.12B, 12.7, 14.11*)

Science (*Figures 6.20B, Opening to Chapter 7, 7.3, 10.2, 11.10, 14.6, 14.12, 14.13, Table 7.1*)

Scientific American (*Figure 14.10*).

Society for Neuroscience (*Figures 4.12, 6.9, 10.16, 10.18*)

Springer-Verlag (*Figures 2.7, 3.1, 5.3B, 6.15, 10.5, 10.6, 10.8B, 10.15, 11.9, Table 2.2, 2.3, 3.1, 6.1*)

Warner Brothers, Inc., representative Paula Allen informally agreed to the use of a Pinky and The Brain drawing and quote (*Opening to Chapter 6*)

Karl Herrup, Case Western Reserve University, used the phrase "What's Wrong With My Mouse?" as the title of a 1996 Short Course sponsored by the Society for Neuroscience. Dr. Herrup and the Society for Neuroscience kindly approved the use of "What's Wrong With My Mouse" as the title for the present book.

David Kupfer, University of Pittsburgh, used the phrase "give your mouse a physical" (Chapter 3), at a workshop on mouse behaviors sponsored by Dr. Norman Anderson, Director of the Office of Behavioral and Social Sciences Research, National Institutes of Health, Bethesda, MD, July 13, 1998.

Personal thanks to Ann Boyle, Robert Harington, Luna Han, and Thomas Moore, at John Wiley & Sons, Inc.

Very special thanks to Barry Wolfe, for his infinite kindness and understanding.

1

Designer Mice

The disease is inherited. Family pedigrees indicate an autosomal dominant gene. Linkage analyses reveal one strongly associated chromosomal locus. Mapping identifies the gene. The cDNA for the gene is sequenced. The anatomical distribution of the gene is primarily in the brain. The symptoms of the disease are neurological and psychiatric. There is no effective treatment. The disease is ultimately lethal.

Your mission, should you choose to accept it, is to develop a treatment for the disease. Replacement gene therapy is the best hope. But you don't know the gene product, you don't know its function, and you don't know if gene delivery would be therapeutic. Where do you start?

These days, you may choose to start with a targeted gene mutation, to generate a mutant mouse model of the hereditary disease. A DNA construct containing a mutated form of the responsible gene is developed. The construct is inserted into the mouse genome. A line of mice with the mutated gene is generated. Characteristics of the mutant mice are identified in comparison to normal controls. Salient characteristics relevant to the human disease are quantitated. These disease-like traits are then used as surrogate markers for evaluating the effectiveness of treatments. Putative therapies are administered to the mutant mice. A treatment that prevents or reverses the disease traits in the mutant mice is taken forward for further testing as a potential therapeutic treatment for the human genetic disease. Gene therapy, based on targeted gene replacement of the missing or incorrect gene in the human hereditary disease, is described in Chapter 14. In the future, medicine may shift emphasis from treating the symptoms to administering replacement genes that efficiently and permanently cure the disease.

Targeted gene mutation in mice has revolutionized biomedical research. *Transgenic* mice have extra copies of a gene, or copies of a new gene, inserted into the mouse genome. Mice with additional copies of a normal gene enable the investigation of the functional outcomes of the overexpression of the gene product. Additions of a

What's Wrong With My Mouse? By Jacqueline N. Crawley
Copyright © 2007 John Wiley & Sons, Inc.

new gene that is not normally present in the mouse genome, such as the aberrant form of a human gene linked to a disease, enable the investigation of the functional outcomes of the expression of the disease gene. For example, the mutated form of the human *huntingtin* gene is added to the mouse genome to generate a mouse model of Huntington's disease. *Knockout* mice have a disrupted gene that is nonfunctional. The *null mutant homozygous* knockout mouse is deficient in both alleles of a gene; the *heterozygote* is deficient in one of its two alleles for the gene. The *genotype* designation $-/-$ represents the null mutant, $+/-$ represents the heterozygote, and $+/+$ represents the *wildtype normal control*. The *phenotype* is the set of observed traits of the mutant line of mice. Phenotypes include pathological, biochemical, anatomical, physiological, and behavioral characteristics.

Targeted mutations of genes expressed in the brain are revealing the mechanisms underlying normal behavior and behavioral abnormalities. Mouse models of human neuropsychiatric diseases, such as Alzheimer's disease, amyotrophic lateral sclerosis, anorexia, anxiety, ataxias, autism, bipolar disorder, depression, drug and alcohol addiction, Huntington's disease, obesity, Parkinson's disease, and schizophrenia, are characterized in part by their behavioral phenotypes.

This book will introduce the novice to the rich literature of behavioral tests in mice. Careful readers will learn how to optimize the application of these tests for behavioral phenotyping of their line of mutant mice. Based on our experiences, our laboratory is working toward a unified approach for the optimal conduct of behavioral phenotyping experiments in mutant mice. Recommendations are offered in Chapter 13 for a three-tiered sequence of behavioral tests, applicable to the range of behavioral domains regulated by genes expressed in the mammalian brain.

SCOPE

This book is designed as an overview of the field of behavioral neuroscience, as it can be applied to the behavioral phenotyping of transgenic and knockout mice. Molecular geneticists may browse through the chapters relevant to their gene, to get ideas for possible tests to try. Behavioral neuroscientists who have no experience with mutant mice may wish to read about the latest genetic technologies, the behavioral tests that have been used to study mice with targeted gene mutations, and some of the successful experiments published in the genetics literature. Behavioral geneticists who are expert in one area of mouse behavior may find the chapters describing other areas of mouse behaviors to be useful for expanding their research repertoire.

Chapters are organized around behavioral domains, including general health, neurological reflexes, developmental milestones, motor functions, sensory abilities, learning and memory, feeding, sexual and parental behaviors, social behaviors, and rodent paradigms relevant to fear, anxiety, depression, schizophrenia, reward, drug addiction, neurodevelopmental disorders such as autism, and aging diseases such as Alzheimer's. Each chapter begins with a brief history of the early work in the field and the present hypotheses about mechanisms underlying the expression of the behavior. A list of general review articles and books is offered for each topic, encouraging the interested reader to gain more in-depth knowledge of the relevant literature. Chapter lengths vary, reflecting the number of established tests available for mice in each behavioral domain, and the current number of publications of mutant mice displaying interesting behavioral phenotypes within each behavioral domain.

Standard tests are then presented in detail. Highlighted are those tasks that have been extensively validated in mice. Demonstrations of genetic components of task performance are described, including experiments comparing inbred strains of mice (strain distributions), naturally occurring mutants (spontaneous mutations), chemical mutagenesis (ENU), quantitative trait loci linkage analysis (QTL), and gene expression patterns (DNA microarrays). Experimental design and specific behavioral tasks are presented as simply as possible. Extensive references are included for each behavioral test, to encourage extensive reading of the primary experimental literature from experts on each topic.

Illustrations are provided for the most frequently used behavioral tasks. Photographs of the equipment or diagrams of the task accompany the text. Samples of representative data are shown. The data presentation is designed to indicate the qualitative and quantitative results that can be expected when the task is properly conducted.

Each chapter includes the results of elegant experiments in which these tasks are successfully applied to characterize transgenic and knockout mice. Examples are limited to the behavioral phenotypes. A more global discussion including physiological, anatomical, and neurochemical phenotypes is beyond the scope of this book.

Classics in the neuroscience literature are listed at the end of this chapter. Background literature for individual behavioral domains is referenced at the end of each chapter. These chapter lists offer good books, special issues of journals, review articles, and primary literature focused on the hypotheses and current body of knowledge for each behavioral domain. A wealth of original publications, describing specific methods and results in detail, is referenced within the text and listed at the end of the book. The referenced literature is designed to provide helpful examples but does not attempt to compile a comprehensive survey. The field is moving too fast. As this second edition goes to press for a 2006 publication date, only articles appearing by late 2005 are included.

The concept behind each chapter is to present simple descriptions of the abundance of behavioral paradigms, developed over the long, illustrious history of behavioral neuroscience. The secondary goal is to provide access to the primary literature and selected reviews. The tertiary goal is to highlight some of the excellent behavioral neuroscience laboratories conducting these types of experiments. Many of these behavioral neuroscientists may be willing to provide intensive training, and/or to engage in collaborative research, to phenotype a new transgenic or knockout mouse.

MESSAGE TO MOLECULAR GENETICISTS

Welcome to the world of behavioral neuroscience! Animal behavior has fascinated human observers throughout history. Systematic studies of animal behavior began over 100 years ago with Darwin's observations on finches of the Galapagos Islands, in which feeding behavior correlated with beak shape and ecological niche (Darwin, 1839). The 1973 Nobel Prize in Medicine was awarded to Konrad Lorenz, Nikolaas Tinbergen, and Karl von Frisch for their elegant ethological studies of naturalistic animal behavior (von Frisch, 1967; Lorenz, 1974; Tinbergen, 1974). Behavioral genetics grew out of observations of species and strain differences. Present-day behavioral neuroscience and behavioral genetics focus on mechanisms underlying observed behaviors, and on the interaction of genetic, anatomical, physiological, biochemical, and environmental factors. Behavioral neuroscientists are found in universities, research institutes, biotechnology and pharmaceutical companies, usually within departments of neuroscience, psychology, biology, zoology, pharmacology, and psychiatry. At the annual meetings of the Society for Neuroscience in the United States, which are attended by more than 30,000 neuroscientists, approximately 15% of the lectures and poster presentations represent behavioral neuroscience research.

Caveat to Molecular Geneticists

This is not a "how-to" manual. Animal behavior is too complex and requires too much training to be attempted for the first time directly from this book alone. As in any field of science, behavioral research has evolved proper experimental designs and controls that must be correctly applied for the data to be interpretable. Little things, such as how to handle the mouse to reduce stress, can greatly affect the results of a behavioral experiment. Like microinjecting an oocyte or operating a DNA sequencer, the tricks of the trade are best learned from the experts. You don't want to waste your time reinventing the wheel. Setting up the tasks described without a behavioral neuroscientist collaborator is not recommended.

Instead, this book is a brief overview of what is available. The descriptions of behavioral tests and experimental design are provided as an initial framework for your thinking and planning. After the orientation provided in these chapters, you will be knowledgeable about the wide range of behavioral tests available for investigating your mice.

The first step is to define your hypotheses about the gene of interest. The second step is to choose the relevant tests. The third step is to develop a collaboration with a reputable behavioral neuroscience laboratory. With a few phone calls, you are likely to find a good behavioral neuroscientist in your geographic area who is willing to work with you on behavioral testing of your knockouts. The names of some of the best laboratories appear within the referenced primary publications in each chapter. Many universities have mouse behavioral phenotyping core facilities available to their investigators. Or you may prefer to contract with a company that conducts rodent behavioral tests on a fee-for-service basis. Contact information for companies is listed in Chapter 2. In addition a set of good review articles and books is provided for each behavioral domain in each chapter. These can help you to identify optimal potential collaborators to contact.

The collaboration is often arranged such that the behavioral neuroscience laboratory or core facility conducts all of the behavioral phenotyping experiments independently. Intellectual input from the molecular genetics group generates the hypotheses to address, suggests the behavioral tests to conduct, and contributes to the interpretation of the results. In this model, a scientific collaboration is set up along the lines of a program project or a center grant, with equitable agreements to share funding and authorships. Alternatively, your molecular genetics group may wish to learn the behavioral techniques from a good behavioral neuroscience laboratory, and then set up the experiments in your own laboratory. In this model the senior molecular geneticist visits the behavioral neuroscience laboratory and observes the techniques directly. Postdoctoral fellows and graduate students in the molecular genetics laboratory then spend several weeks or months in the laboratory of the behavioral neuroscience collaborators, learning the intricacies of conducting the experiments. Courses in mouse behavioral research that may provide additional useful training are listed at the end of this chapter.

This approach ensures that the behavioral concepts and techniques employed in the molecular genetics laboratory are consistent with the highest standards of behavioral neuroscience research, and that the methods used will be acceptable for publication in the best journals. This book will provide you with sufficient background for informed discussion of behavioral tests in the course of your first conversations with potential behavioral neuroscience collaborators.

MESSAGE TO BEHAVIORAL NEUROSCIENTISTS

Behavioral neuroscience has a new tool. Transgenic and knockout mice with mutations in genes linked to human diseases provide excellent models of the behavioral traits characterizing many human genetic disorders. Behavioral abnormalities in these mouse models offer quantitative surrogate markers for the symptoms of the human disease. Reversal or prevention of the behavioral abnormalities in the mouse model can be used as a powerful preclinical endpoint, to assess the efficacy of new treatments for a genetic disease.

Targeted gene mutation offers an alternative approach to investigating endogenous mechanisms underlying behavior. Gene mutations may provide information that complements findings based on electrolytic or neurochemical lesions. For example, null mutation of a gene critical for the functioning of one hippocampal layer may be a superior alternative to an electrolytic hippocampal lesion. Deletion of the gene for a neurotransmitter receptor is analogous to administering a receptor antagonist. This technique is especially useful in understanding the behavioral relevance of a receptor subtype for which no pharmacological antagonists are available. Disruption of a developmental gene can test hypotheses about the role of a neurotrophic factor in neuronal migration or synapse formation. Addition of extra copies of a gene for a hypothalamic neuropeptide can address the outcome of excessive quantities of that neuropeptide on feeding behaviors and the etiology of obesity. Single gene mutations can be used to explore the interactions of genetic and environmental factors, such as social and parental behaviors in mutants missing receptors for olfactory signaling cues. As the technology advances, gene mutations are targeting specific brain structures at specific periods of development or during discrete experimental time points in the adult mouse.

This book is designed to provide you with an introduction to the transgenic and knockout technology, and approaches to optimize its application to behavioral neuroscience research. Chapter 2 presents descriptions of the molecular genetics of the gene targeting vector, generation of the chimera, and breeding strategies. Chapter 14 discusses quantitative trait loci analysis, DNA microarray analysis, chemical- and radiation-induced mutagenesis, viral vector gene delivery, RNA silencing, conditional mutations expressed only in one tissue type, and inducible mutations expressed at controlled time periods. Illustrations and diagrams were chosen to clarify these complex molecular technologies. References to primary publications, review articles, and books are offered at the end of the chapter for each topic, chosen for the interested reader wishing to learn more about molecular genetics. A list of Web sites for further information on the mouse and human genome projects, databases of mouse phenotypes, sources of mutant mice, and overviews of mouse behavioral genetics are provided at the end of Chapter 2.

This book is further designed to give concrete examples of experimental design, methods, and optimization of specific tasks for behavioral phenotyping of mutant mice. Tasks originally developed for mice, and rat tasks that have been successfully adapted for mice, are highlighted. Strategies for converting rat tasks to mouse tasks are included. Cases in which rat tasks apparently cannot be performed by mice are stated. Examples are given of proper and improper behavioral methods that have been used to characterize transgenic and knockout mice.

For behavioral neuroscientists working on one type of behavior, who wish to learn more about tasks for another type of behavior, extensive references are provided within each chapter for each behavioral domain. These publications can be used to obtain specific details of experimental methods. Your excellent behavioral neuroscience laboratories may be highlighted, as suggested contacts for geneticists to consult for more information or to obtain additional training in a specific behavioral task.

Caveat to Behavioral Neuroscientists

This is a guidebook, not a comprehensive or scholarly review of behavioral neuroscience methods. Please understand that this is not a textbook, nor a review, nor an in-depth analysis of theory. Rather, the goal of this book is a brief introduction to behavioral neuroscience for the novice. Chapters are designed as overviews of the behavioral domains that are most likely to be useful in behavioral phenotyping of mutant mice. The behavioral tests chosen for presentation are primarily those that work well with mice, and have been used with some success in some cases to characterize mutant mice. Many interesting and important behavioral paradigms are not discussed because they have not yet been fully validated for mice, or replicated across several laboratories, or have limited applicability to experiments with mutant lines of mice. Chapters are of unequal lengths, reflecting the abundance of good tests for mice in some behavioral domains and the dearth of well-characterized tests for mice in other behavioral domains.

Of course, this book cannot replace a thorough reading of the many excellent books and review articles representing the breadth and depth of behavioral research. Instead, recommended textbooks and review articles are listed in each chapter. Small samplings of the multitude of original data papers are referenced in the text.

Finally, the descriptions of the methods are intentionally superficial. Behavior is not a cookbook discipline. In the opinion of this author, molecular geneticists are best advised to seek training or collaboration with a good behavioral neuroscience laboratory, such as yours, rather than set up the behavioral tasks independently.

BACKGROUND LITERATURE

Alcock J (1989). *Animal Behavior*. Sinauer Associates, Sunderland, MA.

Buccafusco JJ (2000). *Methods of Behavioral Analysis in Neuroscience* CRC Press, Boca Raton, FL.

Baker C (2004). *Behavioral Genetics*. American Association for the Advancement of Science and The Hastings Center, Washington, DC.

Becker JB, Breedlove SM, Crews D (1993). *Behavioral Endocrinology*. MIT Press, Cambridge.

Bloom FE, Young WG (1995). *Brain Browser*. Academic, San Diego.

Bolivar V, Cook M, Flaherty L (2000b). List of transgenic and knockout mice: Behavioral profiles. *Mammalian Genome* **11**: 260–274.

Campbell IL, Gold LH (1996). Transgenic modeling of neuropsychiatry disorders. *Molecular Psychiatry* **1**: 105–120.

Crawley JN, Ed. (1997 and ongoing supplements). Chapter 8 Behavioral Neuroscience. In *Current Protocols in Neuroscience*, J Crawley, C Gerfen, M Rogawski, D Sibley, P Skolnick, W Wray, R McKay, Eds. Wiley, New York (print version and CD-ROM).

Crusio WE, Gerlai RT, Eds. (1999). *Handbook of Molecular-Genetic Techniques for Brain and Behavior Research*. Elsevier, New York.

Darwin C (1839–1843). *Zoology of the Voyage of the H.M.S. Beagle, Part 3 Birds*, Smith Elder and Company, London.

Dowling JE (1998). *Creating Mind: How the Brain Works*. Norton, New York.

Foster G (1997). *Chemical Neuroanatomy of the Prenatal Rat Brain*. Oxford University Press, New York.

Franklin BJ, Paxinos G (1996). *The Mouse Brain in Stereotaxic Coordinates*. Academic, San Diego.

Goldowitz D, Wahlsten D, Wimer RE, Eds. (1992). *Techniques for the Genetic Analysis of Brain and Behavior: Focus on the Mouse*. Elsevier, Amsterdam.

Grandin T (1998). *Genetics and the Behavior of Domestic Animals*. Academic, San Diego.

Griffin DR (1985). Animal consciousness. *Neuroscience and Biobehavioral Reviews* **9**: 615–622.

Jacobowitz DM, Abbott LC (1998). *Chemoarchitechtonic Atlas of the Developing Mouse Brain*. CRC Press, Boca Raton, FL.

JAX Notes (2003). The importance of understanding substrains in the genomic age. The Jackson Laboratory 491: 1–3.

Jones BC, Mormede P (1999). *Neurobehavioral Genetics: Methods and Applications*. CRC Press, Boca Raton, FL.

Kandel ER, Schwartz JH, Jessel TM (2000). *Principles of Neural Science*, 4th ed. McGraw-Hill, New York.

Kaufman MH (1992). *Atlas of Mouse Development*. Academic, San Diego.

Lorenz K (1971). *Studies in Animal and Human Behavior*. Harvard University Press, Cambridge, MA.

Lorenz KZ (1974). Analogy as a source of knowledge. *Science* **185**: 229–234.

Lorenz KZ (1981). *The Foundations of Ethology*. Springer, New York.

Mak RW (1998). *The Gene Knockout FactsBook*. Academic, San Diego.

Marler P, Hamilton WJ, Eds. (1968). *Mechanisms of Animal Behavior*. Wiley, New York.

Nelson RJ (2005) *An Introduction to Behavioral Endocrinology*. Sinauer, Sunderland, MA.

Nelson RJ, Young KA (1998). Behavior in mice with targeted disruption of single genes. *Neuroscience and Biobehavioral Reviews* **22**: 453–462.

Paxinos G (1985). *The Rat Nervous System*. Academic, New York.

Paxinos G, Franklin KBJ (2000, 2003). *The Mouse Brain in Stereotaxic Coordinates*. Academic, San Diego.

Plomin R, DeFries JC, McClearn GE, Rutter M (1998). *Behavioral Genetics*. Freeman, New York.

Schambra UB, Lauder JM, Silver J (1991). *Atlas of the Prenatal Mouse Brain*. Academic, New York.

Silver LM (1995). *Mouse Genetics*. Oxford University Press, New York.

Squire LR, Bloom FE, McConnell SK, Roberts JL, Spitzer NC, Zigmond MJ (2002). *Fundamental Neuroscience*, 2d ed. Academic Press, New York.

Swanson, L (1998). *Brain Maps: Structure of the Rat Brain*. Elsevier Science, New York (print version and CD-ROM).

Tinbergen N (1974). Ethology and stress diseases. *Science* **185**: 20–27.

Von Frisch K (1967). *The Dance Language and Orientation of Bees*. Belknap, Cambridge, MA.

Wahlsten D, Crabbe JC (2006). Behavioral testing. In *The Mouse in Biomedical Research*, volume III, in press.

Wahlsten D, Metten P, Phillips TJ, Boehm SL, Burkhart-Kasch S, Dorow J, Doerksen S, Downing C, Fogarty J, Rodd-Henricks K, Hen R, McKinnon CS, Merrill CM, Nolte C, Schalomon M, Schlumbohn JP, Sibert JR, Wenger CD, Dudek BC, Crabbe JC (2003). Different data from different labs: Lessons from studies of gene-environment interaction. *Journal of Neurobiology* **54**: 283–311.

Whishaw IQ, Kolb B (2004). *The Behavior of the Laboratory Rat: A Handbook with Tests*. Oxford University Press, Oxford.

Special Issues of Journals

Behavior Genetics **26**(3), May 1996 Special Issue: *Molecular Genetic Approaches to Mammalian Brain and Behavior*.

Behavior Genetics **27**(4), July 1997 Special Issue: *The Genetics of Obesity*.

Behavioural Brain Research **95**(1), September 1998 Special Issue: *Behavioural Neurogenetics*.

Brain Research **835**, 1999 Special Issue: *Knockouts and Mutants: Genetically Dissecting Brain and Behavior*.

Hormones and Behavior **31**(3), June 1997 Special Issue: *Single Gene Mutations, Gene Knockouts and Behavioral Neuroendocrinology*.

Learning and Memory **5**(4–5), September/October 1998 Special Issue: *Transgenic/Knockout Approaches in Neurobiology*.

Nature Neuroscience **7**(5), May 2004: *Scaling up Neuroscience*.

Neuropeptides **36**(2–3), April/June 2002 Special Issue: *Transgenics and Knockouts with Mutations in Genes for Neuropeptides and Their Receptors*.

Psychopharmacology **132**(2), July 11, 1997, Special Issue: *Behavioral and Molecular Genetics*.

Psychopharmacology **174**(4) August 2004 Special Issue: *Molecular Genetics and Psychopharmacology*.

Neuroscience Methods Series

Current Protocols in Neuroscience. Wiley.

Journal of Neuroscience Methods. Elsevier.

Methods in Neuroscience. Academic Press.

Neuroscience Protocols. Elsevier.

Techniques in the Behavioral and Neural Sciences. Elsevier.

Courses in Mouse Behavior

Cold Spring Harbor Laboratory Courses in Mouse Behavioral Analysis. Summer 2003 course was organized by Stephen Anagnostaras and Mark Mayford, held at Cold Spring Harbor Laboratory, New York, USA. *http://meetings.cshl.org/2003/2003c-maze.htm*

EMBO/FENS Practical Course in Mouse Transgenics and Behavior. July 2003 course was organized by David P. Wolfer, Hans-Peter Lipp, and Richard G. M. Morris, held at the University of Zurich, Switzerland. *http://www.dpwolfer.ch/mouse-course*

International Summer School on Behavioral Neurogenetics. August 2004 course was organized by Robert W. Williams, Douglas Matthews, Byron Jones, and Dan Goldowitz, held at the University of Tennessee, Memphis, Tennessee, USA. *http://tnmouse.org/documents/BehNeurog-SummerSchool04.pdf*; and Second Annual Experimental Neurogenetics Mouse Workshop, organized by Dan Goldowitz and members of the Tennessee Mouse Genome Consortium, May 2005, Memphis, Tennessee, USA.

JAX Neurogenetics Conference, organized by Wayne Frankel. June 2004 course held at The Jackson Laboratory, Bar Harbor, Maine, USA. *http://www.jax.org/courses/events/coursedetails.do? id=28*

The Laboratory Mouse in Vision Research, Organized by John Macauley, Maureen McCall, and Patsy Nishina, October 2004 course held at The Jackson Laboratory, Bar Harbor, Maine, USA *http://www.jax.org/courses/events/coursedetails.do?id=42*

Walk This Way: Gait Dynamics in Rodent Models of Human Diseases. August 2005 workshop organized by Mouse Specifics, Inc., Cambridge Life Sciences Center, Cambridge, MA.

Mice expressing the rat metallothionen-growth hormone fusion gene grew significantly larger than their litter-mates. *(From the cover of Nature, Vol. 300, December 16, 1982; Palmiter et al., 1982.)*

2

Of Unicorns and Chimeras

Targeted gene mutation technologies began in the 1980s (Jaenisch, 1976, 1988; Costantini and Lacy, 1981; Gordon and Ruddle, 1981; Harbers et al., 1981; Wagner et al., 1981a, 1981b; Jaenisch, 1988; Pascoe et al., 1992; Doetschman, 1991; Smithies, 1993; Bronson and Smithies, 1994; Smithies and Kim, 1994; Capecchi, 1989, 1994). Building on manipulations of yeast and fruitfly genomes, mouse mutations multiplied. The first big success in detecting a phenotype relevant to behavior in a transgenic mouse appeared in 1982. The cover illustration of the December 16 issue of *Nature*, shown on the opposite page, excited the popular imagination with the dramatic results of the elegant experiments by Richard Palmiter and co-workers at the University of Washington (Palmiter et al., 1982). A growth hormone overexpressing transgenic mouse was much larger than normal littermate control mice of the same age and gender, as a result of more rapid weight gain. Technical advances in targeted gene mutations in mammals raised hopes that the new technology could be applied to discovering the role of individual genes in normal and abnormal behavioral processes. This dream moved into the realm of reality over the past decade. Many excellent books and review articles describe the techniques for generating transgenics, knockouts, knock-ins, conditional mutations, inducible mutations, and further elegant genome manipulations (Bradley et al., 1992; Hogan et al., 1994; Accili, 2000; Joyner, 2000; Jackson, 2000; Gossmann et al., 2000; Rülicke and Hübscher, 2000; Nestler et al., 2001; Hofker et al., 2002; Nagy et al., 2002; Wolfer et al., 2002; Tecott and Wehner, 2001; Tecott, 2003; Tenenbaum et al., 2004).

The present chapter provides a brief overview of the types of targeted gene mutations in current use. The primary focus of this chapter is on the steps following the generation of the first founder mouse. Breeding strategies tailored to the needs of behavioral phenotyping are presented. Background strains that have been effectively used to breed mutant lines are recommended for various behavioral domains. Housing, transportation, group size, group composition, and animal welfare requirements are described, to meet the special demands of behavioral phenotyping. Original literature cited in the text, and review articles cited at the end of this chapter, provide more in-depth information for the interested reader. The listings at the end of this chapter include background readings on DNA constructs, embryonic stem cell lines, breeding strategies, mouse handbooks, companies that provide breeding and genotyping services, companies that design and manufacture behavioral test equipment, and academic organizations and companies that generate and phenotype mutant mice. Web sites relevant to these topics are included.

GENERATING A TARGETED GENE MUTATION

The process of developing a transgenic or knockout mouse begins with an identified gene. If the gene has not yet been sequenced, a useful targeting vector cannot be designed.

Transgenic is defined by the insertion of a gene. Transgenic mice may have a new gene added, for example, the human gene for a hereditary disease such as Huntington's (Carter et al., 1999), or extra copies of an existing gene, for example, the corticotropin releasing factor gene in order to investigate excessive expression of this stress-related hormone and neurotransmitter (Stenzel-Poore et al., 1994). Transgenic techniques involve microinjection of the DNA construct containing the transgene into the pronucleus of a fertilized mouse oocyte. The DNA construct also contains a reporter gene, such as β-galactosidase (lacZ), with a nuclear localization signal (nls). The reporter gene is simultaneously driven by the promoter for the transgene. LacZ positive cells indicate the presence of the transgene in the cell. Concentrations of the reporter gene, the transgene, and the gene product are assayed to determine the overexpression level of the gene in the tissue of interest. Anatomical mapping of the localization of the reporter gene, the transgene, and the gene product is used to describe the distribution of the transgene in the brain, or to more precisely delineate the specific neurons expressing a known gene product during stages of development (Jacobowitz and Abbott, 1997; Itoh et al., 1998). Transgenic methods are diagrammed in Figure 2.1. Each mouse that develops from a microinjected egg is a potential *founder* of a mutant line. The success rate of the technique is proportional to the number of eggs injected, because insertions through homologous recombinations are random, infrequent events. The more lottery tickets you buy, the better your chances of winning.

Knockout mice represent a loss of function or a null mutation, meaning a mutated gene that does not synthesize its protein. Knockout mice are generated by a different set of techniques (Wynshaw-Boris et al., 1999; Ledermann, 2000). Instead of a gene insertion, knockouts have a mutation introduced into a carefully chosen exon of the cDNA of the gene. The mutation is usually a selective deletion of a portion of DNA that is critical for the expression of the gene product. The gene for resistance to an antibiotic drug, such as the neomycin resistance gene (Neo^r), is inserted within the DNA construct as a marker. The deletion and insertion usually shift the reading frame for the DNA, rendering incorrect reading of the triplet base pair codes for the amino acids comprising the gene product. A typical targeting vector is shown in Figure 2.2.

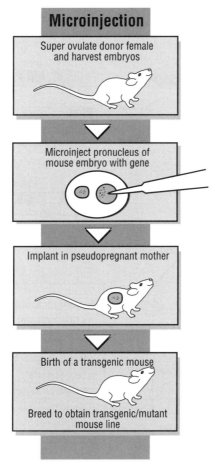

Figure 2.1 *Transgenic mice are generated by inserting a foreign gene, or extra copies of a gene, into the pronucleus of a fertilized egg.* [From Taconic Farms, Inc. (1998), *Research Animal Review* **1**(5): 2.]

The targeted gene construct is inserted into the genome of embryonic stem (ES) cells. ES cell lines used for generating knockout mice are most frequently derived from the 129 inbred strains of mice. Several 129 substrains produce ES cells that grow well in culture, remain viable through the electroporation and implantation processes, and colonize a large proportion of the developing embryo, as compared to ES cells from other mouse strains or from rats strains (Simpson et al., 1997). Knockout methods are diagrammed in Figure 2.3.

The first pup born from a successful gene knockout is called a *chimera* because it contains cells from two independent sources.* The appearance of the fur may be one useful early marker of a successful mutation. The C57BL/6 strain has black fur. The 129 strain has agouti (light grayish-brown) fur. When the coat color of pups appears as

---

*In Greek mythology the Chimera, a fire-breathing monster, was a mixture of a lion's head, a goat's body, and a serpent's tail. Bellerophon, a Greek hero aided by the the winged horse Pegasus, killed the Chimera by thrusting lead down its throat. Its fiery breath melted the lead, but the molten lead trickled down into the stomach, killing the Chimera and saving the Kingdom of Lycia (D'Aulaire and D'Aulaire, 1962).

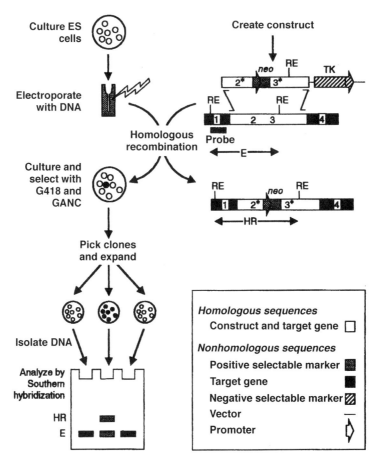

Figure 2.2 *Production, selection, and identification of targeted gene disruption by homologous recombination. An example of a restriction enzyme (RE) and hybridization probe that can be used to identify cells in which homologous recombination has occurred (shaded colony) is shown. The predicted size of the restriction fragment generated from an unaltered target gene (E) and a target gene that has undergone homologous recombination (HR) are shown. If equal amounts of DNA are present in the lanes of the Southern blot, the intensity of each of the two hybridization fragments from the DNA of a homologous recombinant clone will be half the intensity of the hybridizing fragment from unaltered clones.* [From Ausubel (1995), p. 9. 16.2.]

grayish brown, or sometimes a mosaic of black and grayish-brown patches or stripes, chimeras are visually apparent. If no ES cells were incorporated at the blastula stage, the pup will show a black coat color. Genotyping remains the essential method for definitively identifying chimeras.

FOUNDER LINES

If the mutation incorporated into cells of the blastula develop into germ-line gametes, that is, eggs and sperm, then the targeted mutation is transmitted to the next generation of the offspring of the chimera. When incorporation is only in the somatic cells that

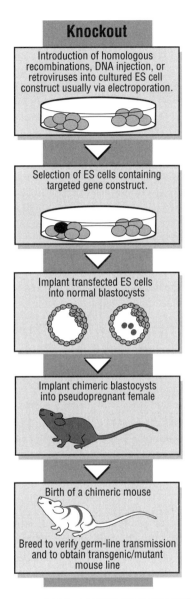

Figure 2.3 *Knockout mice are generated by inserting the mutated cDNA into embryonic stem cells, injecting the transfected cells into normal blastocysts and implanting the injected blastocysts into pseudopregnant normal female mice.* [From Taconic Farms, Inc. (1998), *Research Animal Review* **1**(5): 2.]

develop into nonreproductive tissues, the original chimeras express the mutation, but their offspring do not. The scheme for generating chimeras and their offspring is shown in Figure 2.4.

To detect germ-line transmission, a test cross is conducted. The chimera is bred to a mouse of a normal inbred strain, such as C57BL/6J. The F_1 offspring of the test cross are genotyped for expression of the mutation. An F_1 offspring that receives the

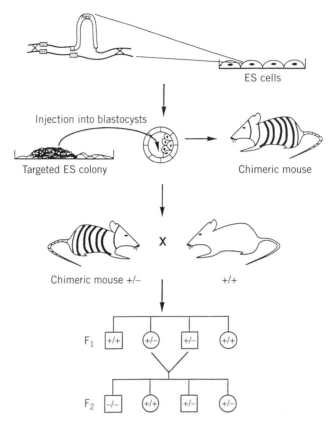

Figure 2.4 *Chimeric mice grow from the blastulas containing embryonic stem cells expressing the targeted gene construct. When the mutation is expressed in germ-line cells of the chimera, the mutation is transmitted to subsequent generations.* [From Wehner and Silva (1997), p. 244.]

mutated gene from the chimeric progenitor parent is heterozygous for the mutated gene. Southern blot analysis or polymerase chain reaction assay is performed on a small tissue sample, usually from the tail of the offspring, to identify positive heterozygotes. Each positive heterozygote can be used to generate a line of mutant mice.

Identified F_1 heterozygote offspring are mated with each other to produce an F_2 generation. Theoretically the F_2 population will follow the principles of Mendelian segregation, resulting in one-fourth (1/4) homozygous mutants ($-/-$), one-half (2/4) heterozygotes ($+/-$), and one-fourth (1/4) homozygous wildtype controls ($+/+$). If the gene is lethal, the homozygotes will not survive. If the gene is located on the X or the Y chromosome, gender issues influence the ratio of males and females for each genotype.

The genotypes of the F_2 mice are tested by Southern blot for the presence of the normal gene in the $+/+$ mice, the presence of half the normal complement of the gene in the $+/-$ mice, and the absence of the gene in the $-/-$ mice. The expression of the gene product, when the gene product is known, is assayed by an appropriate technique, such as high-pressure liquid chromatography for an enzyme or radioimmunoassay for

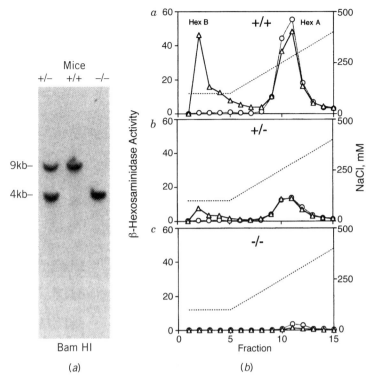

Figure 2.5 *Confirmation of the mutation in a hexosaminidase B knockout mouse model of Sandhoff disease. (a) Southern blot analysis of tail snips digested with BamH1 endonuclease, from Hexb heterozygote, wildtype, and null mutant mice, showing the absence of a band at 9 kb in Hexb −/− mice deficient in the HexB gene. (b) High-pressure liquid chromatograph showing* concentrations *of the HexB gene products, the enzymes β-hexosaminidase A and β-hexosaminidase B. Peaks for both enzymes are seen in the wildtype control mice, top panel. Both peaks are absent in the null mutants, bottom panel, representing the complete lack of β-hexosaminidase A and B enzyme activity. Heterozygotes, shown in the middle panel, have small amounts of the enzymes, indicating a gene-dose effect.* [From Sango et al. (1995), p. 171.]

a peptide. The heterozygotes (+/−) may express half the gene product, reflecting the presence of half the gene dose, or may express variable amounts of the gene product in some cases. The homozygous mutant mice (−/−) should show no gene product. These confirmed −/− mice are termed the *null mutants*. Techniques for confirmation of the mutation are illustrated in Figure 2.5.

Conventional gene disruption technology is now widely used in laboratories around the world. Modifications that increase specificity are increasingly available. Tissue-specific *conditional mutations* allow the investigator to insert the mutation only into specific cell types. Conditional mutations solve two problems: (1) lack of control over the integration site of a transgene into the chromosome and (2) ubiquity of a knockout mutation throughout the body. Linking the transgene to a tissue-specific promoter limits the site of integration to the chromosomal locus homologous with the specific promoter. Linking a knockout to a tissue-specific promoter ensures that the modified DNA construct is expressed only in tissues that normally express that promoter gene.

For example, the CaMKIIα promoter confers regional specificity of a targeted gene mutation to neurons of the forebrain (Mayford et al., 1996). A promoter specific to the CA1 pyramidal cells of the hippocampus was developed using a phage P1-derived Cre/loxP recombination system (Tsien et al., 1996).

Inducible mutations allow the investigator to turn the mutation on and off at a desired developmental stage or period of life (Mansuy and Bujard, 2000). Temporal specificity is conferred by a regulatory expression system that is activated or deactivated by a real-time drug treatment. Tetracycline-controlled transactivator and reverse tetracycline transactivators are respectively activated or deactivated by doxycycline, an antibiotic administered at low doses in the drinking water (Mansuy et al., 1998, 1999; Sakai et al., 2002). Tamoxifen-inducible Cre-ERT recombinase is another inducible that can be designed with a tissue-specific promoter (Weber et al., 2001).

Knock-in mutations are point mutations, targeting a single nucleotide in a gene to produce a single amino acid substitution in its protein product (Giese, 1999; Tecott and Wehner, 2001; Tecott, 2003). The function of the gene may be altered, not eliminated. The alteration may produce a loss of function or a gain in function. For example, a point mutation in the GABA-A receptor that substitutes an arginine for a histidine at amino acid 101 of the α4 and α6 subunits retains most of the functions of the GABA-A receptor but reduces sensitivity to diazepam on the rotarod test (McKernan et al., 2000; Burt, 2003). A point mutation in the enzyme catechol-O-methyl transferase that converts amino acid 158 from methionine to valine results in impaired executive function mediated by the dorsolateral prefrontal cortex and appears linked to schizophrenia (Egan et al., 2001; Blasi et al., 2005). Knock-ins for nicotinic receptor subunits are elucidating the receptor subunit composition that mediates smoking (Champtiaux and Changeaux, 2004). Knock-in mice with elongated polyglutamine tracts within the correct context of the mouse *Hdh* gene develop abnormal activity levels that may mirror the progression of Huntington's disease (Hickey and Chesselet, 2003).

Knockdown mutations include technologies that manipulate gene expression with more precise anatomical specificity. Viral vectors such as adeno-associated virus (AAV), described in Chapter 14, are engineered to deliver a specific gene directly into a brain region (Hommel et al., 2003; Tenenbaum et al., 2004). A suspension of the AAV containing the gene is microinjected into a brain region. Stable transfection of the target tissue allows the gene product to be synthesized from that time point on, in that tissue only, conferring both temporal avoidance of the early stages of development, and anatomical specificity to the site of microinjection. For example, AAV delivery of the gene for the inhibitory neuropeptide galanin, microinjected in to the inferior colliculus, attenuated focal seizures (Haberman et al., 2003). RNA interference (RNAi) is a new technique, described in Chapter 14, that neutralizes specific gene sequences to shut down expression (Hommel et al., 2003; Lavery and King, 2003). Single silencer RNA sequences (siRNA), siRNA libraries, and delivery systems using an siRNA transfection agent or an siRNA expression vector are commercially available (e.g., Ambion Inc., *www.ambion.com*; Qiagen, *www.qiagen.com*; Stratagene, *www.stratagene,com*). Viruses expressing shRNAs that target a distinct sequence of the gene for tyrosine hydroxylase, along with enhanced green fluorescent protein, were microinjected into the mouse midbrain (Hommel et al., 2003). For example, fluorescent immunostaining displayed the fluorescent protein specifically in the substantia nigra zona compacta. Reductions in tyrosine hydroxylase were found two weeks later in the terminal region of the nucleus accumbens, and amphetamine-induced hyperloco-motion was attenuated (Hommel et al., 2003). Conditional, inducible, knock-in and knockdown technologies are discussed at length in Chapter 14.

BREEDING STRATEGIES

To obtain large numbers of offspring for functional studies, heterozygote transgenic or knockout mice containing the desired gene mutation in the germ line are bred with normal mice. Several breeding strategies have been successfully employed. A variety of inbred and outbred strains of mice have been utilized for breeding. The goal of the breeding strategy is to optimize the expression of the mutation, maximize the number of viable offspring, and minimize the potential confounding influence of background genes from the breeder parents.

Recommendations for the choice of inbred mouse strains for breeding the targeted mutation have been discussed at length in many excellent reviews (Silver, 1995; Keverne et al., 1996; Gerlai, 1996; Crawley, 1996; Lathe, 1996; Crusio, 1996; Zimmer, 1996; Crawley et al., 1997; Silva et al., 1997; Markel, 1997; Frankel, 1998; Choi 1997; Nelson and Young, 1998; Dubnau and Tully, 1998; Bolivar et al. 2000a; Cook et al. 2002; Bothe et al. 2004). As described below, there is no "best" strain that is ideal for all behavioral tests (Frankel, 1998). C57BL/6J is a good choice for many, but not all mutations, as it shows average performance in most, but not all, behavioral domains (Wehner and Silva, 1997; Crawley et al., 1997). One domain in which C57BL/6J is unusual is its high propensity to consume ethanol and self-administer cocaine (Crabbe et al., 1994b; Berrettini et al., 1994; Miner, 1997; discussed further in Chapter 11). Another problem with C57BL/6J is progressive deafness in its genetic background, discussed in Chapter 5. As C57BL/6J is a reasonably good breeder, and is readily available from The Jackson Laboratory, it is a good compromise candidate to use for breeding many transgenics and knockouts (Crawley et al., 1997; Banbury Conference, 1997). This author highly recommends that the growing literature on inbred strain distributions be consulted before you make a choice, at the very earliest stages of commitment, to a breeding strategy for a new transgenic or knockout. Choosing a breeding strain that will not immediately confound the interpretation of your mutant phenotype will avoid some of the disastrous consequences discussed throughout this book. Choosing a breeding strain that is internationally available will allow replication and extension of your discoveries by other laboratories.

Various methods for conducting the generational breeding are possible. The best approach for breeding a targeted mutation of a gene relevant to a behavioral phenotype is the *congenic* breeding strategy. A congenic line is created by successive backcrosses into one inbred strain (Silver, 1995; Wehner and Silva, 1997). First, the mutant founder is mated to a normal C57BL/6J recipient mouse. Offspring are genotyped. Individuals receiving the mutation are used for future matings. C57BL/6J mice are then consistently used as breeding partners in alternate generations. Heterozygote brother and sister matings generate batches of mice with the Mendelian 1:2:1 ratio of $-/-$, $+/-$, and $+/+$ littermates for the phenotyping experiments. The congenic breeding strategy maintains a well-defined and behaviorally characterized inbred genetic background, while minimizing genetic drift. Systematic backcrosses between the null mutant and the standard inbred strain effectively retain the single gene mutation on a fixed genetic background (Silver, 1995; Crawley et al., 1997). Flanking or hitchhiker genes, inadvertently included in or downstream from the targeting vector, are bred out during the successive backcrosses (Wolfer et al., 2002). As discussed below, *speed congenic breeding* distills the population containing the phenotype of interest more quickly by selecting breeding males on the basis of genetic markers and phenotypic expression of the mutation (Markel et al., 1997; Wong, 2002).

The wrong solution is to continuously inbreed the $+/+$, $+/-$, and $-/-$ littermates (Wolfer et al., 2002). Mutated "hitchhiker" or "passenger" genes, introduced by the targeting vector and the disrupted reading frame, will remain in the mutant line. Random segregation events, occurring over many generations, may generate new, unusual alleles for background genes and dramatically influence the behavioral phenotype of the mutation. These extraneous mutated genes are transmitted to each future generation, introducing false positives or false negatives into the interpretation of the phenotype of the mutation. The phenotype of the mutant line may in fact be caused by the continued presence of a flanking gene, or a randomly transmitted allele, rather than the targeted gene mutation. Deleterious allele combinations carried through successive inbreeding can suppress reproduction and survival (Banbury Conference, 1997).

Avoid homozygous matings at all costs (and the costs are high). Generating one line of $-/-$ and another independent line of $+/+$ is likely to create serious artifacts for behavioral phenotyping. Littermates are the only true comparisons of genotypes. This is because environmental conditions directly affect mouse behaviors. Parental care, cagemate social interactions, cage changing, room temperature, ambient light level, building noise, season of the year, and many other environmental factors will influence scores on behavioral tasks. For example, emotional reactivity of reciprocal F_1 hybrids from a cross of C57BL/6 and BALB/c was closer to the strain of the foster mother than to the strain of the genetic mother (Calatayud and Belzung, 2001). Even the in utero environment can contribute to the behavioral phenotype, as B6 mice cross-fostered prenatally into BALB/cJ dams showed behaviors more like their Balb/cJ foster mothers on elevated plus-maze, Morris water maze, and open field behaviors (Francis et al., 2003). There is no practical way to keep all environmental conditions constant across breeding cages and breeding generations (Wahlsten et al., 2003c). The only valid solution is to compare treatment animals and control animals living in the same environment. Since one litter cannot produce enough mice for behavioral experiments, the best strategy is to generate groups of littermates from many cages and across several litters. Ensure that approximately equal numbers of each relevant genotype are represented within an experiment.

Breeding a double mutation involves additional challenges in generating sufficient numbers of mice for behavioral phenotyping. In a single gene knockout for gene "A," heterozygotes (Aa) are bred together to produce three genotypes: homozygous mutant aa, heterozygote Aa, and homozygous wildtype AA littermates. The yield will approximate the Mendelian ratio of 1 aa : 2 Aa : 1 AA. Thus, to obtain a minimum N of 10 $-/-$ (aa) and 10 $+/+$ (AA) of each sex, approximately 80 pups are born, requiring about 10 breeding pairs. A double gene knockout could be analogously generated by breeding a male heterozygote for one gene mutation (Aa) with a female heterozygote for the second gene mutation (Bb). Nine genotypes result: AABB, AaBB, aaBB, AABb, AaBb, aaBb, AAbb, Aabb, and aabb. The approximate Mendelian yield would be only 1 homozygous double mutant (aabb) and 1 homozygous double wildtype out of 16 pups born. Thus, considerably more mating is required. To obtain a minimum of $N = 10$ aabb and $N = 10$ AABB of each sex, approximately 320 pups are born, requiring about 40 breeding pairs. Triple mutations multiply these demands by another fourfold. For example, breeding together three lines of mutant mice to replicate several pathological features of Alzheimer's disease, such as overexpression of the amyloid precursor protein, a mutant presenilin gene, and a mutant tau protein gene, would generate 64 genotypes, with a theoretical Mendelian yield of only 1 triple homozygous mutant out of 64 pups. The need for double and triple mutant lines is growing, because inducibles and conditionals require mating a transgene on a tissue-specific Cre/lox promoter with a tetracycline-inducible element.

Animal breeding and holding space are major issues for double and triple mutant lines. The costs multiply for housing and genotyping. An open question is the choice of control genotypes for the phenotyping experiments. Ideally all 9 genotypes of a double knockout cross would be tested on behavioral tasks. When these Ns are impractical or impossible, which are the required control genotypes to compare against the double mutants? For the first experiments, a logical approach is to compare the double mutations to the double wildtypes, that is, aabb is the treatment group and AABB is the control group. The heterozygote combination of AaBb may be the logical heterozygote control. However, in many cases the scientific hypothesis behind the experiments addresses gene dose effects. Phenotypes of the heterozygotes are of interest. Investigators will decide on the heterozygote gene combination(s) that best address their specific scientific questions.

Certainly it is tempting to employ a breeding strategy for double and triple mutants that generates two independent breeding pools, such as a line of double or triple null mutants (aabb or aabbcc) and a line of full wildtype controls (AABB or AABBCC). Unfortunately, as explained above, this is the wrong breeding strategy for the purposes of behavioral phenotyping. Littermates are necessary to avoid confounds of environmental, home cage, and parental factors that directly influence mouse behaviors.

Background Gene Considerations

Background genes are a major issue. All inbred strains of mice were probably derived from one panmictic ancestral population (Van Oortmerssen, 1971). The house mouse, *Mus musculus*, is a highly adaptable species that successfully colonized natural and human-made habitats throughout the world (Silver, 1995). Unusual looking mice appealed to animal lovers in Asia and became "objects of fancy" in Europe (Sage, 1981; Silver, 1995). Miss Abbie Lathrop of Granby, Massachusetts, who bred fancy mice for sale as pets, provided some of these unusual individuals to Harvard University and the University of Pennsylvania between 1910 and 1918 (Morse, 1978; Silver, 1995). Many of the common inbred strains of mice available from commercial suppliers, including C57BL/6, originated from Ms. Lathrop's breeding farm. An elegant family tree of 102 mouse strains was generated by Petko Petkov and co-workers at The

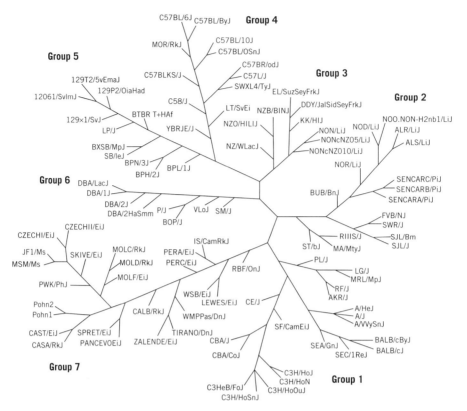

Figure 2.6 *Mouse family tree, constructed from the results of SNP marker analysis of 102 inbred and wild-derived strains.* [Courtesy of Petko Petkov, The Jackson Laboratory, Bar Harbor, ME. From Petkov et al. (2004), p. 1810.]

Jackson Laboratory, using single polynucleotide polymorphism (SNP) analysis (Petkov et al., 2004; Figure 2.6).

Defined inbred strains are maintained by reputable commercial breeders, including Charles River, Harlan, The Jackson Laboratory, and Taconic in the United States. A strain is considered inbred when it is maintained for at least 20 sequential generations by brother/sister matings (Silver, 1995). However, random mutations and genetic drift introduce new alleles into the genotype of an inbred strain. New alleles may appear at each generation and be carried into subsequent generations, resulting in new substrains maintained by each supplier. Independent colonies of inbred strains of laboratory animals vary at many chromosomal loci. All are distinct from the genotypes of the original wild house mouse populations. Thus many genes differ among inbred strains of mice, among substrains of mice, and between the same inbred strains and substrains from different independent colonies and commercial suppliers.

Remarkable variability in behavioral scores is evident in different strains of mice tested on the same behavioral task. A comparison of a phenotype across many strains is termed a *strain distribution*. The literature on strain distributions is rapidly growing for many behavioral domains. Considerable information has been assembled on the behavioral phenotypes of inbred, outbred, and wild strains of mice, demonstrating strain differences on measures of locomotion, learning and memory, aggression, sexual and

TABLE 2.1 Strain Distributions of Inbred Strains of Mice on Two Learning and Memory Tasks Commonly Used for Behavioral Phenotyping of Transgenic and Knockout Mice

Superb Learners	Adequate Learners	Impaired Learners	Visually Impaired
	Spatial selectivity of mouse strains on Morris water task[a]		
B6D2F1	C57BL/6J	129/SvJ	A/J
B10C3F1	C57BL/10J	DBA/2	SJL/J
129B6F1	LP/J	BALB/cByJ	C3H/Ibg
FVB129F1	BALB129F1		FVB/NJ
FVBB6F1	B6SJLF1		BuB/BnJ
129/Svev			
	Contextual fear-conditioning in mouse strains[b]		
C57BL/6J	C3H/Ibg	FVB/NJ	A/J
C57BL/10J		DBA/2	
129/SvJ		BuB/BnJ	
129/Svev			
SJL/J			
BALB/cByJ			
LP/J			
BALB129F1			
FVB129F1			
FVBB6F1			
B6D2F1			
B6SJLF1			
B10C3F1			
129B6F1			

Source: From Wehner and Silva (1997), p. 245.

[a]Animals were trained in the hidden and visible platform versions of the Morris water task using 12 trials per day for 3 days as described previously (Wehner et al., 1990). Using a probe trial to assess spatial selectivity, a preference score was calculated (Wehner et al., 1990) for site crossings. This score is calculated as the mean of crosses at the trained site minus the mean of crosses at the three other platform sites. Mice are designated as "superb learners" if they had a preference score of 3.5 or greater; mice are designated as "adequate learners" if they had a preference score of 2.0 or greater; mice are designated as impaired if they had a score under 2.0 but could perform the visible platform task; mice are designated as visually impaired if they could not perform the visible platform task.

[b]Performance on contextual fear conditioning was measured as described by Paylor et al. (1994b). Animals that performed well on the task could discriminate the context from an altered context as indicated by a greater percent freezing in the same context than in the altered context. Impaired animals freeze equally in both the same context and altered context. A/J, a generalized freezer, exhibited high levels of baseline freezing in the chamber prior to conditioning.

parental behaviors, sleep, vision, hearing, startle, prepulse inhibition, taste conditioning, latent inhibition, anxiety-related behaviors, depression-related behaviors, and responses to drugs such as ethanol, nicotine, cocaine, morphine, antidepressants, antipsychotics, anxiolytics, psychostimulants, and convulsants (Wehner and Silva, 1996; Crawley et al., 1997a; Logue et al., 1997; Paylor and Crawley, 1997; Phillips et al., 1999; Gould and Wehner, 1999; Johnson et al., 2000; Briebel et al., 2000; Bolivar et al., 2000a; Lucki et al., 2001; Miczek et al., 2001; Cook et al., 2002; Holmes et al., 2002; Broadbent et al., 2002; Bouwknect and Paylor, 2002; JAX Notes, 2003; Koehl et al., 2003; Crabbe et al., 2003; Balogh and Wehner, 2003; Ripoll et al., 2003; Danciger et al., 2003; Bothe et al., 2004; Brooks et al., 2004; Mohajeri et al., 2004). The Mouse Phenome Project, maintained by Dr. Molly Bogue at The Jackson Laboratory, offers databases of various phenotypes of selected inbred strains (Bogue and Grubb, 2004).

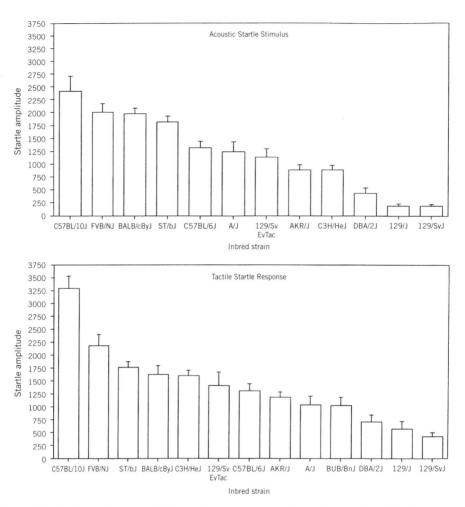

Figure 2.7 *Strain distributions of inbred strains of mice on tests of acoustic and tactile sensory reflexes.* [From Crawley and Paylor (1997), p. 176.]

Table 2.1 gives strain distributions for several learning and memory tasks. Figure 2.7 shows strain distributions for acoustic startle and tactile startle. Table 2.2 presents a strain distribution for an anxiety-related behavior. Table 2.3 describes a strain distribution for self-administration of nicotine.

Strain distributions are not a charming little provincial corner of behavioral genetics. Background gene differences in inbred strains of mice are central to the latest findings in functional genomics. Why does one person with a breast cancer mutation develop malignancy and another does not? Modifier genes in the individual's genome may serve as protective factors. Why does one person with the short form polymorphism of the serotonin transporter develop severe depression and another does not (Holmes and Hariri, 2003)? Multiple genes in the background may contribute protection and susceptibility factors that interact with each other and with environmental life events. For example, Miriam Meisler and co-workers at the University of Michigan discovered a modifier gene, *SCNM1*, that is mutated in the C57BL/6J inbred strain and enhances the

TABLE 2.2 Strain Distribution of Mouse Strains on an Anxiety-Related Task

Mouse Strain	Number of Light ↔ Dark Transitions	
	Baseline	Response to Diazepam
C57BL/6J	49 ± 3	91 ± 11[a]
Swiss Webster/NIH	38 ± 2	67 ± 4[a]
DBA	21 ± 2	27 ± 7
CF-1	20 ± 4	26 ± 3
Swiss Webster/Harlan	14 ± 6	24 ± 3
A/J	9 ± 1	10 ± 2
BALB/cJ	8 ± 3	7 ± 5

Source: From Crawley et al. (1997a), p. 117, adapted from Crawley and Davis (1982).
[a]$p < 0.05$ compared to vehicle.

TABLE 2.3 Strain Distribution of Mouse Strains on Self-administration of Nicotine

	Threshold Tolerance Dose	Nicotine Consumption Max Dose	Nicotine Consumption IC_{50}
A	2.32 ± 0.31	5.7 ± 0.3	42.8 ± 8.9
BUB	3.52 ± 0.60	6.2 ± 0.8	72.0 ± 17.2
C3H	3.93 ± 0.44	4.8 ± 0.5	40.2 ± 7.6
C57BL/6	1.12 ± 0.46	11.7 ± 1.1	114.1 ± 20.2
DBA/2	2.73 ± 0.25	8.2 ± 0.6	89.7 ± 12.3
ST/b	—	2.8 ± 0.3	32.4 ± 7.9

Source: Adapted from data by Allan Collins and co-workers in Crawley et al. (1997a), p. 115.

lethality of a targeted mutation in a sodium channel gene, *medJ* (Buchner et al., 2003). As discussed in Chapter 14, discovery of genes that confer resistance or susceptibility to a disease are likely to enrich our understanding of disease processes, and enhance the development of novel therapeutics with pharmacogenomic customization to the background genes of individual patients.

One reasonable approach for choosing a mouse breeding strain is to review the pertinent literature and choose the inbred strain demonstrating a phenotype with average, middle-of-the-road performance on the behavioral tasks of interest. This approach allows detection of either an increase or a decrease in the behavioral trait in the transgenic or knockout mice. Unusually high or unusually low behaviors in the parental strain on the behavioral domain of interest are likely to affect the performance of the mutant mouse on that behavior. For example, a strain of mouse with the retinal degeneration gene in its genome will develop blindness and will therefore be a poor choice to use as the breeding strain for a mutation in a gene hypothesized to regulate vision. A breeding strain with very high scores on anxiety-related behaviors could produce a "ceiling effect" on a stressful task. If the hypothesized outcome of the gene mutation is to increase anxiety, the effects of the mutation on anxiety-related behaviors may be undetectable because the normal anxiety level of the parental strain is already maximal. If the goal is to discover a treatment for epilepsy-induced neurodegeneration, FVB/N may be a better choice than C57BL/6, since neuronal cell death in the hippocampus after pilocarpine-induced seizures was greater in FVB/N than in C57BL/6 mice (Figure 2.8; Mohajeri et al., 2004).

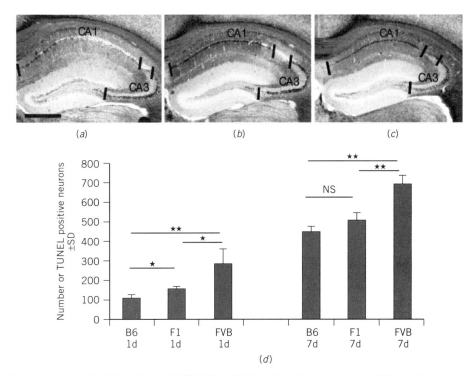

Figure 2.8   *C57BL/6JIbm (B6) and FVB/NJIbm (FVB) strains of mice displayed differential responses to pilocarpine-induced seizures. Neuronal cell loss in pyramidal cells of the hippocampal CA1 and CA3 areas was greatest in FVB (c), least in B6 (a), and intermediate in their F_1 hybrids (b), post-seizure survival times of 1 day (d) and 7 days (e). [From Mohajeri et al., (2004).]*

Further, an unusual allele in the breeding strain may directly interact with the mutated gene, either within the chromosome due to genetic factors or biochemically due to gene product factors. Complex epistatic gene interactions occur between background gene alleles and the mutated gene within the targeted locus (Banbury Conference, 1997; Choi, 1997). Expression of the phenotype depends on which of these background genes are expressed in the strain used to breed the mutant mice (Bailey et al., 2006). For example, mutant mouse models of Alzheimer's disease were designed to overexpress amyloid precursor protein, the molecule cleaved to generate the β-amyloid 1–42 peptide that aggregates to form neuritic plaques in Alzheimer's disease. Dramatic differences in phenotype were found in lines bred with FVB/N, C57BL/6J, and C3H strains (Carlson et al., 1997). When the transgene for amyloid precursor protein was bred into the FVB/N background, the mutation was lethal. When the transgene for the amyloid precursor protein was bred into a C3H background, the mutant offspring survived and memory deficits were detected. Further, transgenic mice overexpressing the β-amyloid precursor protein expressed more severe defects in the corpus callosum, the forebrain commissure connecting the right and left cerebral cortex hemispheres, when bred into 129/SvEv and 129/Ola genetic backgrounds, as compared to a C57BL/6J background (Magara et al., 1999). In other examples, null mutants for the p53 tumor suppressor gene exhibited resistance to kainate-induced neuronal cell death when bred into a C57BL/6J background but lacked resistance to kainate-induced neuronal cell death when bred into a 129/SvEMS background (Schauwecker and Steward, 1997). Decreased sensitivity to the sedative-hypnotic effects of ethanol, and failure to develop chronic tolerance to ethanol, were discovered in γ-protein kinase C knockout mice bred onto a mixed C57BL/6J × 129/SvJ background and backcrossed into a C57BL/6J × 129/SvEvTac background. No genotype difference was detected after the mutation was introgressed into a C57BL/6J background for six generations (Bowers et al., 1999). Anxiety-like phenotypes were detected when serotonin transporter knockout mice were bred onto a C57BL/6J background, but not when the mutation was bred onto a 129SvEv/Tac background (Holmes et al., 2003d). Dopamine transporter knockout mice demonstrated differing amounts of hyperactivity in a novel environment when the mutation was bred onto C57BL/6JOrl, DBA/2JOrl, and the F_1 hybrid backgrounds (Morice et al., 2004).

Thus the choice of strain for the blastula donor and breeders may strongly influence the phenotype obtained for a given gene mutation. The informed investigator chooses a breeding strain based on the hypothesized outcome of the experimental mutation. Strains with normal sensory and motor functions are desirable for most behavioral tasks. Strains with good learning and memory abilities will allow detection of cognitive impairments in mice with mutations in genes necessary for learning and memory. A strain with high levels of anxiety-like behaviors may be ideal for a targeted gene mutation likely to have anxiety-reducing effects. A strain with a high propensity to self-administer addictive drugs will be useful as the background for a genetic mutation hypothesized to reduce drug addiction.

Embryonic Stem Cells Lines

Many strains are available for blastula donation and breeding. However, 129 inbred strains are most extensively used as embryonic stem cell lines for delivering the targeting vector. Figure 2.9 diagrams the pedigree of the many available substrains of 129, indicating the probable family tree of some of the 129 substrains used for the generation of embryonic stem cell lines employed in the knockout technology (Simpson et al., 1997).

The choice of 129 ES line is best determined by knowing the behavioral phenotype of the specific 129 substrain. For example, the 129/J, BTBR $T+$ tf/tf, and the BALB/cWahl substrains fail to develop a normal corpus callosum, the major axonal fiber bundle between the two cortical hemispheres (Wahlsten, 1972, 1982; Livy and Wahlsten, 1997; Wahlsten et al., 2003b). The 129/J, a substrain of 129, shows a dramatic absence of corpus callosum and poor performance on memory tasks (Wehner and Silva, 1996; Montkowski et al., 1997), as shown in Table 2.4. If one wants to study a gene relevant to learning and memory, it is a mistake to use 129/J and some of the other ES cells, since a complex interactions between the corpus callosum background genes and the targeted gene mutation are likely to complicate the interpretation of a learning deficit. Luckily most embryonic stem cell lines were derived from 129/SvJ, 129/SvEvTac, and 129/Ola. The 129/Sv and 129/Ola substrains show normal corpus callosa and normal performance on memory tasks (Wehner and Silva, 1996; Montkowski et al., 1997), as illustrated in Figure 2.10. However, the 129/SvEvTac substrain displays absent or reduced areas of the corpus callosum and deficits in several learning and memory tasks (Balogh et al., 1999; Wahlsten et al., 2003b).

Breeding for behavioral phenotyping would be simple if embryonic stem cell lines for C57BL/6J were available (Wehner and Silva, 1997; Crawley et al., 1997a). The

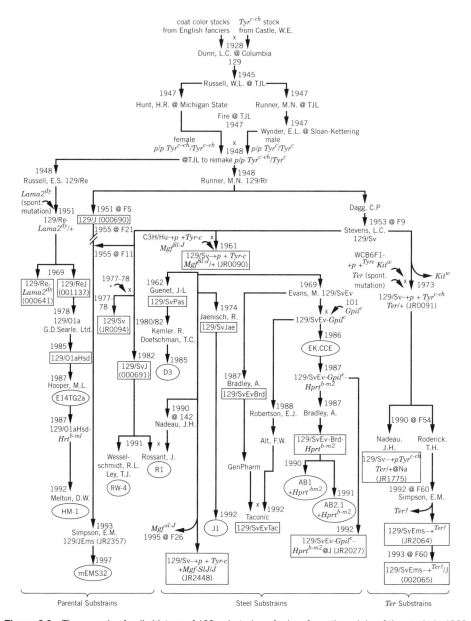

Figure 2.9   *The complex family history of 129 substrains of mice, from the origin of the strain in 1928 to the derivation of commonly used embryonic stem cell lines in 1996.* [From Simpson et al. (1997), p. 21.]

background genotype could then be completely C57BL/6J, rather than including a variable genetic complement from a 129 embryonic stem cell line. C57BL embryonic stem cells have been developed (Lederman and Burki, 1991; Wiles and Keller, 1991; Cheng et al., 2004). Some successes have been reported with generating targeted gene mutations in C57BL/6 embryonic cell lines (Cheng et al., 2004; Seong et al., 2004), although applications are not yet widespread. Another useful approach is to breed into

TABLE 2.4   Complete or Partial Absence of the Corpus Callosum, the Massive Axonal Fiber Bundle Connecting the Right and Left Hemispheres of Cerebral Cortex, in Inbred Mouse Substrains[a]

Strain	% CC Absences	% CC Defect	Initial HC Crossing (g)
	Frequency of callosal absence and defect in relation to the time of initial crossing by hippocampal axons		
B6D2F$_2$	$0^{c,e}$	$0^{c,e}$	0.350
C129F$_2$	$24^{d,g}$	$33^{d,g}$	0.440
BALB/cWahl	20^{c}	$55^{a,c}$	0.470
129/J	16.67^{f}	70^{b}	0.520
RI-1	$100^{d,g}$	$100^{d,g}$	0.750

Source: From Livy and Wahlsten (1997), p. 3.
[a]Absent or defective corpus callosa were discovered in the 129/J and BALB/cWahl substrains. B6D2F$_2$ are the F$_2$ generation from the cross of C57BL/6J × DBA/cJ. C129F$_2$ are the F$_2$ generation of the cross of BALB × 129. The RI-1 is one of the recombinant inbred lines from the 129 × BALB strains.

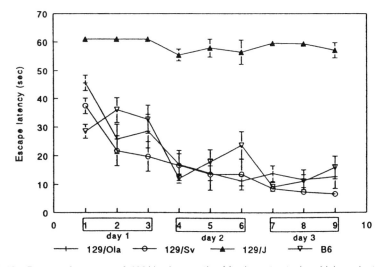

Figure 2.10   *Poor performance of 129/J mice on the Morris water task, which evaluates learning and memory, as compared to two other 129 substrains and the C57BL/6J (B6) inbred mouse strain.* [From Montkowski et al. (1997), p. 14.]

the same inbred substrain of 129 used for the embryonic stem cell line (Lijam et al., 1997), thereby obtaining a complete 129 background genotype.

Breeding Records

Complete and thorough records are necessary for maintaining a breeding colony. Record-keeping systems for breeding a mutation in mice are elegantly described by Lee Silver at Princeton University (1995; *http://www.informatics.jax.org/silver/*). The *mating unit* system records each mating pair and all of its offspring. The *animal/litter* system records each litter and each individual animal. Within either system the researcher or animal care-taker records the number, genotype, gender, birth date, identity of parents, identity of litter, number of pups born to the litter, cage number, and any unusual events or characteristics of the birth and individuals. Commercial software packages for maintaining breeding records of mutant mice include Locus Technology Inc. (*www.locustechnology.com*), Topaz (*www.topazti.com*), and Progeny (*www.progeny2000.com*). The Jackson Laboratory's Applied Genomics Training Center in Bar Harbor, Maine, offers a three-day Colony Management Course (*http://jaxmice.jax.org/library/notes/487m.html* and *http://www.jax.org/courses/events/current.do*). Dr. Lee Silver wrote Animal House Manager (AMAN for the PC and MacAMAN for the MacIntosh computer), a computer software package with specialized database programs for record keeping of animals, genotypes, parents, litters, cages, and weaning dates. His outstanding book, *Mouse Genetics* (Silver, 1995), is online at *http://www.informatics.jax.org/silver/*.

Identification is permanently affixed to the mouse by ear punch pattern, tattoo, toe clip, numbers engraved on metal ear tags, or bar code pattern on a subcutaneously implanted chip. Tattoo is appropriate when identification and genotyping are needed for very young pups, and is best completed by age 5 days. Toe clip should be completed by age 7 days. Commercial sources for bar code chips and scanners include AVID, 3179 Hamner Avenue, Norco, California, 91760, 909-371-7505, *www.avidmicrochip.com*, and BioMedic Data Systems, Inc., 1 Silas Road, Seaford, Delaware 19973, 800-526-2637, *www.bmds.com*. For extra insurance two identification methods may be used in each animal because ear tags occasionally fall off, ear punch patterns may become ragged, and chips occasionally migrate to an inaccessible body region.

HOUSING

Our animal facility at the National Institutes of Health, and many other laboratory animal facilities, houses mice in plastic "shoebox" cages with wire lids and filter-paper-covered plastic microisolator tops. Many facilities maintain these cages in ventilated racks to maximize cleanliness. However, the higher noise levels generated by the ventilation system can impair breeding, and can affect performance on some behavioral tasks. In addition, fluorescent lightbulbs emit ultrasonic vocalizations that can affect breeding and behavior. It is better to locate breeding cages on the lower shelves of large housing racks, as far as possible from fluorescent lights in the ceiling. A maximum of five adult mice per standard shoebox cage provides good animal care and is cost-effective. Food and water are available ad libitum. Nesting material is provided in breeding cages. Temperature and humidity are controlled. A circadian light cycle of 12 hours light/12 hours dark is maintained in the housing room.

Housing is generally arranged by gender and genotype. Each mating pairs is housed in a standard cage in the vivarium. Another approach is harem mating, with one male and two or three females per breeding cage. Opinions vary on the optimal number of breeding females to house with a breeder male, and whether it is best to raise litters separately or together. Pregnant females are often removed from the harem and placed in individual cages with nesting material for parturition. Pups are weaned from the mother between ages 3 and 4 weeks. First estrus for female mice usually begins at 5 to 6 weeks of age (Hedrick and Bullock, 2004). Separation of juvenile males and females into gender-specific home cages at weaning will ensure that no unwanted matings occur.

Separate caging of +/+, +/−, and −/− mice may be necessary if the +/+ mice are larger, healthier, or more aggressive than the −/− mice, to prevent the +/+ individuals from hoarding food or attacking the −/− mice. If severe fighting is ongoing in a cage, it may be necessary to remove the dominant male and house him in an individual cage to prevent severe wounding and social stress. However, isolation is also a stressor in mice. Single housing elevates aggression in male mice, as discussed in Chapter 9. Therefore, if some mice must be individually housed, then ideally all mice for that experiment are individually housed. Unfortunately, individual housing may be limited by costs and the total caging space available to the experimenter. Instead of complete individual housing, a single highly aggressive mouse can be removed from the home cage and removed from the experiment. This is a common occurrence, since many strains of mice display social hierarchies characterized by a single dominant, aggressive male. Records kept of the highly aggressive individuals will reveal whether they are all of one genotype, suggesting further investigation with specific tests for an aggressive phenotype caused by the targeted gene mutation (see Chapter 9). If there are no problems with competition or aggression in the home cage, then the genotypes can be mixed in the home cage.

GROUP SIZE AND COMPOSITION

For meaningful statistical interpretations of behavioral phenotyping experiments, the minimum recommended number of animals is in the range of 10 mice of each genotype. This means at least $N = 10+/+$ wildtype mice, $N = 10+/−$ heterozygous mutant mice, and $N = 10−/−$ null mutants. The first experiments include both males and females of each genotype. Gender analysis of the first results will determine whether a gender difference is detected for the mutation. If the comparison of the male null mutants versus the male wildtypes differs from the comparison of the female null mutants versus the female wildtypes, gender is likely to be a determining factor in phenotyping the mutation. If a gender effect is detected, $N = 10$ of each sex of each genotype is needed.

These numbers of mice may sound like a lot of mice when the null mutants are first being developed and breeding is going slowly. Sizable Ns are usually necessary to satisfy the criteria for appropriate statistical analysis of behavioral data. Doug Wahlsten at the University of Edmonton, Alberta, Canada, offers suggestions for choosing Ns that will provide sufficient power to detect small, medium, large, and very large genotype effects (Chapter 13, Figure 13.1). Appropriate statistical tests include t-tests when comparing only two groups on one behavioral measure, Analysis of Variance when comparing three or more groups on one measure, and Repeated Measures Analysis of Variance when the same mice are used repeatedly within a task. Factors analyzed within a Multiple or Repeated Measures ANOVA may include genotype, sex, time points, or treatments, and combinations such as genotype × treatment. Significant ANOVA values justify follow-up post-hoc tests to compare individual means. For example, the post-hoc comparisons between +/+, +/−, and −/− will reveal whether the heterozygote phenotype differed from the wildtype phenotype on a behavioral test, and whether the null mutants differed from the heterozygotes. Choice of post-hoc tests, such as Fisher's PLSD, Newman-Keuls, Dunnett's, Scheffe, and Bonferroni's, depend on the properties of the data set, such as normal distribution, equal or unequal Ns, multiple tests conducted on the same mice, and missing data cells. Numbers of mice needed to reach sufficient statistical power to detect small genotype effects are illustrated in Figure 13.1. $N = 20$ mice or more per genotype are often needed to complete the first set of behavioral tests in a new line of mutant mice. Another $N = 20$ may then be needed to replicate the first findings for publication. Therefore, the best approach is to establish sufficient breeding pairs to generate full sets of mice dedicated to the behavioral experiments.

The age of the mice is important for behavioral experiments. "Neonatal" is from birth to almost 3 weeks of age. "Juvenile" is usually 3 to 7 weeks after weaning and before sexual maturity. "Adult" is between ages 2 and 12 months. "Aged" is between 13 and 24 months. Most strains of laboratory mice have lifespans of about 2 years (Silver, 1995). Depending on the goals of the experiment, mice can be chosen within the desired age category, for example, older mice to study genes relevant to aging, as described in Chapter 12. For standard adult mice to test in behavioral experiments, it is best to use mice between 3 and 6 months of age. This 4-month age span provides a relatively homogeneous age-matched distribution across genotypes. Behavioral experiments are likely to take several months to complete. The mice will not reach the "aged" status during the course of the behavioral experiments if testing begins at around 3 months of age. Chapter 12 discusses specific testing issues for neurodevelopmental and aging studies.

Sometimes the full Ns for each genotype within the recommended age range cannot be obtained all at once. This problem arises when the mutation impairs survival, when breeding is not prolific (Cheng et al., 1998), or when sufficient cage space is not available to breed large numbers of mice simultaneously. Partial Ns can be tested as subgroups, as long as similar Ns of each genotype are tested concurrently within each round of experiments. Second, third, and fourth rounds of mice from each genotype are tested at later dates, as new subgroups become available. Statistical analyses comparing test dates will reveal any differences across subgroups of mice for each behavioral test conducted. If no differences are detected across test dates, then the data for the several sets of mice can be combined to reach the required total Ns for a given experiment.

TRANSPORTATION

When the mice are developed and bred at one site, and the behavioral phenotyping experiments are to be conducted at another site, issues of transportation, quarantine, and acclimation arise. Two types of vivaria are commonly maintained. (1) Specific-pathogen-free (SPF) facilities are maximally clean of parasites, bacteria, and viruses that infect mice. (2) Conventional facilities ensure that the mice are generally healthy by minimizing and containing pathogens. Once the animals arrive, a quarantine period of up to several months may be required before the mice are introduced into the housing rooms. SPF facility managers will require extensive serology and pathology reports to ensure that incoming animals do not inadvertently introduce parasites, viruses, or diseases. If mice are developed at a conventional animal facility, it may be impossible to import them into a specific-pathogen-free facility. Instead, the genotypes may be rederived by caesarian section and cross-fostered into females housed in the SPF vivarium. If mice are developed at an SPF facility and the behavioral phenotyping is conducted at a conventional housing facility, there is usually no problem with transfer from the very clean to the less clean facility.

Transportation of the mice conforms to the requirements of the facility, in accordance with institutional, national, and international animal care and use guidelines. After the mice have reached their final housing destination, they need to acclimate to their new surroundings for at least one week before the start of behavioral experiments. The stress of transportation and adjusting to a new living environment can directly affect behavior. After about a week, the stress effects have usually diminished, and behavioral baselines have stabilized.

CAVEATS

In an elegant comparison of strain distributions of mice in several behavioral tests across three geographically unrelated sites, Crabbe et al. (1999b) used identical strains, suppliers, breeding, housing conditions, behavioral test equipment, and precisely standardized methods. A housing system of two mice per cage, housed by genotype and gender, was employed for the small Ns of these experiments. This Multi-Center Trial of a Standardized Battery of Tests of Mouse Behavior, conducted by John Crabbe at the Oregon Health Sciences University at Portland, Oregon, Doug Wahlsten at the University of Edmonton in Edmonton, Alberta, Canada, and Bruce Dudek at State University of New York at Albany in Albany, New York, with the support of the National Institutes of Health (NIH) Office of Behavioral and Social Sciences Research, can be viewed in detail at Web site address *http://www.albany.edu/psy/obssr*. The authors were extremely careful about standardizing every component of their methods across the three test sites. Good agreement in the qualitative findings between mouse strains was seen for most of the behavioral tests, analogous to the level of agreement across laboratories for other techniques such as receptor binding assays, microdialysis, and anatomical cell counts. However, as with other kinds of biological assays, variability in the precise numerical results was evident across the three sites. Further, in the case of a null mutant behavioral phenotype, some of the tests in this study showed both qualitative and quantitative differences between sites. Careful in-depth follow-up analyses confirmed that the major differences between inbred strains on most behavioral tasks were replicated across the three laboratories (Wahlsten et al., 2003c).

Thus, no one single set of methods for breeding, housing, and testing will ensure absolutely consistent results across laboratories studying the phenotype of a mutation in mice. At present, the use of wildtype littermates as the comparison group represents the best internal control for all of the variables that may interact with the mutation and influence the behavioral phenotype. Multiple replications with independent sets of mice, independent founder lines, and in independent laboratories, will help to definitively confirm or disprove the findings, as occurs in the normal incremental process of scientific discovery in every field.

BEFORE YOU START

Before starting any behavioral phenotyping experiments, you will need to obtain approval for the behavioral protocols from your Institutional Animal Care and Use Committee or national regulatory organization. All behavioral procedures described in this book conform with the requirements of the NIH Guide for the Care and Use of Laboratory Animals, and the US Public Health Service Policy on Humane Care and Use of Laboratory Animals. Each institution and country has analogous guidelines that must be followed. Web site addresses that contain useful information include *http://dels.nas.edu/ilar_n/ilarhome* and *http://www.aphis.usda.gov/ac/*. Guidelines and standard operating procedures are offered for the specialized needs of behavioral neuroscience, including food and water restriction protocols, types of cage enrichment that will not interfere with behavioral tasks, and specialized methods for cleaning of behavioral test equipment.

Congratulations! You are now ready to start your behavioral phenotyping experiments. Preliminary observations, general tests for sensory and motor functions, and constellations of specific tests for the range of behavioral categories are described in Chapters 3 through 13.

BACKGROUND LITERATURE

DNA Constructs

Good reviews that describe current methods for developing a targeting vector and inserting the mutation into the mouse genome include:

Burt DR (2003). Reducing GABA receptors. *Life Sciences* **73**: 1741–1758.

Campbell IL, Gold LH (1996). Transgenic modeling of neuropsychiatric disorders. *Molecular Psychiatry* **1**: 105–120.

Capecchi MR (1994). Targeted gene replacement. *Scientific American* **270**: 52–59.

Champtiaux N, Changeux JP (2004). Knockout and knockin mice to investigate the role of nicotinic receptors in the central nervous system. *Progress in Brain Research* **145**: 235–251.

Current Protocols in Molecular Biology (1995 and ongoing supplements). *Introduction of DNA into Mammalian Cells*. Wiley, New York, Chapter 9.

Current Protocols in Neuroscience (1995 and ongoing supplements). *Gene Cloning, Expression, and Mutagenesis*. Wiley, New York, Chapter 4.

Doetschman TC (1991). Gene targeting in embryonic stem cells. *Biotechnology* **16**: 89–101.

Giese KP (1999). The use of targeted point mutants in the study of learning and memory. In *Handbook of Molecular-Genetic Techniques for Brain and Behavior Research*, WE Crusio, RT Gerlai, Eds. Elsevier Science, Amsterdam, pp. 305–314.

Goldowitz D, Wahlsten D, Wimer RE, Eds. (1992). *Techniques for the Genetic Analysis of Brain and Behavior: Focus on the Mouse*. Elsevier, Amsterdam.

Hofker MH, Van Deursen J, Sklar HT (2002). *Transgenic Mouse: Methods and Protocols*. Humana Press, Totowa, NJ.

Jaenisch R (1988). Transgenic animals. *Science* **240**: 1468–1472.

Ledermann B (2000). Embryonic stem cells and gene targeting. *Experimental Physiology* **85**: 603–613.

Mayford M, Bach ME, Huang YY, Wang L, Hawkins RD, Kandel ER (1996). Control of memory formation through regulated expression of a CaMKII transgene. *Science* **274**: 1678–1683.

Mansuy IM, Bujard H (2000). Tetracycline-regulated gene expression in the brain. *Current Opinions in Neurobiology* **10**: 593–596.

Mansuy IM, Mayford M, Kandel ER (1999). Regulated temporal and spatial expression of mutants of CaMKII and calcineurin with the tetracycline-controlled transactivator (tTA) and reverse rTA (rtTA) systems. In *Handbook of Molecular-Genetic Techniques for Brain and Behavior Research*, WE Crusio, RT Gerlai, Eds. Elsevier Science, Amsterdam, pp. 291–304.

McGuffin P, Owen MJ, Eds. (2002). *Psychiatric Genetics and Genomics*. Oxford University Press, Oxford.

Morse HC (1978). *Origins of Inbred Mice*. Academic Press, New York. Adapted for the Web by JAX NIAID NIH, http://www.informatics.jax.org/morsebook.

Papaioannou VE, Behringer RR (2005). *Mouse Phenotypes: A Handbook of Mutation Analysis*. Cold Spring Harbor Laboratory Press, Cold Spring Harbor, NY.

Rülicke T, Hübscher U (2000). Germ line transformation of mammals by pronuclear microinjection. *Experimental Physiology* **85**: 589–601.

Sedivy JM, Joyner AL (1992). *Gene Targeting*. Freeman, New York.

Silver LM (1995). *Mouse Genetics*. Oxford University Press, New York.

Smithies O (1993). Animal models of human genetic diseases. *Trends in Genetics* **9**: 112–116.

Steele PM, Medina JF, Nores WL, Mauk MD (1998). Using genetic mutations to study the neural basis of behavior. *Cell* **95**: 879–882.

Tecott LH (2003). The genes and brains of mice and men. *American Journal of Psychiatry* **160**: 646–656.

Tecott LH, Wehner JM (2001). Mouse molecular genetic technologies. *Archives of General Psychiatry* **58**: 995–1004.

Tenenbaum L, Chtarto A, Lehtonen E, Velu T, Brotchi J, Levivier M (2004). Recombinant AVV-mediated gene delivery to the central nervous system. *Journal of Gene Medicine* **6**: S212–S222.

Tsien JZ, Chen DF, Gerber D, Tom C, Mercer EH, Anderson DJ, Mayford M, Kandel ER, Tonegawa S (1996). Subregion- and cell type-restricted gene knockout in mouse brain. *Cell* **87**(7): 1317–1326.

Winshaw-Boris A, Garrett L, Chen A, Barlow C (1999). Embryonic stem cells and gene targeting. In *Handbook of Molecular-Genetic Techniques for Brain and Behavior Research*, WE Crusio, RT Gerlai, Eds. Elsevier Science, Amsterdam, pp. 259–271.

Breeding Strategies and Background Strains

Once the mutation is inserted into the genome of a mouse, the breeding of subsequent generations is designed to retain the mutation in the germ line of the offspring. Some of the pioneers in the development of breeding strategies have published strain distributions and background strain recommendations:

Bogue MA, Grubb SC (2004). The mouse phenome project. *Genetica* **122**: 71–74.

Bolivar V, Cook M, Flaherty L (2000b). List of transgenic and knockout mice: Behavioral profiles. *Mammalian Genome* **11**: 260–274.

Banbury Conference (1997). Mutant mice and neuroscience: Recommendations concerning genetic background. *Neuron* **19**: 755–759.

Cook MN, Bolivar VJ, McFadyen MP, Flaherty L (2002). Behavioral differences among 129 substrains: Implications for knockout and transgenic mice. *Behavioral Neuroscience* **116**: 600–611.

Crawley JN, Belknap JK, Collins A, Crabbe JC, Frankel W, Henderson N, Hitzemann RJ, Maxson SC, Miner LL, Silva AJ, Wehner JM, Wynshaw-Boris A, Paylor R (1997). Behavioral phenotypes of inbred mouse strains: Implications and recommendations for molecular studies. *Psychopharmacology* **132**: 107–124.

Holmes A, Harari AR (2003). The serotonin transporter gene-linked polymorphism and negative emotionality: Placing single gene effects in the context of genetic background and environment. *Genes, Brain and Behavior* **2**: 332–335.

Jackson IJ, Abbott CM (2000). *Mouse Genetics and Transgenics: A Practical Approach*. Oxford University Press, Oxford.

Jones BC, Mormede P (1999). *Neurobehavioral Genetics: Methods and Applications*. CRC Press, Boca Raton, FL.

Joyner AL (2000). Gene *Targeting: A Practical Approach*. Oxford University Press, Oxford.

Lyon MF, Rastan S, Brown SDM (1996). *Genetic Variants and Strains of the Laboratory Mouse*. Oxford University Press, New York.

Nagy A, Gertsenstein M, Vintersten K, Behringer R (2002). *Manipulating the Mouse Embryo: A Laboratory Manual*. Cold Spring Harbor Laboratory Press, Cold Spring Harbor, NY.

Petkov PM, Ding Y, Cassell MA, Zhang W, Wagner G, Sargent EE, Asquith S, Crew V, Johnson KA, Robinson P, Scott VE, Wiles MV (2004). An efficient SNP system for mouse genomic scanning and elucidating strain relationships. *Genome Research* **14**: 1806–1811.

Silver LM (1995). *Mouse Genetics*. Oxford University Press, New York. *http://www.informatics. jax.org/silver/*.

Wehner JM, Silva A (1996). Importance of strain differences in evaluations of learning and memory processes in null mutants. *Mental Retardation and Developmental Disabilities Research Reviews* **2**: 243–248.

Mouse Handbooks

Useful information about laboratory mouse biology and husbandry is found in several major books:

Green EL, Ed. (1966). *Biology of the Laboratory Mouse*. McGraw-Hill, New York.

Crispens CG (1975). *Handbook on the Laboratory Mouse*. C.C. Thomas, Springfield, IL.

Foster HL, Small JD, Fox JG (1982). *The Mouse in Biomedical Research*. Academic, New York.

Fox JG, ed (2006). *The Mouse in Biomedical Research*. Academic, New York.

Green MC, Witham BA (1991). *Handbook on Genetically Standardized JAX Mice*. Jackson Laboratory, Bar Harbor, ME.

Hedrick H, Bullock G (2004). *The Laboratory Mouse*. Academic Press, San Diego.

Jackson IJ, Abbott CM (2000). *Mouse Genetics and Transgenics: A Practical Approach*. Oxford University Press, New York.

Jacobowitz DM, Abbott LC (1998). *Chemoarchitectonic Atlas of the Developing Mouse Brain*. CRC Press, Boca Raton, FL.

Lyon MF, Searle AG (1989). *Genetic Variants and Strains of the Laboratory Mouse*. Oxford University Press, Oxford.

Silver LM (1995). *Mouse Genetics*. Oxford University Press, New York.

Ward JM, Mahler JF, Maronpot RR, Sundberg JP (1999). *Pathology of Genetically-Engineered Mice*, Blackwell, Ames, Iowa.

Companies That Manufacture Behavioral Test Equipment

Some of the companies that manufacture high-quality behavioral equipment include:

Actimetrics, 1024 Austin Street, Evanston, IL 60202 USA, phone 847-922-2643, fax 847-589-8103, *http://www.actimetrics.com*.

AccuScan Instruments, Inc., *www.accuscan-usa.com*, 5090 Trabue Road, Columbus, OH 43228 USA, phone 1-800-822-1344, 1-614-878-6644, fax 1-614-878-3560, *sales@accuscan-usa.com*.

Biobserve, *www.biobserve.com*, 2125 Center Avenue, Suite 500, Fort Lee, NJ 07024, phone 1-201-302-6083, fax 1-201-302-6062, *info@biobserve.com*.

Clever Sys, Inc. 11480 Sunset Hills Road, Suite 210 W, Reston, VA 20190 USA, 1-703-787-6946, *www.cleversysinc.com*, *sales@cleversysinc.com*, *support@cleversysinc.com*.

Columbus Instruments, *www.colinst.com*, 950 North Hague Avenue, Columbus, OH 43204-2121 USA, phone 1-800-669-5011, 1-614-276-0861, fax 1-614-276-0529, *sales@colinst. coin.*

Computervision Laboratory, University of California, San Diego, USA, *http://vision.ucsd.edu/ smart_vivarium.*

Coulbourn Instruments, 7462 Penn Drive, Allentown, PA 18106 USA, phone 1–800 424-3771, fax 610-391-1333, *www.coulbourn.com*.

Hamilton-Kinder, *www.hamiltonkinder.com*, phone 1-858-679-1515, fax 1-858-679-4811, *mkinder@hamiltonkinder.com*, *chamilton@hamiltonkinder.com*.

IITC Life Science, 23924 Victory Blvd, Woodland Hills, CA 91367, USA, phone 1-888-414-4482, 1-818-710-8843, fax 1-818-992-5185, *http://www.iitcinc.com*.

Lafayette Instrument, *www.lafayetteinstrument.com*, 3700 Sagamore Parkway North, PO Box 5729, Lafayette, IN 47903 USA, phone 1-800-428-7545, 1-765-423-1505, fax 1-765-423-4111, *tehard@lafayetteinstrument.com*, *eval@lafayetteinstrument.com*, *yvelez@lafayetteinstruments.com*.

MED Associates Inc., *www.med-associates.com*, PO Box 319, St. Albans, VT 05478 USA, phone 1-802-527-9724, fax 1-802-527-5095, *www.med-associates.com*, *medas@med-associates.com*.

Metris BV, Saturnusstraat 12, 2132 HB Hoofddorp, PO Box 3023, 2130 KA Hoofddorp, The Netherlands, 31-(0)-23-554-2250, FAX 31-(0)-23-557-1069, *info@metris.nl*, *www.metris.nl*.

Mouse Specifics, Inc. 28 State Street, Suite 1112, Boston, MA 02109 USA, phone 1-617-821-6687, *www.mousespecifics.com*.

NewBehavior AG Intellicage, Hardturmstrasse 76, CH-8005 Zürich, Switzerland, 41-44-272-73-44, fax 41-44-440-03-81, *www.newbehavior.com*, *info@newbehavior.com*.

Noldus Information Technology, *www.noldus.com*, Costerweg 5, PO Box 268, 6700 AG Wageningen, The Netherlands, phone 31-317-497677, fax 31-317-424496, *info@noldus.nl*; Sasbacher Strasse 6, D-791 11 Freiburg, Germany, phone 49-761-4701600, fax 49-761-4701609, France 33-389-697323, *infonoldus.de*; 751 Miller Drive, Suite E-5, Leesburg, VA 20175-8993 USA, phone 1-703-771-0440, fax 1-703-771-0441, *info@noldus.com*.

PanLab SL, C/Energia 112, 08940 Cornella, Barcelona, Spain, phone 34-934-750-697, fax 34-934-750-699, *info@panlab-sl.com*, *http://www.panlab-sl.com/English/Letica_Products/Behaviour/Behaviour.htm*.

RAPC Bridgekey Corporation, 2000 Winton Rd S # 5–103 Rochester, NY 14618-3922 USA, phone: (585) 240–6012, fax: (585) 240–6011, *www.rapc.us*.

San Diego Instruments, *www.sd-inst.com*, 7758 Arjons Drive, San Diego, CA 92126-4391 USA, phone 1-858-530-2600, fax 1-858-530-2646, sales@sd-inst.com.

Stoelting Company, *www.stoeltingco.com/physio*, 620 Wheat Lane, Wood Dale, IL 60191 USA, phone 1-630-860-9700, fax 1-630-860-9775, *physiology@stoeltingco.com*.

TSE Systems, Technical and Scientific Equipment GmbH, Siemensstr.21, 61352 Bad Homburg, Germany, phone 49 (0) 6172 789 0, fax 49 (0) 6172 789 500, *www.TSE-Systems.de*, *info@TSE-Systems.de*.

Vicon Peak Performance Systems, 9 Spectrum Pointe Drive, Lake Forest, CA 92630, USA, phone 1–949-472-9140, *http://www.peakperform.com/*.

Companies Generating and/or Distributing Mutant Lines of Mice

Several academic consortia and commercial organizations offer transgenic and knock-out mouse production services, embryonic stem cell lines containing gene mutations, and/or sell donated lines of mutant mice or frozen embryos. Examples include:

BayGenomics, *http://baygenomics.ucsf.edu*, in collaboration with the National Institutes of Health National Center for Research Resources *http://www.ncrr.nih.gov/*, Mutant Mouse Regional Resource Center *http://www.mmrrc.org/index.html*

Biocon, Inc., 15801 Crabbs Branch Way, Rockville, MD 20855, phone 301-762-3202.

Charles River Laboratories, 251 Ballardvale Street, Wilmington, MA 01887.

Chrysalis DNX Transgenics, 303B College Road East, Princeton Forrestal Center, Princeton, NJ 08540, phone 609-520-0300.

Deltagen, 1031 Bing Street, San Carlos, CA 94070 USA, phone 650-569-510, fax 650-569-5280, *http://www.deltagen.com*.

International Gene Trap Consortium, *http://www.igtc.ca*.

JAX Mice, The Jackson Laboratory, 600 Main Street, Bar Harbor, ME 04609 USA.

Knockout Mouse Project, National Institutes of Health, Bethesda, MD, USA, described in Austin et al., (2004) *Nature Genetics* **36**(9): 921–924.

Lexicon Genetics Inc., 4000 Research Forest Drive, The Woodlands, TX 77381, phone 281-364-0100, *www.lexgen.com*.

Neuromice, Northwestern University Center for Functional Genomics, 2205 Tech Drive, Evanston, IL 60208, phone 847-467-4686, *www.neuromice.org*.

PolyGene, the Swiss Transgenic Mouse Services Company, Riedmattstrasse 9, 8153 Ruemlang, Switzerland, phone 41-(0)44/828 63 86, *www.polygene.ch*.

Taconic, 273 Hanover Avenue, Germantown, NY 12526 USA, *http://www.taconic.com*.

Companies Offering Software for Mouse Breeding Colony Management:

Big Bench Software, *http://www.bigbench.com*

Circusoft Instrumentation, Hockessin, DE, USA, *http://www.circusoft.com/gmouse.html*

The Jackson Laboratory, Bar Harbor, ME, USA, *http://jaxmice.jax.org/library/notes/487m.html* Colony management course at The Jackson Laboratory and *http://jax.org/courses/archives/2003/colf03_wray_breeding.pdf* Charles G. Wray, Ph.D., The Jackson Laboratory, lecture notes on breeding strategies

Locus Technology Inc., Annapolis, MD, USA, *http://www.locustechnology.com*

Software for maintaining mutant mouse breeding colony records:

LAMS, Medical Research Council, Edinburgh, UK, *http://www.hgu.mrc.ac.uk/Softdata/Lams/*

Progeny Software, South Bend, IN, USA, *http://www.progeny2000.com*

Topaz Technologies, Austin, TX, USA, *http://www.topazti.com*

Transgenic Software Inc., Nampa, IN, USA, *http://www.transgenic-software.com*

Organizations and Companies Conducting Mouse Behavioral Phenotyping

Academic institutions, consortium organizations, and companies established to conduct behavioral phenotyping of transgenic and knockout mice, as well as biochemical, histological, and other aspects of phenotyping, include:

EUMORPHIA: The European Union consortium conducting mutagenesis, phenotyping and informatics, toward understanding human disease through mouse genetics. *www.eumorphia.org*

Mouse IQ: Neurobehavioural Research: Dr. Richard E. Brown, Psychology Department, Dalhousie University, Halifax, Nova Scotia B3H 4J1, phone 902-494-3647; fax 902-494-6585.

Neuro-Bsik Mouse Phenomics: Consortium of Dutch neuroscience institutes and corporations, Eric Meijer, Consortium Manager, phone 31-10-4087571, fax 31-10-4089457, *e.j.c.meijer@erasmusmc.nl.*

NeuroDetective, Inc.: Drs. Ian Q. Whishaw and Brian Kolb, University of Lethbridge, 4401 University Drive, Lethbridge, Alberta, Canada, T1K 3M4; phone 403-329-2235, fax 403-329-2555. Dr. Forrest Haun, 1757 Wentz Road, Quakertown, PA 18951, phone 215-536-8757; *fhaun@neurod.com*, and *www.neurodetective.com.*

Neurofit: rue Jean Sapidus, Parc d'Innovation, 6700 Illkirch, France, phone 33-388-651606, fax 33-388-651622, *neurofit@tpgnet.net.*

PsychoGenics, Inc.: 4 Skyline Drive, Hawthorne, NY, *www.psychogenics.com.*

Genetics Web Site Addresses

Alphabetical list of knockouts, their description and references, *http://www.bioscience.org/knockout/knochome.htm*

Bay Genomics embryonic stem cell lines with insertional mutations, *http://baygenomics.ucsf.edu/*

BiomedNet database of mouse knockouts and mutations, *http://research.bmn.com/mkmd*

BioMedNet Mouse Knockout Database, *http://biomednet.com/db/mkmd*

DNA Data Bank of Japan, *http://www.ddbj.nig.ac.jp*

European Bioinformatics Institute, *http://www.ebi.ac.uk*

GenBank database maintained by the National Center for Biotechnology Information, includes Mouse Genome Resources, access to LocusLink, UniGene, Mouse Blast, *http://www.ncbi.nlm.nih.gov*

Genetics Home Reference: Your guide to understanding genetic conditions. National Library of Medicine, USA, *http://www.ghr.nlm.nih.gov/ghr/page/Home*

Guide for the Care and Use of Laboratory Animals, *http://www2.nas.edu/ilarhome/240a.html*

Human Genome Project, site maintained by the US Department of Energy, *http://www.ornl.gov/hgmis*

International Behavioural and Neural Genetics Society/Useful links, *http://www.ibngs.org*

International Behavioral Neuroscience Society, *http://www.ibnshomepage.org*

Knockouts and Mutants: Genetically Dissecting Brain and Behavior, *http://elsevier.com/locate/bri98*

Mammalian Genetics Unit, Medical Research Council, Harwell, UK, *http://imsr.har.mrc.ac.uk/*

Mouse Genome Database, including single nucleotide polymorphisms database across inbred strains, maintained by The Jackson Laboratory, *http://www.informatics.jax.org/*

Mouse Phenome Database, *http://www.jax.org/phenome*

National Institutes of Health Mouse Repository, *http://mouse.ncifcrf.gove*

Neurosciences on the Internet, *http://www.neuroguide.com*

Oak Ridge National Laboratory Mutant Mouse Database, *http://bio.lsd.ornl.gov/mouse*

Perlegen database of single nucleotide polymorphisms in inbred strains of mice, *http://mouse.perlegen.com/mouse/download.html*

Reference list of gene knockouts in mice that affect nervous system phenotype and functions. Index list maintained by Dr. Jonathan Pollock, National Institute on Drug Abuse, National Institutes of Health, *http://www.nida.nih.gov* and *http://165.112.78.61/genetics/ko/ko-index.html*

The Jackson Laboratory search engine for genetically modified mouse strains, *http://www.jax.org*

The Society for Neurosciences, *www.sfn.org*

Trans-NIH Mouse Initiative, *http://www.nih.gov/science/mouse/*

Transgenic/Targeted Mutation Data Base, maintained by The Jackson Laboratory, *http://tbase. jax.org*

Whole Mouse Catalog—Genome Databases and Genome Maps, *http://www.rodentia.com/wmc/ index.html*

3

General Health

GIVE YOUR MOUSE A PHYSICAL

The first step in behavioral phenotyping is to check your mice for gross abnormalities that will obviously interfere with behavioral testing. A very sick mouse will lie immobile on the bottom of the home cage and fail to respond to most behavioral challenges. Hyperactivity to handling, including biting the hand of the investigator trying to pick up the mouse from its home cage, often reflects illness. A mouse that cannot walk will do poorly on many behavioral tasks. The overwhelming majority of behavioral tasks require motor coordination and locomotor activity. A blind mouse is useless for any sort of visual discrimination task. A deaf mouse cannot be cued with an auditory tone. An anosmic mouse shows no response to olfactory cues.

The goal of preliminary observations is to eliminate artifacts. Preliminary behavioral observations of mutant mice in their home cage, along with a set of simple reflex tests, allow the investigator to detect overt, serious dysfunctions. You would much rather know from the beginning that your mouse is blind than to spend considerable time on training in the Morris water maze, a spatial learning task based on distant visual cues, and then falsely conclude that your mouse is learning impaired when it merely failed to see the signs. Foreknowledge allows the investigator to design alternative tasks that are unaffected by the physical defect. For example, blind mice can be tested for learning and memory abilities using an olfactory discrimination task (described in Chapter 6).

Preliminary observations are analogous to the tests done in a routine physical examination. Your physician may check your body weight, body temperature, complexion, blood pressure, pulse, heartbeat, vision, hearing, neurological reflexes, and ask about your mood, work, social life, drinking, eating, and sleeping patterns as measures of your general health. If any of these are abnormal, the physician will subsequently pursue the problem in depth. If you are fine on these superficial measures, but came in

What's Wrong With My Mouse? By Jacqueline N. Crawley
Copyright © 2007 John Wiley & Sons, Inc.

for a specific complaint, the physician then has a greater level of confidence that there is not a more general reason for the specific symptom, and will proceed with more sophisticated tests to diagnose and treat your illness.

OBSERVATIONAL BATTERIES TO EVALUATE GENERAL HEALTH

Perhaps the first ethological observations of mouse mutant lines were conducted by van Abeelen at the University of Nijmegen in The Netherlands (van Abeelen, 1963). Samuel Irwin at the University of Oregon systematized one of the earliest comprehensive assessment protocols (Irwin, 1968). Many of the presently used screens derive from the original list of tests defined by Irwin, shown in Table 3.1. Scoring scales from 0 to 8 are employed for each of 50 categories of observations. Assessment begins with observations of the undisturbed animal's body position, respiration, gait, and locomotor activity, along with any bizarre or stereotyped motor behaviors, tremors, convulsions, excessive salivation, urination, or diarrhea. The experimenter records body tone, body temperature, pupil size, pelvic elevation, tail elevation, piloerection, exophthalmos, skin color, and gait. Responses elicited to startle, tail pinch, tail suspension, and handling are noted. Righting reflex, grip strength, provoked biting, and freezing are scored. The Irwin battery used at the Schering Corporation, a pharmaceutical company in Bloomfield, New Jersey, was one of the first behavioral pharmacology approaches to characterize drug responses. For example, chlorpromazine treatment reduces behavioral arousal and muscle tone and induces hypothermia, measures easily detected with the Irwin screen.

Behavioral neurotoxicologists developed excellent test batteries, such as the comprehensive sequence designed by Virginia Moser and colleagues at the US Environmental Protection Agency in Research Triangle Park, North Carolina (Moser et al., 1995). Observations are organized by behavioral domains. Six domains are analyzed: autonomic, neuromuscular, activity, sensorimotor, excitability, and physiological measures. Table 3.2 shows the individual parameters that are quantitated in this battery.

The SHIRPA protocol was originally designed by Derek Rogers and co-workers at SmithKline Beecham Pharmaceuticals in Harlow, England (*http://www.mgu.har.mrc. ac.uk/facilities/mutagenesis/mutabase/shirpa_summary.html*). The SHIRPA Consortium in the United Kingdom is analyzing phenotypes of mutant mice generated by chemical mutagenesis, described in Chapter 14. Three levels of behavioral testing are defined. The first level is a set of preliminary behavioral observations (Rogers et al., 1997). As shown in Figure 3.1, categories include muscle and lower motor neuron functions, spinocerebellar functions, sensory functions, and autonomic functions. The second level of tests includes exploratory locomotion, feeding, analgesia tests, histology, and biochemistry. The third level represents more complex behavioral and physiological approaches, including learning and memory, anxiety, prepulse inhibition, neurophysiology, and magnetic resonance imaging. Several other behavioral batteries to screen chemical mutagenesis progeny are in current use (Tarantino et al., 2000; Sayah et al., 2000; Keays and Nolan, 2003; O'Brien and Frankel, 2004).

A battery of neurological reflexes was defined by Royle and co-workers (Royle et al., 1999) to compare four strains of mice: 129/Ola, BALB/c, C57BL/6, and FVB/N. Core temperature, righting reflex, corneal reflex, salivation, and grip strength were not significantly different among the strains. Body weight was significantly lower in

TABLE 3.1 Irwin Observational Test Battery

I. In viewing jar

Body position
- 0 Completely flattened
- 1 Lying on side
- 2 Lying upright
- 4 Sitting up
- 6 Standing on *HL* (rearing)
- 8 Repeated vertical leaping
  - *P* = Palpebral closure (0–8)

Locomotor activity
- 0 None; resting
- 2 Casual scratch, groom/slow spatial
- 4 Vigor. scratch, groom/mod. spatial
- 6 Vigor. mvmt.; sl. sharp, rapid/dart
- 8 Ext. vigor. mvmt.; ext. sharp, rapid/dart
- *S* Scratch; *R* Restless; *W* Writhe

Bizarre behavior (0–8)

HF	Head flick	*P*	Prancing
HS	Head search	*UW*	Upright walk
H	Hallucinatory	*AW*	Aimless wander
B	Compulsive bite	*C*	Circling
SB	Self-destr. bite	*W*	Waltzing
L	Compuls. licking	*R*	Retropulsion
L	Compuls. licking	*R*	Retropulsion
		D	Spat. disorient.

Exophthalmos
- + Present

Resp. rate/arrhythmia
- 0 Arrest

2 60/min	*L*	Labored
4 120/min	*R*	Retching
6 180/min	*D*	Dyspneic
8 240+/min	*G*	Gasping

Tremors
- 0 None
- 2 Sl. fine body tr. (1.5 mm)
- 4 Mod. coarse (3 mm) w. sl. impair. locom.
- 6 Mkd. coarse (4.5 mm) w. mod-mkd impair.

Locomotion
- 8 Ext. coarse (6 mm) w. locom. impossible

E	Exertion tr.	*R*	Rest tr.
H	Head only	*T*	Tail only

Twitches

0–8 Sl., mod., mkd., ext. (magnitude/freq.)

Convulsions

SB	Self-destr. bite	*AW*	Aimless wander
C	Clonic	*T*	Tonic
Cs	Clonic, symmet.	*Tf*	Tonic flexion
Rn	Running excit.	*Op*	Opisthotonus
Ch	Champing	*Em*	Emprosthotonus
P	Popcorn	*Rr*	Rock & roll
A	Asphyxeal	*Su*	Sitting up
D	Terminal death	*Pr*	Praying

Transfer arousal (appearance)
- 0 Coma/sl. vibrissae mvmt. only
- 1 Mkd. dulled; slow, sl. mvmt.
- 2 Mod. dulled; slow, mod. mvmt.
- 3 Sub-alert; sl. dec. mvmt.
- 4 Alert; active mvmt./slow freeze
- 5 Hyperalert; rap. mvmt./act. frze
- 6 Sl. excit.; sl. sharp/dart mvmt.
- 7 Mod. excit., sharp/dart mvmt. ("hypomanic")
- 8 Ext. excit., sharp/dart mvmt. ("hypomanic")
- *C* Catalepsy

Spatial locomotion
- Duration (0–4) times speed of movement factor:
- Slow = 1
- Active = 1.5
- Rapid = 2

Palpebral closure
- 0 Eyes wide open
- 2 1/4 closed
- 4 1/2 closed
- 6 3/4 closed
- 8 All closed

Piloerection
- 0–8 Sl., mod., mkd., ext.

Startle (jerk)
- 0 None
- 2 1/4 cm
- 4 1/2 cm
- 6 3/4 cm
- 8 1 cm or more

Alley progression (cm)
- *D* 4+ sec. delay; *E* Early exit
- → No exploration

Pelvic elevation
- 0 Mkd. flattened
- 2 Barely touches
- 4 3 mm elev. (1/8″)
- 6 6 mm elev. (1/4″)
- 8 12 mm elev. (1/2″)
- *C* Crouched; *H* Abnormal head position

Tail elevation
- 0 Flattened
- 2 Horizontally extended
- 4 Diagonally elevated (45°)
- 6 Vertically upright (90°)
- 8 Diagonally retrograde (135°)

(Continued)

TABLE 3.1 *(continued overleaf)*

II. In arena

Finger-approach
 0 None
 2 Head mvmt. only; at distance
 4 Mvmt. To finger; no contact
 6 Contact; partially on finger
 8 Completely on & explores
Finger-withdrawal
 0 None
 2 Mkd. eye-squint only
 4 Squint w. sl. head & body retract.
 6 Squint w. mod. withdrawal
 8 Continuous withdrawal
  B Biting; *F* Freeze
Touch-escape
 0 None
 2 Slow esc./sl. frze (firm stroke)
 4 Mod. rapid esc./act. frze (light stroke)
 6 Vigor., rap. esc. (light stroke)
 8 Ext. vigor. run-escape (barely touch)
Ataxic gait (lurch)
 0 None
 2 Slight, definite
 4 Considerable/w/o fall
 6 Marked; fall every 4–6 steps
 8 Barely moves w/o fall
Hypotonic gait
 0 None

 2 Sl. w. sub-norm. pelvis; *HL* sl. post.
 4 Mod. w. low pelv.; *HL* sl. lat.
 6 Mod. w. flat pelv.; *HL* mod. lat./post.
 8 Ext. w. flat abdom.; *HL* ext. lat./post./drag
Impaired gait, other
 St Steppage
 Sp Spastic
 W Waddling
 Ds Dysmetric
 Dk Duck-walk
 Sc Scissor
Limb rotation
 0–8 Sl., mod., mkd., ext.
 A Anterior
 P Posterior
 L Lateral
 M Medial
  F Forelimbs only
Total gait incapacity
 0–8 Sl., mod., mkd., ext.
Positional passivity (no struggle)
 2 Held by neck
 4 Held supinely
 6 Held by foreleg
 8 Held by hindleg

III. Tail suspension

Visual placing (nose distance)
 0 None, even after nose contact
 1 After nose contact
 2 After mkd. vibris. contact (6 mm)
 4 After sl. vibris. contact; active (12 mm)
 6 Before vibris. contact; active (18 mm)
 8 Early vigor. extension, incl. *HL* (25 mm)
Grip strength (grid-grip resistance)
 0 None
 2 Sl. grip; semi-effective
 4 Mod. grip; effective
 6 Active grip; effective
 8 Unusually effective
Body tone
 0 Complete flaccid.; no return cavity normal
 2 Sl. flaccid.; rapid or v. sl. slowed return
 4 Sl. resistance
 6 Mod. resistance
 8 Ext. resistance; board-like
Hypothermia
 + Present
Pinna
 0 None

 6 Very brisk flick
 8 Hyperact., repet. flick barely touch
  W Body withdrawal response
Cornea
 0 None
 2 Sluggish closure
 4 Active single eye-blink
 6 Double eye-blink
 8 Triple eye-blink
Toe-pinch
 0 None
 2 Sl. withdrawal
 4 Mod. rapid withd.; not brisk
 6 Brisk, rapid withdrawal
 8 Very brisk w. repet. extens. & flex.
Positional struggle
 0–8 Sl., mod., mkd., ext.
  (Mean pinna-corneal & vis. placing)
Wire maneuver
 0 Actively grasps w. *HL*
 2 Mod. diff. grap w. *HL*
 4 Unable grasp w. *HL*; sl. raises
 6 Unable lift *HL*; falls 6–10 sec

(Continued)

TABLE 3.1 (*continued overleaf*)

2 Mod. retract./sl. brisk flick	8 Falls immediately
4 Active retract./mod. brisk flick	*B* Behavioral fall

IV. Supine restraint

Skin color
  0 Ext. blanching
  2 Mod. blanch; mod. pink tone
  4 No blanch; sl.-mod. dusky rose
  6 Deep dusky rose
  8 Bright, deep red flush
    C Cyanosis (0–8)
Diarrhea
  + Present
Limb tone
0–8 Sl., mod., mkd., ext. resist.
Abdominal tone
  0 Completely flaccid; no return cavity normal
  2 Sl. flaccid; rapid or v. sl. slowed return
  4 Sl. resistance
  6 Mod. resistance
  8 Ext. resistance; board-like
Pupil size (dilatation)
  0 Pin-point
  0.5 1/16th
  1 1/8th
  2 1/4
  4 1/2
  6 3/4

  8 Fully
  Op Opacity; *N* Nystagmus
Light-pupil
  0 Absent
  2 Mkd. sluggish
  4 Mod. sluggish
  6 Sl. sluggish
  8 Very active
Lacrimation
  + Present
  C Chromodacryorrhea
Salivation
  0 None
  2 Very sl. wet margin sub-maxillary area
  4 Wet zone 1/4 sub-max. area
  6 Wet zone 1/2 sub-max. area
  8 Wet zone entire sub-max. area
Provoked biting
  0 None
  2 Slight; weak
  4 Mod. active; not vigorous
  6 Vigorous; not immediate or cont.
  8 Ext. vigor., intense & continuous
    (Mean dowel & tail-pinch)

V. In arena

Tail-pinch
  0 No response
  1 Very sl. mvmt., sl. freeze/vocal.
  2 Sl. bite, escape/mod. frze/mkd. vocal.
  4 Mod. bite, escape/abrupt active freeze
  6 Vigorous biting/escape
  8 Ext. vigor. biting/escape
Righting reflex
  0 No impairment

  1 On side 1–2 times
  2 On side 3–4 times
  3 On side 5 times
  4 On back 1–2 times
  5 On back 3–4 times
  6 On back 5 times
  7 Sluggish when placed on back
  8 Absent when on back & tail pinched

VI. Throughout handling

Grasp irritability (biting tendency)
  0 None
  1 Slight
  2 Moderate
  3 Marked
    (Sum scores body tone, pos. passivity &)
    supine restraint)

Provoked freezing
  0 None
  1 Slight
  2 Moderate

  3 Marked, abrupt freeze
    (Sum scores transf. arousal, touch-esc. &
    tail-pinch)
Vocalization
  f Number during handling
Urination-defecation

  f Number during handling
Acute death
  + Present
  C Convulsions; *R* Respir. depress.
  Code *w.* with; *w/o* without;/or;
  HL hindlimbs; *mvmt.* movement

Source: From Irwin (1968), pp. 226–227.

TABLE 3.2 Moser Neurobehavioral Toxicology Test Battery Organized According to Domains of Neurological Function

Autonomic	Activity	Excitability
Lacrimation	Rearing	Ease of removal
Salivation	Motor activity counts	Handling reactivity
Palpebral closure	Home cage posture	Arousal
Pupil response		Clonic movements
Urination		Tonic movements
Neuromuscular	*Sensorimotor*	*Physiological measures*
Gait score	Tail-pinch response	Body weight
Righting reflex	Click response	Body temperature
Forelimb grip strength	Touch response	Piloerection
Hindlimb grip strength	Approach response	
Landing foot splay		

Source: From Moser et al. (1995), p. 176.

the BALB/c strain. Rotarod performance was significantly lower in the 129/Ola and BALB/c strains.

The fields of neurodevelopment and developmental neurotoxicology contribute some of the best observational tests for neonatal and juvenile mice (Spear, 1990; Stanton and Spear, 1990; Moser et al., 1995, 1997; Bignami, 1996; File, 1998; Meyer, 1998; Le Roy et al., 1999; Cory-Slechta et al., 2001). Functional categories of sensory, motor, arousal, motivation, cognitive, and social traits are designed for comparability across mice, nonhuman primates, and humans (Stanton and Spear, 1990). Developmental

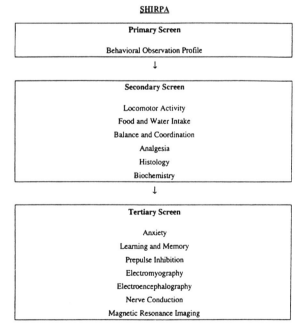

Figure 3.1 *SHIRPA protocol for comprehensive phenotype assessment.* [From Rogers et. al. (1997), p. 712.]

disabilities are evaluated in young mice through a set of observations of simple behaviors representing *developmental milestones* during early postnatal development (Fox, 1965b; Altman and Sundarshan, 1975; Spear, 1990; Spear and File, 1996; Gozes et al., 1997; Chapillon et al., 1998; Le Roy et al. 1999). Charles Heyser at Franklin and Marshall College in Lancaster, Pennsylvania contributed an excellent step-by-step series of protocols for conducting behavioral tests of neurodevelopmental milestones in mice (Heyser, 2003). More information on behavioral tests useful for very young mice are described in Chapter 12.

W. M. Fox of The Jackson Laboratory in Bar Harbor, Maine, assembled a comprehensive battery of developmental milestones in mice (Fox, 1965b). Figure 3.2 shows representative data over days 1–21 for rat pups tested on 23 reflex responses. Developmental tests include sensory measures of olfaction, tactile responses, thermal, vestibular, auditory, visual, and taste. Motor abilities are quantitated on measures of crawling, walking, pivoting, running, foot placing, cliff avoidance, probing, wall climbing, rearing, face washing, grooming, suckling, and eating solid food. Social behaviors are quantitated for developmental milestones over the time periods 1–21 days, 22–70 days, and 71–113 days in rats, including observational analyses of huddling, fighting, grooming, activity, and sexual interactions (Day et al., 1982). Age-dependent emergence of these elements in normal pups can be compared to age of appearance of the developmental milestones in mutant mice. Altered timing of appearance, or complete absence of normal developmental milestones, may reflect anatomical abnormalities such as altered cortical migration (Berger-Sweeney and Hohmann, 1997). The role of early experience and early social deprivation on developmental milestones and subsequent adult behaviors has been extensively studied (Hofer and Shair, 1987; Brunelli et al., 1989; Meaney, 2001) and is described in Chapters 8 and 13.

Richard Brown and colleagues at Dalhousie University in Halifax, Nova Scotia, Canada, assembled a standardized set of behavioral tests that include developmental measures of sensory and motor functions, learning and memory tasks, and ethological tests. Entitled the "Mouse IQ Test," the procedures are listed in Table 3.3. Another excellent battery of observational ratings evaluates responses to handling and degree of wildness versus placidity (Wahlsten et al., 2003a). Fifteen behavioral tests were validated in two expert laboratories. This battery has been applied to 21 strains of mice in the Mouse Phenome Project organized by The Jackson Laboratory (Wahlsten et al., 2003a). Standard batteries of behavioral tests are offered by companies that conduct mouse behavioral phenotyping on a contract basis, including Neurodetective and Psychogenics, as listed at the end of Chapter 2.

TABLE 3.3 Mouse IQ Test[a]

Developmental Tests	Learning and Memory	Ethological
Righting reflex	Eight-arm radial maze	Nest building
Grasp reflex	Step-down passive avoidance	Parental care of pups
Rooting and orienting reflex	Olfactory discrimination	Mating
Acoustic startle reflex	Hebb-Williams maze	Resident-intruder
Ultrasonic vocalizations	Morris water maze	Visible burrow
Forelimb grip strength	Odor preferences	Elevated plus-maze

[a]Standardized battery developed by Dr. Richard E. Brown and co-workers, Psychology Department, Dalhousie University, Halifax, Nova Scotia, Canada.

Development of Reflex Responses in 45 Litters of 214 Normal Mice

Response	1	2	3	4	5	6	7	8	9	10	11	12	13	14	15	16	17	18 – 21 days in age
Righting	1/50	5/75	5/100	5/100	9/75	9/100	9/100											
Crossed extensor	1/100	5/100	9/100	5/75	1/25	0/0	0/0											
Forelimb placing	1/100	5/75	5/100	9/100	9/100													
Hindlimb placing	0/0	1/100	5/75	5/100	9/50	9/100	9/100											
Acceleration righting	–	–	–	–	–	–	–	–	0/0	0/0	1/75	1/100	5/50	5/75	5/200	9/75	9/100	9/100
Postural flexion	9/100	9/100	9/75	5/75	5/50	1/50	0/0	0/0										
Postural extension	0/0	0/0	0/0	1/25	1/50	5/50	1/100	1/25	0/0	0/0								
Normal posture	0/0	0/0	0/0	0/0	0/0	0/0	1/25	5/100	9/75	9/100	9/100							
Forelimb grasp reflex	0/0	1/100	5/100	5/75	5/100	5/100	9/75	9/100	9/100				100					(hindlimb grasp appears later, 9 by 12-14 days)
Pivoting	1/100	5/100	9/100	9/100	9/75	5/25	1/25	0/0	0/0									
Swimming	Circling until 6-8 days, then swim straight																	
Straight-line walking	0/0	0/0	0/0	0/0	1/25	5/75	9/80	9/100	9/100					(straight crawl, can later run, hop and jump)				
Hyperkinesias	Affecting head, limbs and trunk, reach a peak between 6-12 days, then gradually disappear																	
Rooting	1/100	9/75	9/100	9/100	9/100	9/75	5/75	1/75	0/0	0/0								
Vibrissa placing	0/0	0/0	0/0	1/100	1/100	1/100	5/75	5/100	5/100	9/75	9/100	9/100						
Visual placing	0/0	0/0	0/0	0/0	0/0	0/0	0/0	0/0	0/0	0/0	0/0	1/25	1/100	5/75	9/75	9/75	9/100	9/100
Negative geotropism	0/0	1/100	1/100	1/100	5/75	9/75	9/100	9/100 (pivoting predominates)						(less stereotyped at 14-16 days; explore the slope)				
Bar-holding ability	0/0	0/0	0/0	0/0	0/0	0/0	0/0	0/0	0/0	1/20	1/100	1/100	5/75	5/100	9/75	9/75	9/100	9/100 (motor co-ordination)
Cliff drop aversion	1/50	1/100	1/100	1/100	5/75	5/100	9/100	9/100										
Eyes open	0/0	0/0	0/0	0/0	0/0	0/0	0/0	0/0			5/100	9/75	9/100	9/100				
Auditory startle	0/0	0/0	0/0	0/0	0/0	0/0	1/25	1/25	1/50	1/75	1/100	1/100	5/75	5/100	9/75	9/75	9/100	9/100
Generalized pain response	5/100	5/100	9/100	9/100	9/100	9/75	5/100	5/80	1/25	0/0	0/0							
Overgeneralized sensory response	0/0	0/0	0/0	0/0	0/0	0/0	0/0	1/100	5/50	5/75	9/100	9/100	9/100	9/100	9/100			(disappears by 21-28 days of age)

Figure 3.2 *Reflex responses over postnatal days 1–21, in C3H/HeJ and C57B1/6 mice.* [From Fox (1965b), p. 236.]

Our laboratory continues to refine the set of preliminary behavioral observations that best detect fundamental abnormalities in mutant mice. Richard Paylor, while a postdoctoral fellow in our laboratory, designed a set of simple observations that work well for many transgenic and knockout mice (Crawley and Paylor, 1997; Paylor et al., 1998, 2001; Bowers et al., 2005). The Irwin screen was refined and adapted for mutant mice and their littermate controls. Approximately 40 measures are scored; they evaluate physical characteristics, spontaneous behaviors, and a battery of sensory, motor, and sensorimotor reflexes. Measurements take approximately 20 minutes per mouse. A well-trained behavioral neuroscientist can learn to conduct this battery in an hour or two. Novice high school student interns in our laboratory have become proficient after two days of intensive training. Table 3.4 presents the list of tests conducted, along with representative data from the Paylor screen. We routinely run this battery of preliminary

TABLE 3.4 Paylor Screen for General Motor and Sensory Responses, as Applied to *Acra7* Deficient Mice, Missing the Nicotinic Cholinergic Receptor Subunit7

	Wild Type	*Acra7* Deficient
Physical characteristics		
Weight	26.5 ($\pm$ 1.2)	28.1 ($\pm$ 1.3)
Body temperature	37.4 ($\pm$0.1)	37.3 ($\pm$ 0.1)
Whiskers (% with)	50	80
Bald patches (% with)	18	13
Palpebral closure (% with)	0	0
Exophthalmos (% with)	0	0
Piloerection (% with)	0	0
General behavioral observations (% subjects displaying response)		
Wild running	0	0
Freezing	0	0
Sniffing	100	100
Licking	0	0
Rearing	100	100
Jumping	0	0
Defecation	6	9
Urination	6	0
Move around entire cage	100	100
Sensorimotor reflexes (% subjects displaying "normal response")		
Cage movement	100	100
Righting	100	100
Whisker response	100	100
Eye blink	100	100
Ear twitch	100	100
Hot-plate test		
Latency to first hind-paw response (sec)	6.42 ($\pm$0.69)	5.95 ($\pm$0.61)
Motor responses		
Wire suspension time (sec)	40 ($\pm$6)	48 ($\pm$5)
Pole test score	6 ($\pm$2)	6 ($\pm$1)
Tail suspension (% with normal response)	100	100
Elevated platform		
Latency to edge (sec)	1 ($\pm$0)	1 ($\pm$0.3)
Exploratory nose pokes	9 ($\pm$1)	8 ($\pm$1)

Source: From Paylor et al. (1998), p. 307.

observations to characterize a new transgenic or knockout line (Paylor et al., 2001; Steiner et al., 2001; Holmes et al., 2001, 2002d, 2003b; Kinney et al., 2002; Wrenn et al., 2004).

In all of the behavioral test batteries described above, the full series of preliminary observations is conducted sequentially on each individual mouse. In most cases the tests are simple, noninvasive, and show no apparent overlap between tests. The exact order of testing does not appear to be critical in most cases (McIlwain et al., 2001; Voikar et al., 2004). One noteworthy exception is the need to run the elevated plus-maze before the battery of preliminary observations, due to the sensitivity of this anxiety-related task to prior experience, as discussed in Chapter 10. Investigators with some training in rodent handling can easily learn to conduct the series of observations with accuracy and good interobserver reliability. Equipment requirements are generally simple and inexpensive.

BEHAVIORAL TESTS

General Appearance

The physical appearance of the mouse often reveals problems with general health. A healthy mouse is a well-groomed mouse. Rodents routinely engage in grooming behaviors in which the body hair is licked and smoothed. Self-grooming and social grooming are ongoing behaviors in mice, occurring sporadically during the day and intensively after eating. A sick animal often stops grooming. Yellow, gray, and/or sticky fur indicates insufficient grooming. Bald patches, in which fur is missing, reflect aberrant grooming. Sparse, short, or absent whiskers may reflect inadequate diet, premature weaning, aberrant self-grooming, or excessive whisker barbering by a cagemate (Strozik and Festing, 1981; David D. Myers, DVM, Ph.D., Director Laboratory Animal Health, Task Force on Dermatitis in C57BL/6J and Related Strains, The Jackson Laboratory, described in November 27, 1996, *Memo to Users of JAX Mice*; Sarna et al., 2000).

Pink coloration of the ear pinna and footpads is a sign of good health. Hypothermia produces pale or purple coloration of the pinna and footpads. Subcutaneous protrusions may mask spontaneous tumors, which are common at older ages in some strains of mice. Very large or very small body size may represent abnormal growth, feeding, or metabolism. Dark red or yellow crustiness around the nostrils or eyes is a sign of respiratory infection. Dark red lesions on the feet or tail indicate a recent wound. Bleeding or dark red scabs on the tail, rump, and back indicate severe fighting behavior in the home cage. Observations of such traits should be recorded in the health records, treated appropriately, and considered carefully before behavioral analyses are pursued.

Body Weight

Mice are easily weighed on scales or animal balances (Figure 7.2). The animal is placed in the basket of a standard rodent triple-beam balance, and the weight is recorded. Inexpensive electronic pan balances can be readily adapted for weighing mice. A glass jar, Plexiglas cylinder, or wooden box is placed on the pan, and the balance tare is set at zero. The mouse is then placed in the container and the weight recorded.

A single measurement of body weight in adult mutant mice and littermate controls is often sufficient as an index of general health. If the gene under investigation is relevant to feeding behaviors, growth, or metabolism, then it may be useful to measure body weight daily. If a developmental gene is mutated, weekly body weights, beginning at birth, provide one good measure of development.

Body Temperature

Body temperature is measured in mice by a rectal thermistor. Mouse core temperature equipment is commercially available (e.g., Columbus Instruments, Columbus, OH; Physiotemp Instruments, Clifton, NJ). The investigator holds the mouse loosely by the tail and gently inserts the thin, flexible probe into the rectum. Temperature is automatically recorded and displayed digitally in about 5 seconds.

Posture and Gait

Normal mice show a hunched posture, with feet tucked under the body, when resting in the corner of the home cage. Mice locomote in the horizontal plane with all four feet moving and the ears and whiskers twitching. In a novel environment the back is kept relatively horizontal, with frequent long extensions as the mouse stretches forward and pauses for sniffing. Vertical postures occur at frequent intervals during normal exploration of the walls of the cage or test chamber. Gait is generally in a straight line for short lengths. Mice stop frequently and change direction often during spontaneous exploratory locomotion. During pauses the head and whiskers may move quickly and erratically, as the mouse sniffs and touches the environment. Bouts of complete immobility may represent fear-induced freezing behavior (described in Chapters 6 and 10).

Home Cage Activity Patterns

Home cage behaviors are usually analyzed by videotaping several 24-hour periods in the animal housing facility, and subsequent scoring of the videotape by two independent observers. An infrared camera is set up in the housing room to videotape the cages for later scoring. Alternatively, an observer scores home cage behaviors in real time through a videofeed or window. Standing or sitting in the room at a distance from the cages is usually unproblematic. Turning off the room lights during the daytime circadian phase, or turning on the room lights during the night phase, may disrupt home cage behavior patterns, and influence subsequent results on behavioral tasks. Therefore, during the dark cycle it is better to observe home cage behaviors by keeping the room lights off. A lamp with a dim red bulb provides sufficient illumination for the investigator, while being relatively undetectable to mice. Some animal test rooms are built with overhead red light bulbs on a separate room lighting switch. Digital videorecorders are available that provide good images under red lighting conditions.

Dedicated equipment and software for monitoring locomotor activity in specialized home cages are commercially available for mice housed in a standard vivarium. Columbus Instruments has assembled a Comprehensive Lab Animal Monitoring System (Figure 3.3) that records weight, activity, feeding, drinking, and other measures for up to 32 mice (*http://www.colinst.com/brief.php?id=61*), Accuscan Instruments, Inc.,

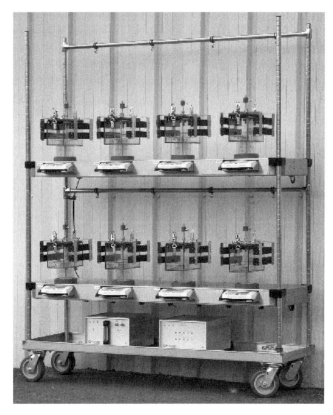

Figure 3.3 *Automated 24-hour monitoring of activity, food and water consumption, and metabolism is built into the cages and rack of a vivarium monitoring system. Sold as the Oxymax Comprehensive Laboratory Animal Monitoring System by Columbus Instruments, http://www.colinst.com.*

in Columbus, Ohio, has developed an analyzer that detects infrared light beam breaks in the home cage. Mounting hardware can attach one, two, three, or four beams per cage, for increasing sensitivity. The Smart Frame Rack by Hamilton-Kinder (Figure 3.4) is one of the systems for continuously measuring home cage locomotion by placing the normal mouse housing cage within a frame containing a small number of infrared photocell beam emitters and detectors. Hamilton-Kinder is developing a linked system, MotorMonitor, that will offer open field, cued and contextual fear conditioning, light ↔ dark and elevated plus-maze anxiety-related tests, place preference, hole board, forced swim, novel object recognition, runway, and social interaction analyses using a unified software package (Michael Kinder, personal communication). The animal care program and engineers at the University of California, San Diego have developed a "smart vivarium" that includes automated monitoring of several mouse behaviors in the home cages to facilitate health assessment (*http://research.calit2.net/smartvivarium*). Fully automated home cages designed to measure components of locomotion, circadian rhythms, feeding, approach-avoidance anxiety-related behaviors, learning, etc., include PhenoLab by Noldus Information Technology, a patented cage monitoring system by Psychogenics, Home Cage Activity by Coulbourn Instruments, Laboratory Animal Behaviour Observation Registration and Analysis System by Metris, and Intellicage

Figure 3.4 *Home cage locomotion is automatically recorded for mice living in normal housing cages, through infrared photocell beams in the outer frame. Sold as the SmartFrame Cage Rack by Hamilton-Kinder, http://www.hamiltonkinder.com.*

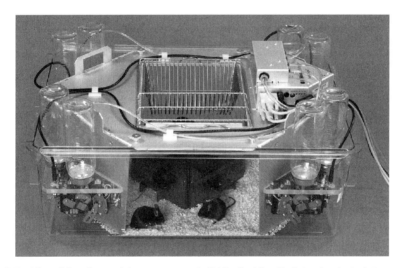

Figure 3.5 *Mice living in social groups are individually identified by implanted transponders. Detectors located at feeding, drinking, and nesting stations, and other key sites in the complex environment, record the location of each identified mouse across programmed time periods. Microprocessors and software analysis convert the locations into behavioral measures including activity, feeding, drinking, nesting, and social interactions. Sold as Intellicage by New Behavior, http://195.129.94.217/web/nb/products/ic.*

by NewBehavior. Intellicage (Figure 3.5), designed by Hans-Peter Lipp, David Wolfer and co-workers at the University of Zurich in Switzerland, offers a complex, naturalistic home cage environment. Transponders implanted in individual mice activate detectors placed at key locations in the cage, to automatically record measures of feeding and drinking, locomotion, sleeping patterns, nesting, and social interactions (Galsworthy et al., 2005). Automated home cage monitoring has considerable appeal as a first screen for measures of general health. Practicality, accuracy, and applications of these various automated home cage systems are presently being tested by several behavioral neuroscience laboratories.

Mice are a nocturnal species. Home cage activities are higher during the dark phase of the circadian cycle than during the light phase. Activity levels generally rise immediately before the room lights go off, and remain high for several hours or until the room lights go on. Methods for quantitating circadian activity in the home cage are described in Chapter 4.

Exploratory activity in the home cage occurs spontaneously. Bouts of exploration alternate with bouts of resting and sleeping. Feeding and drinking similarly occur in bouts, at times consistent with general activity. Mating occurs primarily during the dark phase.

Sleeping patterns in the home cage generally resemble a group huddle. All of the mice are in close physical contact and/or partially on top of each other. If nesting material is provided in the home cage, mice will fluff up the nesting material and sleep in a group huddle within the nest, as described in Chapters 8 and 9.

Abnormal Spontaneous Behaviors

The mouse is placed in an empty standard mouse cage for observation of any abnormal spontaneous behaviors. Examples of aberrant actions include wild running, constant circling, excessive grooming, excessive stereotyped sniffing, excessive stereotyped head bobbing, hunched body posture, and frozen immobility. The investigator records the presence of such behaviors over a three-minute observation period. A stopwatch is used to quantitate cumulative duration of excessive running, circling, stereotyped grooming, sniffing, freezing, and any other unusual postures or bizarre actions.

Neurological Reflexes

The *righting reflex* is evaluated by turning the mouse onto its back. Normal mice will immediately turn themselves over, to right themselves onto all four feet. The *postural reflex* is evaluated by placing the mouse in an empty cage and shaking the cage from side to side and up and down. The normal postural reflex is the extension of all four legs, to maintain an upright, balanced position. The *eye blink reflex* is tested by approaching the eye with the tip of a clean cotton swab. The *ear twitch reflex* is tested by touching the ear with the tip of a clean cotton swab. The *whisker-orienting reflex* is tested by lightly brushing the whiskers of a freely moving animal with a small paint brush. Normal mice will stop moving their whiskers when they are touched, and may turn the head to the side on which the whiskers were touched.

Vision

One simple measure to evaluate visual abilities is the response to light. The beam of a penlight or a small flashlight is directed at the eye. *Pupil constriction* occurs immediately, followed by *pupil dilation* when the light is removed. However, it is difficult to see the pupils in many strains of mice. Another method is based on the natural tendency of mice to move away from the light into a dark environment. Latency to exit a brightly lit open area and enter a dark area is timed. The visual placing or forepaw reaching response is another simple and useful test (Heyser, 2003). When held by the tail above a surface, mice generally hold their paws retracted close to the body. As the head approaches the table surface, the mouse will extend the paws toward the surface, to reach for a soft landing. Blind mice will not see the surface coming, and will not extend the paws. The visible platform component of the Morris water maze test is often used to verify visual abilities. Further measures of visual abilities are described in Chapter 5.

Olfactory

Simple olfactory tests are used to detect *anosmia*, a total absence of the sense of smell. A strong, pleasant odor is painted onto the side of a clean, bare test cage. Vanilla, lemon, or peppermint extract, a smear of cheese, or urine from a mouse of the opposite sex provide attractive olfactory cues for mice. Latency to the first sniff and time spent sniffing the odorous spot are quantitated with a stopwatch. Alternatively, an aromatic food source is buried in the test cage litter. Foods such as a cube of Swiss cheese or a chocolate chip provide positive olfactory cues. Latency for a hungry mouse to dig up the food is a measure of olfactory ability. Usually it is necessary to introduce the mouse to the new food source by placing it into the home cage a few days before the olfactory test, to avoid food neophobia. More sophisticated tests of olfactory discrimination are described in Chapter 6.

Tactile

Response to touch is measured with Von Frey hairs. The mouse stands on a wire mesh platform. Fine wires are inserted through the holes from below, touching the footpads. Foot withdrawal to the wire poke is the measure of touch sensitivity. Wires of increasing diameter are tested to determine the minimum threshold for foot withdrawal.

Reflex responses to pain are quantitated with two standard tests, *tail flick* and *hot plate*. Both measure the time it takes for the mouse to move a body part away from intense heat. This test is analogous to the reflexive withdrawal of the hand from a hot stove in humans. Methods for analgesia testing are described in Chapter 5. It may be necessary to assay pain perception in the preliminary screen in cases where the hypothesized behavioral phenotype will involve nociceptive stimuli that must be correctly perceived. For instance, if the mutation causes a complete loss of pain perception, then a learning task based on avoiding a footshock will yield results indicating that the animal failed to learn the task. Interpreting the results as indicating a learning deficit would constitute a false positive, since the underlying problem was a failure in a procedural ability; that is, the mouse did not detect the footshock on which the learning task was based.

TRANSGENICS AND KNOCKOUTS

In some cases, the gross abnormality defines the expected phenotype. An obvious example is seen in null mutants deficient in the receptor for fibroblast growth factor (FGF) receptor-3. FGFs are neurotrophic proteins that bind to a family of four transmembrane tyrosine kinases (Deng et al., 1996; Colvin et al., 1996). The FGFR-3 receptor subtype mRNA is localized in the cartilage rudiments of bones and appears to regulate bone growth and development. Mutations in the coding sequence for the *FGFR-3* gene in humans causes skeletal bone abnormalities, including achondroplasia, hypochondroplasia, and thanatophoric dysplasia (Muenke and Schell, 1995). *FGFR-3* knockout mice showed abnormal bone development. Progressive bone dysplasia was accompanied by prolonged endochondral bone growth (Deng et al., 1996; Colvin et al., 1996). Skeletons of *FGFR-3* knockout mice displayed bone overgrowth, crooked tails, kyphotic bends along the vertebral column, larger knee joints, and femur bones that were curved and longer than in wildtype controls. Kyphoscolisis or curved backs were easily observed in the null mutants (Deng et al., 1996), as seen in Figure 3.6. Deficits on specialized motor tasks characterize *FGFR-3* knockout mice (McDonald et al. 2001), as shown in Figure 3.6 and further described in Chapter 4.

Body weight is a simple preliminary indicator for evaluating genes regulating growth or contributing to obesity. The illustration at the beginning of Chapter 2 shows the dramatic increase in body size in transgenic mice overexpressing growth hormone. Transgenic mice overexpressing insulin-like growth factor type 1 (IGF-1) are 30% overweight by early adulthood, leading to the discovery of postnatal brain overgrowth and higher numbers of neurons and synapses in the hippocampus at juvenile ages (D'Ercole et al. 2002). Genes regulating developmental growth are described in Chapter 11. The illustration at the beginning of Chapter 7 shows the high levels of obesity in the *ob/ob* mouse. Behavioral phenotypes of transgenic and knockout mice with mutations in genes regulating ingestive behaviors are described in Chapter 7.

Developmental milestone behaviors proved useful in studying the role of SNAP proteins in synaptogenesis during mammalian development. SNAP-25 is a presynaptic nerve terminal protein that is a component of synaptic vesicle docking and fusion with the presynaptic terminal membrane, necessary for the process of exocytosis during neuro-transmitter release (Bark and Wilson, 1994). SNAP-25 may also play a major developmental role in neurite outgrowth before synaptogenesis (Bark et al., 1995). Deletion of the *SNAP-25* gene was performed by neutron irradiation-induced gene deletion of the distal portion of mouse chromosome 2, the coloboma mutation *Cm* (Heyser et al., 1995). This mutation is homozygous lethal. Null mutant *Cm* mice die at an early stage of embryonic development. Heterozygote *Cm* mutant mice have SNAP-25 mRNA and protein levels that are 50% of the levels in wildtype littermate controls. The *Cm* heterozygotes show behavioral deficits in developmental milestones from postnatal day 2 to postnatal day 20 (Heyser et al., 1995). The heterozygote mutant mice had significantly lower body weights, deficits in wire hang, and deficits in the righting reflex. Other developmental milestones were normal, including crossed extensor disappearance, forelimb placing, hindlimb placing, rooting reflex disappearance, grasp reflex, and eye opening. Hyperreactivity to touch was seen at postnatal days 8–10. Heterozygote pups showed exaggerated jumping or freezing behaviors when their skin was stimulated. The postnatal period during which the developmental milestones were delayed in *Cm* mutant mice corresponds to the postnatal age at which there is a shift in the dominant expression of the SNAP-25a isoform to the SNAP-25b isoform (Bark et al., 1995).

RECOMMENDATIONS ON ORDER OF OPERATIONS

Observations of general health and neurological reflexes are a good place to start for all behavioral phenotyping experiments. Our laboratory first conducts the preliminary

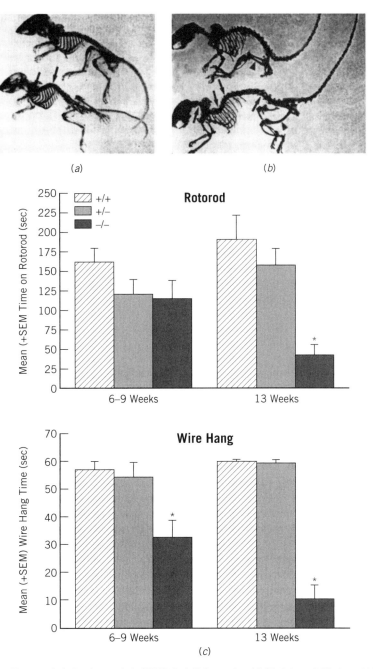

Figure 3.6 *Abnormal skeletal growth in FGFR-3 deficient mice (a) Skeleton of 15-day-old heterozygote (upper) and 15-day-old homozygote (lower) mutants. (b) Skeleton of 207-day-old heterozygote (upper) and 190-day-old homozygote (lower) mutants. Arrows indicate kyphotic bends in the spine. Triangles indicate elongated femur bones* [From Deng et al. (1996), p. 914.] *(c) Motor abnormalities in FGFR-3 knockout mice.* [Adapted from McDonald et al. (2001).]

battery of observations described in this chapter, and then the motor and sensory tests described in Chapters 4 and 5, to confirm that the test mice are able to perform the various procedures necessary for the protocols and equipment involved in testing more complex behavioral domains. False positives are thus avoided. If no major problems are detected in general health, reflexes, motor functions, or sensory abilities, then constellations of tests are chosen within each behavioral domain relevant to the hypothesized function(s) of the gene of interest (Crawley and Paylor, 1997). Three different tests within each behavioral domain are recommended, each with different task demands. False negatives are thus avoided.

The order in which the behavioral experiments are done does not appear to be critical for most of the observations and quick tests described in Chapters 3, 4, and 5. None of these procedures is particularly stressful. None have been reported to produce long-lasting effects that might confound the interpretation of future tests. The main advantage of checking more general behaviors first is that the hypothesis-driven tasks can be chosen to circumvent any physical problems detected. For example, if motor functions are not optimal, learning tests involving vigorous swimming or running can be avoided in favor of learning tests conducted in small operant chambers that require minimal movement by the mouse.

Careful studies of the optimal order of experiments have been rigorously tested in only a few publications (McIlwain et al., 2001; Voikar et al., 2004), as discussed by Crabbe, Wahlsten, and Dudek (Crabbe et al., 1999b). Therefore no specific recommendations are offered in this book for a fixed sequence of tests. However, instances in which a carryover effect has been documented, or appears to be likely, are discussed at relevant places throughout Chapters 6 to 11, and more comprehensively in Chapter 13. Flowcharts, offering guideline for order of behavioral tests, are shown in Chapter 13.

BACKGROUND LITERATURE

Cory-Slechta DA, Crofton KM, Foran JA, Ross JF, Sheets LP, Weiss B, Mileson B (2001). Methods to identify and characterize developmental neurotoxicity for human health risk assessment. I: Behavioral effects. *Environmental Health Perspectives.* **109** (Suppl 1): 79–91.

Crawley JN, Paylor R (1997). A proposed test battery and constellations of specific behavior paradigms to investigate the behavioral phenotypes of transgenic and knockout mice. *Hormones and Behavior* **31**: 197–211.

Fox WM (1965). Reflex-ontogeny and behavioral development of the mouse. *Animal Behavior* **13**: 234–241.

Heyser CJ (2003) Assessment of developmental milestones in rodents. *Current Protocols in Neuroscience* 8.18.1–8.18.15. Wiley, New York.

Irwin S (1968). Comprehensive observational assessment: Ia. A systematic, quantitative procedure for assessing the behavioral and physiologic state of the mouse. *Psychopharmacologia* **13**: 222–257.

McIlwain KL, Meriweather MY, Yuva-Paylor LA, Paylor R (2001). The use of behavioral test batteries: effects of training history. *Physiology and Behavior* **73**: 705–717.

Moser VC, Tilson HA, MacPhail RC, Becking GC, Cuomo V, Frantik E, Kulig BM, Winneke G (1997). The IPCS collaborative study on neurobehavioral screening methods: II. Protocol design and testing procedures. *Neurotoxicology* **18**: 929–938.

Rogers DC, Fisher EMC, Brown SDM, Peters J, Hunter AJ, Martin JE (1997). Behavioral and functional analysis of mouse phenotype: SHIRPA, a proposed protocol for comprehensive phenotype assessment. *Mammalian Genome* **8**: 711–713.

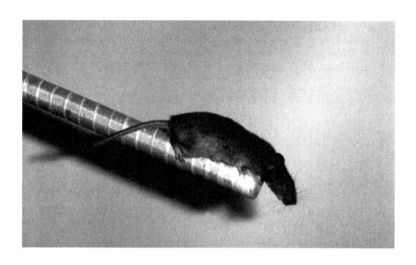

4

Motor Functions

HISTORY AND HYPOTHESES

Almost every behavior requires the animal to move. If motor functions are impaired, your mouse will not be able to perform complex tasks, simply because it cannot walk to the food, initiate social interactions, or swim the maze. Artifacts will obscure any meaningful interpretation of the behavioral phenotype. Before starting on a series of more sophisticated behavioral experiments, the rigorous behavioral phenotyping laboratory will first test a variety of motor functions.

In some cases, the motor deficit defines the expected phenotype. Mutant mouse models of ataxias are described in the last section of this chapter. Mutations in spinal, cerebellar, and vestibular genes result in abnormal motor coordination and balance, as discussed below. Motor abnormalities revealed many of the known natural mutants that originated at The Jackson Laboratory in Bar Harbor, Maine. The aberrant motor patterns observed in the home cage by animal caretakers at JAX led to the discovery of several spontaneous mutations relevant to posture, gait, and balance.

Robert Yerkes completed one of the first comprehensive studies of motor behaviors in mice in 1907 at the Harvard University Psychology Laboratory. Yerkes described a natural mutant known as "The Dancing Mouse," published in the *Classics in Psychology* (Yerkes, 1973). Also known as the Japanese dancing mouse and as the waltzing mouse, this inbred strain is characterized by its running in circles around the perimeter of the cage, and whirling in place with remarkable rapidity. Yerkes quantitated the number, direction, frequency, footprint pattern, circadian rhythm, and ontogeny of the whirling activity, meticulously documenting the phenotype of this remarkable mutant. Normal neuromuscular abilities of the dancing mice were verified in a separate set of motor tasks, including pole climbing and balance-beam walking. Further, Yerkes investigated potential mechanisms underlying the "dancing" motor dysfunction.

What's Wrong With My Mouse? By Jacqueline N. Crawley
Copyright © 2007 John Wiley & Sons, Inc.

The anatomy and physiology of the inner ear was abnormal in some dancing mice, indicating a vestibular basis for the phenotype. Yerkes's historic publication is perhaps the first behavioral phenotyping study of a natural line of mutant mice that encompasses analyses of hearing, vision, olfaction, learning, social, aggressive, and parental behaviors. The 290-page monograph (Yerkes, 1973, p. 277) concludes with wise words to which this author can deeply relate:

I have presented...a program rather than a completed study. To carry out fully the lines of work which have been suggested by my observations and by the presentation of results would occupy a skilled observer many months. I have not as yet succeeded in accomplishing this, but my failure is not due to lack of interest or of effort.

Motor behavior is the final common behavioral output. A defect in a gene mediating a single aspect of the neuronal cell body, axon, dendrite, neurotransmitter, receptor, transducer, muscle, tendon, or bone may impair motor functions. Proceeding backward through the chain of command, bones are moved by muscles, muscle cells are innervated by motor neurons, spinal and cranial motor neurons are innervated by hindbrain neurons in the medulla and cerebellum, hindbrain neurons are innervated by midbrain and forebrain descending pathways, and the motor cortex contributes the highest level of integration. Sensory and associative neurons send their input to the motor pathways. Therefore there are a large number of genes, working at many sites, that influence motor behaviors. Mutations in any of these genes can produce an abnormal motor behavior phenotype.

BEHAVIORAL TESTS

Open Field Locomotion

The most standardized general measure of motor function is spontaneous activity in the open field. Early open fields were constructed of large wooden boxes in which the surface of the floor was marked off into small squares. Original open fields were large, circular (Hall, 1934), or square areas (Broadhurst, 1961; Denenberg and Morton, 1962; Henderson, 1967) in which the spontaneous locomotor activity of the rodent was manually quantitated by human observers. For example, Broadhurst (1961) popularized a 45×45 in.^{2} wooden arena, 18 in. high, painted flat black on all inner surfaces. Thin white stripes were painted across the floor to create 25 blocks, each 9 in.^{2} Ambulatory locomotion in the horizontal plane was quantitated by an observer who scored the number of lines crossed, or squares entered, within a predetermined time period, for example, 5 minutes or 1 hour.

Square, rectangular, and circular equipment is presently in common use. The size of the open field varies considerably across laboratories. Larger open fields, greater than 1 m^2, provide more opportunity for long-distance movements and are more likely to detect components of fear-related behavior. Rats and mice have been tested in open fields of the same size. However, to ensure sufficient surface area to measure exploratory activity and detected anxiety-like tendencies, a larger open field is better for larger animals. In nonautomated open field tests, the test session is videorecorded to allow quantitation at a later time. Scoring of videotaped sessions avoids the confounding variables introduced by the presence of the observer in close proximity to the test animal.

Automated open fields have largely replaced the observer-scored wooden box. Automated open fields now routinely used in behavioral neuroscience laboratories are equipped with either photocell beams or videotracking cameras and software. Both types of automated systems calculate a useful range of locomotor parameters. The Digiscan system by Accuscan Instruments shown in Figure 4.1 employs an array of photocell beams that detect movement. The mechanism is similar to the photocell light beam control system that detects a person approaching the entrance to the supermarket and opens the automatic door.

Automated locomotor systems generally have smaller open fields. The reduced size of the exploratory environment increases the rate at which the mouse habituates to the novelty of the open field. The Digiscan or Versa Max open field (Figure 4.1a) is a Plexiglas box, 40×40 cm. Eight photocell emitters send out horizontal beams of infrared light from the dark red panel on one side, which are detected by eight photocell receptors on the opposite side (x-coordinate). Similarly another set of eight photocells is located in the adjacent two panels (y-coordinate). A third set of photocells is located in two additional panels at the higher elevation along two of the walls, raised 4 cm above the floor (z-coordinate). When the animal moves through a beam, the beam path is broken and the photocell analyzer records the beam break. The computer-assisted analyzer tallies the number of beam breaks for each set of photocells. Beam breaks of the lower 16 sets of photocells (x-y) constitute horizontal activity, because the animal is moving in the horizontal plane on all four feet when it crosses these photobeams. Beam breaks of the top 8 sets of photocells (z) constitute vertical activity, since the animal stands upright in a vertical posture, usually exploring the side of the wall, when it crosses these photobeams. Juvenile mice may be too small to break the standard photobeams, as their body size falls below the beam height. A solution is to raise the floor surface by adding a layer of litter, or to raise the open field arena with respect to the outer photobeam panels by placing the open field on top of a square piece of Plexiglas of the same dimensions.

The computer software calculates a large number of relevant variables over the preset time period. Useful variables derived from the x, y, z beam breaks include horizontal activity, vertical activity, total distance traversed, total number of movements, and time spent in the center of the open field versus time spent in the perimeter of the open field. The measure of stereotypy derived by the automated software is misleading and is not used by behavioral neuroscientists, because automated stereotypy scores generally do not correlate well with the stereotypy scores obtained with standard observational methods that tally grooming, sniffing, and repetitive movements (Creese and Iverson, 1973; Kelley, 1998).

Photocell-equipped open field systems are remarkably accurate and reliable for many years if properly cared for. Diagnostics are provided in the software to inform the investigator when a set of photocell beams is not functioning correctly. The usual cause of beam blockage is litter, dirt, or dust on the red panels, which can be easily wiped clean.

The Plexiglas test box is usually thoroughly washed with detergent and water and wiped dry at the end of each daily use. In addition, the open field usually requires some cleaning after each individual test session to prevent the next mouse from being influenced by the odors deposited in the urine and feces of the previous mouse. This is particularly important if the Plexiglas box is used with its floor surface bare. Alternatively, clean cage litter can be placed in the floor and changed between animal

(a)

(b)

Figure 4.1 (a, b) Digiscan photocell-equipped automated open field. Parameters measuring exploratory locomotor activity are quantitated over time by a computer-assisted analyzer. The software tallies spatially identified beam breaks and derives measures such as horizontal activity, vertical activity, total distance traversed, time spent in the center of the open field, and time spent in the periphery or margins of the open field. Photographs contributed by the author. (c) Representative data illustrating hyperlocomotion in knockout mice deficient in the dopamine transporter (black circles and black bars), as compared to normal spontaneous locomotion in heterozygotes and wildtype littermate controls. [From Giros et al. (1996), p. 609.]

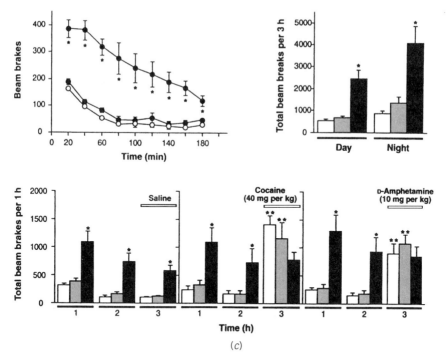

Figure 4.1 *(Continued)*

sessions. A thin layer of litter will not block the lower sets of photocells. Rationale and methods for cleaning the open field box between test sessions vary across laboratories. Many behavioral neuroscience laboratories recommend meticulous cleaning with ethanol between test animals to remove all olfactory cues. Others wipe gently with a damp paper towel and/or only remove any urine and feces, on the assumption that a constant layer of olfactory cues will be less stressful than a very clean environment. Guidelines for cleaning behavioral test equipment may require full cleaning with detergent and water at regular intervals, such as at the end of the test day or the end of the experiment.

For the preliminary assessment of general motor activity, a 5-minute test session in an open field is usually sufficient to evaluate gross abnormalities in locomotion. Testing may be performed under standard room lighting conditions during either the light or the dark phase of the circadian cycle, or within a darkened environmental chamber during either the light or dark phase of the circadian cycle. Quantitative and qualitative differences in locomotion are seen across these four conditions (Figure 4.1*b*).

A *reverse light cycle*, where the animal housing room is maintained on a circadian lighting cycle with lights off during the day and lights on at night, may have advantages for mouse behavioral testing. The house mouse used in laboratory research, *Mus musculus*, is a nocturnal species. The reverse light cycle allows the investigator to test the mouse during its night phase, when mice are most active, rather than during its day phase, when mice normally sleep. Twenty-four-hour circadian light cycle controllers are commercially available to regulate the daily lighting system. A red light bulb in the room lamp, and/or in a handheld flashlight, enables the investigators and the animal caretakers to see the animals in the dark. Red light is probably invisible to rodents. Practical considerations often prohibit maintenance of a mouse colony on a reverse light cycle, especially in a large, multiuser vivarium. Although the mouse behavior literature includes many studies in which the mice live in a reverse light cycle environment, and many more studies in which the behavioral tests are conducted during the mouse's normal lights-on sleeping time, there are few systematic comparisons between behavioral data obtained from mice housed in standard versus reverse lighting schedules across various behavioral domains.

A 5-minute test session in a novel open field provides a measure of exploratory locomotion in a novel environment. Five minutes is usually sufficient to detect highly significant hyperactivity or behavioral sedation. More detailed analysis of motor behaviors is conducted by examining the time course of locomotor parameters. Depending on strain, mice generally habituate to the novelty of the open field environment within 30 minutes to 2 hours. Each measure obtained from the software is summed over a fixed interval, for example, 5-minute intervals for a total of 2 hours. Typical data illustrating habituation to a novel open field over a 180-minute test session is shown in (Figure 4.1*c*). Daily test sessions in the open field provide further measures of habituation (Rubinstein et al., 1997). Familiarity over repeated test sessions in the identical open field may reflect a form of recognition memory. Decreases in perimeter time with concomitant increases in center time during habituation are sometimes considered an index of reductions in anxiety (see Chapter 10). Motor activity across 24-hour periods can be quantitated by placing the entire home cage inside the Plexiglas open field and setting the software for appropriate intervals and session lengths. Home cages should be checked to be sure that they are made of plastic that transmits the infrared light beams, and that home cage components do not block the beams.

Inbred strains of mice display very different levels of spontaneous open field activity and habituation (Flint et al., 1995; Logue et al., 1997b; Crawley et al., 1997a; Bolivar et al., 2000a; van Gaalen and Steckler, 2000; Cabib et al., 2002). Figure 4.2 illustrates the pathways traversed by 8 inbred strains in a large circular open field (Lipkind et al., 2004). Quantitative trait loci analysis of recombinant inbred strains and F_2 offspring from two strains that differ on open field locomotion revealed chromosomal loci that may contain the genes responsible for strain differences in exploratory locomotion (Plomin et al., 1991b; Mathis et al., 1995; Gershenfeld et al., 1997; Kelly et al., 2003; Henderson et al., 2004). Included are investigations of locomotion in 129/SvEv-Tac and C57BL/6J, two inbred strains commonly used for embryonic stem cells and breeding, respectively. Since background genes influence the phenotype of a mutation, baseline levels of open field locomotor activity in transgenic and knockout mice are best compared to wildtype littermate controls, as described in Chapters 2 and 13. This is because open field activity in the wildtype controls may be influenced by a large number of background genes from the multiple parental strains used to generate the mutant mice, as described in Chapter 2. In cases where an unusual strain is employed to generate the mutation, it is useful to test that parental strain for baseline open field locomotion, to know the approximate levels of locomotion to expect before beginning the experiments with the null mutants, heterozygotes, and wildtype littermates.

Locomotor abnormalities are seen in many natural mutants (Van Daal et al., 1987). For example, hyperactivity is seen in the *weaver* mouse, which displays dopamine deficiency (Schmidt et al., 1982). SNAP-25 (synaptosomal-associated protein-25 kDa) heterozygote mutant mice showed elevated locomotor activity at early dark and mid-light

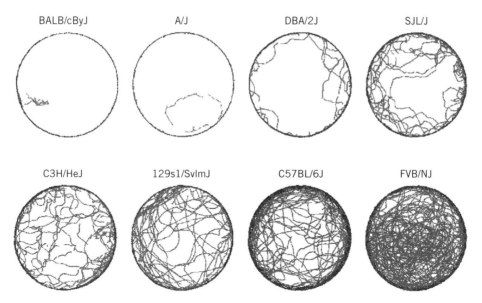

Figure 4.2 *Plots of the path traced by 8 inbred strains of mice in a large open field, 250 cm in diameter, during a 30 minute open field session.* [Kindly contributed by Dina Lipkind, Yoav Benjamini, and Ilan Golani, Tel Aviv University, Israel; data derived from Lipkind et al. (2004).]

periods of the circadian cycle (Hess et al., 1992). SNAP-25 is localized in presynaptic nerve terminals and participates in neurotransmission. This mutation was identified in the spontaneously hyperactive mouse mutant *coloboma*, which expresses an ocular deformation.

Hole board exploration takes advantage of the tendency of mice to poke their noses into holes in the wall or floor (Bradley et al., 1968; File and Wardill, 1975a, b; Lister, 1987a; Rao et al., 1999; van Gaalen and Steckler 2000). The hole board apparatus is a wooden or Plexiglas box, approximately $40 \times 40 \times 27$ cm, with four or more holes, each 3 cm in diameter, spaced along the floor and/or walls (File et al., 1998). Head dipping into the hole interrupts infrared photocell beams just below the edge of each hole. Locomotor activity is measured by the number of beams broken over a 5- or 10-minute period. Automated versions of the hole board test are available, often as add-ons to automated open fields.

Rotarod

Motor coordination and balance are measured by performance on the rotarod (Jones and Roberts, 1968; Sango et al., 1995, 1996; Chapillon et al., 1998; Carter et al., 1999; Rustay et al., 2003). As illustrated in Figure 4.3, the rotarod is a rotating cylinder, approximately 3 cm in diameter. The mouse must continuously walk forward to keep from falling off the rotating cylinder. Constant-speed rotarods have been used for many years. The mouse is given one or two practice trials and then placed on the rotating cylinder. Latency to fall off is the dependent variable. Most mice are easily able to maintain balance on the rotarod for many minutes at standard speeds of 5 revolutions per minute (rpm) or higher. A 1-minute cutoff maximum per session is often used.

Automated rotarod systems include a timer connected to a treadle switch in the floor panel onto which the mouse falls, or a photocell beam just above the floor, to measure the latency to fall. The drop to the floor is in the range of 15 cm below the rotating rod, to motivate the mouse to walk forward on the rotating rod rather than jumping down.

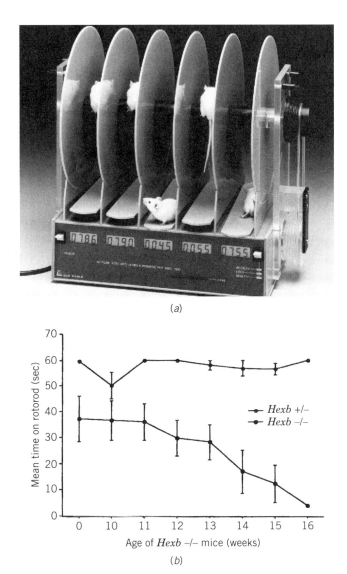

Figure 4.3 *(a) Rotarod apparatus for measuring motor coordination and balance. The Ugo Basile/Stoelting rotarod is based on a rotating cylinder, 3.2 cm in diameter, covered with grooved plastic. Each section is 6.0 cm wide, allowing five mice to be tested simultaneously, one per section. Mice walk forward on the cylinder as it rotates at a constant speed, or at speeds increasing from 4 to 40 rpm over a 5-minute test session. (Photograph kindly contributed by Charles W. Scouten, Stoelting Company, Wood Dale, Illinois.) (b) Representative data showing progressive loss of rotarod performance in Hexb knockout mice, as compared to normal rotarod scores in heterozygote littermate controls. This mouse model of Sandhoff disease, a human neurodegenerative disease, shows an analogous accumulation of gangliosides in cerebellar neurons. [From Sango et al. (1995), p. 172.]*

Accelerating automated rotarods are now in common use (Barlow et al., 1996; Liu et al., 1997b; Crabbe et al., 1998; Carter et al., 1999; Wrenn et al., 2004a). The mouse is placed on the cylinder at a slow rotational speed, for example, 4 rpm. Rotational speed gradually increases over a 5-minute test session, according to a predetermined program, ramping up to a maximum rotational speed of 40 rpm. Latency to fall off is thus a more stringent measure of ability to maintain balance, since the requirements for running forward are gradually increasing. Mice with deficits in motor coordination or balance will fall off the rotarod well before the end of the 5-minute test session, usually during the first minute. Performance on the rotarod task over days of repeated training is quantitated as a measure of motor learning, as described in Chapter 6.

Most inbred strains of mice used for targeted gene mutation perform well on the rotarod task (Cook et al. 2002; Rustay et al., 2003; Brooks et al., 2004). However, strain differences on rotarod performance have been reported (Homanics et al. 1999; Tarantino et al. 2000; McFadyen et al. 2003; Wahlsten et al. 2003). C57BL/6J mice show much longer latencies to fall from the rotarod than 129/SvJ mice (Homanics et al., 1999). Both strains improved over a course of eight daily rotarod tests, reaching the same asymptote of remaining on the rotarod for 3 minutes over an acceleration from 3 to 19 rpm (Homanics et al., 1999) Visually impaired C3H/HeJ mice remained on the rotarod longer than a C3 congenic line without retinal degeneration (McFadyen et al., 2003), indicating that vision is not essential. Some strains of mice frequently grasp the surface of the rotating rod, flatten themselves across the rod, and ride the rod around, instead of walking forward to stay on top (Wahlsten et al., 2003). Altering the diameter of the rod and its surface properties may reduce this alternate behavioral strategy in mice. Doug Wahlsten at the University of Alberta in Edmonton and John Crabbe at Oregon Health and Science University constructed a covering of fine grade sandpaper, wrapped around the rod, to reduce the riding around strategy, but discovered that any gap where the edges of the sandpaper met would allow mice to grasp the juncture and hang on (Wahlsten et al., 2003). While no simple solution has been found for the "riding around" complication, many labs take the mouse off and start the session again. Similarly, if a mouse falls off the rotarod immediately, it is often given a second chance. Pretraining at a low rotational speed, either the day before or a few minutes before the test session, may give the mouse sufficient familiarity and practice with the task to avoid hanging on or immediately falling off.

Cerebellar defects result in performance deficits on the rotarod test (Mason and Sotelo, 1997). *Staggerer, hot-foot*, and *lurcher* are natural mutants with known defects in the neuroanatomy of the cerebellum (Lalonde et al., 1996). *Staggerer* mutant mice lose inferior olive, cerebellar granule, and Purkinje cells. *Hot-foot* mutant mice show defective innervation of Purkinje cells and some necrotic cerebellar granule cells. *Lurcher* mutant mice show degeneration of neurons in the cerebellar cortex. All three mutants perform poorly on a rotarod task, with latencies to fall off in the range of 0 to 100 seconds, as compared to latencies of 300 to 400 seconds for wildtype controls (Lalonde et al., 1996).

Balance-Beam Test

Motor coordination and balance is also evaluated by the ability of mice to traverse a graded series of narrow beams to reach an enclosed safety platform (Carter et al., 1999). Figure 4.4 diagrams the balance-beam apparatus designed by Steve Dunnett

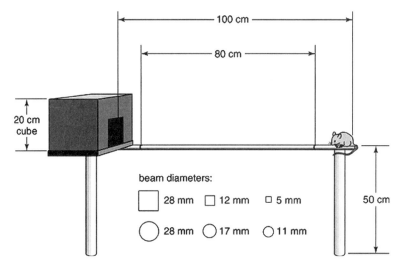

Figure 4.4   Balance-beam apparatus for measuring motor coordination and balance. Latency to cross from the elevated open platform to the dark enclosed box is measured. Foot slips off the beam are counted. Decreasing diameters of square and round beams are used as further challenges, increasing the sensitivity of the task. [From Carter et al. (2003), Current Protocols in Neuroscience, p. 8.12.4.]

and colleagues at Cambridge University and Cardiff University in the UK. Beams are horizontal and 50 cm above the table. A bright light illuminates the start area. An enclosed escape box, 20 cm^2, is at the end of the beam. Mice are trained to traverse square and/or circular beams of decreasing diameter, for increasing difficulty. One sequence is a 28-mm^2 beam, then a 12-mm^2 beam, then a 5-mm^2 beam. The mice are then trained to traverse a 28-mm round beam, a 17-mm round beam, and an 11-mm round beam, on two consecutive trials per beam. Latency to traverse each beam and number of times the hind feet slip off each beam are recorded.

Vertical Pole Test

The *vertical pole test* is a measure of motor coordination and balance that requires minimal equipment (Figure 4.5). A metal or plastic pole, approximately 2 cm in diameter and 40 cm long, is wrapped with cloth tape for improved traction. The mouse is placed in the center of the pole, which is held in a horizontal position. The pole is then gradually lifted to a vertical position. Latency to fall off the pole is the dependent variable. Normal mice will stay on the pole and may walk up or down the length of the pole. Deficits in motor coordination and balance will be detected by the mouse falling off the pole, usually before the pole reaches a 45° angle.

Hanging Wire

Neuromuscular abnormalities can be detected with simple measures of motor strength. Balance and grip strength are required for a mouse to hold its body suspended. A standard wire cage lid is used. Masking tape or duct tape placed around the perimeter of the lid prevents the mouse from walking off the edge. The hanging wire test is performed by placing the mouse on the top of a wire cage lid. The investigator shakes

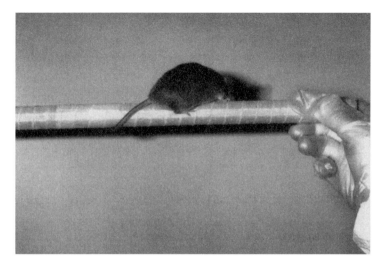

Figure 4.5 *Vertical pole test for grip and motor coordination. A wooden or metal dowel is covered with masking tape and gradually inclined up to 45°. A normal mouse is able to grip and climb the pole.*

the lid lightly three times to cause the mouse to grip the wires, and then turns the lid upside down. The upside-down lid is held at a height approximately 20 cm above the cage litter, high enough to prevent the mouse from easily climbing down but not high enough to cause harm in the event of a fall. The investigator uses a stopwatch to quantitate the latency to fall off the wire lid. Normal mice can hang upside down for at least one minute. A 60-second cutoff time is used for the standard test session. Figure 4.6 illustrates the hanging wire test and shows representative data (Sango et al., 1996). Figure 4.7 illustrates an automated grip test.

Posture and Gait Observations

Severe abnormalities in posture and gait can often be detected by the observant investigator during the 5-minute session in the open field, the hole board test, and the rotarod test. Curvature of the spine, muscle weakness, poor limb coordination, and the like, will result in the mouse wobbling, stumbling, and weaving. Vertical activity in the open field is sometimes a reasonably sensitive measure of postural and gait abnormalities, as the vertical upright position requires more intricate coordination of limbs and muscles than horizontal locomotion.

Footprint Pattern

Ataxia and gait abnormalities are quantitated from the footprint pattern of a mouse (Crawley and Paylor, 1997; Barlow et al., 1996; Carter et al., 1999). The hindpaws are dipped in nontoxic ink or paint. The mouse is then placed at the brightly lit end of a tunnel or runway, which is dark at its far end. The bottom surface of the tunnel is lined with white paper. Tunnel dimensions used can be 9 cm wide × 35 cm long × 6 cm high or 10 cm wide × 50 cm long × 10 cm high. The mouse walks down the tunnel, leaving a set of black footprints on the white paper. The paper is then removed and the footprint pattern analyzed. Distance between each stride, variability in stride length, variability

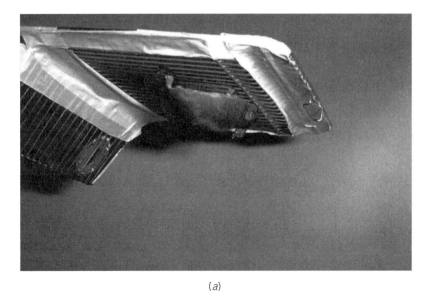

(a)

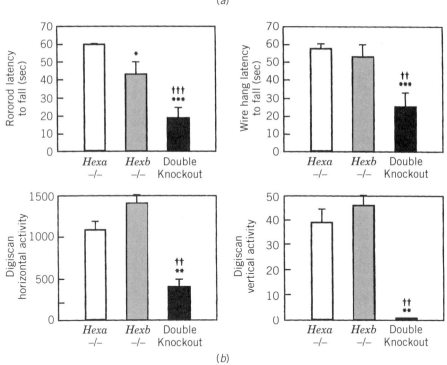

(b)

Figure 4.6   Hanging wire grip test for muscle strength. (a) Normal mice can easily hang upside down from the wire cage lid for 60 seconds or longer (Photograph contributed by the author.) (b) Hexa −/− and Hexb −/− mice are generally able to maintain their grip in the hanging wire test at eight weeks of age. Double knockouts of Hexa and Hexb at the same age fall from the wire during the 60-second test. [From Sango et al. (1996), p. 351.]

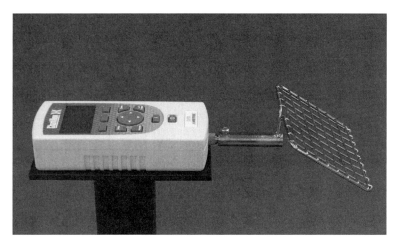

Figure 4.7 *Automated grip test. Neuromuscular function is assessed by a sensor that measures the peak amount of force an animal applies in grasping a pull bar assembly.* [From Columbus Instruments, http://www.colinst.com.]

around a linear axis, hindbase width between left and right hindpaws, frontbase width between left and right forepaws, and forepaw/hindpaw overlap, provide measures of the ability of the mouse to walk in a straight line, with regular, even steps. Ataxic gait is represented by highly variable stride length and path (Barlow et al., 1996). Jenny Morton and Steve Dunnett at the University of Cambridge use purple paint for the hindpaws and orange paint for the forepaws, to maximize the sensitivity of the footprint analysis. Footprint patterns and calculations are shown in Figure 4.8.

Staircase Test

Coordinated paw reaching is sensitive to unilateral lesions of motor structures in the brain, such as the striatum in lesion models of Parkinson's disease (Baird et al., 2001). Climbing up a staircase requires motor coordination and paw reaching. A staircase walking test measures the number of steps that a mouse can climb up, to reach food pellet reinforcers placed on the steps. This task has been used to separate locomotion from anxiety-related behaviors in response to benzodiazepines (Belzung et al., 1989; Pruhs and Quock, 1989; Weizman et al., 1999) and in response to neuroactive steroids (Pick et al., 1996). Lafayette Instruments manufactures a rodent motility staircase apparatus, incorporating a food pellet reinforcer for staircase climbing (Figure 4.9).

Stereotypy

Repetitive, invariant, perseverative motor patterns that do not appear to be directed toward a purposeful action are termed stereotyped behaviors (Kelley, 1998; Lewis, 2004). Stimulant drugs such as amphetamine and cocaine induce a pattern of stereotypy that includes head dipping, excessive sniffing, and intense grooming (Iversen and Iversen, 1981). Standardized rating scales (Creese and Iversen, 1973) and scoresheets (Kelley, 1998; Turner et al., 2001) are employed to precisely quantitate the type and magnitude of stereotyped behaviors over the test session. No automated systems developed to date give accurate measures of true stereotyped behaviors.

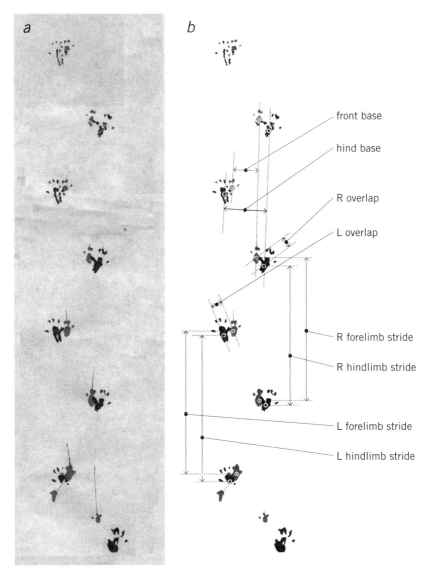

Figure 4.8 Illustration of (a) footprints and (b) the standard measurements that can be collected from them. The darker footprints are those of the hindlimbs. Overlap is the mean of the right and left overlaps, and stride length is the mean of the right (R) and left (L) forelimb or hindlimb strides. [From Carter et al. (2003), Current Protocols in Neuroscience, p. 8.12.7.]

Circling

Rotational locomotion is a robust consequence of unilateral lesions of the nigrostriatal dopaminergic pathway (Ungerstedt, 1971; Mura et al., 1998) and of some neurological mutant mice such as *whirler* (Sackler and Weltman, 1985). Circling in a round chamber is highly biased toward the lesioned side, for example, counterclockwise when the left nigrostriatal bundle is lesioned with 6-hydroxydopamine. Drug treatments with dopaminergic agonists such as dopamine, apomorphine, amphetamine, and quinpirole

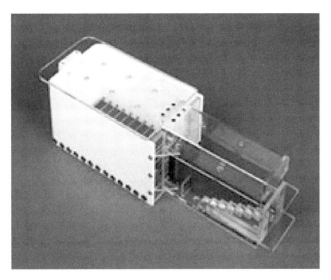

Figure 4.9 *Staircase test. Staircase climbing and paw reaching for foot pellets placed on each step provide measures of motor abilities relevant to nigrostriatal lesions and focal ischemia. [From Lafayette Instruments, http://www.licmef.com/animalspecialty.htm.]*

induce contralateral circling, away from the lesioned side, due to receptor denervation supersensitivity after the unilateral lesion. A good rat model of Parkinson's disease is based on unilateral 6-hydroxydopamine lesions combined with L-DOPA, the original treatment for Parkinson's disease (Ungerstedt, 1971; Mura et al., 1998). Rotational behavior after striatal lesions has been characterized primarily in rats (Dziewczapolski et al., 1998). Figure 4.10 illustrates an automated rotational monitoring system for rats.

Circadian Wheel Running

Circadian rhythms are well known for many biological functions. Hormone release, feeding, and sleep show reproducible activity patterns that vary across the 24-hour daily cycle (Richter, 1967; Marler and Hamilton, 1968; Takahashi, 1995; Dunlap, 1996). Daily rhythmicity of home cage activity in hamsters, rats, and mice can be measured by temporal analysis of running wheel activity. Strain differences in circadian wheel running, correlated with body temperature and brain EEG activity, have been reported for female mice (Koehl et al., 2003). Commercially available home cage running wheels for mice and representative data are shown in Figure 4.11.

Sleep

Electroencephalogram (EEG) electrodes implanted onto the dura of the cerebral cortex and connected to a polygraph recorder measure neurophysiological correlates of slow wave sleep, paradoxical sleep, and wakefulness in rodents. Spontaneous diurnal sleep—waking cycles were measured in mice using standard polygraphic sleep monitoring equipment (Boutrel et al., 1999). Consistent with diurnal rhythms of running wheel activity, wakefulness was highest during the lights-off nighttime period, and slow wave sleep was highest during the early hours of the lights-on period. Paradoxical sleep in mice,

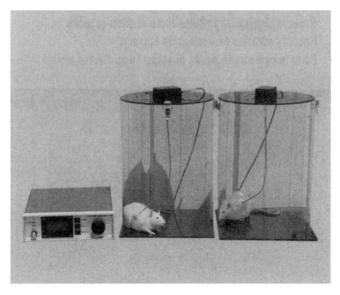

Figure 4.10   Rotational behavior monitoring system. The animal is harnessed to a light, flexible tether. The tether is attached to a sensor that counts the number of clockwise and counterclockwise revolutions of rotation as the rat circles in one direction in the cylindrical chamber. (Photography kindly contributed by Ken Kober, Columbus Instruments.)

analogous to rapid eye movement sleep in humans during dreaming, occurred during the same periods as slow wave sleep. Amount of paradoxical sleep increased following sleep deprivation (Boutrel et al., 1999). Inbred strains of female mice show different patterns of rapid eye movement and nonrapid eye movement sleep across the light/dark cycle, a genotype effect that was more pronounced than sleep pattern differences at different phases of the estrous cycle (Koehl et al., 2003). Inbred strain differences have been reported with an automated sleep recording system (Veasey et al., 2000).

Seizures and Vestibular Dysfunctions

Ataxia, immobility, twitching, "waltzing," "reeling," "tottering," and convulsions are motor abnormalities that may reflect dysfunctions in vestibular structures in the inner ear, or abnormal electrical discharges during seizures in the brain. Natural mutants with unique motor abnormalities are known to have specific mutations and/or structural defects in the labyrinth of the inner ear, in spinal motor neurons, and/or in the cerebellum. Severe incoordination of the hindlimbs is seen in the *reeler* mouse (Myers, 1970). Rapid, erratic running is seen in the *dancing* mouse (Yerkes, 1973). The *staggerer* mouse falls from the rotarod and fails to grip in the wire hang test (Steinmayr et al., 1998). Developmental delays in wire hang and rapid deterioration of rotarod performance are seen in the *twitcher* mouse (Olmstead, 1987).

Many mouse models of epilepsy arose from phenotypes detected in natural mutant mice observed to be seizing in the home cage. The *stargazer* mouse shows an unusual posture with the head angled upward, gazing into the sky, along with bouts of freezing behavior. Prolonged, generalized spike-wave cortical discharges accompany the stargazing behavior (Noebels et al., 1990). Spontaneous seizures begin at 4 to 6 weeks

(a)

Wildtype

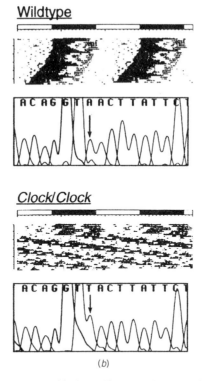

Clock/Clock

(b)

Figure 4.11 *Circadian wheel running. (a) Mice will run inside a wheel attached to the home cage. Wheel running follows a daily rhythm that is quantitated over time by an automated data analysis system. The resulting actagram data are presented as consecutive 24-hour periods. (From David Osgood, Mini-Mitter Company.) (b) Mice with mutations in the* clock *gene show irregular wheel running activity patterns over 24-hour periods, as compared to normal actagrams from wildtype control mice.* [From King et al. (1997b), p. 647.]

of age in *tottering/leaner* mice (Noebels, 1984; Abbott et al., 1990; Hess and Wilson, 1991; Austin et al., 1992). A loud tone induces clonic seizures in audiogenic seizure susceptible mice, including the epilepsy-prone mouse derived from a BALB/c line, and DBA/2J mice at juvenile ages peaking at 21 days (Seyfried, 1982; Chapman et al., 1996; Banko et al., 1997). Differential susceptibility to drug-induced seizures in inbred strains of mice have been identified for drug treatments including pilocarpine, methyl-β-carboline-3-carboxylate and other γ-amino butyric acid (GABA) receptor antagonists (Kosobud and Crabbe, 1990; Martin et al., 1991; Chapouthier et al., 1991, 1998; Mohajeri et al., 2004). Genes mediating these aberrant motor manifestations, spontaneous seizures, and susceptibility to seizures are under investigation.

Swimming

Ability to swim in deep water is measured in a specialized swim tank (Figure 4.12), or in the circular pool during visible platform training in the Morris water maze (Figure 6.3). Motor functions involved in swimming differ from motor functions involved in locomotion, wheel running, balance beam, rotarod, and other tasks described above (Carter et al., 2003; Brooks et al., 2004). Each motor ability may require a different set of muscles, spinal reflexes, and brain regions. Choice of appropriate motor tasks will depend on the predicted deficits in a mouse model of a motor dysfunction, such as amyotrophic lateral sclerosis versus ataxias.

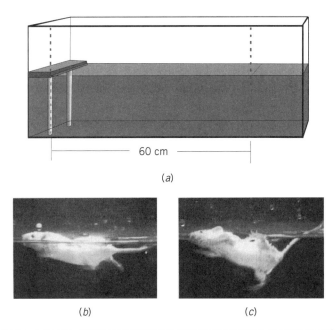

(a)

(b)   (c)

Figure 4.12 *Horizontal swim chamber measures rodent motor abilities specific to swimming.* (Kindly contributed by Dr. Stephen Dunnett, Cardiff University, Cardiff, Wales, taken from Figure 2, "Schematic illustration of the swim tank for rating of swimming performance," in Assessment of Motor Impairments in Transgenic Mice by Stephen B. Dunnett, Short Course II, Mouse Behavioral Phenotyping, with permission from the Society for Neuroscience, Copyright 2003.)

TRANSGENICS AND KNOCKOUTS

Single gene mutations have been linked to cerebellar ataxias, motor neuron diseases, Huntington's disease, and familial amyotrophic lateral sclerosis (reviewed in Lee et al., 1996b). Motor behavior tests provide straightforward methods to evaluate motor dysfunctions in mouse models of these inherited human diseases. Where a significant motor phenotype is detected, the mutant mouse model provides a powerful research tool for evaluating new therapeutic approaches, including gene therapy (discussed in Chapter 14).

Motor deficits have been characterized in several mouse models of human ataxias, including ataxia telangiectasia, spinocerebellar ataxia type 1 and type 7, and episodic ataxia type-1 (Barlow et al. 1996; Cummings et al. 2001; Herson et al. 2003; Michalik et al. 2004). Motor coordination and balance are impaired in many mutant mouse models of cerebellar dysfunction. Mutations in genes expressed specifically in Purkinje cells of the cerebellum, such as calbindin, the neural recognition molecule NB-3 and the *Grid2* gene for glutamate receptor GluRδ2, lead to severe deficits on the rotarod and other motor coordination tasks (Takeda et al., 2003; Barski et al., 2003; Lalonde et al., 2003). Disruption of a prion gene, *Prp*, resulted in an extensive loss of cerebellar Purkinje cells, progressive ataxia, and impaired rotarod performance (Sakaguchi et al., 1996; Katamine et al., 1998).

Ataxia telangiectasia is a disease caused by a mutation in the *Atm* gene. In humans, this is an autosomal recessive disorder leading to neurologic degeneration, cerebellar ataxia, choreoathetosis, oculocutaneous telangiectasias, growth retardation, endocrine abnormalities, extreme sensitivity to ionizing radiation, and lymphatic cancers (Boder, 1975; Barlow et al., 1996). Null mutations for the *Atm* gene resulted in mice that showed an ataxic gait on the footprint test, characterized by irregular footprint pathway and shorter stride length (Barlow et al., 1996). Rotarod performance was impaired. Activity in the Digiscan open field was significantly reduced on both horizontal and vertical beam breaks. Motor dysfunctions in *Atm* knockout mice are analogous to the neurological deficits seen in humans suffering from ataxia telangiectasia. Further, a selective loss of dopaminergic neurons was detected in the substantia nigra and ventral tegmentum of *Atm* deficient mice (Eilam et al., 1998), consistent with the known role of the nigrostriatal and mesolimbic dopamine pathways in normal locomotion. This knockout mouse model provides an example of motor deficits that mimic a primary phenotype of the human mutation.

Fibroblast growth factor receptor-3 (FGFR-3) is a subtype of the family of neurotrophic receptors that is expressed at high levels in cartilage and mediates bone ossification (Deng et al., 1996). A point mutation of *FGFR-3* appears to be the cause of a common form of human dwarfism, achondroplasia. Knockout mice deficient in *FGFR-3* show severe overgrowth of skeletal bones, resulting in kyphotic bends in the vertebral column (Deng et al., 1996; McDonald et al. 2001), as described in Chapter 3. *FGFR-3* null mutants show major motor deficits on the rotarod and wire hang tasks (see Figure 3.3c).

Dopamine is a catecholamine neurotransmitter in the nigrostriatal and mesolimbic pathways of the basal ganglia in mammals. Dopamine contributes to spontaneous and evoked motor activity. Dopaminergic D_2 receptor agonists increase open field locomotion, while dopaminergic D_2 receptor antagonists reduce open field locomotion to the point of catalepsy. Knockout mice deficient in tyrosine hydroxylase, the synthetic enzyme for dopamine, show very low levels of activity in the open field (Zhou and Palmiter, 1995). Knockout mice deficient in the D_2 dopamine receptor subtype have low scores on horizontal activity, horizontal distance, and vertical activity in the Digiscan open field (Kelly et al., 1998) and impaired performance on the rotarod test (Balk et al., 1995). Similarly null mutant mice deficient in the D_{1A} receptor subtype showed lower open field activity (Smith et al., 1998a). One line of knockout mice deficient in the D_3 dopamine receptor subtype demonstrated higher locomotor activity and rearing behavior (Accili et al., 1996b). Another line of D_3 receptor knockout mice showed normal or reduced locomotion in both null mutants and heterozygotes, as compared to wildtype controls (Boulay et al., 1999), raising questions about the consistency of the D_3 phenotype across different DNA promoter constructs and different parental genetic backgrounds. Dopaminergic D_4 receptor subtype knockout mice showed reductions in open field horizontal distance on both the first and second days of open field activity testing, as compared to wildtype littermate controls (Rubinstein et al., 1997). D_5 receptor knockout mice showed normal baseline locomotion (Holmes et al. 2001) but less hyperlocomotion in response to cocaine (Elliott et al. 2003). Direct comparison of the same behavioral parameter, open field horizontal activity, on receptor subtype null mutants, can be a useful tool for discovering differential roles of each receptor subtype for a neurotransmitter. Challenges with agonist treatments can further elaborate the selective actions of each subtype.

The dopamine transporter is a presynaptic protein that actively transports released dopamine from the synapse back into the presynaptic axon terminal, thereby terminating synaptic transmission and replenishing intracellular stores of dopamine. Dopamine transporter mutations result in mice with higher baseline locomotion in an open field (Giros et al., 1996), as illustrated in Figure 4.1c. Drug responses in dopamine transporter mice are discussed in Chapter 11.

Mutant mice deficient in genes that participate in dopaminergic transmission, including RIIβ-protein kinase A (Adams et al., 1997; Brandon et al., 1998), *Nurrl* (Zetterstrom et al., 1997), retinoid receptors (Krezel et al., 1998), and DARPP-32 (Nally et al., 2004), show impairments in either baseline or drug-induced locomotor behaviors. The calcium binding protein calbindin D28k is expressed in the dopaminergic nigrostriatal pathway, as well as in the cerebellar cortex, Purkinje cells, and inferior olive neurons (Airaksinen et al., 1997). Null mutant mice deficient in calbindin D28k showed aberrant gait by the footprint test and poor grip strength (Airaksinen et al., 1997).

Serotonin is a monoamine neurotransmitter in the brain and spinal cord that contributes to a variety of behaviors, including feeding, sleep, and motor activity (Jacobs, 1999). Serotonin receptor subtype and serotonin transporter knockout mice are discussed in Chapters 7 and 10. While motor behaviors appeared to be normal in mice with mutations in several of the serotonin receptor genes, the serotonin receptor subtype 5-HT$_{5A}$ knockout mouse showed reduced exploratory activity in a novel open field (Grailhe et al., 1999). Stimulatory actions of lysergic acid diethylamide (LSD) on exploratory activity was attenuated in the 5-HT$_{5A}$ knockout mice (Grailhe et al., 1999).

Epileptic seizure susceptibilities have been reported in transgenic and knockout mice. Seizure susceptibility results in death in serotonin receptor 5-HT$_{2C}$ mutant mice (Applegate and Tecott, 1998). Sound-induced audiogenic seizures were reported for serotonin 5-HT$_{2C}$ receptor knockout mice (Brennan et al., 1997) and Fyn kinase knockouts (Miyakawa et al., 1995). *Fyn* null mutants also show greater sensitivity to convulsant drug treatments, including pentylenetetrazol, picrotoxin, bicuculline, kainic acid, and *N*-methyl-D-aspartate (NMDA) (Miyakawa et al., 1996a, 1996b). Heterozygote mice with mutations in the glutamatergic GluR-B receptor subunit developed seizures and died by age 3 weeks (Brusa et al., 1995). Recurring episodes of excessive jumping and running were observed in *GluR-B* mutants, along with postmortem evidence of neuronal degeneration in the hippocampal CA3 field (Brusa et al., 1995). Mutation of the *E6-AP* ubiquitin ligase gene, a mouse model of Angelman's syndrome, produced mice with reduced body weight, reduced brain weight, poor performance on the rotarod, defective contextual fear conditioning, and severe tonic-clonic seizures in response to audiogenic seizure stimuli (Jiang et al., 1998b).

Narcolepsy appears to be caused by mutation in the gene for hypocretin/orexin in some cases (Sakurai, 2005). Mice with a mutation in the gene for orexin or the orexin-2 receptor displayed disrupted wakefulness and unusual bouts of non-REM sleep (Willie et al., 2003; Mieda et al., 2004).

Huntington's disease is a human autosomal dominant hereditary disorder, linked to an identified gene, *HD*, encoding a 348-kD protein, huntingtin (Nasir et al., 1995; Reddy et al., 1998). Loss of neurons in the striatum and cortex leads to motor restlessness and chorea, a movement disorder, as well as cognitive decline, psychiatric symptoms, and death (Zeitlin et al., 1995; Reddy et al., 1998). Triplet repeats of the nucleotides CAG characterize the Huntington's phenotype, as well as other neurological disorders such as spinocerebellar ataxias (Burright et al., 1997). Although the human motor phenotype of Huntington's disease has not been precisely duplicated in *huntingtin* transgenics, motor abnormalities have been reported (Zeitlin et al., 1995; Nasir et al., 1995: Burright et al., 1997; Reddy et al., 1998; File et al., 1998; Carter et al., 1999). A phase of hyperactivity in the home cage, followed by hypoactivity and death, were observed in several Huntington's transgenic lines (Reddy et al., 1998; Menalled et al. 2002). In another line, Huntington's disease transgenic mice showed reduced exploratory activity on the hole board test (File et al., 1998). Reductions in locomotor activity on this task grew progressively worse, beginning at 8 weeks of age. A third transgenic line expressing the *HD* gene carrying a 141–157 CAG repeat showed progressive declines in swimming, balance-beam walking, footprint patterns, and rotarod performance at 33 to 44 rpm, beginning at 5 weeks and declining severely at 11 weeks of age (Carter et al., 1999). The R6/2 Huntington's mice are proving useful in evaluating therapeutics that reverse components of the behavioral phenotype (Morton et al., 2005a).

Tay-Sachs and Sandhoff diseases are degenerative neurological diseases caused by one of several mutations in the genes necessary for the synthesis of β-hexosaminidase (Sandhoff et al., 1971; Gravel et al., 1995; Sango et al., 1995, 1996). β-hexosaminidase, a heterodimer enzyme, composed of α- and β-subunits, metabolizes gangliosides in the brain. Gangliosides are normal constituents of neuronal cell membranes. Insufficient metabolism results in ganglioside accumulation in the cytoplasm and cell death. Mutations in the *HEXA* gene, which codes for the α-subunit, result in Tay-Sachs syndrome. Mutations in the *HEXB* gene, which codes for the β-subunit, result in Sandhoff disease. The infantile form of both Tay-Sachs and Sandhoff diseases is characterized by ganglioside accumulation in the brain, neurodegeneration, ataxia, muscle wasting, mental retardation, and death by age 4 years. Knockout mice with null mutations for the *Hexa* gene show normal performance on motor tasks (Yamanaka et al., 1994). Knockout mice with null mutations for the *Hexb* gene show severe impairments on the rotarod task, as shown in Figure 4.2, indicating deteriorating motor coordination (Sango et al., 1995).

Progressive loss of motor coordination on the rotarod begins at 9 weeks of age in *Hexb* −/− mice and continues to decline until death at 16 to 20 weeks.

Double knockouts of *Hexa* and *Hexb* show much more severe deficits on rotarod performance than mice with mutations in *Hexb* alone (Sango et al., 1996). At 8 weeks of age, *Hexa* −/− *Hexb* −/− double-knockout mice showed major deficits on rotarod performance, significantly worse than *Hexb* −/− single-knockouts at 8 weeks of age (Figure 4.3*b*). Complete absence of all forms of lysosomal β-hexosaminidase in the homozygote *Hexa/Hexb* null mutants resulted in early ganglioside accumulation and accelerated lethality (Sango et al., 1996).

The *Hexb* knockout mouse model of a lethal human neurodegenerative disease provides a model system to test potential treatments for Tay-Sachs and Sandhoff diseases. The rotarod deficit in motor coordination provides a simple, fast, and sensitive measure of efficacy. Treatment of *Hexb* −/− null mutants with bone marrow transplants from +/+ wildtype control mice extended the time course of motor functions (Norflus et al., 1998). Rotarod performance deficits showed a delayed onset, appearing first at 13 weeks of age (Figure 14.13). Further, survival time was extended twofold, from 120 to 240 days (Norflus et al., 1998). Lentivirus delivery of the *Hexb* gene reduced ganglioside accumulation and inflammation in *Hexb* mice (Kirkanides et al., 2005). These experiments validate the use of mutant mice as a research tool, providing a proof of principle that a simple motor behavior phenotype of a knockout mouse can serve to screen novel therapies for a human genetic disorder.

BACKGROUND LITERATURE

Bolivar VJ, Caldarone BJ, Reilly AA, Flaherty L (2000a). Habituation of activity in an open field: A survey of inbred strains and F1 hybrids. *Behavior Genetics* **30**: 285–293.

Brooks, SP, Pask T, Jones L, Dunnett SB (2004). Behavioral profiles of inbred mouse strains used as transgenic background: 1. Motor tests. *Genes, Brain and Behavior* **3**: 206–215.

De Zeeuw CI, Strata P, Voogd J (1998). *The Cerebellum: From Structure to Control*. Elsevier Science/North Holland, Dordrecht.

Kelley AE (1998). Measurement of rodent stereotyped behavior. In *Current Protocols in Neuroscience*. Wiley, New York, pp. 8.8.1–8.8.13.

Klockgether T, Evert B (1998). Genes involved in hereditary ataxias. *Trends in Neuroscience* **21**: 413–418.

Marler P, Hamilton WJ (1968). *Mechanisms of Animal Behavior*. Wiley, New York, pp. 25–72.

Pierce C, Kalivas P (1997). Locomotor behavior. In *Current Protocols in Neuroscience*, Wiley, New York, pp. 8.1.1–8.1.8.

5

Sensory Abilities

HISTORY AND HYPOTHESES

We filter the world through our vision, hearing, smell, taste, and touch. How does a mouse use its five senses? Psychophysics is the field of investigation that measures sensory abilities in humans (Sekuler, 1974; Lloyd and Appel, 1976; Bartoshuk and Beauchamp, 1994). Some of the same principles and practices, such as quantitating the "just noticeable difference" between two sensory stimuli, can be applied to mice. However, in many cases, the sensory tasks used for humans cannot be adapted for mice. This chapter will describe some of the available tests for sensory abilities in mice. Yes-or-no measures of gross sensory abilities are presented. Sensitive neurophysiological measures of visual and auditory acuity are described. However, available tests for rapid, accurate evaluation of sensory modalities in mice are presently insufficient. More efforts toward the development of quick but comprehensive behavioral screens and more intensive, graded behavioral analyses of sensory functions in mice are greatly needed.

Olfaction is the best developed sense in most rodents (Getchell et al., 1984; Doty, 1986; Margolis and Getchell, 1988; Farbman, 1994; Beauchamp and Yamazaki, 2003; Brennan and Keverne, 2004). Receptors in the olfactory epithelium of the nasal cavity, and sensory neurons in the vomeronasal organ of the accessory olfactory system, receive information about smells (Luo et al., 2003; Brennan and Keverne, 2004). Bipolar sensory neurons send the information through the olfactory nerve to the olfactory bulb, which is large in comparison to brain size in rodents. Mice use the sense of smell to detect food, predators, territorial scent markings, colony odors, sexual receptivity, social dominance, and many other aspects of their environment (Liebenauer and Slotnick, 1996; Beauchamp and Yamazaki, 2003). Major urinary proteins appear to convey individual-specific scent marks in laboratory mice (Hurst et al., 2001; Nevison et al.,

2003). Genes for a very large number of olfactory receptors have been discovered in rodents (Dulac and Axel, 1995; Breer et al., 1996; Carver et al., 1998; Buck, 2004; Mombaerts, 2004). Activation of olfactory receptors induces the production of intracellular cyclic adenosine monophosphate (cAMP), which stimulates a cAMP-activated ion channel (Lowe et al., 1989; Lin et al., 2004). Sensitive tests for olfactory abilities in mice are designed to quantitate the acuity of the animal, that is, the detection limit for a given odorant. Other tests measure discrimination between odors, that is, the ability of the mouse to distinguish two different smells.

The sense of taste is mediated by a chemical transduction process similar to olfaction (Schiffman and Erickson, 1971; Simon and Roper, 1993). Gustatory receptors are located in the taste papillae and taste buds on the surface of the tongue. The gustatory receptors detect sweet, salty, umami, sour, and bitter tastes to determine the identity and the quality of food sources (Zhao et al., 2003; Matsunami and Amrein, 2004; Mombaerts, 2004). Taste discrimination is a well-developed sense in many rodents, including mice. Rats and mice will eat food with a familiar taste but not with an unfamiliar taste, as measured by food neophobia tests. Members of a colony of rats appear to learn which foods are safe and which are not by observing and sniffing other colony members during and after feeding (Galef and Wigmore, 1983; Bunsey and Eichenbaum, 1995). Grooming the food from another's whiskers appears to provide information about acceptable food sources. Taste preference and taste aversion tests measure the ability of mice to distinguish among tastes (Riley and Freeman, 2004). Two bottle choice tests for taste preference are used to quantitate the threshold for detection of a new or unusual taste. Chromosomal loci linked to sweet and bitter tastes have been described (Whitney and Harder, 1986; Berrettini et al., 1994; Lush, 1995).

Vision is a fairly well-developed sense in rodents, but poor in comparison to species such as hawks and humans. The gross anatomy of the visual system is similar across mammals (Glickstein, 1969; Van Essen, 1979; Deeb, 1993; Dowling, 1998). Retinal rod cells detect black and white images; cone cells detect colors. Absorption of a photon of light, by a sensory neuron in the retina, generates an amplified neural signal that is transmitted to higher order visual neurons. Photoactivated rod and cone receptors activate a guanosine triphosphate (GTP)-binding protein, transducin, that stimulates a cyclic guanosine monophosphate (cGMP) phosphod (Baylor, 1996). Retinal photoreceptors then send information to second-order cells and ganglion cells, which convey information to the optic nerve. Visual information is relayed to the lateral geniculate, superior colliculus, and visual cortex. Visual acuity in mice is generally measured neurophysiologically by electroretinogram, or behaviorally with a visual discrimination task, using an appetitive or aversive reward.

The sense of hearing is transduced by sensory hair cells of the cochlea. Information about pitch and loudness are conveyed through the auditory nerve to the olivary complex, lateral lemniscus, inferior colliculus, medial geniculate, and auditory cortex (Altschuler et al., 1991). Hearing is a well-developed sense in mice (Willott, 2001). High-frequency ultrasonic vocalizations are a major form of social communication between pups and parents. Distress vocalizations appear to convey information about aversive events to conspecifics in some rodent species. Simple measures of hearing can be assessed with threshold tests for acoustic startle (Logue et al., 1997b; McCaughran et al., 1999). Auditory acuity across frequencies is measured by the auditory-evoked brainstem response, a well-established neurophysiological test that has been adapted for mice (Erway et al., 1996). The auditory-evoked brainstem response and transient evoked otoacoustic emissions are used to measure hearing in adult and infant humans (Finitzo et al., 1998).

The sense of touch is remarkably keen in the whiskers, or vibrissae, of rodents. Watching a mouse in a novel environment, the astute human observer immediately notices the way the mouse constantly uses its whiskers to touch objects and surfaces. Whisker barrels are specialized regions of the cerebral cortex that map the activity of individual whisker hairs (Welker et al., 1996; Melzer and Smith, 1998). Mice show a strong startle response to a puff of air directed at the face or body. Compared to rats, mice are more sensitive to handling. Pain sensitivity has been well characterized in mice, with several standardized tests available for evaluation of pain threshold and the actions of analgesic drugs, as described below. Genetic influences on pain thresholds and sensitivity to analgesic drugs have been demonstrated in strain distributions of inbred and recombinant inbred strains of mice (Mogil et al., 1996; Pick et al., 1991; Quock et al., 1996; Elmer et al., 1998; Mogil and McCarson, 2000; Gaveriaux-Ruff and Kieffer, 2002), and across strains of rat (Hoffmann et al., 1998).

Interoceptive sensory information is perceived through sensory feedback neurons from the vagus nerve to the brainstem, as discussed in Chapter 7. Gastrointestinal sensations created by consumption of a large meal contribute to the cessation of feeding (Smith 1998a, 2000). Internal stimuli produced by food that makes the animal ill, or by a drug such as cocaine or morphine administered systemically, are detected by rats and mice and can be used as interoceptive discriminative stimuli (Riley, 1998).

BEHAVIORAL TESTS

Olfactory Acuity

Olfactory detection of smells is measured in simple tests in which the mouse locates a hidden object by odor. For example, a chocolate chip, a piece of Fruit Loop cereal, or a small amount of cheese or peanut butter is buried in the cage litter. A mouse that smells the odor of a familiar buried food will quickly burrow in the litter and uncover the source of the odor. Latency to locate the buried food is measured as the index of olfactory ability. Sensitive, quick tests for buried food generally involve overnight food deprivation, followed by placing the subject mouse in a cage, aquarium, or box containing clean litter or sand. A small piece of a familiarized palatable food, such as a cookie, chocolate chip, potato chip, or food pellet, is buried 1 to 4 cm below the surface (Nelson et al., 1995; Takeda et al., 2001; Bakker et al., 2002; Luo et al., 2002; Wersinger et al., 2002). An observer uses a stopwatch to determine the time it takes for the mouse to locate the food. This quick test gives a yes or no answer to whether the mouse lacks the ability to smell a strong odor. Buried food detection confirms that the mouse is not anosmic, that is, lacking a sense of smell entirely. Latency to locate the same food placed on the surface is a good control for the visual recognition of an olfactory stimulus (Nathan et al., 2004). A positive control to ensure that olfactory deficits are detectable by the methods used is to induce anosmia, the complete lack of smells, by infusing zinc sulfate into the nasal cavities of a normal mouse (Takeda et al., 2001), or perform a surgical olfactory bulbectomy (Trinh and Storm, 2003). Chemical or surgical anosmia should block detection of the buried food, confirming that the test will be sensitive to a mutation in the olfactory system.

Another simple approach is to quantitate time spent sniffing an attractive, novel odor. As described in Chapter 3, an odorant is painted onto the side of the test cage or soaked into a cotton ball or cotton swab (Winslow and Insel, 2002; Wrenn et al., 2003; Figure 5.1). A smear of cheese, a drop of an aromatic extract such as vanilla, almond, peppermint, or banana, a swipe of the bottom of a cage occupied by other mice, or an aliquot of urine from a mouse of the opposite gender are olfactory cues that are generally attractive to mice. The odorant located on the side of the test cage will stimulate bouts of directed sniffing at the location of the odor by the test mouse. Cumulative time spent sniffing the odorant is quantitated by the observer with a stopwatch. Cumulative time spent sniffing a neutral location, such as a drop of water painted onto an adjacent side of the test cage, is the control comparison. Smells such as banana and almond, soaked onto cotton swabs, elicit sniffing behaviors that can be quantitated and evaluated for habituation and dishabituation as a sensitivity measure of olfactory acuity (Luo et al., 2002; Wrenn et al., 2003; Figure 5.1). Time spent sniffing volatile urinary odors measures pheromone detection, more ethologically relevant to the smells used by mice in social communication. Urine from estrous females is often presented to male subject mice. Collected urine is presented on filter paper behind a mesh barrier (Baum and Keverne, 2002), at one end of a three chambered apparatus (Isles et al., 2001), at one location in a four chambered maze (Roman et al., 2002), on a sponge (Capone et al., 2002), plastic resin pad (Macknin et al., 2004), or via soiled litter (Wersinger and Rissman, 2000; Wiedenmayer et al., 2000).

More sophisticated tests for olfactory ability require the mouse to choose between two odors to obtain a food reward (Staubli et al., 1985; Eichenbaum et al., 1988; Zhang et al., 1998; Doty et al., 1999). Figure 6.20 in Chapter 6, illustrates an odor discrimination apparatus developed by Ursula Staubli and Gary Lynch at the University of California, Irvine (Staubli et al., 1985). The apparatus is a form of radial maze, containing two or more ports, through which distinct odors are delivered in a jet of forced air. The animal pokes its nose into the odor outlet tube, breaking a photocell beam, to receive a food or water reward. The choice may be designed to detect very different odors or different concentrations of the same odor. Automated olfactometers based on this principle have been elaborated for operant training in olfactory choice tasks to measure sensitivity to smells, and to investigate olfactory learning (Bodyak and Slotnick, 1999; Schellinck et al., 2001, Larson et al., 2003, Slotnick and Restrepo, 2005, and described in Chapter 6). Olfactometers that are now commercially available deliver volatile chemicals such fragrances and alcohols, through an airflow port located above a lever or nose-poke or water bottle spout. Commercially manufactured olfactometers enhance the comparison of results across laboratories.

Choice tasks require the animal to learn the procedures necessary to perform the task and to remember these procedures, as described in Chapter 6. Therefore it is possible to obtain a false positive on an olfactory discrimination if the deficit is related to learning or memory rather than to olfactory abilities. Motivation, attention, and general health deficits can similarly contribute to a false positive. Many rodent olfactory tests that were designed for rats require further optimization for mice.

Visual Acuity

Simple tests to determine blindness are commonly used in mice (Pinto and Enroth-Cugell, 2000). Reflex responses to visual stimuli are described in Chapter 3. These

(a)

(b)

Figure 5.1 *Olfactory habituation/dishabituation. (a) The mouse sniffs a cotton swab inserted into the cage lid. Time spent sniffing is scored by an observer with a stopwatch. A series of cotton swabs are sequentially inserted. Mice with normal olfaction will habituate to repeated exposures to the same smell, and will dishabituate and sniff more when presented with a new smell. (Photograph by Katrina Cuasay and Dr. Joanna Hill, contributed by the author.) (b) Habituation to three exposures to water-soaked swabs, dishabituation to a banana extract-soaked swab, habituation across three exposures to banana-soaked swabs, dishabituation to an almond extract-soaked swab, and habituation across three exposures to almond-soaked swabs. Both wildtype (WT) and galanin overexpressing transgenic mice (GAL-tg) displayed similar habituation/dishabituation, indicating normal olfaction in GAL-tg. This olfactory test was conducted as a control for the social transmission of food preference test for olfactory learning and memory, described in Chapter 6.* [From Wrenn et al. (2003), p. 28.]

include the blink reflex and pupillary constriction in response to a strong beam of light directed at the eye. The *visual placing test* involves holding the mouse by its tail at a height of about 15 cm from a table surface and noting the limbs held close to the body. As the mouse is gradually lowered to the table surface, it extends its forepaws for a "soft landing" on the horizontal surface (Heyser, 2003). A blind mouse will not see the approaching surface and will not extend its forelimbs until the whiskers or nose touch the table (Pinto and Enroth-Cugell 2000). The ability to discriminate light from dark is easily determined in mice by providing an environment consisting of two visually distinguishable areas, such as a well-lighted versus a dark region of a large cage or a Y-maze. As typical nocturnal rodents, mice prefer the dark and will quickly enter the darker area of the environment. Since a large, well-lighted, open area is aversive to mice, there is a component of fear or anxiety in light/dark choice tests, as described in Chapter 10.

The *visual cliff test* provides an appealing measure of gross visual ability (Fox, 1965a; Sloane et al., 1978; Rader et al., 1980; Bertenthal and Campos, 1984; Salinger et al., 2003). The visual cliff apparatus evaluates the ability of the mouse to see the drop-off at the edge of a horizontal surface. As shown in Figure 5.2, a cardboard or wooden box is built with a horizontal plane, connected to a vertical drop, connected to a second horizontal plane at a lower level. The vertical drop is approximately

Figure 5.2 *Visual cliff. Checkerboard contact paper lines the horizontal surface and the vertical drop-off. A sheet of clear Plexiglas resting on the horizontal surface extends out over the drop-off. If the mouse can see, it will stop at the edge of the horizontal surface and explore the edge. If a mouse is blind, it will not see the visual illusion of the drop-off but will feel the continuing horizontal surface. A blind mouse is more likely to walk forward across the Plexiglas without stopping. (Photograph contributed by the author.)*

0.5 meter. A black and white checkerboard pattern accentuates the vertical drop-off. The checkered pattern can be applied by painting the surfaces or by covering the surface with checkered adhesive paper. A sheet of clear Plexiglas is placed across the top horizontal plane, extending across the cliff. Thus there is the visual appearance of a cliff, but in fact the Plexiglas provides a solid horizontal surface. The mouse is placed onto the center ridge, and the time taken for the mouse to move off the ridge onto the horizontal surface is recorded. Both sides need to be sufficiently illuminated so that the checkerboard pattern is clearly visible. White walls around the edges will reduce reflections on the Plexiglas surface. If the animal sees the "cliff," it will stop at the edge of the vertical drop-off. A blind animal may move forward across the "cliff" without pause.

The original visual cliff developed by Fox (1965a) includes a ridge of aluminum, 1 inch wide and 1.5 inches thick, at the edge of the cliff. The mouse was placed on the aluminum block at the start of each of 10 consecutive trials. Fox (1965a) analyzed the visual cliff data as "safe" or positive when the mouse chose to step down onto the horizontal checkered surface and "negative" when the mouse stepped down onto the vertical-appearing surface. A 50/50 ratio of responses indicated that the mouse moved off the ridge onto the two sides with equal frequency. Strain comparison with this method showed 80–90% safe responses in the A/J, Sm/J, C57BL/6, and C57BL/10 strains of mice, as compared to only 52% safe responses in the C3H/HeJ mice, which are a blind strain (Fox, 1965a).

The visual cliff task is confounded by the ability of mice to use other senses for navigation. The mouse may sniff and explore the Plexiglas sheet with the whiskers for several seconds before venturing forward across the "cliff." If the animal is blind, it may successfully navigate on the basis of nonvisual senses, including feedback from the whiskers and paws, detecting the tactile feel of the continuous Plexiglas floor. It is possible to eliminate the sensory information from the whiskers by shaving off the whiskers before the visual cliff test. Removal of the whiskers will interfere with performance on many other behavioral tasks. If whisker removal is chosen for the visual cliff test, this experiment should be done at the end of all other behavioral testing.

Behavioral tests for vision in mice require the animal to make a *visual discrimination*, i.e. a reinforced choice between visual cues. Mice are trained to press a lever in the presence of a small light on the wall of the cage, to obtain a food reward. Visual acuity is assessed by a visual discrimination between graded levels of illumination or colors of the light. Sensitive visual discrimination tasks require the rodent to make a choice between two complex visual stimuli. Either an appetitive or an aversive reinforcement accompanies the choice. For example, the subject must choose the lever below a checkerboard pattern, rather than the lever below a solid color, to receive a food reward. Or the mouse must enter the arm of a Y-maze with horizontal stripes, rather than the arm of the Y-maze with vertical stripes, or an illuminated versus a dark arm of a Y-maze, to avoid a footshock. These tasks are limited by the relatively poor visual abilities of mice and the requirement that the mouse must be able to learn the choice procedure for the task. The cognitive components of these tasks are described in Chapter 6.

Circadian locomotor rhythms are sensitive to brief pulses of light, allowing circadian running wheel patterns to be used as a measure of progressive photoreceptor degeneration. If a light pulse does not re-set the circadian activity rhythm in an animal housed in continuous darkness, blindness is one likely explanation. Aged mice with retinal degeneration showed deficits in visual and circadian responses to light (Provencio et al., 1994).

Electrophysiological recording from the visual cortex during presentation of visual stimuli remains the most sensitive and accurate measure of visual acuity in mice. Electroretinogram (ERG) recordings measure responses of retinal cells to a flash of bright light (Pinto and Enroth-Cugell, 2000; Peachey and Ball, 2003). Visual thresholds in mice have been quantitated with single-unit neurophysiological recording from retinal ganglion axons and from cells in the superior colliculus, and correlated to behavioral performance in a water maze test (Hayes and Balkema, 1993). Absolute visual threshold and the retinal image of a monochromatic light source have been measured with a double-pass imaging apparatus (Green et al., 1994; Artal et al., 1998). This method is used to calculate contrast sensitivity functions in rats. Visual evoked potentials to pattern stimuli, such as a horizontal striped grating, have been recorded from the binocular visual cortex of awake mice (Porciatti et al., 1999).

Spontaneous mutations in the visual system are common in laboratory mice. At least 16 naturally occurring retinal degeneration mutations in the photoreceptors of the retina were discovered at The Jackson Laboratory (Chang et al., 2002). Retinal degeneration genes are common in the genetic backgrounds of many inbred strains used to generate and breed transgenic and knockout mice, including C57BL/6J, FVB/NJ, C3H, and CBA/J (Chang et al., 2002). Age dependence of the loss of vision complicates the phenotypic analysis of mutant mouse models of aging and Alzheimer's disease, as described in Chapter 12. For example, the presence of the retinal degeneration gene in three Alzheimer's transgenic lines produced profound impairments on spatial memory tasks, including the Morris water maze, radial arm water maze, and platform recognition, conducted when the mice were 5 to 8.5 months old (Garcia et al., 2004).

Auditory Acuity

The *acoustic startle test* provides a good measure of gross hearing ability and of auditory threshold. Acoustic startle is a reflex in most mammals. A sudden, loud noise causes the subject to flinch. Human acoustic startle experiments measure the startle response to a sudden, loud noise by the eye blink reflex. Mouse acoustic startle experiments measure amplitude of whole body flinch. The Preyer reflex is a flinch response to the sound of a clicker or a loud hand clap (Henry and Willott, 1972; Huang et al., 1995). The Preyer reflex is measured as a yes-or-no response by the human observer. Automated startle equipment, shown in Figure 5.3a, is commercially available. Automated software delivers uniform tones across a range of loudnesses, in a random order. The mouse is placed in a small cylinder. Cylinders of various sizes are available, designed for rats, adult mice, and young mice. An optimal cylinder diameter is chosen to reduce the stress of restraint, being wide enough to allow the animal to turn around. A five-minute habituation period before the start of the session is usually sufficient for a mouse to adjust to the cylinder and sit quietly during presentations of sounds. White noise is present at a background level of 65 to 75 decibels (dB). Brief tones varying from 75 to 120 dB are randomly delivered through a small speaker in the wall of the chamber. Normal mice of many inbred strains (Figure 2.6) will respond to tones of 100 dB and louder by a whole body

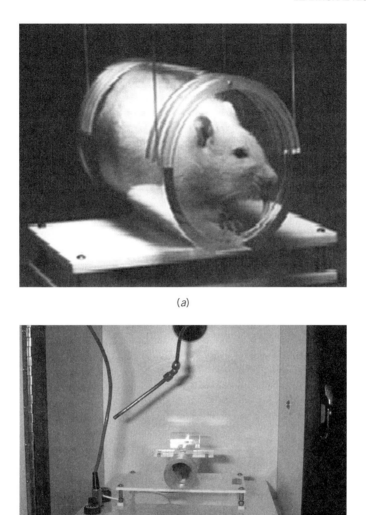

(a)

(b)

Figure 5.3 *Acoustic startle test. (a) The mouse is placed in the plastic cylinder, which is enclosed in (b) a sound-attenuating environmental chamber. A sudden, loud noise is introduced through a speaker within the chamber. This startle stimulus produces a flinch reflex response in the normal mouse. Amplitude of the whole body flinch is measured by a sensor under the cylinder. The startle response can be tested across a range of decibel levels, 70 to 120 dB, to evaluate acoustic threshold.* (Photograph (a) kindly contributed by Dr. Stephen Dunnett, Cardiff University, UK. Photograph (b) contributed by the author.)

twitch (Geyer and Swerdlow, 1998). As illustrated in Figure 2.7, strain distributions document inbred strains of mice that are deficient in acoustic startle (Logue et al., 1997b; Paylor and Crawley, 1997; McCaughran et al., 1999; Willott et al., 2003). The equipment measures flinch amplitude through an electrostatic sensor located immediately under the chamber in which the mouse is resting. Acoustic startle threshold is defined as the minimum decibel level that elicits a flinch. A deaf mouse will not flinch until the tone is in the range of 120 dB. At this extreme loudness, the mouse is probably responding to vibrations through the sense of touch, rather than hearing the sound. A deaf mouse may not respond to startle stimuli below 120 dB. A mouse with impaired hearing shows an elevated acoustic startle threshold to startle stimuli below 120 dB. Prepulse inhibition of the acoustic startle response, described in Chapter 10, may provide an even more sensitive measure of hearing, as mice appear to detect lower decibels of the prepulse, 90 dB or below, which inhibits the 120 dB startle stimulus. However, startle and prepulse inhibition are not pure measures of hearing. Emotional and motor components of the startle reflex circuitry and of the pathways mediating sensorimotor gating contribute to performance. Mutations affecting any of these underlying mechanisms will confound a direct interpretation of a hearing deficit in a mutant mouse line. The neural circuitry mediating the acoustic startle reflex has been elegantly elucidated (Davis et al., 1994; Geyer et al., 2002; Yeomans et al., 2002). Genes linked to deficits in acoustic startle have been revealed in linkage studies with recombinant inbred strains with high-frequency hearing loss (Hitzemann et al., 2001).

The *auditory-evoked brainstem response* (ABR) is the most sensitive measure of auditory threshold that has been extensively applied to mice. Synchronized neural discharges from the auditory nerve and brainstem are emitted in response to short acoustic stimuli, such as a series of clicks or tones (Altschuler et al., 1991; Steel, 1995; Erway et al., 1996). This method requires specialized neurophysiological recording equipment, including a precise auditory stimulus input, and an evoked potential detector. A commonly used system is the Intelligent Hearing System unit (IHS, North Miami, FL). The equipment is best used within a sound-controlled environment such as the NIOSH noise exposure facility (Davis and Franks, 1989). Anesthetized mice are acutely implanted with stainless steel electrodes, inserted subcutaneously below the pinnae, superficial to the auditory nerve (Erway et al., 1996). Click stimuli tones, randomized over 8 to 32 kHz, are presented through earphones over both ears. The loudness of each tone is varied in 5 dB steps. Responses from the auditory nerve are amplified 25,000- to 100,000-fold and filtered through a preamplifier. The ISH system calculates the neurophysiological response profile from the auditory nerve electrodes, and presents the auditory-evoked brainstem response as a wave form for each tone. The auditory threshold is defined as the lowest stimulus intensity at which a normalized ABR wave can be identified.

Strain distribution studies have demonstrated differences in ABR thresholds for inbred strains of mice. Age-related hearing loss was demonstrated in the DBA/2J strain at 3 to 4 weeks of age, in the C57BL/6J strain beginning at approximately 2 months of age, and in the BALB/c strain beginning at approximately 2.5 months (Li and Borg, 1991; Willott and Bross, 1996; Willott et al., 1998; Johnson et al., 2000). Natural mutants with severe deficits in ABR and cochlear function include the Ames waltzer, varitint-waddler, jerker, and Tasmanian devil (Raphael et al., 2001; Erven et al., 2002; Kim et al., 2003; Davis et al., 2003). Noise exposure induces hearing loss in mice, with differential sensitivities to noise-induced deafness seen in different inbred strains detectable by the ABR test (Erway et al., 1993, 1996). Potential deafness due to background genes in these strains must be considered for any task involving an auditory cue. For example, it would be incorrect to interpret a learning and memory deficit if poor performance on auditory cued fear conditioning was actually an artifact of deafness (see Chapter 6).

Taste

The ability to detect tastes is often measured in choice tests (Riley and Freeman 2004). Two bottle choice tests are frequently used. A different taste solution is placed in each of two identical water bottles in the home cage. The location of the two bottles in the cage is randomly switched across test days to avoid place preferences. The volume of each solution that is consumed by the mouse over a fixed time period is measured. For example, a choice between tap water and quinine will measure the ability of the mouse to detect a bitter taste (Berrettini et al., 1994; Wong et al., 1996; Riley, 1998; Ferraro et al., 2005). A choice between water and saccharine will measure the ability of the mouse to detect a sweet taste (Belknap et al., 1992). A choice between different concentrations of a solution can evaluate taste acuity. Some of the species-typical responses to aversive tastes are shown in Figure 5.4. Detailed methods to measure consumption of solids and liquids are described in Chapter 7.

Touch

The ability to respond to tactile stimuli can be measured in simple reflexes, as described in Chapter 3. A mouse will twitch its whiskers in response to a light touch of the whiskers by a small paintbrush. A gap-crossing test may be useful for assessing the active use of the vibrissae in locomotion (Barneoud et al., 1991). Tactile startle can be measured in the same automated equipment used for acoustic startle, as illustrated in Figure 5.3b. The whole body *flinch response* to a puff of compressed air, directed at the body, is measured in the tactile version of the startle test. Inbred strain distributions for tactile startle are similar to inbred strain distributions for acoustic startle (Figure 2.6). Discrimination between the arms of a Y-maze in which different textures cover the floor (e.g., cloth toweling on the floor versus wire mesh flooring) can provide further information about touch sensitivity of the paws. The foot will be withdrawn in response to acute pressure. Commercially available equipment measures pressure sensitivity by the application of force to the paw or tail and measurement of a flinch response.

Pain Sensitivity

Several good tests for pain sensitivity have been validated for responses to analgesic drugs in inbred strains of mice (Gonzales-Rios et al., 1986; Pick et al., 1991; Richardson et al., 1998; Malmberg and Bannon, 1999). The stimulus is a brief, intense exposure to a focused heat source in some tasks, or a noxious chemical injected subcutaneously in other tasks. Spinal reflex pain is measured with the tail flick test, illustrated in Figure 5.5. Centrally mediated acute pain reflexes are measured with the hot plate test, illustrated in Figure 5.6. Delayed responses to tissue damage are measured with the formalin test. Stress-induced analgesia, induced by restraint, footshock, or cold-water swim, is a well-characterized phenomenon in rodents (Moskowitz et al., 1985; Rubinstein et al., 1996; Mogil and Belknap, 1997). To avoid the confounding effects of novelty and handling stressors, prior handling of the mice and habituation to the test equipment is advisable. A 5-minute session for exploration of the equipment, without presentation of the stimulus, will minimize stress-induced analgesia during the subsequent test session.

	Solution			
Response	0.1 M Sucrose	0.1 M NaCl	0.01 M HCl	0.0005 M QHCl
Ingestive				
Mouth movement	77.8 (±10.8)	45.6 (±8.5)	86.1 (±16.8)	60.8 (±10.6)
Tongue protrusion	43.2 (±13.1)	31.1 (±8.7)	46.6 (±14.9)	44.7 (±14.8)
Aversive				
Gape	3.8 (±2.1)	9.3 (±4.5)	6.4 (±3.6)	8.2 (±4.9)
Passive drip	1.5 (±0.7)	1.2 (±0.5)	1.7 (±0.8)	1.7 (±0.9)
Head shake	1.3 (±0.5)	1.9 (±0.5)	3.3 (±1.1)	3.6 (±1.0)
Forelimb flail	6.7 (±2.0)	17.1 (±3.6)	17.7 (±7.1)	15.1 (±5.2)
Fluid expulsion	0.2 (±0.2)	0.3 (±0.2)	1.3 (±0.6)	1.1 (±0.6)
Neutral				
Paw licking	9.6 (±3.4)	4.6 (±2.3)	4.7 (±1.6)	8.7 (±2.5)

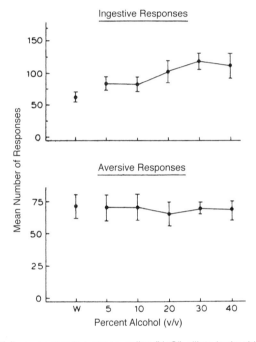

Figure 5.4 *Taste reactivity responses to sucrose, saline (NaCl), dilute hydrochloric acid (HCl), quinine (QHCl), water (W), and alcohol, infused into the mouths of mice.* [From Kiefer et al. (1998), p. 1149.]

Tail flick is a spinal reflex in which the mouse moves its tail out of the path of a strong beam of focused light (D'Amour and Smith, 1941; Hole and Olsen, 1993; King et al., 1997c; Malmberg, 1999). Several automated tail flick systems are commercially available. As shown in Figure 5.5*a*, the mouse is gently placed on a platform with its tail extended in a straight line along a narrow groove. The mouse is gently held by a gloved hand resting on the mouse's back, or lightly wrapped in a soft cloth towel, with the tail extending out of the towel along the groove. A high-intensity, heat-producing, narrow beam of light is directed at a small, discrete spot in the groove, about 15 mm from the tip of the tail. The tail lies in the path of the beam. After a few seconds, the

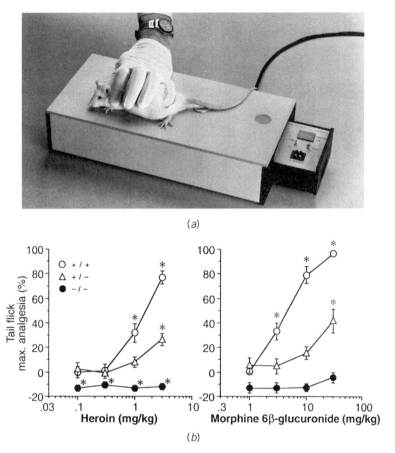

(a)

(b)

Figure 5.5 *(a) Tail flick test measures sensitivity to a high-intensity light beam focused on the tail of the mouse. Latency to flick the tail out of the path of the light beam is the measure of pain sensitivity or analgesia. (Photograph kindly contributed by Dr. Charles Scouten, Stoelting Company, Wood Dale, IL.) (b) See legend to Figure 5.6b.*

heat of the light beam on the tail becomes slightly painful. The mouse will move its tail out of the path of the beam. Latency to flick the tail away from the light beam is the dependent variable. Latencies can be recorded manually by an observer with a stopwatch, or by timers within the automated tail flick equipment. The intensity of the light beam is adjusted by the experimenter to produce tail flick latencies of 4 to 6 seconds in control mice. Age and body weight change the properties of the tail, requiring recalibration of the light beam intensity for each control group. Treatment with an analgesic drug, such as morphine, will increase the latency to flick the tail out of the path of the light beam. A cutoff time of 10 to 30 seconds is generally used, after which the mouse is removed from the apparatus. A brief cutoff time is necessary to avoid severe pain and tissue damage to the tail. Accuracy is improved by averaging three separate determinations spaced over intervals of at least 5 minutes. Different areas of the tail are in contact with the light beam on each trial.

The *Hargreaves test* (Hargreaves et al., 1988; Clark and Tempel, 1998; Richardson et al., 1998) is similar to the tail flick test. A focused beam of light is directed to the

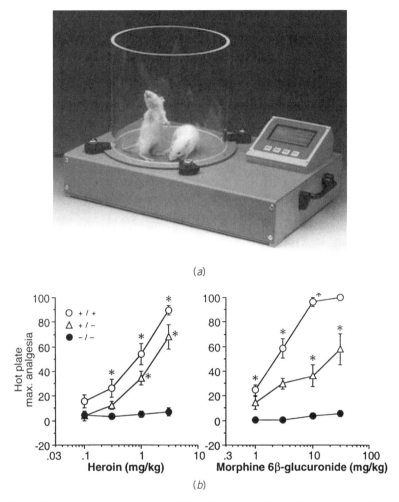

Figure 5.6 *(a) Hot-plate test measures sensitivity of the paws to the hot temperature of the surface of the floor. Latency to lift and lick a paw, vocalize, or jump is the measure of pain sensitivity or analgesia. (Photograph kindly contributed by Dr. Charles Scouten, Stoelting Company, Wood Dale, Illinois.) (b) Knockout mice deficient in the µ-opioid receptor (black circles) lack the analgesic responses to heroin and morphine on the hot-plate and tail flick tests, as compared to heterozygote (white triangles) and wildtype controls (white circles).* [From Kitanaka et al. (1998), p. R2.]

skin of a hindpaw rather than the tail. Latency for the mouse to withdraw the paw is measured. The *hot-water tail flick* and *cold-water tail flick* tests measure the latency for the mouse to withdraw the tail when immersed in a beaker of water maintained at 52.5°C or in ice water, respectively (Gonzales-Rios et al., 1986; Pick et al., 1991; Fairbanks and Wilcox, 1997).

The *hot-plate* test measures a reflex that requires circuitry in the brain as well as in the spinal cord (O'Callaghan and Holtzman, 1985; Bannon et al., 1995; Malmberg, 1999). The mouse is placed on a horizontal surface that is heated to 52–55°C. The hot-plate temperature is calibrated by the experimenter to produce a response within about 10 seconds in control mice. A tall plastic cylinder or square is placed on the surface, around the mouse, to prevent the animal from walking off the surface or jumping out of the test apparatus. After a few seconds the heat of the surface becomes slightly painful to the paws. The mouse will lift a paw and lick the ventral surface of the paw. Latency to lick a rear paw is a reliable indicator of discomfort. Latencies are timed by an observer with a stopwatch. Some mice will jump up and some will vocalize. Many investigators accept jumping and vocalizing responses as equivalent to hindpaw licking. Automated versions of the hot-plate test measure the latency to jump. A cutoff time of 30 to 60 seconds is generally used, after which the mouse is removed from the apparatus to avoid tissue damage that can occur with prolonged exposure to the hot surface. Mice treated with an analgesic drug such as morphine, or with a mutation that induces a hyperalgesic condition, will show increased hot-plate latencies. The cutoff time is critical in hot-plate experiments designed to test analgesic drugs or mutations that block pain transmission, to prevent injury to the paws.

Commercial equipment that delivers a well-controlled temperature to the horizontal surface is available, as shown in Figure 5.6. The choice of temperature is made on the basis of the responses of the control mice. Larger, older mice may have thicker fat pads or denser skin in the ventral surface of the paws, increasing resistance to the heat stimulus to some extent. A hot-plate temperature of 55°C may be optimal for adult mice and rats. For young mice, a lower hot-plate temperature such as 52.5°C may be optimal. Wildtype littermates are the best controls for a pilot study to decide on the hot-plate temperature for a given experiment with mutant mice. Age- and weight-matched mice from the parental strains are an alternative approach.

The *formalin test* (Dubuisson and Dennis, 1977; Wheeler et al., 1990; Clark and Tempel, 1998; Malmberg, 1999) measures the response to a noxious chemical injected into a hindpaw. A volume of 20 µl of a 1% formalin solution is injected through a fine-gauge needle subcutaneously into the dorsal surface of one hindpaw. Licking and biting of the hindpaw is quantitated as cumulative number of seconds engaged in the behaviors. A rating scale is often used in which 1 = the formalin-injected paw rests lightly on the floor, bearing less weight; 2 = the injected paw is elevated; 3 = the injected paw is licked, bit, and/or shaken (Dubuisson and Dennis, 1977; Malmberg, 1999). Two phases of responses are seen in the formalin test. Phase 1 begins immediately after injection and lasts about 10 minutes, representing the acute burst of activity from pain fibers. Phase 2 begins about 20 minutes after injection and continues for about an hour. This phase appears to represent responses to tissue damage, including inflammatory hyperalgesia.

Another test for touch, which can be used to measure pain threshold, employs *Von Frey hairs* (Pitcher et al., 1999; Fuchs et al., 1999). As illustrated in Figure 5.7, Von Frey hairs are a set of very fine gauge calibrated metal wires. Withdrawal threshold to mechanical stimulation is measured. The animal stands on an elevated platform in which the surface is a wide gauge wire mesh. The Von Frey hair is inserted from below, up through the holes in the mesh, to poke the undersurface of a hindpaw. At threshold, the mouse responds by quickly flicking its paw away from the hair, generally followed by raising the paw, licking the paw, and/or vocalization. Mechanical withdrawal threshold is defined as the minimum gauge wire stimulus that elicits withdrawal reactions in two out of three consecutive trials (Robertson et al., 1997a).

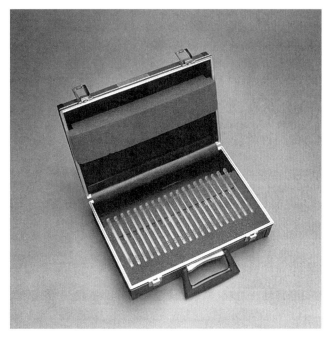

Figure 5.7 *Von Frey hairs.* (Photograph kindly contributed by Dr. Charles Scouten, Stoelting Company, Wood Dale, IL.)

TRANSGENICS AND KNOCKOUTS

Odors occupy specific receptors that activate G_{olf}, a G protein expressed at high levels in olfactory cilia, which stimulates membrane-bound adenylate cyclase (Bakalyar and Reed, 1990). Targeted mutation of the G_{olf} gene generated mice with major reductions in electrophysiological responses of primary olfactory sensory neurons to odors (Belluscio et al., 1998). Most of the null mutants did not nurse and died within 2 days of birth. Surviving homozygous mutants had lower body weights and were hyperactive on an open field test. Female knockout mothers failed to nest and crouch over the pups, and did not retrieve the pups, which may reflect an inability to detect the pup odors. Olfactory impairments included inability to locate a buried food pellet at ages up to 10 weeks, lack of habituation to identical odors presented on cotton swabs, and lack of dishabituation to new odorants presented sequentially on cotton swabs (Luo, 2002).

Investigations of genes mediating olfactory abilities are increasingly employing operant training in airflow delivery olfactometers to analyze olfactory discrimination of similar smells or serial dilutions of the same smell. The *CNGA4* cyclic nucleotide gated channel gene determined odor detection threshold for a discrimination between cineole and a background odor of mineral oil, or between 1-octanol and mineral oil, without affecting choice accuracy between more similar odors such as 1-octanol and 1-heptanol (Kelliher et al., 2003). *AC3* adenylate cyclase knockout mice failed an olfactory passive avoidance response to citralva versus lilial (Wong et al., 2000), and an habituation test to volatile odorants (Trinh and Storm, 2003). *CNGA2* knockout mice, deficient in cyclic nucleotide gated channel subunit A2, showed chance levels of choice accuracy for the detection of lilian and geraniol, while responding normally to ethyl acetate (Lin et al., 2004).

Olfactory recognition underlies many social sexual, parental, and aggressive behaviors in mice (Hurst et al., 2001; Brennan and Keverne, 2004). Deletion of an approximately 600 kilobase region containing 16 vomeronasal receptor genes resulted in female mice showing less maternal aggression and male mice initiating fewer sexual mounts (Del Punta et al., 2002). Estrogen receptor alpha null mutant males spent less time investigating females and did not show the normal preference for soiled bedding from females, although their neuroendocrine responses to female-soiled bedding remained intact (Wersinger and Rissman, 2000). Male mice deficient in the gene for aromatase, the enzyme that converts testosterone into estradiol, showed reduced copulation (Aste et al., 2003), and less investigation of odors from females in a Y-maze (Bakker et al., 2002). Reduced olfactory abilities have been reported in Alzheimer's disease patients (Kier and Molinari, 2003). Mouse models of Alzheimer's disease, discussed in Chapter 11, may be useful for evaluating the role of specific pathological symptoms in the olfactory deficit. For example, transgenic mice overexpressing the human Tau protein failed to show odor habituation, as compared to their wildtype controls (Macknin et al., 2004). ApoE knockout mice performed poorly on locating a buried food pellet and odor avoidance tests, while showing normal latencies to locate a visible food pellet (Nathan et al., 2004).

Retinal degeneration genes have been extensively documented in natural mutants. Spontaneous retinal degeneration has been described for the *rd/rd, rds/rds, nr/nr, pcd/pcd, spastic*, and *vitiligo* strains of mice, which lose photoreceptors at older ages (Yazulla et al., 1997; Evans and Smith, 1997; Pieke-Dahl et al., 1997; LaVail et al., 1997, 1998; Hopp et al., 1998). The genes responsible for the apoptotic loss of retinal neurons in these lines of mice are being investigated for their relevance to human retinal degeneration. *Rd* appears sporadically in the genome of several strains of inbred mice, requiring careful testing for progressive blindness in older animals. For example, chromosomal loci linked to iris atrophy and glaucoma in DBA/2J mice have been identified at the *ipd* locus on mouse chromosome 6 and the *isa* locus on mouse chromosome 4 (Chang et al., 1999).

Mutations for photoreceptor genes have been analyzed with both behavioral and neurophysiological techniques. Mice deficient in rods and cones were generated by a triple knockout of the rod gene *Gnat1*, the cone gene *Cnga3*, and the melanopsin gene (Hattar et al., 2003). Null mutants showed no pupillary constriction to light, and no circadian photoentrainment, while heterozygote triple mutants were similar to wildtype controls. Transgenic mice expressing a human cone photopigment gene displayed enhanced spectral sensitivity in electroretinogram responses to square wave pulses of light, and the small number of transgenics that were tested behaviorally showed an expanded spectral range in a visual discrimination task (Jacobs et al., 1999).

Lowe syndrome is a human hereditary disease, inherited as an X-linked trait from a mutation of the *OCRL1* gene, that presents as congenital cataracts, mental retardation, and kidney failure (Charnas and Nussbaum, 1994). Knockout mice deficient in *Ocrl1* did not develop cataracts, and showed normal performance on open field and passive avoidance tasks (Jänne et al., 1998). A second gene, *Inpp5b*, encodes a product highly homologous to the *Ocrl1* gene product, inositol bisphosphate 5-phosphatase. Double knockout mice deficient in both *Ocrl1* and *Inpp5b* were extremely impaired and inviable as compared to either knockout alone, suggesting mutually complementary and compensatory roles of these two genes in mice (Jänne et al., 1998).

An interesting rescue experiment reported beneficial effects of ciliary neurotrophic factor gene transfer by an adenoviral vector into the *rds/rds* mutant mouse (Cayouette et al., 1998). Continuous administration of this neurotrophic factor is suggested as a treatment for some forms of human retinal degeneration. Recent gene delivery therapies in mutant mice are described in Chapter 14.

Genes for deafness in mice are being mapped for investigations of their relevance to human deafness (Robertson et al., 1997b). Natural spontaneous mutations in some of these genes have been seen in mice (Steel, 1995; Raphael et al., 2001; Erven et al., 2002; Kim et al., 2003; Davis et al., 2003; Johnson et al., 2003; Libby et al., 2003). Hereditary cochlear deafness is well described in a natural mutant mouse, *deafness (dn/dn)*. Sound responsiveness was tested in these mice by the simple Preyer startle reflex. Auditory brainstem response testing in these mutant mice detected thresholds 30 to 50 dB higher for the mutant mice (Huang et al., 1995). The Ames waltzer (*av*) mouse is another natural mutation characterized by deafness. This autosomal recessive deafness mutation was mapped to loci on chromosome 10, using the pinna orienting reflex to a loud noise (Zobeley et al., 1998). The *deafwaddler (dfw)* mouse is a natural mutant with a point mutation that causes a substitution of a glycine to a serine in a conserved amino acid of a protein in the hair cells of the cochlea (Street et al., 1998). Mouse models of Usher syndrome, an autosomal recessive human syndrome of deafness and retinitis pigmentosa blindness (Petit, 2001), were discovered by mapping genes in natural mutants including deaf circler, deafwaddler, and waltzer (Pieke-Dahl et al., 1997; Johnson et al., 2003; Noben-Trauth et al., 2003; Libby et al., 2003).

Knockout mice showing loss of hearing and/or anatomical defects in the auditory system have been described (Agnostopoulos, 2002). The autosomal recessive deafness gene *DFNB29* was investigated by targeted gene mutation in *Claudin 14* knockout mice and found to show cochlear hair cell degeneration (Ben Yosef et al., 2003). *TrkB* and *TrkC* tyrosine kinase receptor knockout mice showed loss of sensory neurons (Silos-Santiago et al., 1997); however, functional studies of hearing remain to be performed. Elevated sensitivity to sound-induced seizures was detected in serotonin receptor $5\text{-}HT_{2C}$ knockout mice (Brennan et al., 1997). However, careful neurophysiological analysis of responses of inferior colliculus neurons in these mice revealed elevated thresholds for high-frequency sounds in both the wildtype and knockout genotypes, consistent with the moderate age-related hearing loss common in C57BL/6J mice. Mice with mutations in the adrenergic α_{2C} receptor show unusual responses on acoustic startle and prepulse inhibition (Sallinen et al., 1998). Amplitude of the startle reflex to a 118 dB loud tone was greater in the α_{2C} knockout mice as compared to wildtype control mice. Less prepulse inhibition was seen in the α_{2C} knockout mice to a 75, 78, 81, or 87 dB prestimulus tone, 100 msec before the 118 dB acoustic startle tone. Knockout mice deficient in the gene for fibroblast growth factor receptor 3 show severe abnormalities in bone growth and development of the cochlea, with evidence for deafness as measured by the acoustic startle reflex and the auditory-evoked brainstem response (Deng et al., 1996; Colvin et al., 1996; McDonald et al., 2001). Mice with mutations in *dishevelled-1*, a developmental gene in *Drosophila* that is expressed in mouse brain, showed reduced responses on both acoustic startle and prepulse inhibition to acoustic startle as compared to wildtype littermates (Lijam et al., 1997).

Genes for taste receptors have been identified in experiments with inbred strains of mice and knockout mice (Reed et al., 2004; Mombaerts, 2004; Matsunami and Amrein, 2004). Mice deficient in gustducin, a G protein in taste receptor cells, showed reduced responses to bitter tastes (Wong et al., 1996). A two-bottle taste preference test was used to evaluate taste responses to sweet, salty, sour, and bitter solutions. Knockouts showed normal preferences to a series of concentrations of sodium chloride, the salty taste, and hydrochloric acid, the sour taste. Aversive responses to higher concentrations of two bitter solutions, denatonium benzoate and quinine sulfate, were significantly lower in knockouts as compared to wildtype littermates. Preference for two sweet solutions, sucrose and SC45647, were also reduced in the knockout mice. Taste receptors T1R1, T1R2, and T1R3 were found to confer the ability to detect the tastes of sweet and umami (the taste of monosodium glutamate) in mice, using choice tests with mouse strains containing polymorphisms of the Tas1r3 gene (Reed et al., 2004; Inoue et al., 2004), and in knockout mice deficient in subunits of the T1R receptor (Zhao et al., 2003).

Mutations in genes relevant to pain transmission produce marked changes in pain threshold and sensitivity to analgesic drug treatments. Mice with null mutations in the μ-opioid receptor gene do not respond to morphine on the tail-flick and hot-plate tests (Matthes et al., 1996; Sora et al., 1997; Gaveriaux-Ruff and Kieffer, 2002). Similarly μ-opioid receptor knockout mice were normal on responses to Von Frey hair stimulation but did not display the dose-dependent reduction in withdrawal responses to Von Frey hair stimulation after morphine treatment (Fuchs et al., 1999). While the analgesic effects of morphine on the tail-flick and hot-plate tests, and morphine-induced lethality, were dramatically reduced in μ-opioid receptor knockout mice, normal responses were seen to ligands of other opioid receptor subtypes, using the δ receptor subtype agonist DPDPE and the κ receptor subtype agonist U50488 (Loh et al., 1998). Null mutants deficient in pre-proenkephalin showed normal baseline responses on the tail-flick, hot-plate, and formalin tests and normal stress-induced analgesia (König et al., 1996). Null mutants deficient in the gene for β-endorphin exhibited normal analgesic responses to morphine but did not exhibit stress-induced analgesia following swim stress (Rubinstein et al., 1996). Baseline and pharmacological responses on analgesia tests in mice lacking opioid receptors are summarized by Brigitte Kieffer in Table 5.1.

Substance P is a neuropeptide transmitter synthesized in small-diameter sensory pain fibers (Liu et al., 1997a). Transgenic mice overexpressing substance P displayed hyperalgesia, as measured by faster reaction time on tail withdrawal in a modified tail-flick test (McLeod et al., 1999). Treatment with a substance P antagonist reversed the increased responsivity of the substance P overexpressing transgenics on the tail-flick test, confirming that the phenotype was due to the substance P overexpression (McLeod et al., 1999). Using the opposite strategy, null mutant mice deficient in a substance P receptor were generated. Substance P receptor knockout mice were normal on the tail-flick and hot-plate tests (De Felipe et al., 1998). However, substance P receptor knockout mice did not show normal nociceptive responses on the formalin test and had a significantly reduced response to morphine on the tail-flick test. Further, substance P receptor knockout mice did not demonstrate stress-induced analgesia following swim at 4°C, and did not show normal intensity coding or "wind-up" neurophysiological responses to electrical or noxious mechanical stimuli, which normally activate unmyelinated C fibers. The authors conclude that substance P orchestrates the response

TABLE 5.1   Summary of Behavioral Phenotypes of Several Opioid Receptor Knockout Mice on Measures of Analgesic Responses to Opioid Treatments

Opioid (Selectivity)	In vivo Responses in Wildtype Mice	MOR −/−	DOR −/−	KOR −/−
Mice lacking μ-(MOR −/−), δ-(DOR −/−) and κ-(KOR −/−) opioid receptors: Responses to opioids				
Morphine (μ)	Spinal analgesia	Abolished	Maintained	Maintained
	Supraspinal analgesia	Abolished	−	−
	Reward	Abolished	−	Maintained
	Withdrawal	Abolished	−	Decreased
	Respiration depression	Abolished	−	−
	Inhibition of GI transit	Abolished	−	−
	Immunosuppression	Abolished	−	−
	Hyperlocomotion	Abolished	−	−
M6G (μ)	Spinal analgesia	Abolished	−	−
	Supraspinal analgesia	Abolished	−	−
Endomorphin 2 (μ)	Spinal analgesia	Abolished	−	−
	Supraspinal analgesia	Abolished	−	−
DPDPE (δ)	Spinal analgesia	Decreased or maintained	Abolished	−
	Supraspinal analgesia	Decreased or maintained	−	−
Deltorphin II (δ)	Spinal analgesia	Decreased	−	−
	Supraspinal analgesia	Maintained	−	−
U50488H (κ)	Spinal analgesia	Maintained	−	Abolished
	Supraspinal analgesia	Maintained	−	Abolished
	Hypolocomotion	−	−	Decreased
	Dysphoria	−	−	Decreased

Source: From Kieffer (1999), p. 21.
DPDPE, *cyclic*[D-penicillamine[2], D-penicillamine[5]]enkephalin; GI, gastrointestinal; M6G, morphine-6-glucuronide.

of the mouse to major stressors. Mutations in the *tachykinin 1* gene, producing lack of both substance P and substance K, showed longer latencies on the hot-plate test and lack of responses to formalin (Zimmer et al., 1998). Table 5.2 summarizes the behavioral phenotypes detected on pain sensitivity tests in lines of mice with mutations in substance P and neurokinin 1 systems (Woolf et al., 1998).

Galanin is another neuropeptide transmitter that is implicated in endogenous nociception. Localized in sensory and spinal dorsal horn neurons, galanin inhibits C-fiber pain transmission in rodents (Xu et al., 2000). Galanin overexpressing transgenic mice displayed reduced spinal cord sensitization to C-fiber stimuli during wind-up conditioning (Grass et al., 2003). Responses after nerve damage were attenuated in another line of galanin transgenics in which overexpression was restricted to primary afferent dorsal root ganglion neurons (Holmes et al., 2003). Baseline responses were similar in galanin transgenics as compared to wildtype controls in standard hot-plate and tail

TABLE 5.2 Summary of Behavioral Phenotypes of Substance P and Neurokinin 1 Receptor Knockout Mice on Various Tests for Pain Sensitivity and Analgesia

		PPT-A/Tac 1 KO (Zimmer)	PPT-A/Tac 1 KO (Basbaum)	NK1 KO (Hunt)	NK1 Antagonists (Subataxic Doses)
Nociception					
	Thermal				
	Tail flick[a]	+		+	+
	Hot plate	− @ 52°C	+ @ 52.5°C and 58.5°C	+	+
	Mechanical				
	Von Frey threshold	− −	− @ 55.5°C	+	+
	Chemical				
	Acetic acid writhing[a]	+	−	+	−
	Formalin: Early phase[a]	− −	− −	+ +	+
Central sensitization	Formalin: Late phase[a]	− (20 ul, 5%)	+ (10 µl, 1.2%)	(0.6 and 2%)	− (25 µl, 2%)
Inflammatory hyperalgesia	Mechanical		+ (20 µl, 5%)	+ (10 µl, 2%)	− (25 µl, 2%)
Miscellaneous	Stress-induced analgesia[a]	+		−	−
	Neurogenic extravasation		−	−	−

Source: From Woolf et al. (1998), p. 1065.
Legend: [a]Apparent discrepancy: +, normal; −, reduced; − −, abolished.

flick tests (Kinney et al., 2002), consistent with the frequency coding hypothesis that neuropeptides are released only under conditions of high neuronal activity (Hökfelt et al., 1999). Similarly galanin receptor subtype GAL-R1 knockout mice were normal on baseline measures of pain sensitivity, but displayed greater mechanical and heat sensitivity after sciatic nerve injury (Blakeman et al., 2003).

Muscarinic and nicotinic cholinergic receptors mediate some types of analgesic responses, as seen in m2 receptor subtype knockout mice (Gomeza et al., 1999). The nonselective muscarinic agonist, oxotremorine, produced a dose-dependent analgesic response on both the tail-flick and hot-plate tests in wildtype control mice and m2 heterozygotes. M2 null mutants showed a right shift in the dose–response curve to oxotremorine on both tail-flick and hot-plate responses, indicating reduced sensitivity to the analgesic actions of the muscarinic agonist. Analgesic response to morphine on the tail-flick test was unaffected by the mutation. Knockout mice deficient in the beta 2 and alpha 4 subunits of the nicotinic receptor had reduced responses to nicotine on the hot-plate and tail flick tests (Cordero-Erausquin et al., 2000).

Cannabinoid receptors are implicated in some forms of antinociception (Pertwee, 2001). Cannabinoid receptor CB1 knockout mice showed normal responses on hot-plate and tail immersion, but did not develop the normal stress-induced analgesia following forced swim (Valverde et al., 2000). Hypoalgesia was seen in the formalin test (Zimmer et al., 1999).

Mutations in several ion channel genes produced unusual phenotypes on pain sensitivity. The glycine receptor subunit GlyR α3 is expressed in the superficial layers of the dorsal horn. GlyR α3 knockout mice showed reduced responsivity to Von Frey hairs and reduced latency to paw withdrawal on the Hargreaves test (Harvey et al., 2004). Hyperalgesia was detected in knockout mice deficient in the *Kv1.1* potassium channel gene (Clark and Tempel, 1998). The null mutants showed reduced latencies on the hot-plate test, the Hargreaves paw flick test, and the formalin test. Blunted analgesic responses to morphine were detected in the homozygous mutants on paw flick latency. Mutations in the calcium ion channel TRPV4 gene reduced the sensitivity to tail pressure, while normal heat avoidance, taste, and olfaction were reported (Suzuki et al., 2003).

Mice deficient in the G protein G_o were hyperalgesic on the hot-plate test (Jiang et al., 1998a). These mice were also hyperactive, unable to perform the rotarod task, ran in circles on the open field test, showed generalized tremor, and died at early postnatal ages, complicating the interpretation of the hyperalgesia finding. The G_o knockout provides an illustrative example of the need to measure general health, sensory, and motor functions, as described in Chapters 3, 4, and 5, to avoid an overinterpretation of one component of a broader behavioral phenotype.

Normal baseline responses for opiate receptor knockout mice and substance P receptor knockout mice on some pain tests raises the interesting question of compensatory genes, discussed in Chapter 14. Since the μ-opioid receptor and substance P receptor knockout mice show reduced or absent responses to morphine on pain tests, it is reasonable to assume that pain transmission is not completely normal in these null mutants. Perhaps the tail-flick and hot-plate tests are insufficiently sensitive to detect differences in baseline pain thresholds. Alternatively, substance P, the endogenous opiate peptides, and other endogenous neurotransmitters such as acetylcholine, galanin, and cholecystokinin, which are present in pain transmission pathways, may interact synergistically or redundantly under normal circumstances to mediate sensitivity to pain sensation.

Mutation of one component in a complex circuitry may be compensated by changes in other elements of the circuitry. It is interesting to speculate that critical sensory modalities, such as pain, olfaction, and hearing, may be multiply determined, with many redundant neurotransmitters and receptors contributing to normal performance on sensory tasks. Null mutants deficient in only one element may show normal phenotypes because of these compensatory processes in redundant systems. These null mutants provide interesting models to investigate compensatory mechanisms. The developmental biology of redundant regulatory elements is a research field to which the transgenic and knockout mouse technology can make unique contributions. The interested researcher may choose to breed double knockouts, deficient in two potentially redundant genes, such as substance P and the μ-opioid receptor. Another approach is to assay for upregulation of the hypothesized compensatory mechanism, for example, increased synthesis of messenger RNA for the compensatory neuropeptide or its receptor.

BACKGROUND LITERATURE

Altschuler RA, Bobbin RP, Clopton BM, Hoffman DW (1991). *Neurobiology of Hearing: The Central Auditory System*. Raven, New York.

Brennan PA, Keverne EB (2004). Something in the air? New insights into mammalian pheromones. *Current Biology* **14**: R81–R89.

Doty RL (1986). Odor-guided behavior in mammals. *Experientia* **42**: 257–271.

Farbman AI (1992). *Cell Biology of Olfaction*. Cambridge University Press, Cambridge.

Fraser HF, Harris LS (1967). Narcotic and narcotic antagonist analgesics. *Annual Review of Pharmacology* **7**: 277–300.

Gaveriaux-Ruff C, Kieffer BL (2002). Opioid receptor genes inactivated in mice: The highlights. *Neuropeptides* **36**: 62–71.

Glickstein M (1969). Organization of the visual pathways. *Science* **164**: 917–926.

Johnson KR, Zheng QY, Erway LC (2000). A major gene affecting age-related hearing loss is common to at least ten inbred strains of mice. *Genomics* **70**: 171–180.

Kieffer BL (1999). Opioids: First lessons from knockout mice. *Trends in Pharmacological Sciences* **20**: 19–26.

Margolis F, Getchell TV Eds. (1988). *Molecular Neurobiology of the Olfactory System*. Plenum, New York.

Malmberg AB, Bannon AW (1999). Models of nociception: Hot-plate, tail-flick, and formalin tests in rodents. In *Current Protocols in Neuroscience*. Wiley, New York, pp. 8.9.1–8.9.16.

Martin WR (1967). Opioid antagonists. *Pharmacological Reviews* **19**: 463–521.

Moskowitz AS, Terman GW, Liebeskind JC (1985). Stress-induced analgesia in the mouse: Strain comparisons. *Pain* **23**: 67–72.

Pinto LH, Enroth-Cugell C (2000). Tests of the mouse visual system. *Mammalian Genome* **11**: 531–536.

Simon SA, Roper RD, Eds. (1993). *Mechanisms of Taste Transduction*. CRC Press, Boca Raton, FL.

Van Essen DC (1979). Visual areas of the mammalian cerebral cortex. *Annual Review of Neuroscience* **2**: 227–263.

Willott JF (2001). *Handbook of Mouse Auditory Research: From Behavior to Molecular Biology*. CRC Press, Boca Raton, FL.

Willott JF, Tanner L, O'Steen J, Johnson KR, Bogue MA, Gagnon L (2003). Acoustic startle and prepulse inhibition in 40 inbred strains of mice. *Behavioral Neuroscience* **117**: 716–727.

Pinky and the Brain™ cartoon characters.

6

Learning and Memory

They're Pinky and the Brain
One is a genius
The other's insane
They're laboratory mice
Their genes have been spliced
They're Pinky and the Brain, Brain, Brain, Brain . . . *

HISTORY AND HYPOTHESES

What is a memory? The truth is, we really don't know. The biological nature of an individual memory remains mysterious. The concept of memory is operationally defined as *the retention of experience-dependent internal representations over time* (Dudai, 1989). Many excellent books and reviews have been written on the subject (Hebb, 1949; Dudai, 1989; Cohen and Eichenbaum, 1993; LeDoux, 1996; Martinez and Kesner, 1998; Eichenbaum 2002; Gallagher 2002; LeDoux 2002). The interested reader will find a rich history of theories and experiments in the overviews offered at the end of this chapter.

One compelling current hypothesis is that a given memory is a highly specific set of associations, which reside in a network of neural connections with self-sustaining feedback loops (Bhalla and Iyengar, 1999). A favorite example is Marcel Proust's *Remembrance of Things Past*. The taste of a crumb of madeleine cookie, soaked in lime-flower tea, released a flood of memories from Proust's childhood in "Combray" (Proust,

---

*From the theme song to Steven Spielberg's *Pinky and the Brain*, a children's cartoon show, 1995–1998, produced by Amblin Entertainment and Warner Brothers Studios.

1956). You can try this trick yourself. Remember your high school girlfriend/boyfriend? Or lack thereof? Does the name/image release a host of memories, which unfold, labyrinth-like, from connection to connection, along the network of your adolescence? The same process might be elicited by your favorite old song (if you're a baby boomer like me, try the Beatles' *Sgt. Pepper's Lonely Hearts Club Band* album), or by driving down the road to the beach, or walking into your college hangout bar. A standard question to check for memory loss in older Americans is, "What were you doing when you heard that [US President John F.] Kennedy was shot?"

Two underlying types of memory have emerged (Figure 6.1). The first is the specific, individual memory for a fact or an event. Originally called the engram (Lashley, 1950), this type of memory has been variously described as "declarative," "explicit," or "episodic" memory (Morris, 2001). Remembering a specific item of information, for example, your lab's telephone number, the answer to a multiple-choice question about state capitals, or exactly where you left your keys, represents explicit memory. Experiments designed to understand explicit memory address the question, *What is a memory?*

In contrast, there are the memory processes underlying the ability to acquire, store, retain, retrieve, and extinguish memories. This concept is termed "procedural," "implicit," or "semantic" memory. Remembering how to use a telephone, the right way to fill in the little ovals on the multiple-choice answer sheet for the exam, or the process of turning a key in a lock, represent implicit memory of underlying procedures or general knowledge. Implicit memory includes the learning of skills, habits, and complex reflexes. Most of the current knowledge of the neurochemical, neurophysiological, and neuroanatomical mechanisms underlying learning and memory concerns the procedural components of acquiring, storing, and retrieving memories. These experiments address the question, *What is memory?* or *How does memory work?*

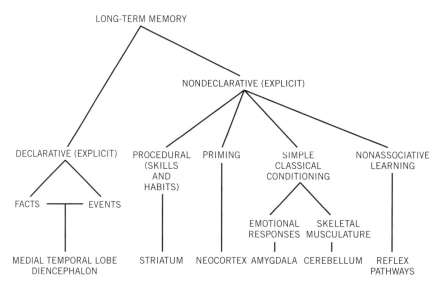

Figure 6.1 *Explicit and implicit forms of memory with suggestions of the brain structures mediating each kind of long-term memory. [From Milner et al. (1998), p. 451.]*

There have been many theories over the years about the physical basis for learning and memory. Early thinkers on the subject, such as Aristotle in the fourth century BC, held that mental functions were located in the heart (Dudai, 1989). The writings of Decartes in the 1600s recognized nerves mediating muscle reflexes, and suggested a role for the pineal gland and the brain in volition (Dudai, 1989). Careful research into the nature of reflexes was conducted by Ivan Petrovich Pavlov at the Institute of Experimental Medicine in St. Petersburg, Russia, for which the Nobel Prize was awarded in 1904. Pavlov expanded on the physiology of reflexes, initiating some of the earliest formal experiments on learning. Dogs were presented with an unconditioned stimulus, food in the mouth, which elicited an unconditioned response, salivation. A conditioned stimulus, a tone, was then paired with the unconditioned food stimulus. Subsequent presentation of the conditioning stimulus tone (without the food) elicited the unconditioned response, salivation (Pavlov, 1927). This form of learning was termed *classical conditioning*, and is a well-characterized form of associative learning across species and tasks (Fanselow and Poulos, 2005).

Various experimental approaches have been in vogue over the years. Technical advances enabled researchers to address hypotheses with increasingly sophisticated techniques. In the late nineteenth and early twentieth centuries, memory deficits in people suffering from brain damage were reported by neurologists, including Ribot in France, Korsakoff in Russia, and Jackson in England. Integration of neurology and psychology, with the brain as the focus of learning and memory, was championed by Hebb (1949). Neuroanatomical lesion studies in animals were initiated by Lashley (1929). More selective lesioning techniques employ a stereotaxic apparatus to locate small brain regions from a three-dimensional brain atlas (e.g., Paxinos and Watson, 1982). Focused electrolytic lesions revealed a major role for the hippocampus in spatial learning (Cohen and Eichenbaum, 1994). Postmortem correlations from patients with lesions in discrete regions of the hippocampus and cerebral cortex revealed selective neuronal loss that correlated with specific types of memory (Damasio et al., 1994). The most famous case is H.M., a man who lost all ability to retain new information (anterograde amnesia) after surgical bilateral hippocampal lesions for treatment of his intractable epileptic seizures (Scoville and Milner, 1957).

In vivo positron emission tomography and functional magnetic resonance imaging (fMRI) techniques now allow investigators to identify specific human brain regions mediating performance on specific types of memory tasks (Posner and Raichle, 1998). Elegant fMRI studies map site–selective activation patterns in brain pathways and structures including the cerebral cortex, hippocampus, or cerebellum of normal human volunteers, while they are engaged in diverse tasks such as picture recognition (Squire et al., 1992; Ungerleider, 1995; Robbins, 1996), word comprehension (Buckner et al., 1996; Abdullaev and Posner, 1998), word lists (Alkire et al., 1998), maze learning (Van Horn et al., 1998), cerebellar motor tasks (Turner et al., 1998), error detection (Carter et al., 1998), pain processing (Derbyshire et al., 1997), facial emotion recognition (Morris et al., 1998), musical processing (Zatorre et al., 1998), playing a videogame (Koepp et al., 1998), retrieval of names for faces (Zeineh et al., 2003), fear acquisition and extinction (LaBar et al., 1998), and gambler's regret (Camille et al., 2004). Imaging techniques reveal a characteristic pattern of hypometabolism and reduced receptor occupancy by neurotransmitters in many brain regions of patients suffering from Alzheimer's disease (Minoshima et al., 1997; Gilman, 1998; Meltzer et al., 1998).

Neurochemical mechanisms have been explicated for components of short-term and long-term memory. In the 1960s Louis Flexner and co-workers at the University of Pennsylvania discovered that inhibition of protein synthesis completely blocked the ability of mice to acquire long-term memory of a Y-maze avoidance task (Flexner et al., 1963). Further, an RNA synthesis inhibitor, actinomycin D, blocked one-trial passive avoidance learning in mice (Barondes and Jarvik, 1964). Elegant experiments from several laboratories supported the interpretation of temporal phases of memory formation with differential requirements for protein synthesis (McGaugh, 1966). Short-term memory is established during the first few minutes after training and is unaffected by protein synthesis inhibition. Long-term memory is consolidated over the course of several hours after training and is blocked by treatment with protein synthesis inhibitors such as puromycin, cyclohexamide, and anisomycin (Quartermain and McEwan, 1970; Flood et al., 1978; Davis et al., 1980; Rainbow et al., 1980; Figure 6.2).

Neurotransmitters that contribute to memory formation include glutamate, GABA, acetylcholine, norepinephrine, dopamine, enkephalin, vasopressin, oxytocin, and galanin (Martinez et al., 1981; Nakanishi, 1992; Dunnett, 1993; Richter-Levin et al., 1995; McDonald and Crawley, 1996; Izquierdo and Medina, 1997; McGaugh and Cahill, 1997; Everitt and Robbins, 1997; Vizi and Kiss, 1998; Lynch, 1998; Ungerer et al., 1998; Sauerberg et al., 1998; Monks et al. 2003). Synaptic docking proteins such as syntaxin, and postsynaptic phosphorylated proteins such as CREB, are just a few of the molecular mechanisms in the chain of events leading to synaptic strengthening relevant to memory formation (Davis et al., 1996; Silva et al., 1998; Bartsch et al., 1998; Kandel, 2001; Bozon et al., 2003).

Neurophysiological recording during performance of memory tasks reveals brain regions and pathways participating in learning and memory processes (Desimone, 1992). *Long-term potentiation* (LTP) and *long-term depression* (LTD) have emerged as strong candidates for some of the neurophysiological mechanism underlying the cellular basis of synaptic strengthening. LTP is the enduring enhancement of synaptic efficacy, which occurs in a specific synapse following bursts of high-frequency

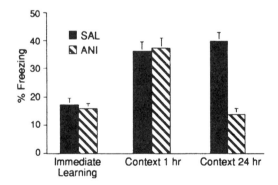

Figure 6.2 *Protein synthesis is an essential step in the establishment of a long-term memory. Anisomycin, a protein synthesis inhibitor, given during training, blocks contextual fear conditioning when mice are tested 24 hours after the aversive stimulus but not when mice are tested 1 hour after the aversive stimulus.* [From Abel et al. (1997), p. 620.]

electrical stimulation (Lynch, 1986; Bliss, 1998). Considerable debate surrounds the question of whether LTP is necessary and sufficient for spatial learning and memory in rodents (Barnes, 1995). Transgenic and knockout mice with mutations in genes thought to be key synaptic elements of LTP are providing useful tools to sort out the conflicting hypotheses.

Targeted mutations have been generated for genes involved in many aspect of the neuroanatomy, neurophysiology, and neurochemistry of synaptic transmission during LTP and memory formation (Thomas and Huginar, 2004). The knockout approach is revealing critical components of the biological mechanisms underlying implicit and explicit memory.

BEHAVIORAL PARADIGMS

The Morris Water Task

The Morris water maze is currently the most frequently used paradigm to evaluate learning and memory abilities in transgenic and knockout mice. This is a spatial navigation task in which the animal swims to find a hidden platform, using distant visual landmarks in the environment to locate the platform. Escape from the water is the positive reinforcement. The task is based on the principle that rodents are highly motivated to escape from a water environment by the quickest, most direct route. Step-by-step standard procedures for this task, and methods for troubleshooting, were described by Gary Wenk, Ohio State University (Wenk, 1997).

The paradigm was originally invented by Richard G. M. Morris at the University of St. Andrews in Scotland (Morris, 1981, 1984). Professor Morris, now at the University of Edinburgh, continues to develop interesting new modifications and applications for his apparatus (Chen et al., 2000). Variously called the water maze. Morris water task, and the Morris swim test, the task was originally used to investigate anatomical brain structures required for spatial learning and memory in rats. One of the most consistent findings is that hippocampal lesions impair acquisition of the Morris water task in both rats (Morris et al., 1982; Eichenbaum et al., 1990) and mice (Logue et al., 1997a). Rats show impaired performance on the Morris water task after electrolytic and immunotoxin lesions of brain regions, including the hippocampus, medial septum/diagonal band, and entorhinal/perirhinal cortex, and also after pharmacological treatments with cholinergic and glutamate receptor antagonists (Decker and Majchrzak, 1992; McNamara and Skelton, 1993; Nilsson and Gage, 1993; McAlanon et al., 1995; Nagahara et al., 1995; Roullet and Sara, 1998). Aged rats show dramatic impairments in performance on this task (Gallagher et al., 1993; Chouinard et al., 1995; Geinisman et al. 2004).

Performance on the Morris water task is highly influenced by genetic background (Crusio et al., 1995; Wehner and Silva, 1996; Whishaw and Tomie, 1996; Crawley et al., 1997a; Wahlsten et al., 2003a; Table 2.1). Some of the 129 substrains perform well on the Morris water task, some poorly (Wehner and Silva, 1996; Montkowski et al., 1997; Clapcote and Roder, 2004). Luckily the 129 substrains used as sources of embryonic stem cells for gene targeting, the 129/Sv, 129/SvEv, 129S6/SvEvTac, and 129/Sv-Ola substrains, perform well on the Morris water task (Wehner and Silva, 1996; Montkowski et al., 1997; Clapcote and Roder, 2004). Poor performance seen in the 129/J (Montkowski et al., 1997) and 129/SvEvTac (Balogh et al., 1999) appears to reflect incomplete development of the corpus callosum in these 129 substrains (Livy and

Wahlsten, 1997; Wahsten et al., 2001; Wahsten et al., 2003). Hippocampal mossy fiber densities may account for some strain differences (Crusio et al., 1995). Reduced visual abilities in albino mice may impair performance on this visually cued task (Wehner and Silva 1996; Montkowski et al., 1997; Wahlstsen et al., 2001). Visual deficits due to retinal degeneration genes in the background of some inbred strains impair performance on the Morris tasks (Wehner and Silva, 1996; Wahlsten et al., 2001). Remarkably, blind rats have been reported to learn the Morris water task, although their rate of acquisition is impaired (Lindner et al., 1997), indicating that nonvisual cues such as odors, sounds, vibration, tactile contact, and temperature gradients may also be used for spatial navigation. Anecdotal observations indicate idiosyncratic strategies employed by various inbred strains of mice. Unusual problems for some strains in the Morris water task include floating or sinking rather than swimming, immediately jumping off the platform back into the water, and circling the perimeter. It seems possible that some strains enjoy swimming, rather than hurrying to get out of the water, especially if the water temperature is warm. In addition, some strains may be more sensitive to the stressful component of this vigorous swimming task, especially if the water temperature is cold. Inbred mouse strain differences in performance on the Morris water task are shown in Table 2.1 and Figure 2.8.

Performance of mice on Morris water tasks is often inferior to the performance of rats, although mice are equal to rats in spatial tasks on dry land (Whishaw and Tomie, 1996). This species difference is perhaps indicative of better innate swimming abilities of rats, a species that appears to be better adapted to water and swimming in their natural environment. Male rats and mice have been reported to acquire the hidden platform location faster than female rats and mice, thought to be related to higher responses to stress in females, as measured by greater elevations in corticosterone after water maze training (Beiko et al., 2004).

The pool is circular, with a diameter varying from 80 to 200 cm and a height varying from 30 to 50 cm in different laboratories (Wahlsten et al., 2003a). For mice, the smaller diameter pool will speed the acquisition of the task. Water depths of 20 to 50 cm are used. Originally, horse feeding troughs were used as the swimming pools. Sturdy white plastic tanks can now be purchased, along with an automated videotracking system and software, from several behavioral equipment companies. The pool is filled to a depth of 20 to 30 cm, so that the mouse can neither escape over the edge of the tank nor balance its tail on the bottom of the tank. Tap water is used to fill the pool, usually with a hose from the sink to the tank. The water reaches room temperature overnight or is warmed to approximately $25°C$ with aquarium heaters. Nontoxic white paint or milk powder is mixed into the water to yield an opaque liquid. The white coloring makes a white or clear platform invisible in a white pool, from the point of the view the mouse swimming on the surface of the water. Alternatively, a black pool with a black platform may be used when testing white mice, in which case there may be no need for coloring the water. For hygiene purposes, the water is changed daily or at the end of the week, depending on how many mice are tested each day. The water is pumped out of the pool at the end of each test day, using tubing and an inexpensive commercial electric sump pump, emptying into a nearby sink. Some tanks have drainage taps to facilitate emptying. A floor drain in the room expedites emptying the water directly out of the pool. The inner surface is then cleaned and the pool is refilled. Refilling at the end of the day allows time for temperature equilibration before the next test session the following morning.

The hidden platform is usually clear or white Plexiglas, 10 to 12 cm^2 or in diameter. Larger platforms will accelerate the rate of acquisition, especially in pools of large diameter (Richard Morris, personal communication). Grooves near the edge assist the mouse in climbing aboard. The platform assembly is heavy enough to remain upright when its base rests on the bottom, or attaches to an assembly at the bottom of the tank. The hidden platform is submerged, such that its upper surface is 1 to 2 cm below the surface of the water.

The visible platform is clear, white, or black Plexiglas, of the same dimensions as the hidden platform. The visible platform is raised approximately 5 cm above the surface of the water. Alternatively, a flag or a dark-colored object is mounted on the hidden platform, to extend above the surface of the water, for easy visibility. The visible platform provides a large "you can't miss it" proximal cue within the swimming area

Videotracking systems are based on visual contrast between the animal and its background. The color scheme as described is useful for targeted mutations in mice or rats with dark coat colors. If white mice or rats are used, a black tank, black Plexiglas platforms, and plain tap water will provide the needed contrast. Lamps are arrayed at a distance, at a height less than the surface of the pool, to avoid light reflecting from the water surface that could interfere with the videotracking software. Dim illumination is sufficient for most videotracking cameras.

Highly visible room cues are strategically placed around the pool. Large, high-contrast geometrical patterns are mounted on the walls of the room, for example, dark-colored construction paper against a white wall to form a checkerboard on one wall. Alternatively, patterns are placed on drapes or room-divider walls located around the circumference of the pool. These distal cues are sufficiently far away to serve as distant spatial landmarks that the mouse learns to recognize as it forms a cognitive map of the environment. If the cues are too close, they become proximal internal signposts like the visible platform within the pool. A videocamera is mounted in the ceiling to track the swimming behavior.

Figure 6.3 illustrates the Morris tank, videotracking camera, and room cues used in our laboratory. The camera mount, the illumination gradient of the room lighting fixtures, the computer, equipment, cart, racks, air vents, temperature gradients, circulation patterns, noises, vibrations, and many other features of the test room, all comprise salient environmental cues. It is critical that all cues remain in a fixed location over the course of an experiment. Importantly, the same experimenter(s) must handle and test the animals over the course of an experiment. We ourselves are powerful olfactory, visual, and tactile cues. Visitors, unfortunately, are welcome during the conduct of the Morris water task only if they are willing to be present constantly throughout the training week. A two-way mirror or window into the Morris water task test room is a useful luxury to accommodate guest observers, if your laboratory is on the high-profile tour route.

Excellent automated videotracking and data analysis systems are available. Noldus (Wageningen, The Netherlands), San Diego Instruments (San Diego, California), HVS Tracking System (Hampton, England), CPL Systems (Cambridge, England), Columbus Instruments (West Lafayette, Indiana), Actimetrics/Coulbourn (Wilmette, Illinois), Hamilton Kinder (San Diego, California), and others, sell systems that have been widely used by behavioral neuroscientists. As seen in Figure 6.3a, a videocamera is mounted on the ceiling, centered above the tank. The camera lens is focused to encompass the diameter of the pool. Room lighting is adjusted to yield good contrast between

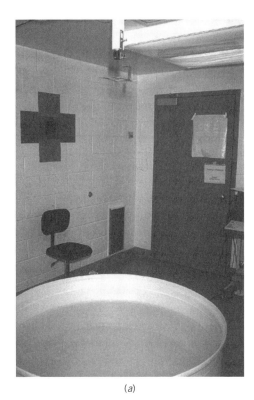

(a)

(b)

Figure 6.3 *Examples of equipment for the Morris water tank. (a) Room cues, swim tank, and videotracking camera. The white plastic pool is filled with clean tap water and made opaque by the addition of nontoxic white tempera paint. The clear Plexiglas hidden platform is not visible to the mouse at the surface of the water. Visual objects in the room, such as patterns on the walls, furniture, computer, and door, provide cues for spatial navigation. (Photograph contributed by the author.) (b) The mouse swims to the platform and climbs aboard. (Photograph contributed by the author.) (c) Schematic illustration of a rat reaching the hidden platform. [From Wenk (1997), p. 8.5.5.] (d) Visible platform, with intramaze cues in the perimeter of the pool. [From Sweeney et al. (1988), p. 143.] (e) San Diego Instruments tracking software.* (Photograph courtesy of Dr. Richard Butcher, San Diego Instruments Inc., San Diego, California)

(c)

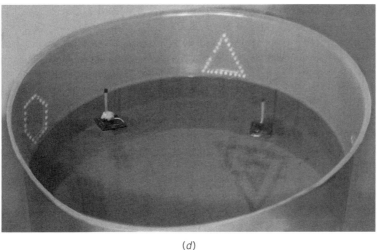

(d)

Figure 6.3   (Continued)

the color of the mouse and the color of the water. Commercially available software receives tracking data from the camera and automatically records the location of the mouse, approximately 10 times per second. Software packages calculate the pathway of swimming (Figure 6.4), time spent in each of four imaginary quadrants (Figure 6.5), average swim speed and path length (Figure 6.6), and latency to reach the platform (Figure 6.7). These data are stored for each trial and used in subsequent calculations. Statistical packages analyze the data in terms of individual animals and treatment groups, over consecutive training sessions and during probe trials.

There are many different protocols for using the Morris swim tank equipment to evaluate various aspects of learning and memory. The description below is one method commonly used in mice. Minimum standard training on the Morris water task for mice

(e)

Figure 6.3 (Continued)

consists of daily training trials on the *visible platform task*, followed by the *hidden platform task*, followed by a *probe trial* at the end of the last hidden platform training session.

The first step of the procedure is *pretraining*. The mouse is gently introduced to the pool and the platform. The experimenter places the mouse on the visible platform for several seconds. Generally, the mouse will jump off the platform after a few seconds. The mouse is allowed a short swim, for example, 15 seconds, then gently guided back to the platform. A single pretraining trial is sufficient for some inbred strains of mice; three pretraining trials are common for many strains of mice; more pretraining trials are needed for a few strains. If mice are singly housed, they may need to be kept warm between training blocks by adding a cloth towel to the holding cage and/or a warming light above the holding cage. *Visible platform trials* test the ability of the animal to conduct the procedures of the task, particularly visual ability to see the room cues and motor ability to swim in the pool (Figure 6.6). However, the visible platform task requires the mouse to see a large proximal local cue at close range, whereas the hidden platform task requires the mouse to see distant extra-maze cues outside the pool

Wildtype β-APP$_{751}$

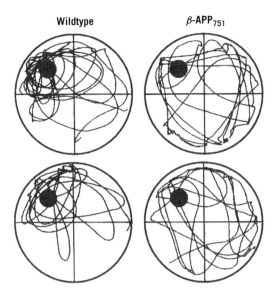

Figure 6.4 *Typical swim path in the Morris water task, illustrating search patterns for wildtype mice and for transgenic mice overexpressing β-amyloid precursor protein. The black circle in the upper left represents the hidden platform.* [From Moran et al. (1995), p. 5344.]

(Morris, 2001). It remains possible that vision is sufficient to see the visible platform but insufficient for distal cues on the walls and ceiling. Latency to reach the visible platform is measured. Swim speed is calculated. Swim path is drawn by some software packages.

Training trials for acquisition of the hidden platform task then commence. Each training session begins with the experimenter taking the mouse from the cage and placing it in the water. Placement is usually at the edge of the pool, facing the wall. Placement location is randomized between quadrants on each successive trial for a mouse. Some laboratories vary the quadrant of the platform across individual mice within a group, although the hidden platform stays in the same place for a given mouse. During each training trial the mouse is allowed 60 seconds to reach the platform and climb up out of the water. After a brief interval of several seconds or minutes, the experimenter places the mouse back in the water for the next trial. One common experimental design is 2 trials per day for 10 days. At the opposite extreme, an intensive training protocol is 12 trials per day in 3 blocks of 4 trials each, for 3 days. In the training block design, after the first block, for example 4 trials, the mouse is returned to its holding cage. The next animal then begins its training trials, until all of the subjects have completed the first training block. The first animal is then started on the next block of trials, and so forth, until all subjects have completed the daily session. The training procedure is repeated on the next day. Another common protocol is 4 trials a day for as many days as it takes for the wildtype group to reach the training criterion.

Over the course of training, normal mice will swim to the hidden platform with an increasingly direct swim pathway and a diminishing latency to reach the platform. A criterion of 15 seconds or less to reach the hidden platform is commonly used. Training days are conducted consecutively until the acquisition criterion is reached by all mice

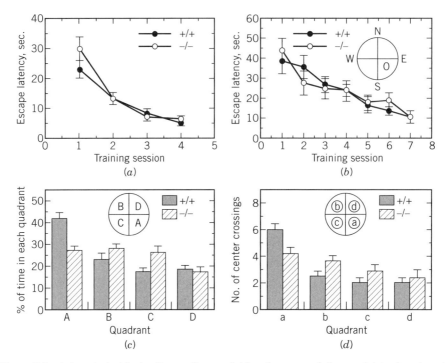

Figure 6.5 *Automated videotracking software divides the area of the pool into four imaginary quadrants. The software calculates the percentage of time that the mouse spends in each of the four quadrants and the number of times the mouse crosses the location of the platform. (a) Visible platform escape latencies. (b) Hidden platform escape latencies. (c) Probe trial, percentage of time spent swimming in previously trained quadrant A as compared to nontrained quadrants B, C, and D. (d) Number of crossings in the center of training quadrant versus nontrained quadrants. Adenylyl cyclase knockout mice (−/−) showed normal acquisition curves on both hidden and visible platforms, but lack of selective quadrant search on the probe trial, as compared to +/+ wildtype controls, indicative of failure learn spatial navigation using distant environmental landmark cues. [From Wu et al. (1995), p. 223.]*

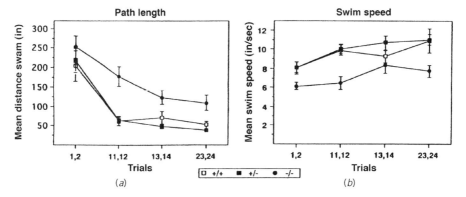

Figure 6.6 *Automated videotracking software calculates (a) the length of the path that the mouse swam to reach the platform and (b) the average speed of swimming for successive trials during visible platform training. Mutant mice deficient in the early immediate gene c-fos (black circles) showed slower swimming and longer distances swum as compared to wildtype controls (white circles). [From Paylor et al. (1994a), p. 279.]*

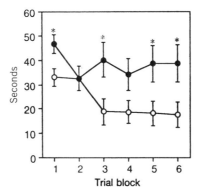

Figure 6.7 *Automated videotracking software calculates the latency for the mouse to reach the platform on each trial. Blocks of several trials may be averaged for ease of data presentation. Normal mice (white circles) showed a gradual reduction in latency over trials, reaching the criterion of 15 seconds or less after several days of training. Mice overexpressing the 751 amino acid isoform of human β-amyloid precursor protein (black circles) did not improve their spatial navigation over successive trials.* [From Moran et al. (1995), p. 5342.]

or only by the wildtype littermate controls. Criterion may be reached in as few as 3 or as many as 14 training days. Number of training days depends on the natural abilities of the wildtype mice to learn the task. Elements of the apparatus and the procedure can be optimized to increase the rate of acquisition. Parameters that may affect acquisition include the temperature of the water, diameter of the pool, the dimensions of the hidden platform, the prominence of distal room cues, room lighting, the number of trials per training block, the number of training days, and the total number of training trials.

If the wildtype littermate controls do not reach the criterion for latency to locate the hidden platform, after many training days, then the Morris water task cannot be used. Other learning and memory tasks must be employed to analyze the behavioral phenotype of the mutation. In some cases, breeding the mutation into another genetic background for at least five generations will solve the problem (see Chapter 2).

The number of animals per genotype that can be tested during an experiment is limited by the number of animals that can be run through the daily regimen within one day. Mice should be tested only during one phase of the circadian cycle, for example, during the 12 hours comprising lights-on. At the end of the training session, mice are returned to their housing room for a good night's rest. The experimenter will need even more rest time than the mouse, after a long week of strenuous aerobic exercise by the investigator.

At the end of the last training day, each mouse is tested on a *probe trial*. Probe trials test the ability of the animal to identify the spatial location that previously contained the hidden platform (Figure 6.8). The platform is removed from the pool. The mouse is then placed in the pool as usual. If the mouse learned the location of the hidden platform by using the distal environmental room cues, it will swim directly to the quadrant that formerly contained the platform and spend the 60-second trial period swimming predominantly in the quadrant that formerly contained the platform. Selective quadrant search indicates that the mouse has formed a cognitive map of the environment during training, and is using the distant spatial landmarks to solve the probe trial challenge.

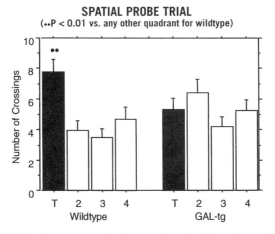

Figure 6.8 *Probe trial. Galanin overexpressing transgenic mice fail to show selective quadrant search in the probe trial. Wildtype littermates displayed normal selective quadrant search, indicating that they learned the Morris spatial navigation task.* [Adapted from Steiner et al. (2001).]

Criteria for successful learning of this version of the Morris water task are based on the latency to find the hidden platform during training trials and selective quadrant search in the probe trial. The true criterion for acquisition of the Morris water task is performance on the probe trial. Search time spent in the trained quadrant must be significantly greater than search time spent in the other three quadrants of the pool. Number of platform crossings, that is, number of the times the mouse swims over the former location of the platform, must be significantly greater for the former platform location in the trained quadrant as compared to the comparable imaginary locations in the other three quadrants of the pool. Normal performance on the visible platform task but impaired performance on the hidden platform task and probe trial is interpreted as a true deficit in learning and memory.

The Morris water task has the advantages of built-in control parameters. Swim speed and swim pathway provide measures of the procedural abilities of the mouse to perform the task. Dysfunctions in visual neurons, motor neurons, cerebellum, and spinal cord, which may impair navigation and swimming abilities, will be reflected in slower swim speeds, random swim pathway, or failure to swim at all on both the visible and hidden platform tasks. If swim speed and/or pathway is abnormal, the experimenter is alerted to a behavioral phenotype relevant to visual and motor functions. In the case of impaired swimming abilities, the Morris water task cannot be used as a measure of learning and memory.

Additional measures of learning and memory using the Morris apparatus are obtained by varying the procedure (Figure 6.9). In cases where the mutants have not reached criterion, the acquisition experiment can be extended by adding more days of training. Long acquisition curves are useful to differentiate slower rates of acquisition from complete inability to learn the task. A second probe trial can be conducted at the end of additional training days to complete the acquisition curve for slow learners. Training and probe trials can be repeated at intervals of several days or weeks to evaluate the rate of relearning. A probe trial without retraining is used to evaluate retention. Repeated probe trials without additional training can be conducted at chosen time points after the

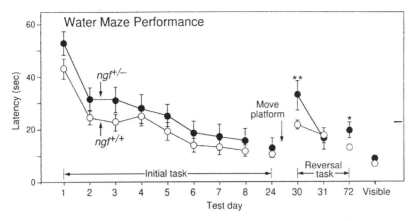

Figure 6.9 *Performance after training on the Morris water task persists after a 16-day delay (training days 1–8, testing day 24). Reversal tasks can be conducted after the initial acquisition and probe trial, to measure the ability of the mouse to learn a new location of the hidden platform, in a different quadrant of the pool. Mice heterozygous for nerve growth factor (black circles) were not significantly different than wildtype controls (white circles) during the first acquisition task but showed poorer performance on the reversal task.* [From Chen et al. (1997a), p. 7291.]

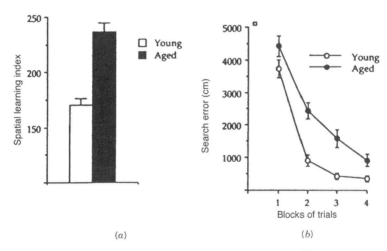

Figure 6.10 *Spatial learning index and cumulative search error scores differentiate young and aged Long-Evan rats on acquisition of the Morris task.* [From Chouinard et al. (1995), p. 958.]

end of training, to evaluate extinction. Search errors provide another sensitive index, for example, to measure the effects of aging (Figure 6.10). The platform can be placed in a new location and the animal retrained for the new platform location, after criterion was reached for acquisition of the original platform location, to evaluate *reversal learning* (Chen et al., 2000). Inbred strain distributions demonstrated reversal learning in the 129S6/SvEvTac strain, often used for embryonic stem cells to generate knockouts, that was better than C57BL/6J on some measures (Clapcote and Roder, 2004). Reversal learning appears to be sensitive to test history in other tasks (Voikar et al., 2004).

The Morris water task is among the most stressful of the tests presented in this chapter. To avoid the confounding long-term sequelae of stress, it is best to conduct the more stressful behavioral tests at the end of the series of behavioral phenotyping experiments. Our laboratory usually runs the Morris task last.

Contextual and Cued Fear Conditioning

Contextual and cued fear conditioning is a fear conditioning task that measures the ability of the mouse to learn and remember an association between an aversive experience and environmental cues (Fanselow, 1980; Fanselow and Tighe, 1988; LeDoux, 1995; Wehner and Radcliffe, 2004; Fanselow and Poulos, 2005). This test is extensively used in behavioral phenotyping of transgenic and knockout mice. Fear conditioning is often chosen as a second independent learning and memory task that complements the Morris water task. Fear conditioning and spatial navigation learning require very different sensory and motor abilities, so their procedural components do not overlap, and the same mice can be tested in both tasks. Contextual and cued fear conditioning requires less elaborate equipment, less laboratory space, less physical exertion by the investigator, and much less training time for the mouse. This is a 2- or 3-day task, requiring about 10 minutes per day per mouse.

Contextual and cued fear conditioning is among the most intuitive of the mouse memory paradigms. Freezing, defined as complete immobility except for breathing, is a common response to sudden fearful situations in many species (Blanchard and Blanchard, 1969; LeDoux, 1996). A colloquial example is the immobile "deer in the headlights" of an oncoming car. Most species acquire and respond to memories of fear-producing stimuli as a fundamental survival mechanism. Simple fear conditioning tests are available for several species in the laboratory environment. Rabbits show fear conditioning of the nictitating membrane reflex, an extension of the third internal eyelid, which has been extensively used by Thompson and co-workers to investigate cerebellar learning (McCormick et al., 1982; Bao et al., 1998). Eyeblink to an auditory cue is a useful fear conditioning paradigm in cats (Woody, 1970), rats (Ivkovich and Stanton 2001), and mice (Kim et al., 1997). The neural circuitry mediating cued and contextual fear conditioning is being elaborated. The amgydala is a major site for cued fear conditioning in rats (Fanselow, 1994; LeDoux, 1995, 1996). The hippocampus, subiculum, anterior cingulate cortex, prefrontal cortex, perirhinal cortex, sensory cortex, and medial temporal lobe appear to mediate components of contextual fear conditioning in rodents (Davis et al., 1994; Squire and Zola, 1996; Eichenbaum et al., 1996; Logue et al., 1997a; Han et al., 2003; Frankland et al., 2004a).

Genetic substrates for cued and contextual conditioning are currently being elucidated. Inbred strain distributions have been extensively demonstrated on components of the cued and contextual conditioning task (Paylor et al., 1994b; Crawley et al., 1997b; Bolivar et al., 2001b; Balogh and Wehner, 2003; Table 2.1). Quantitative trait loci analyses of contextual fear conditioning were conducted by Jeanne Wehner and co-workers at the University of Colorado Institute for Behavioral Genetics in Boulder, and by Lorraine Flaherty and co-workers at the Wadsworth Center in Troy, New York. Highly significant linkages were detected on chromosomes 10 and 16 in an F_2 analysis of BXD mice, with suggestive QTLs on chromosomes 1, 2, and 3 (Wehner et al., 1997). Highly significant linkages were detected on chromosome 1 in an F_2 analysis of C57BL/6J $\times$ C3H mice, with suggestive QTLs on chromosomes 3, 7, 8, 9, and 18 (Caldarone et al., 1997).

Equipment for the cued and contextual conditioning task is illustrated in Figure 6.11. The apparatus is a standard plastic or metal test chamber with an electrifiable grid floor,

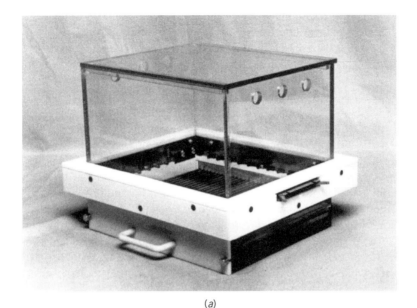

(a)

(b)

Figure 6.11 *Contextual and cued fear conditioning apparatus. Footshock is delivered on day 1, paired with a sound cue. "Freezing," the number of seconds spent in complete behavioral immobility, is scored by a photocell array, videotracking, or by an observer with a stopwatch. Freezing is quantitated under three conditions: (1) 24 hours later in the identical chamber, (2) 48 hours later in a different chamber with different sensory contextual cues such as a solid plastic floor, patterns on the walls, and almond odor, (3) 48 hours later in the novel context chamber with the auditory cue. (a) Contextual chamber. (Photograph kindly contributed by Dr. Richard Butcher, San Diego Instruments.) (b) Novel context triangular white plastic chamber with auditory cue emitter. (Photograph contributed by the author.)*

a sound source, and a calibrated shock generator. The chamber dimensions are not critical, as dimensions varying from 54 cm long × 30 cm high × 27 cm deep (Wehner et al., 1997) to 25 × 30 × 35 cm (Quirk et al., 1997) have been effectively used. Commercially available systems such as the San Diego Instruments Freeze Monitor include automated contextual chambers and software that delivers the auditory cue and footshock, and records freezing behavior across session time. Experience with the Freeze Monitor in our laboratory indicates that defining a bout of freezing by a 2-second minimum provides automated data that is almost as accurate as data scored by a human observer (Nathan Rustay, Laboratory of Behavioral Neuroscience, NIMH, unpublished observations).

Conditioning training on day 1 consists of placing the mouse in the chamber and exposing the animal to a mild footshock paired with an auditory cue. One standard method is described by Wehner and co-workers (Wehner et al., 1997; Radcliffe et al., 1998; Wehner and Radcliffe 2004) with representative data shown in Figure 6.12. The mouse is brought from the home cage to the testing room and placed into the conditioning chamber. It is given 2 minutes to explore the environment. The auditory cue, such as a white noise, 90 dB tone, or 80 dB broadband auditory clicker, is sounded for approximately 30 seconds. A stimulus light in the wall of the chamber may also be illuminated. During the last seconds of the tone the unconditioned aversive stimulus, a mild footshock in the range of 0.25 to 0.5 mA, is administered through the grid floor for 2 seconds. Some laboratories administer more than one pairing of the auditory cue + aversive stimulus, at intervals such as 2 minutes between footshocks, to strengthen the association. The number of seconds spent freezing in the test chamber on the training day is considered the control measure of *unconditioned fear*. The mouse is left in the conditioning chamber for a length of time after the last pairing, such as 30 seconds, during which the association between the aversive stimulus and the properties of the conditioning chamber is further established. The mouse is then returned to its home cage.

Testing on day 2 begins approximately 24 hours after the conditioning session. The mouse is returned to the same conditioning chamber and scored for bouts of freezing behavior. No footshock is administered on day 2. The number of seconds spent freezing in the identical test chamber on day 2 is considered the measure of *contextually conditioned fear*, that is, freezing to the identical context. Freezing is defined as no movement other than respiration. Presence or absence of freezing behavior is generally recorded by a human observer every 10 seconds for 5 minutes, for a maximum total score of 30 freezing bouts. Automated systems sample more frequently or continuously. The mouse is then returned to its home cage. If the scoring is conducted by a human observer with a stopwatch, then it is important that the person conducting the scoring be unaware of the treatment condition or genotype of the mouse. If two or more individuals participate in scoring the freezing bouts, to prevent observer fatigue during a large experiment, it is necessary to achieve a high interrater reliability ($r > 0.95$).

The second phase of testing may begin an hour or more later on day 2, or may be conducted on the next day 3. Another testing chamber with very different properties provides the new context. It is essential to change the sensory cues as much as possible so that the mouse perceives the novel context as unrelated to the conditioning chamber. A smaller, differently shaped box is used, for example, a triangle-shaped test chamber. Visual cues may be taped to the walls of the chamber. Different lighting is used, for example, white light versus red light. Olfactory cues are painted on the walls, such

Mean scores for measures of learning

	Context[a]	Altered context[a]	Auditory cue[a]
B6 (N = 86)	16.6[b] (6.07)	6.33[b] (3.50)	13.03[b] (3.34)
D2 (N = 79)	8.76 (4.87)	3.85 (2.79)	10.13 (3.41)
F$_1$ (N = 77)	15.01 (5.19)	6.63 (3.56)	12.04 (3.05)
F$_2$ (N = 479)	14.14[c] (6.64)	6.12[d] (4.02)	11.53[c,e] (3.97)

[a]Means (S.D.) for bouts of freezing. [b]Significant difference ($P<0.0001$) between B6 and D2 mice, using ANOVA with *post hoc* Tukey-HSD. [c]Significant difference ($P<0.0001$) between both B6 and D2 mice, using ANOVA with *post hoc* Tukey-HSD. [d]Significant correlation with context, $r=0.63$ ($P<0.0001$). [e]Significant correlation with response to context, $r=0.52$ ($P < 0.0001$).

(*a*)

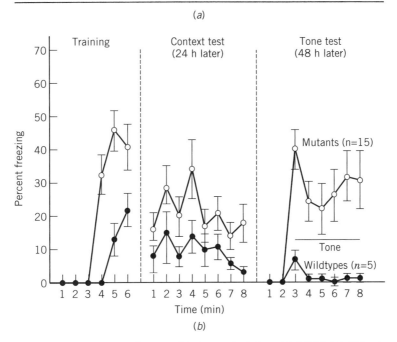

(*b*)

Figure 6.12 *Contextual and cued fear conditioning typical data. (a) C57BL/6J (B6) mice showed more conditioned fear, that is, better memory for the footshock, than DBA/2J mice (D2), as seen in higher freezing scores for same context, and for the novel context plus auditory cue conditions. [From Wehner et al. (1997), p. 331.] (b) Mutant mice deficient in the metabolic enzyme monoamine oxidase A (open circles) showed more conditioned fear than wildtype controls (closed circles), as seen in higher freezing scores during the training session, contextual, and auditory cue tests. [From Kim et al. (1997), p. 5931.]*

as a drop of almond extract or lemon juice. The floor surface texture is composed of a different texture, for example, litter or Plexiglas versus floor bars. The investigator may wear gloves and a lab coat of different texture than on the training day, and/or a different investigator may conduct the novel context testing. Further, it may be useful to conduct the novel context testing in a different room and different time of day. The mouse is placed in the new chamber. No footshock is administered. Freezing behavior is scored for 3 minutes. *Contextual discrimination of fear conditioning* is quantitated by comparing number of freezing bouts in the same contextual environment to number of freezing bouts in the novel contextual environment. Freezing should be minimal in the altered context, since there should be no association between its environmental properties and the previous aversive experience that took place in a very different environmental context.

At the end of the first 3-minutes, the white noise or tone that was presented on training day 1 (and light stimulus cue if used on day 1) is presented in the novel context environment. Freezing behavior is scored for the next 3 minutes in the presence of the sound (and light) cue. *Cued conditioning* is calculated by comparing number of freezing bouts in the novel context environment in the presence of the cue with number of freezing bouts in the novel context environment in the absence of the cue.

Trace fear conditioning is a modification in which the footshock is delivered several seconds after the end of the auditory tone, instead of simultaneously during the end of the tone presentation, on the training day (Moye and Rudy, 1987). The temporal separation increases the difficulty of forming the association. Trace fear conditioning is increasingly used as a more challenging version of the standard delay fear conditioning (Huerta et al., 2000; Crestani et al., 2003; Kinney et al., 2002; Han et al., 2003), and varies across inbred strains (Holmes et al., 2002b).

Long-term fear conditioned memory is evaluated by measuring retention of freezing when the mouse is placed in the contextual environment and in the auditory cued novel environment more than 24 hours after training. Most strains displayed freezing to context and cue at 14 days after training; some strains froze to context and cue 60 days after training (Balogh and Wehner, 2003).

Based on genetic evidence, the cross-species applicability of the task, and its rapid, lower tech methodology, the cued and contextual conditioning task is gaining increasing popularity for behavioral phenotyping of knockout mice. However, consideration of sensory and motor abilities is essential. Differences in pain threshold, hearing, vision, or smell can result in a false positive interpretation of a memory deficit in a mutant mouse. Independent measurement of pain threshold is particularly important for knockout mice that show deficits in fear conditioning. Doug Wahlsten of the University of Edmonton, Alberta, Canada, discovered that BALB/c mice jump remarkably high in response to the conditioning footshock (personal communication). The amount of shock received by these mice is less than expected because the feet are off the floor during the high jumps. To equate the total amount of aversive stimulation received by each mouse, Professor Wahlsten developed a program to measure and correct for actual contact between the feet and the grid floor. Further, motor impairments that produce freezing like behavior, for example, neuromuscular dysfunctions, subthreshold seizures, or sedation, could be overinterpreted as improving fear-conditioned memory.

The advantage of this task lies in the distinction between the contextual and the cued fear conditioning. If an animal (1) freezes to the identical context, (2) freezes in response to the auditory cue, but (3) does not freeze in the contextually altered environment, it is safe to conclude that (4) the animal is normal on sensory and motor abilities, (5) remembers the cues previously paired with the aversive stimulus on day 1, and (6) discriminates the cues not previously paired with the aversive stimulus on day 1. This profile reflects good memory in all components of the task. Impairment or improvement in just one of the components yields information about neuroanatomy, neurotransmitters, and genes regulating emotional components of memory. Examples are described below and illustrated in the data of Figures 6.12 and 6.13.

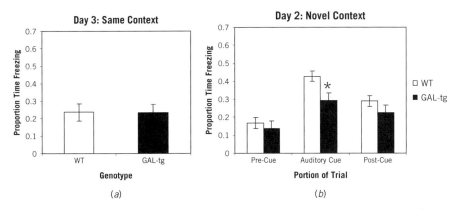

Figure 6.13 *Trace fear conditioning in galanin overexpressing transgenic mice. (a) Similar amounts of freezing behavior were seen in wildtype (WT) and galanin transgenics (GAL-tg) mice in the identical chamber where 24 hours before they had received 4 pairings of auditory cue followed 2.5 seconds later with a footshock. (b) GAL-tg froze less than WT to the auditory cue delivered in the novel context, 48 hours after the original auditory-footshock pairing. Freezing in the novel context before (pre-cue) and after (post-cue) the auditory cue was not significantly different between genotypes. [Adapted from Kinney et al., 2002.]*

Fear conditioning is a relatively stressful task. Our laboratory generally conducts contextual and cued fear conditioning toward the end of a behavioral phenotyping sequence, often just before the Morris water maze task. Overlap between fear conditioning and other tasks involving footshock associations, such as passive avoidance, are likely to confound results if both tasks are conducted in the same animals.

Passive and Active Avoidance

Avoidance tasks are technically similar to cued and contextual conditioning from the point of view of the experimenter—perhaps not from the point of view of the mouse. The stimulus is a mild footshock; the response is avoidance of the location in which the footshock was received. Passive avoidance tasks require the mouse to refrain from entering the chamber in which the aversive stimulus was previously delivered. Active avoidance tasks require the mouse to exit from the chamber in which the aversive stimulus is delivered.

Equipment for avoidance tasks is shown in Figure 6.14. The standard shuttlebox apparatus with a grid floor and shock generator is available from many behavioral equipment manufacturers, including Coulbourn Instruments, Med Associates, and San Diego Instruments. The two chambers of the shuttlebox are the same size and shape but differentiated by their walls. For example, one chamber is black or solid metal; the other is transparent Plexiglas. The floor grid on the dark side receives electrical current; the floor grid on the light side is disconnected and does not receive current. A door in the wall between the two chambers is lowered and raised by the experimenter.

Passive avoidance typically begins by placing the mouse in the light chamber for 10 seconds, then opening the door between the chambers (McGaugh, 1966; Intrioni-Collison and Baratti, 1992; Sarter et al., 1992; Matthis et al., 1994; Kawamata et al., 1997; Wehner and Silva, 1997). Most strains of mice are highly exploratory and prefer the dark over a lighted chamber, so the mouse will quickly enter the dark chamber.

Figure 6.14 *Passive avoidance apparatus. Nocturnal rodents prefer dark, enclosed spaces over open, brightly lit spaces. When the mouse enters the dark compartment, the door between the two chambers is closed and a single footshock is administered. The mouse remains in the dark chamber for 10 seconds, then is returned to the home cage. Twenty-four hours later the mouse is placed in the lighted chamber. Latency to enter the dark chamber is the measure of memory of the aversive experience on the previous day.* (Photograph contributed by the author.)

The door is closed, and a single footshock is delivered through the grid floor. Intensity of the shock is the minimum sufficient to cause flinch and vocalization. A range of shock parameters has been used, depending on the requirements of the experiment and the strain of mouse, from 0.2 to 0.8 mA, at 100 V, 50 to 60 Hz AC, for a duration of 1 or 2 seconds. The mouse remains in the dark chamber for an additional 10 seconds, a period designed to allow strengthening of the association between the properties of the chamber and the footshock. The mouse is then returned to the home cage.

Twenty-four hours later the mouse is taken from the home cage and placed in the lighted chamber, with the door open between chambers. The experimenter records the latency for the mouse to enter the dark chamber. Normal mice will be very slow to enter the dark chamber, often not entering at all, up to a 300-second cutoff latency, presumably because the mouse remembers that a shock was delivered in the dark chamber the day before. Figure 6.15 illustrates typical latencies at various shock intensities. Table 6.1 summarizes drug effects on passive avoidance.

Several variations on avoidance tasks are in use (Sarter et al., 1992; Flood et al., 1993; Schrott and Crnic, 1996; Wehner and Silva, 1997; Kawamata et al., 1997; Kim et al., 1997; Heyser et al., 1997; Flood and Morley, 1998; Sauerberg et al., 1998; Ungerer et al., 1998). *Active avoidance* typically begins with the mouse being placed into the dark chamber where footshock is delivered. The mouse must leave the dark chamber and enter the lighted chamber to escape the footshock. The mouse is then returned to the home cage. Twenty-four hours later the mouse is placed in the dark chamber. Latency to exit into the light chamber is the measure of learning the task. *Step-down avoidance* equipment includes a narrow ledge or small platform, such as a 9-cm diameter platform, elevated over a grid floor. When the mouse steps down, with all four paws on the grid floor, a single footshock is delivered, for example,

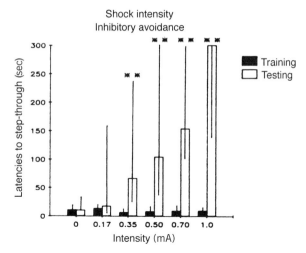

Figure 6.15 *Typical latency to enter the dark compartment at various shock intensities.* [From Brioni and McGaugh (1988), p. 506.]

0.5 mA, 1-second duration. Twenty-four hours later, latency to step down is measured with a 3-minute maximum time on the platform. *Y-maze avoidance* learning employs a three-sided Y-shaped runway, where shock is delivered through the floor grid in only one arm of the Y. Similar equipment in the shape of the letter T is used in *T-maze avoidance*.

Avoidance tasks have been widely used in screening drugs for cognitive enhancement, as summarized in Table 6.1 (Sarter et al., 1992). Retest intervals of 48 hours and 72 hours may be used, to identify drugs that increase latencies to step through at long time intervals after the training day (retention). Genetic components of avoidance task performance are inferred from strain distributions (Bovet et al., 1969; Mathis et al., 1994; Wehner and Silva, 1997).

Procedural components of the task are more difficult to distinguish from declarative memory components in avoidance tasks, as compared to the Morris water task or contextual and cued fear conditioning. The only built-in control parameter in avoidance tasks is latency to enter the dark compartment on day 1. Independent control experiments must be performed in mutant mice to rule out deficits in vision, altered pain threshold, sedation, and motor impairments (using tests described in Chapters 3, 4, and 5), which could interfere with the procedural components of the task.

Shuttlebox avoidance tasks and fear conditioning tasks both employ a mild footshock through a grid floor. This common component is likely to produce a carryover effect if both tasks are administered to the same mouse. To prevent the confounding effects of generalization, these two classes of aversively motivated tasks should not be conducted in the same group of mice.

Conditioned Taste Aversion

Conditioned taste aversion is a well-established classical conditioning paradigm in rodents (Schafe et al., 1996; Lamprecht et al., 1997; Riley, 1998; Riley and Freeman, 2004). Rats are particularly good at avoiding food sources that previously produced

TABLE 6.1 Passive Avoidance Used to Investigate Actions of Neurotransmitters and to Evaluate Potential Cognitive Enhancing Drugs[a]

Drug	Administration Time			Response Latencies[b]			Shock Parameters	
	Pre	Post (min)	Preret	NS	VS (sec)	DS	(sec)	mA
Physostigmine		Im		19.5	19.0	44	1.5	?
Secoverine		Im		13	17	34	1.5	?
Piracetam	60			?	?	+4	1.0	1
Aniracetam	60			?	?	+4	1.0	1
Pramiracetam	60			?	?	+7	1.0	1
Oxiracetam	60			?	?	+6	1.0	1
Piracetam	120			?	10	16	1.0	1
Aniracetam	120			?	10	13	1.0	1
Pramiracetam	120			?	10	18	1.0	1
Oxiracetam	120			?	10	16	1.0	1
Piracetam		Im		?	10	16	1.0	1
Aniracetam		Im		?	10	16	1.0	1
Pramiracetam		Im		?	10	16	1.0	1
Oxiracetam		Im		?	10	16	1.0	1
Aniracetam		Im		?	?	+5	1.0	1
Pramiracetam		Im		?	?	+5	1.0	1
Oxiracetam		Im		?	?	+5	1.0	1
Naloxone		Im		?	16	65	2.0	0.3
Naloxone		Im		?	60	300	2.0	0.7
Epinephrine		Im		?	65	300	2.0	0.7
Hoe 175		Im		?	70	256	2.0	0.6
Etiracetam	30			31	615	963	15	0.25
Physostigmine		1		?	215	570	2.0	0.8
Arecoline		2		?	210	590	2.0	0.8
Oxotremorine		1		?	220	560	2.0	0.8
Nicotine		1		?	270	590	2.0	0.8
4-aminopyridine		1		?	280	590	2.0	0.8
Rolipram		Im		?	220	620	?	0.8
Ro 20–1276		Im		14	316	600	?	0.22
IBMX		Im		17	102	600	?	0.22
Epinephrine		Im		?	60	240	0.35	0.7
MK 801	30			?	?	+15	1	0.2
MK 801	30			?	?	+10	1	1
MK 801		Im		?	?	+14	1	1
AP 7	30			?	?	+5	1	1
D-Cycloserine	60			?	15.3	37	2	0.5
D-Cycloserine		10		?	15.3	34	2	0.5
D-Cycloserine			60	?	15.3	28	2	0.5
Picrotoxin		Im		+6	+50	+120	10	0.35
Bicuculline		Im		+1	+20	+120	10	0.35
Alaproclate			30	23	192	430	?	0.13
Oxotremorine			30	27	192	450	?	0.13
Flumazenil	30			?	+20.2	+124.4	2	0.3
β-CCB	30			?	9.9	+187.9	2	0.3
Ro5–4864		Im		−0.9	+21.0	+180	2	0.35
DHEA		3		?	101	162	1	0.35
Neuropeptide Y		3		?	85	140	1	0.35
Angiotensin II			15	?	70	230	2	0.25

[a]Abbreviations: pre = pretraining; post = post-trail; preret = pre-retest; NS = not shocked; VS = vehicle-treated and shocked; DS = drug-treated and shocked; im = immediately after shock; ? = not indicated; DHEA = dehydroepiandrosterone; 1: maximal effect. Note that most of the values were taken from figures and therefore may not be perfectly accurate. + = change with respect to baseline (absolute data not indicated).

[b]Latency to enter the dark chamber is sensitive to treatments that act at muscarinic cholinergic receptors, glutamate receptors, GABA receptors, enkephalin receptors, and others.

Source: From Sarter et al. (1992), p. 463.

aversive internal stimuli or sickness (Garcia et al., 1955). Acquisition is achieved in one or two trials. The paradigm pairs a pleasant new taste in the drinking water, such as sucrose or saccharin, with an intraperitoneal injection of a malaise-inducing agent, such as a high dose of lithium chloride (Steinert et al., 1980; Houpt ct al., 1996; Lamprecht et al., 1997; Riley 1998; Riley and Freeman, 2004). A subsequent choice test between water and saccharin is then used to assess avoidance of the saccharin solution. Suppression of saccharin drinking is the measure of associative learning. Taste aversion conditioning is remarkably robust, long-lasting, and generally acquired with one or very few trials.

Conditioning is learned even with delays as long as 10 hours between consumption of the novel food and administration of the aversive drug (Riley, 1998). Taste aversion has been useful for investigating brain regions and biochemical mechanisms of extinction and reconsolidation (Bahar et al. 2004). Typical taste aversion data are shown in Figure 6.16.

Eyeblink Conditioning

Classical conditioning of eyeblink is an established paradigm that has been adapted for mice by Richard Thompson and co-workers at the University of Southern California in Los Angeles (Chen et al., 1996c, 1999; Kim et al., 1997; Bao et al., 1998). This task is considered particularly useful as a measure of cerebellar motor learning (Steinmetz, 1998), and is also sensitive to lesions of the dorsal hippocampus (Tseng et al. 2004). The unconditioned stimulus is a 100-msec footshock. The conditioning stimulus is an 80-dB tone. The eyeblink response is measured with an eyelid EMG amplification system (Figure 6.17).

Mazes

The T-maze, radial arm maze, and Barnes maze are variations on spatial maze tasks (Barnes et al., 1990; Wenk 1997). In all cases, the task requires the animal to choose specific arm(s) of the maze to receive a food or water reinforcement or to avoid a footshock. Labyrinth mazes, used in the early 1900s (Small, 1901), remain popular in the public imagination and for elementary school science projects. Researchers have streamlined the labyrinth maze to the simpler configurations of a T, Y, or radial shape to allow larger numbers of training trials and greater precision in experimental design.

An apparatus used for *T-maze delayed alternation* is shown in Figure 6.18. The sides of the T-maze are black Plexiglas or wood. Flooring is metal mesh. Hidden food wells are recessed below the floor, at opposite ends of the T, far enough from the choice point to prohibit the mouse from seeing or smelling the reinforcer. T-maze delayed alternation requires several weeks of training (Hepler et al., 1985; Mastropaolo et al., 1988; Markowska et al., 1989; Wenk, 1997). The animals are maintained on a food-restricted or water-restricted diet. Mice are first habituated to the maze, then shaped to run to the ends of the T to obtain the reinforcer. Alternation training consists of the food or water reinforcer being located at alternating arms on successive trials. Delay training introduces a time delay, in the range of 30 seconds to 5 minutes, between successive trials. The mouse must remember which arm was reinforced last time to make the correct choice on the next trial. Criterion for acquisition of delayed alternation is generally 75–90% correct responses per daily session of 10 to 20 trials per day, on

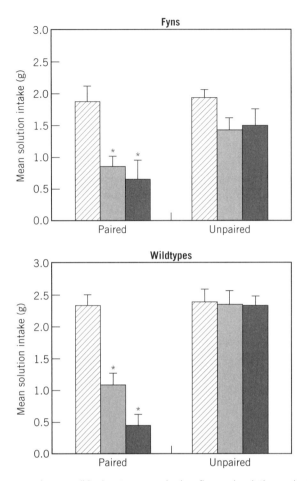

Figure 6.16 *Taste aversion conditioning to a saccharine flavored solution paired with an 0.15 M intraperitoneal lithium chloride injection. Wildtype mice and fyn mutant mice both showed normal drinking of the pleasant-tasting saccharine solution (unpaired) on the conditioning day, first test day, and second test days ("unpaired" three bars). Reduced consumption of the saccharine solution was seen in both genotypes when the pleasant-tasting saccharine solution was paired with the unpleasant interoceptive cues associated with a lithium chloride injection.* [From Schafe et al. (1996), p. 846.]

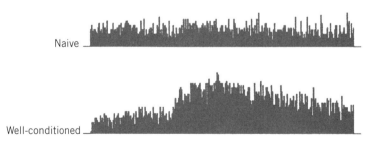

Figure 6.17 *Eyeblink conditioning in BALB/c mice. The blink response is recorded by electromyography from four subcutaneously placed recording electrodes. Temporal pattern of the eyelid reflex response to an aversive stimulus (naïve) and to a tone paired with the stimulus (well-conditioned).* [From Bao et al. (1998), p. 717.]

Figure 6.18 *Typical T-maze. The mouse starts in the lower end of the T on each trial. Food or water reinforcement is located in a recessed cup at the end of either the left or the right arm of the T. For alternation tasks, the location of the reinforcer alternates between the right and the left arm on each successive trial. For the delayed alternation task, a delay of several seconds to several minutes intervenes between successive trials.* (Photograph contributed by the author.)

several consecutive days. This working memory task is used to evaluate both the rate of acquisition and the maximum asymptote reached for choice accuracy. Cholinergic receptor antagonists, and lesions of the cholinergic nucleus basalis magnocellularis medial septum/diagonal band neurons, interfere with performance on T-maze delayed alternation in rats (Hepler et al., 1985; Mastropaolo et al., 1988; Welner et al., 1988). Most strains of mice are considerably slower than rats in learning delayed alternation. A water T-maze, in which the mouse learns intramaze visual cues to find a hidden platform located only at one arm, may be useful for more rapid T-maze acquisition in mice (Gorski et al., 2003).

Radial maze learning requires a different set of behavioral strategies for the mouse (Olton and Samuelson, 1976; Markowska et al., 1989; Crusio et al., 1995; Johnson and Kesner, 1994; Ikari et al., 1995; Wenk, 1997). As shown in Figure 6.19, the apparatus is generally about 200 cm in diameter, composed of 8 or 12 arms, radiating from a central start box. Equipment is built from Plexiglas or wood. Radial mazes, including automated versions, are commercially available from several companies, such as Columbus Instruments and Actimetrics/Coulbourn Instruments. Training takes several days or weeks for rats, many weeks for mice. Dietary restriction to 85% of free-feeding body weight is necessary. Training begins with *habituation* and *shaping* sessions to gradually accustom the animal to the maze and to the presence of food toward the ends of the arms of the maze. Then the maze is baited with food, such as a piece of sweet breakfast cereal, only in the recessed container at the end of each arm. Liquid diet reinforcers are convenient. The mouse is placed in the start box. Correct arm choices are those that are visited once to obtain the reinforcement. Incorrect responses are visitations to the same arm more than once, where the food has already been obtained. Time elapsed between the start of the test session and finding all of the food rewards

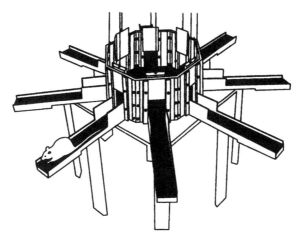

Figure 6.19   An eight-arm radial arm maze. Radial maze tasks measure entries into arms baited with a food or water reinforcer. Entries into previously visited arms are scored as errors. Radial arm maze tasks often bait only some of the arms, with the same arms baited on successive trials. Errors are scored when the rat enters an unbaited arm. [From Wenk (1997), p. 8.5.2.]

is another measure of acquisition. Alternatively, only some of the arms are baited with food, for example, 5 out of 8 arms, requiring the animal to learn the location of the baited versus unbaited arms and also to learn not to return to an arm already visited. Genetic components of performance on radial maze tasks have been described by Wim Crusio at CNRS in Talence, France. Volume of hippocampal mossy fiber projections correlates with radial maze performance among nine inbred strains of mice (Crusio et al., 1995).

The *Barnes maze* was invented by Carol Barnes, University of Arizona. The Barnes maze task has similarities to the Morris water task and the radial maze task but takes advantage of the superior abilities of mice to find and escape through small holes (Barnes, 1979; Markowska et al., 1989; Bach et al., 1995; Mayford et al., 1996; Pompl et al., 1999; Miyakawa et al., 2001). The Barnes maze for rats, illustrated in Figure 6.20, is a brightly lighted circular white platform, 1.2 m in diameter, with a series of 18 evenly spaced holes, 9 cm in diameter, around the outside edge. For mice, a smaller diameter and fewer holes increase the rate of acquisition. The reinforcer is escape from the brightly lit, open platform into a small, dark, enclosed box. One of the holes exits to the dark box, which is recessed under the platform and invisible from the surface of the platform. The mouse learns the spatial location of the correct hole to exit into the escape box. A probe trial is conducted at the end of the training trials, with the escape box removed, to confirm learning based on distal environmental room cues. The entry mode and properties of the escape box appear to be critical for acquisition of the Barnes maze task. Mice appear to sniff the escape hole but avoid jumping straight down into the unfamiliar escape box. A graded ramp from the maze surface into the escape box, and the addition of home cage litter to the escape box, facilitates entries (Sheryl Moy, University of North Carolina, personal communication). The Barnes maze has advantages for behavioral phenotyping of mutant mice. In contrast to T-maze and radial maze tasks, dietary restriction is not necessary. In comparison to the Morris swim task, the stress component is less in the dry-land Barnes maze task.

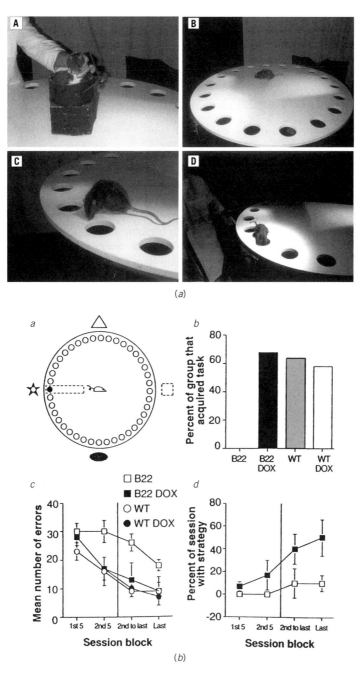

Figure 6.20 *Barnes spatial maze. (a) The mouse explores the brightly lit open field and the holes around its perimeter. One hole leads to an escape tunnel. Number of incorrect holes explored, and latency to locate the hole leading to the escape tunnel, are measured on successive trials.* (Photograph kindly contributed by Dr. Carol Barnes, University of Arizona.) *(b) Mutant mice expressing an activated calcium-independent form of calcium-calmodulin-dependent kinase II (B22) showed impaired acquisition of the Barnes maze task, as compared to wildtype controls (WT). Expression was regulated by a forebrain-specific promoter and a tetracycline transactivator (DOX).* [From Mayford et al. (1996), p. 1680.]

All of these spatial learning and memory tasks measure similar types of spatial navigation based on environmental room cues. Use of two or three of these spatial navigation tasks for evaluating a mutation in separate batches of mice will confirm a significant cognitive phenotype across sensory modalities and motor requirements. Maze tasks have many components in common. It is best to choose only one for a given batch of mice. While careful studies of carryover effects between the radial maze, T-maze, Y-maze, and Barnes maze tasks have not been conducted, it seems likely that some generalization of learning across these similar types of equipment can confound the learning curve if conducted in the same mice.

Olfactory Tasks

Olfaction is a highly developed sense in rodents. Olfactory discrimination tasks are excellent measures of learning and memory in rats and mice (Staubli et al., 1989; Eichenbaum et al., 1988; Zhang et al., 1998; Brennan and Kerverne, 2004). While many olfactory learning and memory tasks are hippocampally dependent, controversy on the role of the hippocampus in some tasks remains (Wrenn et al., 2003). For example, Howard Eichenbaum and co-workers at Boston University demonstrated no effect of hippocampal lesions on a one-trial odor-reward association task for episodic memory in rats (Wood et al., 2004).

An early olfactory learning apparatus is shown in Figure 6.21. Odors are generated by forcing air through water-filled flasks mixed with flavors, odors, and extracts such as amyl acetate, limonene, eugenol, and phenethyl alcohol. Individual odors are delivered through odor-carrying tubes mounted at the ends of radially arranged runways. Odors are removed from the testing area by a fan mounted above the apparatus. The animal pokes its nose into an odor outlet tube, breaking a photodiode beam, which is detected and automatically recorded by computer. In one version of this task, water-restricted rats obtain a water reinforcement by choosing the lower intensity odor in a two-choice discrimination (Staubli et al., 1989). Olfactory learning can also be assessed in a spatial maze apparatus (Staubli, et al., 1985; Lynch, 1986). Rodents are trained to choose one specific odor in the alleys of a maze to obtain a water reinforcement. Pairs or triads of odors are used, with intertrial intervals varying from 1 to 10 minutes. The position of each odor is randomized across trials.

Howard Eichenbaum and colleagues use a successive-cue procedure for olfactory learning in rats and hamsters. Two odors are presented successively on separate trials, and the animal must stay in the odor port of only one of the two odors to receive the food or water reinforcement (Eichenbaum et al., 1988; Otto and Eichenbaum, 1992; Petrulis et al., 2005; Figure 6.22). Social odors delivered through olfactory ports confirmed the ability of hamsters to discriminate individual conspecifics, a naturalistic recognition task (Petrulis et al., 2005). Transverse patterning is a similar but more challenging sequential presentation of three overlapping odor discriminations that has been applied to mice (Rondi-Reig et al., 2001). Robert Mair at the University of New Hampshire employs odorants in an olfactory continuous delayed nonmatching to sample procedure (Figure 6.23) to investigate brain regions mediating olfactory learning in rats (Zhang et al., 1998). Automated olfactory conditioning chambers are commercially available from companies, including Med Associates.

A simple olfactory conditioning test for mice was developed by Keverne and colleagues in the animal behavior subdepartment of the University of Cambridge in

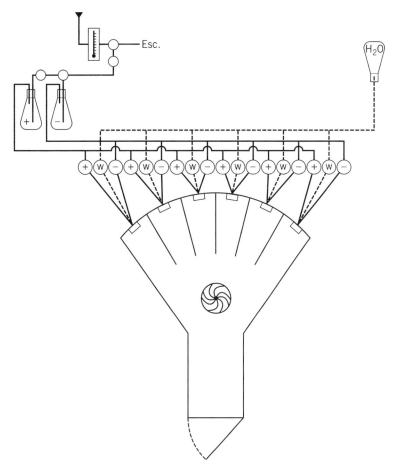

Figure 6.21 *Olfactory discrimination learning apparatus. A wedge-shaped chamber delivers various odors at each of six holes. A ceiling fan rapidly removes the odors from the testing area. The apparatus includes floor plates through which a footshock can be delivered, for one-way active avoidance conditioning to specific odors.* [From Staubli et al. (1989), p. 55.]

England. (Brennan et al., 1998). Mice are maintained on a food restriction regimen. The reinforcer is fragments of sugar cubes. The conditioned stimulus is an odor such as lemon or peppermint. Clean wood shavings in a plastic Petri dish are sprinkled with 50 μl of one of the odors (e.g., peppermint). Each mouse receives 8 days of training sessions, with two 10-minute trials per day. One day after the last training session, each mouse is placed in the middle compartment of a three-compartment chamber. A dish containing the lemon odor is placed in one of the side compartments; a dish containing the peppermint odor is placed in the other side compartment. The total time spent in the compartment containing the previously reinforced conditioned odor (e.g., peppermint) is compared to the total time spent in the compartment containing the nonreinforced unconditioned odor (e.g., lemon). The time spent digging in the wood shavings in each dish provides a second measure of odor conditioning.

Modifications of olfactory discrimination tasks add gradations of difficulty for mice and rats. Scented food can be buried under litter or mixed with sand (Mihalick et al.

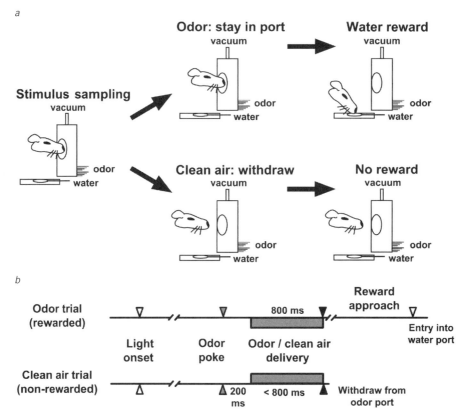

Figure 6.22 Social odor recognition task. Schematic illustrations of the testing apparatus, sequence of behavior and timeline for trials in the task. (a) The illustrations identify the sequence of behaviors in the olfactory detection task. On each trial the hamster must sample the chemosensory stimulus by inserting its nose into an opening in a hollow hemi-cylinder (odor port). Odors and clean air are presented into the hemi-cylinder through separate tubing lines lying below the odor port. Constant application of vacuum dorsal to the odor port pulls odors and clean air past the odor port and is exhausted to a vacuum dump. If the presented stimulus is an odor, then the hamster must keep its nose in the port ($200 + 800$ ms) to receive water reward. If the stimulus is clean air, then the hamster must withdraw earlier than $200 + 800$ ms for a correct response. No reward is presented on clean air trials. (b) The schematic timeline indicates the sequence of behavioral events and the duration of time periods over which electrophysiological data were analyzed: baseline (1000 ms prior to light [trial] onset), trial onset (500 ms after light onset), odor poke (200 ms prior to and 200 ms following entry into odor port); odor/clean air delivery (from 200 ms after odor poke until withdrawal from odor port, calculated for each session), and reward approach (700 ms prior to entry into water port). [From Petrulis et al. (2005), p. 261.]

2000). Multiple two-odor pairings allow measurement of positive transfer across problems (Slotnick et al., 2000). The olfactory discrimination can be based on previous contextual conditioning (Otto and Giardino, 2001). Testing at time points up to 60 days after training allows assessment of long-term memory and forgetting (Schellinck et al., 2001).

Social transmission of food preference is an intriguing olfactory learning and memory task developed by Bennett Galef at McMaster University in Hamilton, Ontario,

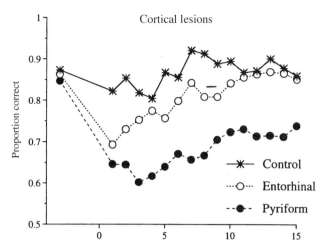

Figure 6.23 *Olfactory continuous delayed nonmatching to sample. Rats in an operant chamber receive a water reinforcement for a nose poke into an odor port containing a specific odorant. Odors, including amyl acetate and eugenol, are rapidly removed by two exhaust fans connected to the operant chamber. Lesions of the entorhinal or pyriform cortex impair performance on this olfactory discrimination task.* [From Zhang et al. (1998), p. 40.]

Canada. Galef and Wigmore (1983) described the wild Norway rat as a social, central-place forager, living in a colony in a burrow. Each member of the colony emerges from the burrow to forage for food. The colony appears to serve as an information center, where conspecifics acquire information about the availability of food and potential feeding sites. This concept was based on the ability of bees to transfer information by complex body movements about the location of distant food sources (von Frisch, 1967).

In the experimental environment, Long-Evans laboratory rats and many strains of mice appear able to convey information to each other about novel foods (Galef and Wigmore, 1983; Bunsey and Eichenbaum, 1995; Wrenn et al., 2003). As elucidated in Galef (2002) and Wrenn (2004) and shown in Figure 6.24, a "demonstrator" is placed in a test cage and given powdered chow mixed with a novel flavor, for example, Hershey's cocoa powder. The demonstrator is then placed in a two-compartment cage with its former cagemates. The cagemate, designated the "observer," sniffs the demonstrator, presumably detecting the smell and/or flavor of the consumed chow on the whiskers or breath of the demonstrator. Later the observer is placed in a test cage containing a choice of cocoa-flavored chow and a different flavor, for example, ground-cinnamon-flavored chow. Observer rats and mice generally show a highly significant preference for the flavored chow previously consumed by the demonstrator cagemate. Bunsey and Eichenbaum (1995) employed a variety of flavor pairings, including cocoa versus cinnamon, clove versus garlic, marjoram versus cumin, and turmeric versus thyme. Retention time between the observation session and the choice session is either zero (immediately thereafter), providing a procedural control, or 24 hours later, providing the delay condition for evaluating memory. Hippocampal lesions severely impair choice accuracy in this task (Bunsey and Eichenbaum, 1995). These findings suggest that olfactory transmission learning may be a useful test of learning and memory in mutant mice, representing a different set of sensory modalities than those employed in other tests. Social transmission of food preference has been successfully adapted for

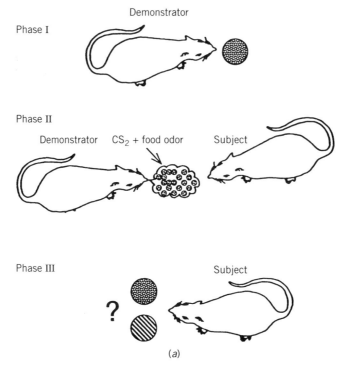

Phase I

Demonstrator

Phase II

Demonstrator   CS$_2$ + food odor   Subject

Phase III   Subject

?

(a)

Figure 6.24 *Social transmission of food preference by olfactory cues. (a) The "demonstrator" rat eats ground rat chow mixed with one of eight distinctly scented spices. The food cup is removed and a subject rat is allowed to sniff the demonstrator rat. Twenty-four hours later the subject rat is given the choice of two foods, one identical to that consumed by the demonstrator rat and one containing a different scented spice. (b) A subject mouse sniffs the mouth of the demonstrator mouse who has just eaten the novel cinnamon-flavored chow. (c) The subject mouse subsequently eats more of the now-familiar cinnamon-flavored chow than a completely unfamiliar cocoa-flavored chow. (d) Subjects will consume more of the food source with the odor previously associated with the demonstrator, as compared to consumption of a food source with a new odor. Rats with lesions of the hippocampus (H) + Dentate gyrus (D) + subiculum (S) chose equally between the two food sources. [(a) and (d) From Bunsey and Eichenbaum (1995), pp. 549 and 552. (b) and (c) Photographs by Craige Wrenn, National Institute of Mental Health, contributed by the author.]*

behavioral phenotyping of mice with mutations in genes mediating memory processes, genes expressed in the hippocampus, and genes for olfactory receptors (Ferguson et al., 2000a; Mayeux-Portas et al., 2000; Giese et al., 2001; Otto et al., 2001; Wrenn et al., 2003, 2004).

Social Recognition

Individual recognition of conspecifics forms the basis of the social recognition paradigm. Originally developed by Robert Dantzer and co-workers in Bordeaux, France (Gheusi et al., 1994), social recognition tasks are increasingly applied as memory tasks for phenotyping mutant mice (Ferguson et al., 2002; Choleris et al., 2004; Winslow and Insel, 2004). Considerable ethological evidence describes the abilities of many species to recognize members of their own group, often according to kin relationship, sexual

(b)

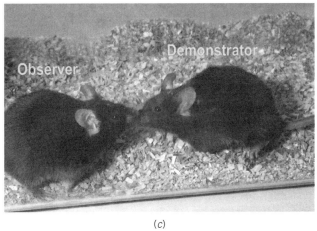

(c)

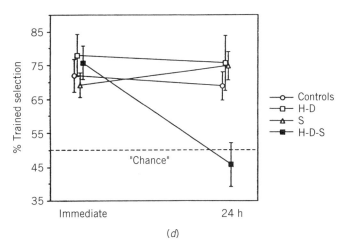

(d)

Figure 6.24   (Continued)

status, dominance hierarchy status, colony identity, and individual identity (Colgan, 1983). Social recognition focuses on degree of familiarity between two individuals. Dantzer's original task for rats is diagrammed in Figure 6.25. The test rat is placed in a test cage with a wire mesh window. A circular cage containing two chambers, each with a wire mesh window, is placed next to the test chamber. In one chamber is a rat who has been paired with the test rat for 5 minutes, termed the familiar conspecific. In the other chamber is a novel rat who has not previously been paired with the test rat. The circular cage is rotated to allow the wire mesh of the test cage to line up with the wire mesh of each of the two chambers, for a 5-minute session. The time spent by the test rat in investigating each of the two conspecifics is recorded. Normal rats will spend more time sniffing the novel rat, as compared to time sniffing the familiar rat. A time interval of hours to days between the exposure presentation and the test session provides a delay component to the task. This task is sensitive to treatments with gonadal steroids and vasopressin, a neuropeptide transmitter in pathways of the rat brain relevant to sexual and social behaviors.

Another version of the social recognition task focuses on the olfactory ability of adult rodents to recognize a familiar juvenile (Bluthe et al., 1993; Engelmann et al., 1995, 1998). In each session a juvenile rat is exposed to an adult female rat over a 4-minute observation period in a test cage. Duration of exploration, sniffing, and licking of the juvenile by the adult is recorded by the experimenter. The juvenile is then removed and placed in a separate cage with fresh bedding, with an interexposure interval of 30, 60, 120, or 180 minutes. The adult is then placed in the cage with both the familiar juvenile and a novel juvenile. Investigatory behaviors by the adult toward each juvenile is recorded. A significant difference between the time spent exploring the novel juvenile versus the familiar juvenile indicates that the adult remembers the familiar conspecific.

Social recognition employs olfactory and other sensory modalities and brain mechanisms that may be ethologically relevant components of learning and memory. Social recognition appears to be mediated by the lateral entorhinal cortex, ventral subiculum, vomeronasal organ, androgens, and hypothalamic neuropeptides including oxytocin and vasopressin and their receptors (Dantzer, 1998; Winslow and Insel, 2002; Wersinger et al., 2002; Kavaliers et al., 2003; Bielsky et al., 2004; Petrulis et al., 2005). Social olfactory recognition is sensitive to lesions of the hippocampus and septum and declines with age (Terranova et al., 1994; Kogan et al., 2000). Social recognition and social memory tasks are successfully used for mice (Terranova et al., 1996; Ferguson et al., 2000b; Winslow, 2003; Figure 9.3). Group housed mice demonstrate significant social memory for at least 7 days after a single interaction (Kogan et al., 2000). An interesting alternative is measuring ultrasonic vocalizations emitted by female mice in the presence of a familiar versus an unfamiliar female (D'Amato and Moles, 2001).

Schedule-Induced Operant Tasks

Schedule-controlled behavior is a classic tool of experimental psychology. B. F. Skinner developed the automated operant chamber, in which a rat learns to press a lever to receive a reward (Skinner, 1938). Standard operant schedules include a fixed ratio (FR), in which every nth lever press is reinforced (e.g., FR 20 in which every 20th lever press delivers a food pellet), and a fixed interval, in which a lever press produces a reward every x seconds (e.g., FI 30 in which a food pellet is delivered every 30th second in

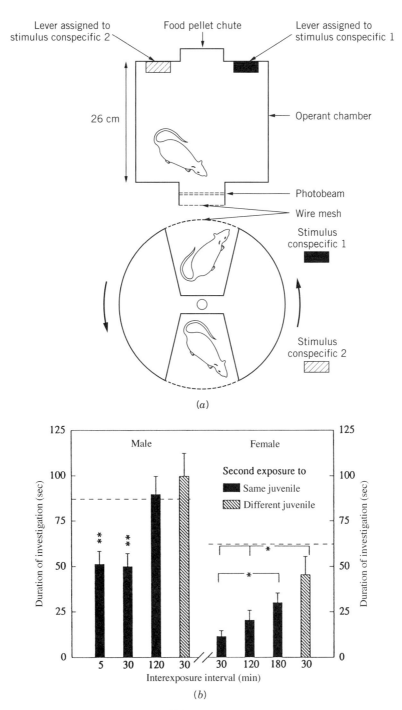

Figure 6.25 *Social recognition memory. (a) Subject rat is exposed to two new rats through a wire mesh screen. The subject is reinforced for pressing one lever in response to the presence of rat 1 and the other lever in response to the presence of rat 2. (b) Duration of investigation of a familiar versus a new rat provides another measure of social recognition. This task is sensitive to gender, age, and interval between social sessions.* [From Gheusi et al. (1994), pp. 67 and 69.]

response to the lever press at the 30-second time point). Variable ratio and variable interval schedules are useful for evaluating many aspects of behavior, including timing, motivational state, and drug responses (Salamone and Correa, 2002). An interesting observation with direct implications for human behavior is that a rat exhibits the largest number of lever presses when the schedule is both variable ratio and variable interval; that is, the most lever presses are elicited when the contingencies appear to be random. Loosely translated, you work hardest when you don't know exactly when you will be rewarded.

Schedule-induced operant tasks have many advantages, including strict experimental control and a high level of precision. The subject reaches stable performance levels after extensive training. Complete automation eliminates the labor-intensive nature of most memory tasks for the researchers. Each subject serves as its own control for drug treatments, lesions, and inducible mutations. A small number of animals may suffice for a large number of experiments. The disadvantage is the long training time required, from several weeks to several months. Although mice generally take longer than rats to train on operant tasks, a rapid autoshaping task for mice helps to reduce the initial time for familiaring the mouse with the operant reinforcement procedures (Barrett and Vanover, 2003).

Lever pressing apparently does not come as naturally to mice as to rats. Nose-poke behavior is more within the natural repertoire of mice and provides a good alternative to lever pressing in the operant chamber (Figure 6.26). A mouse will explore a small, dark hole in the wall. A photocell beam across the hole in the wall detects the choice of a specific hole. A commercially available nose-poke unit, equipped with a photo-cell detector, can be substituted for the lever press unit in a standard operant system (available from companies including Med Associates Inc., St. Albans, Vermont). Jim Barrett and co-workers describe their success with the nose-poke unit for autoshaping of normal CD-1 mice on acquisition for a sucrose liquid reinforcement (Vanover and Barrett, 1998; Barrett and Vanover, 2003; Figure 6.27). Two days of training was sufficient for acquisition of a rate of reinforcement that was sensitive enough to detect deficits induced by scopolamine. Acquisition of a low rate of delayed reinforcement using nose pokes has been achieved with mice (Ripley et al., 1999). In an elegant study of inbred strains of mice by Jeanne Wehner and co-workers at the Institute for Behavioral Genetics at the University of Colorado in Boulder, nose-poke training for sucrose reinforcement was successfully used to investigate impulsivity, the ability to withhold nosepokes until an auditory cue signal, as compared to ethanol preference in a three-bottle choice test (Logue et al., 1998).

Among the best rodent operant learning and memory tasks are delayed matching and delayed nonmatching to position (Dunnett, 1993; Robinson and Crawley, 1993; Dawson et al., 1994; McAlonan et al., 1995; Kirkby et al., 1996; Aura et al., 1997; Higgins et al., 2002). The equipment is a commercially available operant chamber controlled by a software program. The chamber is equipped with two response levers on the front wall and a response lever on the rear wall, each under a stimulus light. A liquid dripper recessed into the front wall dispenses reinforcers of 0.05 ml (50 μl) water to rats on a water-restricted diet. Similarly food dispensers are available that deliver single food pellets in a size suitable for mice. Med Associates sells a smaller size mouse operant chamber with small, sensitive levers for mice. Delayed nonmatching to position has been successfully used with C57BL/6 and DBA/2 mice (Estape and Steckler, 2002)

(a)

(b)

Figure 6.26 *(a) An operant chamber for mice. This adaptation has smaller levers, which are more sensitive to a lighter touch, and smaller food pellets dispensed as reinforcers, as compared to the original operant chamber for rats. A nose-poke detector replaces the response lever in some chambers for mice. (b) Mouse in nose-poke operant chamber.* (Photographs kindly contributed by Dr. Vern Davidson, Med Associates, Inc, were used as concepts for photographs contributed by the author.)

Subjects are first *autoshaped* to press either the left or the right retractable lever or nose-poke hole to obtain the reinforcement. In delayed nonmatch to sample, the first stage of training consists of learning *nonmatch to position*. A cue light above one of the two levers or nose-poke holes is illuminated on the front wall (e.g., sample = left lever). Then both cue lights are illuminated. To earn a reinforcer, the subject makes a correct choice by pressing the lever or nose-poke hole that was not previously illuminated on the sample phase (e.g., the correct nonmatch choice is the right lever). The second stage of training consists of adding a *delay* between the sample and choice phases. Delays are randomly varied from 1 to 20 seconds or longer (i.e., delayed nonmatching to position). The response to the sample initiates the delay phase. The subject must

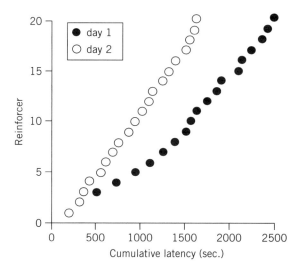

Figure 6.27 *Using an autoshaping program, mice earn access to liquid sucrose reinforcers during the first day of autoshaping (black circles). The learning curve is steeper on the second day (white circles).* [From Barrett and Vanover, *Current Protocols in Neuroscience* (2003).]

press a rear lever after the delay has elapsed to initiate the choice presentation. The longer the delay, the longer the subject must remember the sample position. Choice accuracy declines as the delay lengthens. Delay length is randomized across trials.

Treatments that reduce choice accuracy at all delays are described as *delay-independent*. Delay-independent deficits are thought to interfere with implicit memory or some procedural component of the task. Treatments that have no effect at short delays, but that reduce choice accuracy more as the delay gets longer, are described as *delay-dependent*. Delay-dependent deficits are considered to reflect a specific impairment in explicit short-term working memory. Delayed nonmatching to position is extensively used to evaluate the effects of brain lesions, neurotransmitter receptors, and cognitive enhancing drugs on memory processes (Sarter et al., 1992; Dunnett, 1993; Robinson and Crawley, 1993; Dawson et al., 1994; McAlonan et al., 1995; McDonald and Crawley, 1996; Kirkby et al., 1996; Aura et al., 1997).

Mice are not as good as rats on complex, schedule-induced operant tasks (Figure 6.28). Rats reach high performance criteria on complex schedules, for both appetitive and aversive reinforcement (Figure 6.29). The rat was chosen as an ideal experimental subject, and the operant chamber was designed for rats, because of the high level of performance of this species on these types of operant tasks. Mice are generally not as fast on acquiring schedule-controlled tasks in operant chambers, even with modification including smaller chambers and nose-poke holes. Some folks conclude that mice are stupid. In the opinion of this author, this conclusion is premature. Every species that has evolved and survived is smart in its own way. Mice are more active and exploratory than rats. Operant tasks require the subject to remain in a relatively small, fixed location, and to focus on a small number of stimuli over a long time period. Mice may be better at cognitive tasks that require high levels of exploratory locomotor activity, such as the maze tasks described above.

Successes with training mice on the simpler schedule-induced operant tasks have been reported. Fixed ratio responding has been successfully conducted in mice (Glowa,

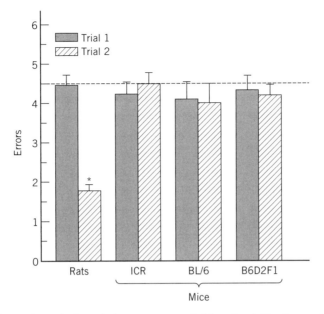

Figure 6.28 *Comparison of rats and mice on an operant delayed matching to sample win-stay task. Sprague-Dawley rats reached criterion on the task after 20 days of training. Three strains of mice did not reach criterion.* [From McNamara et al. (1996), p. 182.]

1986; Glowa et al., 1988; Baron and Meltzer, 2001). Mice were trained on a fixed ratio 30 schedule in which the 30th nose-poke produced access to a 0.025 ml evaporated milk reinforcer. Behavioral effects of ethanol, the ethanol antagonist Ro 15–4513, amphetamine, cocaine, nicotine, and caffeine were reported with this method for schedule-controlled responding in mice. Ts65Dn mice, a model of Down syndrome, acquired a fixed ratio 15 schedule with a milk reinforcer (Wenger et al., 2004). DARPP-32 knockout mice and their wildtype controls also reached criterion on a FG-15 schedule of reinforcement (Heyser et al. 2002). On an FR1 or FR3 reinforcement schedule in a nose-poke operant chamber, some nicotinic receptor mutants on a C57BL/6J or mixed background reached criterion within 10 days (Keller et al., 2005). Mike McDonald in our laboratory succeeded in training mice on several tasks, beginning with a simple acquisition for a food reward (Figure 6.30; McDonald et al., 1998). At the beginning of each trial, a cue lamp was illuminated. Pressing the lever below the cue lamp produced immediate delivery of a food pellet and extinguished the cue lamp. Lever presses while the cue lamp was off produced no consequences. Normal CD-1 mice reached the criterion of earning at least 10 reinforcers in a 40-trial session after 8 daily sessions. These mice were then able to learn a reaction-time task, which is used to assess attention in rats and humans (McGaughy and Sarter, 1995; Zahn et al., 1991). In this task the subject must respond during a brief period when the cue light is illuminated, with the period of time progressively shortened to measure minimum reaction time. Further, a go/no-go task was mastered by many of the CD-1 mice. This type of signal detection paradigm assesses both sustained attention (vigilance) and impulsivity (Zahn et al., 1991). A tone is sounded when the stimulus light is illuminated. Reinforcement is earned by responding when the light is illuminated and by not responding when the light + sound are simultaneously presented. Normal mice performed better on

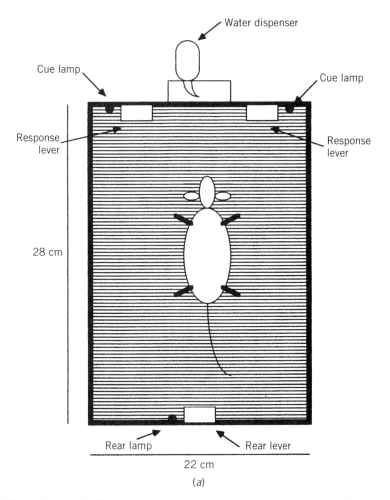

28 cm

22 cm

(a)

Figure 6.29 (a) Schematic diagram of an operant chamber used for a water-reinforced delayed nonmatching to position two lever choice task. (Drawing kindly contributed by Dr. John Robinson, State University of New York at Stony Brook.) (b) Delayed nonmatching to position performance in rats treated with scopolamine, a muscarinic receptor antagonist; with MK-801, a glutamate receptor antagonist; or with galanin, a neuropeptide overexpressed in the basal forebrain in Alzheimer's disease. Dose-dependent, delay-independent reductions in choice accuracy were seen after both treatments. [From Robinson and Crawley (1993), pp. 461 and 462.]

these tasks than transgenic mice with an insertion of a mutant human thyroid receptor gene, a model of human resistance to thyroid hormone syndrome, which sometimes presents symptoms analogous to attention deficit hyperactivity disorder (McDonald et al., 1998).

Automated touchscreens, similar to those used to test human cognition, have been adapted to operant chambers for mice (Bussey et al., 2001; Izquierdo et al., 2006). Figure 6.31 illustrates a touchscreen with two panels presenting illuminated visual cues, a star on the left side and a pattern of dots on the right side. C57BL/6J mice performed well on this visual discrimination and reversal learning task (Izquierdo et al., 2006).

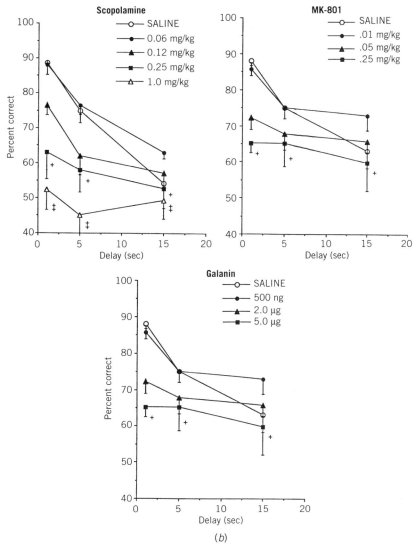

Figure 6.29   *(Continued)*

Five-Choice Serial Reaction-Time Attentional Task

Attentional processes play a significant role in schedule-induced operant behavioral tasks. Trevor Robbins, Barry Everitt, and co-workers at the University of Cambridge developed a selective attentional task for rats (Robbins et al., 1989; Muir et al., 1996; Baunez and Robbins, 1999) that is analogous to continuous performance used in human studies of attentional processes. The rat is required to monitor up to nine spatial locations simultaneously. The wall of an operant chamber is concavely curved, with a series of nine holes and a light above each hole. On a commonly used version of this task, one of five holes is illuminated for 0.5 second on each trial. A food reinforcer is available at that hole for 5 seconds. For mice, the inter-trial interval is critical, to give

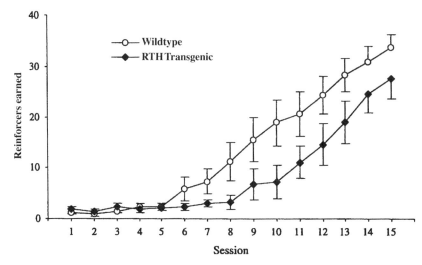

Figure 6.30 *Acquisition of a simple lever press autoshaping task for food reinforcement in mice. Wildtype mice reached criterion after 15 days of training, 40 trials per day. Transgenic mice expressing the human thyroid hormone resistance gene (RTH) acquired the task but at a slower acquisition rate.* [Adapted from McDonald et al. (1998).]

mice sufficient time to eat the food pellet. Liquid diet reinforcers delivering 0.05 ml (50 µl) volumes may shorten the required inter-trial interval. On each successive trial, one of the five lights is lit, in a randomized order, and the food pellet is again available briefly at that hole. A photocell across each hole detects the nose-poke. Photocell beam breaks are analyzed by a software program for time and location of each beam break with respect to the light stimulus presentations.

The animal must monitor all of the holes, requiring attention to be spread over the range of spatial locations. Accuracy and speed of responding measure attentional performance. Gerry Dawson of Merck, Sharp & Dohme in Terlings Park, England, and Lawrence Wilkinson of The Babraham Institute in Cambridge, England, have adapted the five-choice serial reaction time equipment for mice, employing smaller chambers and nose-poke responses, as illustrated in Figure 6.32. Choice accuracy on this attentional task has been demonstrated for C57BL/6J tested with a three-hole and a five-hole task (Bensadoun et al. 2004; Wrenn et al., 2006). Med Associates offers a five-choice panel that inserts into the standard mouse operant chamber.

Motor Learning

Cerebellar learning is a well-established principle that can be measured by repeated testing on a motor task requiring coordination and balance. Learning to walk, swim, ride a bike, or excel at a physical sport are examples of motor learning. Repeated testing on the accelerating rotarod is used to assay motor learning in mice (Lalonde et al., 1995; Gerlai et al., 1996; Le Marec and Lalonde, 1997; Paylor et al., 1998; Hossain et al., 2004). In one standard protocol, the mouse is given daily trials on the rotarod (Figure 4.3*a*), accelerating from 4 to 40 revolutions per minute over a 5-minute period each day. Improvement in performance across training days, as measured by increasing latency to fall from the rotarod, indicates motor learning (illustrated in Figure 6.33).

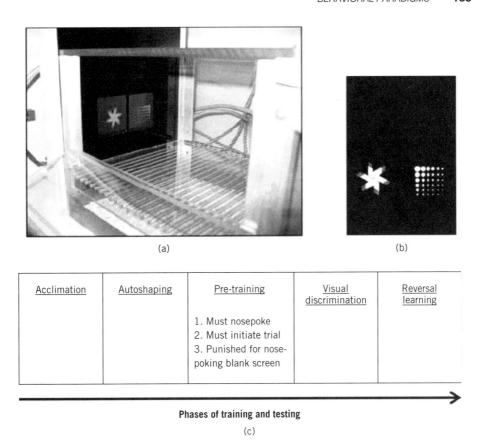

Acclimation	Autoshaping	Pre-training	Visual discrimination	Reversal learning
		1. Must nosepoke 2. Must initiate trial 3. Punished for nose-poking blank screen		

Phases of training and testing

(c)

Figure 6.31 *Operant touchscreen task for mice. (a) A touch sensitive panel replaces the levers or nose-poke holes in a standard Med Associates operant chamber. (b) Equiluminescent stimuli are presented, for example the star on the left window and the pattern of dots on the right window. When the nose of the mouse touches the correct stimulus on the screen, a food pellet is delivered. (c) Training sequence for a visual discrimination and reversal task* [Kindly contributed by Andrew Holmes, from Izquierdo et al. (2006), p. 182.]

Recommendations

The best way to begin assessment of learning and memory in transgenic and knockout mice is to choose several mouse-friendly tasks that require different sets of sensory and motor abilities. This approach allows the investigator to confirm a positive finding in related and unrelated types of tasks, while ruling out false negatives due to simpler explanations, such as motor dysfunctions impairing swimming, or pain threshold differences affecting avoidance learning. An ideal combination of very different tasks includes the Morris water task, contextual and cued fear conditioning, an olfactory discrimination operant task, taste aversion, rotarod learning, and the radial arm maze. These tasks are all well-characterized, have appropriate built-in controls for procedural deficits, employ different sensory modalities, require different motor abilities, and assess different types of learning and memory. Up to three batches of mice may be needed to conduct (1) two or three conceptually different cognitive tasks with the first batch, (2) two or three more tasks with potential carryover effects in the second batch,

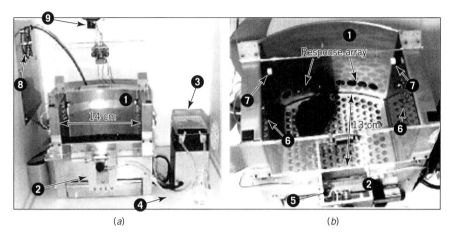

(a)   (b)

Figure 6.32 *Five-choice serial reaction time test of visual attentional function. This attentional task requires the mouse to monitor the light signals over five holes simultaneously. (a) One of the lights is illuminated briefly, and a food pellet is accessible in the corresponding hole for 5 seconds. (b) The mouse must poke its nose through the hole thereby breaking a photocell beam that counts the nose-poke and tallies the correct or incorrect response. On each successive trial, one of the lights is illuminated, in random order. Speed to respond, and number of correct nose-pokes, measure the ability of the mouse to pay attention and make the correct choice.* (Photographs kindly contributed by Dr. Gerry Dawson, Merck Sharp & Dohme Neuroscience Research Centre, Terlings Park, UK, of equipment originally designed and built by Kevin Weston and Rob Barratt, Research Engineering, MSD, updated by Dr. Lawrence Wilkinson, University College London, from *Current Protocols in Neuroscience,* Figures 8.5H1 and 8.5H3.)

and/or (3) replication of a significant genotype difference in the same task on another batch of mice. Positive findings across tasks will strongly corroborate an interpretation of a memory deficit or improvement in a mutant line of mice. Positives may be detected in one or some of these tasks but not others, indicating the type of memory impairment. The insightful investigator then employs additional appropriate tasks to understand the fundamental type of memory affected by the mutation.

TRANSGENICS AND KNOCKOUTS

Transgenic and knockout mice provide an excellent research tool for investigating the anatomical and biochemical mechanisms mediating learning and memory (Silva et al., 1997). The strategy is spoofed in Stephen Spielberg's terrific cartoon television series, *Pinky and the Brain* (illustrated at the beginning of this chapter). Targeted gene mutations have generated many Pinkies, mice deficient in learning and memory, and a small number of cases like The Brain, who appear to show improved learning and/or memory. Individual components of memory processes have been analyzed by mutations in the genes for cortical, hippocampal, and cerebellar development, transcriptional factors, biochemical signaling pathways, proteins involved in synaptic vesicle exocytosis, calcium influx, and synaptic strengthening.

Genes for neurotransmitters and their receptors were mutated in several interesting knockout mice. Glutamate is the major excitatory neurotransmitter in mammalian brain. Glutamate acts at several classes of receptors, some of which participate in

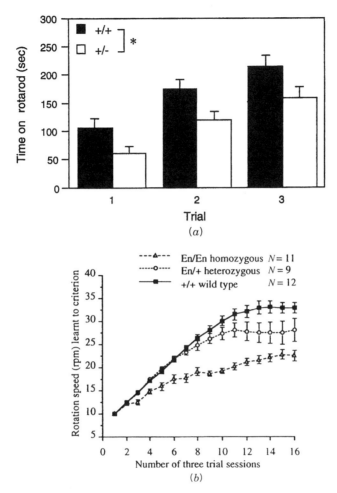

Figure 6.33 *Rotarod motor learning. Performance on the accelerating rotarod task improves over successive test sessions. (a) Both wildtype and mutant mice deficient in lis1, a neuronal migration gene, improved their performance over three consecutive daily 5-minute sessions in which the rotarod speed accelerated gradually from 4 to 40 rpm. Performance by the lis1 mice was significantly poorer than wildtype controls on all three test days, indicating an impairment in cerebellar learning. [Contributed by the author; adapted from Paylor et al. (1999), p. 526.] (b) Improvement in performance on the accelerating rotarod over 16 sessions of 3 trials per session, 2 sessions per day. Performance of mutant mice deficient in the engrailed-2 transcription factor gene En-2 showed less improvement over training sessions as compared to heterozygotes and wildtype controls. [From Gerlai et al. (1996), p. 128.]*

long-term potentiation. The N-methyl-D-aspartate (NMDA) receptor is the primary site for fast excitatory transmission of glutamate actions. NMDA receptor subunit knockout mice showed slower acquisition of the location of the hidden platform in the Morris water task and impaired performance on three probe trials, while performing at close to normal levels in the visible platform task (Tsien et al., 1996b). Knockout mice with NMDA receptor mutations on a promoter specific to the CA1 area of the hippocampus show striking deficits in difficult nonspatial tasks including trace fear conditioning and transverse patterning, while locomotion and feeding were unaffected by the mutation

(Huerta et al., 2000; Rondi-Reig et al., 2001). Transgenic mice overexpressing *NR2B* showed slightly enhanced performance on the Morris water task; however, the Morris visible platform task and other necessary control experiments were not conducted (Tang et al., 1999). Controversy over whether the NR-2B knockouts have increased pain sensitivity emphasizes the importance of conducting the appropriate procedural controls within a batch of mice, in this case for a learning task involving footshock (Tang et al., 2001). Inducible, temporally reversible mutations in the NMDA receptor subtype *NR1* disrupted memory retrieval of fear conditioned stimuli (Cui et al., 2004). Conditioned eyeblink acquisition is impaired in knockout mice with mutations in the NMDA receptor subunit NR2A but not in *NR2C* knockouts, while spontaneous eyeblink and acoustic startle eyeblink were unaffected by the mutations (Kishimoto et al., 1997).

Mice deficient in the metabotropic mGluR2 glutamate receptor performed normally on the Morris water task, although long-term depression (LTD) in the hippocampal mossy fiber-CA3 synapses was abolished (Yokoi et al., 1996). Knockout mice deficient in mGluR5 showed impaired acquisition of the Morris water task and impaired contextual fear conditioning, while auditory cued freezing was normal and locomotor activity in a 5-minute open field test was normal (Lu et al., 1997). Knockout mice deficient in mGluR7 displayed reduced levels of contextual fear conditioning and impaired taste aversion to lithium chloride, while normal phenotypes were observed for pain sensitivity, taste preference to saccharin, and open field activity (Masugi et al., 1999). mGluR8 knockout mice displayed temporally delayed responses on contextual conditioning, and large performance deficits on the Morris water task, that appeared to be caused by visual deficits in the breeding strain, ICR (Gerlai et al. 2002). Knockout mice deficient in the glutamatergic AMPA receptor subunit GluR-A showed loss of associative long-term potentiation but performed normally on the Morris water task (Zamanillo et al., 1999). This growing body of literature illustrates the application of gene mutation techniques to basic research problems such as the elements defining a complex synaptic process. The fundamental hypothesis that long-term potentiation is necessary for learning and memory is thus being tested with the genetic dissection strategy in mice.

In addition to glutamate, other neurotransmitters including acetylcholine, GABA, norepinephrine, serotonin, and galanin play primary or modulatory roles in performance of many memory tasks. One line of muscarinic receptor subtype m1 knockout mice showed performance deficits on a delayed nonmatching to sample working memory task (Anagnostaras et al., 2003). Nicotinic receptor subunit α7 knockout mice were slower to acquire the five choice serial reaction time attentional task and showed higher levels of omitted responses than wildtype littermates (Young et al., 2004). Nicotinic α7 knockout mice, but not other nicotinic receptor subunit mutants, were impaired on acquisition of a fixed ratio appetitive nose-poke operant task when the contingencies increased in difficulty (Keller et al., 2005). Nicotine-enhanced performance on fear conditioning in α7 but not in β2 nicotinic receptor subunit knockout mice (Wehner et al., 2004). Serotonin 5-HT$_{2C}$ null mutants exhibited selective deficits on the Morris water task, along with impaired LTP (Tecott et al., 1998). Null mutants deficient in dopamine β-hydroxylase, the enzyme that synthesizes norepinephrine and epinephrine from dopamine, displayed deficits in some components of contextual fear conditioning and spatial memory (Murchison et al., 2004). Null mutants deficient in monoamine oxidase, the metabolic enzyme that degrades serotonin, norepinephrine, and epinephrine, showed a remarkably selective enhancement of contextual fear conditioning (Figure 6.12*b*), while step-down avoidance learning was normal (Kim et al., 1997). Transgenic mice overexpressing the inhibitory neuropeptide galanin on a dopamine β-hydroxylase promoter were impaired on the more difficult components of learning and memory tasks, including trace fear conditioning, the probe trial of the Morris water maze, and social transmission of food preference (Steiner et al., 2001; Crawley et al., 2002; Kinney et al., 2002; Wrenn et al., 2003).

Nitric oxide, a gaseous neurotransmitter that mediates the actions of GABA and other transmitters, as well as mediating vasodilation, is synthesized by the enzyme nitric oxide synthase. Mice deficient in the gene for endothelial nitric oxide synthase displayed a remarkably faster acquisition of the Morris water task, faster reversal learning, and better long-term retention, while showing normal performance on the visible platform task (Frisch et al., 2000). Heme oxygenase-1, the inducible form of the only known source for carbon monoxide, which is postulated to act as another endogenous gaseous neurotransmitter, was overexpressed in *HO-1* mice (Maines et al., 1998; Morgan et al., 1998). Extensive evaluation of sensorimotor functions in *HO-1* transgenics demonstrated normal body weight, normal pinna flick in response to a loud sound, normal reaching response, normal righting reflex, and normal forepaw grip. Exploratory activity was reduced. No significant differences were detected in performance of a Y-maze exploration task. On the Morris water task, escape latencies remained higher over trial blocks in the transgenics as compared to wildtype controls. Time spent in the training quadrant was significantly less in the transgenics than the wildtype controls. Visible platform testing showed normal swim speed in the transgenics. The authors postulate a developmental role for *HO-1* that affects spatial navigation in adult mice.

Calcium channels open when glutamate binds to NMDA receptors, allowing calcium influx into the postsynaptic neuron. Calcium activates several kinases, including calcium-calmodulin-dependent kinase type II (CaMKII); (Elgersma et al., 2004). Knockout mice deficient in CaMKII exhibit impairments in LTP and in spatial learning on the Morris water task (Silva et al., 1992a, b). Further, αCaMKII knockout mice showed decreased spatial selectivity in a visually cued radial maze, as reflected by decreased selectivity in place cell activity (Cho et al., 1998), and on fear conditioning at a long retention interval of 36 days after training (Frankland et al., 2004a). Transgenic mice expressing a calcium-independent form of mutated CaMKII, generated by the introduction of an aspartate at amino acid 286, failed to learn to navigate in the Barnes maze task (Bach et al., 1995). Using a forebrain-specific promoter combined with a tetracycline transactivator, anatomically and temporally selective mutations in CaMKII produced deficits on the Barnes circular maze (Figure 6.19*b*) and in contextual conditioning (Mayford et al., 1996). These effects were reversed back to normal performance levels when the transgene was suppressed (Mayford et al., 1996).

Calcium influx also activates phosphatases, including calcineurin, a calcium-sensitive serine/threonin phosphatase present in high levels in the hippocampus. Transgenic mice overexpressing a truncated form of calcineurin show impairment in a novel, intermediate form of LTP (Winder et al., 1998). Impairment on visual recognition and spatial tasks was reported for these mice (Mansuy et al., 1998). Mice deficient in calbindin D28K, which regulates the influx of calcium into the postsynaptic neuron, showed impairments in spatial learning and LTP (Molinari et al., 1996). Non-receptor tyrosine kinases, activated by various transmembrane signaling proteins expressed in the hippocampus, include *src, yes, abl,* and *fyn.* Knockout mice with mutations in *fyn* showed impairment on the Morris water task and in conditioned taste aversion learning, while radial maze learning was normal (Grant et al., 1992; Miyakawa et al., 1996; Schafe et al., 1996).

When a neurotransmitter binds to its G-protein coupled receptor, several receptor kinases are activated (Lefkowitz et al., 1992). For example, the NMDA receptor NR2 subunits bind to postsynaptic density-95 protein (PSD-95), a membrane-associated guanylate kinase. Mice with a mutation in PSD-95 were impaired on acquisition of the Morris water task, showing aberrant search patterns and lack of selective quadrant search on the probe trial (Migaud et al., 1998).

Neurotransmitter activation of receptors triggers a sequence of intracellular responses that appear to be necessary for consolidation of long-term memory (Bailey et al., 1996; Abel et al., 1998). Experiments with *Aplysia*, a marine mollusk, *Drosophila*, the fruit fly, and laboratory mice provide evidence for the following cascade of postsynaptic events: (1) neurotransmitter-induced synthesis of cyclic adenosine monophosphate (cAMP) activates (2) the catalytic subunit of protein kinase A (PKA), which phosphorylates (3) cAMP response element-binding protein (CREB) transcription factors, to activate (4) cAMP-inducible genes, initiating (5) further cellular events leading to synaptic growth. Transgenics and knockout mice with mutations in genes regulating each of these events have been tested in learning and memory tasks.

Mutations in protein kinase A (PKA) reveal a critical role for this kinase in long-term memory. Transgenic mice with overexpression of an inhibitory form of the regulatory subunit of PKA show reduced PKA expression in the hippocampus and forebrain, deficits in LTP, deficits in acquisition of the Morris water task, and impaired contextual fear conditioning (Abel et al., 1997). Similarly, *Drosophila* with mutations in PKA showed defects in learning an olfactory avoidance task, with no loss in olfactory acuity or shock reactivity (Goodwin et al., 1997). Mice deficient in MARCKS, the myristoylated alanine-rich C kinase substrate of protein kinase C, showed reductions in acquisition of the Morris water task and reversal learning (McNamara et al., 1998).

A pivotal target protein that is phosphorylated by PKA is cAMP response-element binding protein (CREB). The role of CREB activation in long-term potentiation has been extensively explored in several species (reviewed in Carew, 1996; Silva et al., 1998; Kandel 2001). The CREB/CRE transcriptional pathway is activated during contextual learning (Impey et al., 1998). CREB knockout mice show deficits on several memory tasks, including the Morris water task and social recognition (Bourtchuladze et al., 1994; Kogan et al., 1997). Specific inhibition of CREB in the CA1 region of the dorsal hippocampus produced normal learning of visible and hidden platforms in the Morris water maze, but lack of selective quadrant search on the probe trial (Pittenger et al., 2002). Hippocampal place cell firing during navigation of a visually cued radial arm maze was also impaired in CREB knockout mice (Cho et al., 1998). *Drosophila* expressing a dominant negative CREB transgene did not reach the criterion for long-term memory of the olfactory avoidance task, while olfactory acuity and shock reactivity remained unaffected (Yin et al., 1994). In one amazing experiment, transgenic flies with induced expression of an activator isoform of CREB reached maximum performance after only one training session on an olfactory avoidance task that requires 10 training sessions in wildtype control flies, while olfactory acuity and shock reactivity were similar between genotypes (Yin et al., 1995).

CREB repressor isoforms have been identified that appear to inhibit LTP and may play a role in inhibition of memory formation (De Luca and Giuditta, 1997; Bartsch et al., 1998). Knockout experiments have further investigated the behavioral phenotype of *C/EBP*δ, a transcriptional gene that includes a binding site for CREB on its promoter (Sterneck et al., 1998). Mice deficient in *C/EBP*δ were normal on neurological reflexes, open field behaviors, sensory and motor tasks, the light ↔ dark exploration test for anxiety-related behaviors, and the Morris water task. Contextual conditioning was enhanced in two independent lines of *C/EBP*δ knockout mice. These results indicate a relatively selective increase in fear-related learning in the absence of *C/EBP*δ, and thus that memory can be improved by removal of an endogenous gene product. Targeted gene mutation is providing a tool that may discover ways to improve memory, perhaps by enhancing CREB expression or inhibiting a CREB repressor.

Synaptic proteins mediate many aspects of neurotransmission. Synaptotagmin IV is a synaptic vesicle protein involved in exocytosis. Synaptotagmin IV mutant mice display less freezing to context in a fear conditioning task, and lack of preference for the cued food in social transmission of food preference (Ferguson et al., 2000a). Mutation of the gene for complexion II, another modulator of synaptic vesicle release, resulted in slower acquisition of the Morris hidden platform task but relatively normal selective quadrant search on the probe trial, along with motor deficits including poor rotarod performance and gait abnormalities on the footprint test (Glynn et al., 2003). Synaptic growth and plasticity require morphological and metabolic modifications that may be mediated by serine proteases, including cell adhesion molecules and plasminogen activators. Knockout mice lacking a cell adhesion molecule, *Thy-1*, were normal on the Morris water task, while showing impaired LTP in the dentate gyrus but normal LTP in hippocampal CA1 (Nosten-Bertrand et al., 1996). This finding points to the possibility that cortical input to the dentate gyrus may not be required for some types of spatial learning (Nosten-Bertrand et al., 1996). *Thy-1* knockouts also failed to choose the cued food in the social transmission of food preference test (Mayeux-Portas et al. 2000). Transgenic mice overexpressing urokinase-type plasminogen activator protein were impaired on acquisition of the Morris water task and conditioned taste aversion learning, while swimming and gustatory abilities appear normal (Meiri et al., 1994). This finding is discussed by Dudai and co-workers in terms of the interesting possibility that effective synaptic remodeling depends on a temporal balance in activity of the components of the cascade of events in normal learning (Meiri et al., 1994).

Mutant mouse models of Alzheimer's disease are extensively described in Chapter 13. Memory deficits were detected in many, but not all, lines of mice with mutations that generated brain pathologies similar to Alzheimer's disease, including amyloid overexpression, presenilin mutations, and ApoE variants (Dodart et al., 2002; Higgins et al., 2003; Mineur et al., 2004). Excellent mouse models of Down syndrome and Fragile X syndrome, the most common forms of human mental retardation, are described in detail in Chapter 13.

Multiple, complex biochemical mechanisms mediate learning and memory processes (Kaplan and Abel, 2003). Either too little or too much of one component may disrupt the synaptic cascade and impair the learning process. Considering the dearth of selective pharmacological tools to activate and inhibit each of the components of synaptic transmission and circuitry, knockout mice provide an important resource to study the intracellular events underlying learning and memory. The small sampling of results included above indicates that careful phenotyping of mutant mice can reveal selective functions of genes mediating individual components of cognitive processes.

BACKGROUND LITERATURE

Braveman NS, Bronstein P, Eds. (1985). *Experimental Assessments and Clinical Applications of Conditioned Taste Aversions*. New York Academy of Sciences, New York.

Bushnell PJ (1998). Behavioral approaches to the assessment of attention in animals. *Psychopharmacology* **138**: 231–259.

Cohen NJ, Eichenbaum H (1994). *Memory, Amnesia, and the Hippocampal System*, 2nd ed. MIT Press, Cambridge.

Colgan PW (1983). *Comparative Social Recognition*. Wiley, New York.

Dudai Y (1989). *The Neurobiology of Memory: Concepts, Findings, Trends*. Oxford University Press, New York.

Eichenbaum H (2002). *The Cognitive Neuroscience of Memory: An Introduction*. Oxford University Press, New York.

Elgersma Y, Sweatt JD, Giese KP (2004). Mouse genetic approaches to investigating calcium/calmodulin-dependent protein kinase II function in plasticity and cognition. *Journal of Neuroscience* **24**: 8410–8415.

Fanselow MS, Poulos AM (2005). The neuroscience of mammalian associative learning. *Annual Review of Psychology* **56**: 207–234.

Gallagher M (2002). *Handbook of Psychology, Biological Psychology*. Wiley, New York.

Kaplan MT, Abel T (2003). Genetic approaches to the study of synaptic plasticity and memory storage. *CNS Spectrum* **8**: 597–620.

LeDoux JE (1996). *The Emotional Brain*. Simon and Schuster, New York.

LeDoux J (2002). *Synaptic Self: How Our Brains Become Who We Are*. Viking, New York.

Lipp HP, Wolfer DP (1998). Genetically modified mice and cognition. *Current Opinions in Neurobiology* **8**: 272–280.

Lynch G (1986). *Synapses, Circuits, and the Beginnings of Memory*. MIT Press, Cambridge.

Martinez JL, Kesner, RP, Eds. (1998). *Neurobiology of Learning and Memory*. Academic Press, San Diego.

Martinez JL, Jensen RA, Messing RB, Rigter H, McGaugh JL (1981). *Endogenous Peptides and Learning and Memory Processes*. Academic Press, New York.

Mayford M, Bach ME, Huang YY, Wang L, Hawkins RD, Kandel ER (1996). Control of memory formation through regulated expression of a CaMKII transgene. *Science* **274**: 1678–1683.

Mineur YS, Crusio WE, Sluyter F (2004). Genetic dissection of learning and memory in mice. *Neural Plasticity* **11**: 217–240.

Morris RGM (2001). Episodic-like memory in animals: Psychological criteria, neural mechanisms and the value of episodic-like tasks to investigate animal models of neurodegenerative diseases. *Philosophical Transactions of the Royal Society B: Biological Sciences* 356: 1453–1465.

Schmahmann JD (1997). *The Cerebellum and Cognition*. Academic Press, San Diego.

Seabrook GR, Rosahl TW (1999). Transgenic animals relevant to Alzheimer's disease. *Neuropharmacology* **38**: 1–17.

Silva AJ, Smith AM, Giese KP (1997). Gene targeting and the biology of learning and memory. *Annual Review of Genetics* **31**: 527–546.

Squire LR, Zola SM (1996). Structure and function of declarative and nondeclarative memory systems. *Proceedings of the National Academy of Sciences USA* **93**: 13515–13522.

Steinmetz JE (1998). The localization of a simple type of learning and memory: The cerebellum and classical eyeblink conditioning. *Current Directions in Psychological Science* **7**: 72–77.

Ungerleider LG (1995). Functional brain imaging studies of cortical mechanisms for memory. *Science* **270**: 769–775.

Wenk GL (1997). Learning and Memory. In *Current Protocols in Neuroscience*, JN Crawley, CR Gerfen, R McKay, MW Rogawski, DR Sibley, P Skolnick, Eds. Wiley, New York, Unit 8.5.

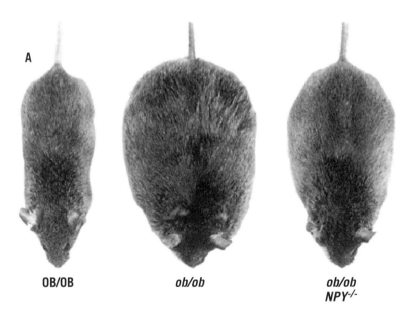

Neuropeptide Y knockout mouse (left), ob/ob *leptin-deficient obese mouse (middle),* $NPY \times$ ob/ob *double mutant mouse (right). Experiment described on page 182 (Erickson et al., 1996b).*

7

Feeding and Drinking

HISTORY AND HYPOTHESES

Feeding is the most basic survival behavior. Every living creature eats, each in its own way. Ingestion of nutrients and water provides the necessary raw materials that allow the organism to maintain its unique structural, biochemical, and physiological integrity against the forces of entropy. Complex mechanisms regulating feeding and drinking have evolved in mammalian species. Specialized brain circuits signal hunger and thirst, initiate foraging behaviors that lead to ingestion, maintain ingestive behaviors until satiation is reached, and then turn off feeding and drinking behaviors. Neurotransmitters regulating ingestive behaviors are multiple and redundant. Unlike Franz Kafka's hunger artist, very few of us are able to give up eating for long.

Circulating levels of insulin and glucose are monitored by chemoreceptors in the liver and hypothalamus that trigger meal initiation (Woods et al., 1998). Nutrients are then absorbed from the ingested food into the bloodstream for subsequent storage and use in cells throughout the body. Energy is mobilized by burning carbohydrates and metabolizing fat stored in adipose tissue.

Neuroanatomical circuitry mediating thirst and appetite includes gastrointestinal feedback signals, hindbrain and hypothalamic nuclei, and their afferent and efferent projections. The first great discoveries explicating brain sites mediating feeding were hypothalamic lesions that triggered or blocked ingestive behaviors (Anand and Brobeck, 1951; Stellar, 1954; Teitelbaum and Epstein, 1962; Epstein, 1971; Danguir and Nicolaidis, 1980). Lesions of the lateral hypothalamus generated hypophagic rats. These animals stopped eating for many days after the lesion, losing weight to the point of death. Lesions of the ventromedial hypothalamus generated hyperphagic rats. These animals increased their meal size and frequency, gaining body weight to the point of severe obesity.

The multiple neurotransmitters that influence ingestion indicate a complex neuropharmacological regulation of thirst and appetite. Many excellent reviews describe the interactions of peripheral gastrointestinal signals with hormones and brain neurotransmitters that integrate hunger, satiety, and energy balance (Blundell, 1977; Morley and Levine, 1982; Stanley and Leibowitz, 1985; Gibbs et al., 1993; Leibowitz, 1995; Crawley, 1995; Rowland and Kalra, 1997; Curzon et al., 1997; Campfield et al., 1998; Woods et al., 1998; Heinrichs et al., 1998; Inui, 1999; Beck, 2000; Ahima and Flier, 2000; Smith, 2000; Rodgers et al., 2002; Butler and Cone, 2002; Friedman, 2002; Segal-Lieberman et al., 2003; Eisenstein and Greenberg, 2003; Drazen and Woods, 2003; Leibowitz and Wortley, 2004; Moody and Merali, 2004; Olszewski and Levine, 2004, Moran, 2004; Woods, 2004). A large number of neuropeptides and classical transmitters modulate food consumption (Table 7.1). Serotonin, cholecystokinin, corticotropin releasing factor, α-melanocyte stimulating hormone, glucagon-like peptide, bombesin, and nociceptin/orphanin FQ are some of the transmitters that inhibit feeding when administered centrally. Neuropeptide Y, galanin, norepinephrine, opiates, ghrelin, and orexins stimulate feeding. Angiotensin regulates drinking and salt consumption (Stellar and Epstein 1991; Schulkin, 1991).

Satiation is the physiological process that underlies the cessation of eating. Satiation is an active inhibitory process that requires complex central neural integration (Richter, 1922; Blundell, 1979; Kulkosky et al., 1982; Stricker and Verbalis, 1990; Smith, 1998b; Moran, 2004). Because so many Americans are overweight, interventions that reduce appetite or promote satiety are of high priority in the United States. Our understanding of underlying mechanisms derives from research into the postprandial stages of satiety (Figure 7.1), especially the elegant studies of Gerard Smith and co-workers in the Bourne Laboratory at Cornell University in White Plains, New York (Gibbs et al., 1993; Smith et al., 1985). First, food entering the digestive tract stimulates the release of digestive enzymes and triggers gastrointestinal contractions. Released neurotransmitters such as cholecystokinin then activate receptors throughout the stomach and upper small intestine (Gibbs et al., 1993; Moran and McHugh, 1988; Smith, 1998b). This sensory information is conveyed to the brain by visceral nerves, particularly the vagus nerve (Smith et al., 1985; Edwards et al., 1986; Moran and McHugh, 1988; South and Ritter, 1988; Moran et al., 1997). Sensory fibers of the vagus nerve innervate the gastrointestinal tract and send information to the area postrema and nucleus tractus solitarius of the hindbrain (Smith et al., 1985; Edwards et al., 1986; Crawley

TABLE 7.1 Some Neurotransmitters That Modulate Feeding[a]

Increase Ingestion	Decrease Ingestion
Norepinephrine	Serotonin
Neuropeptide Y	A-Melanocyte stimulating hormone
Galanin	Melanocortin
Opiates	Corticotropin releasing factor, urocortin
Orexin-A	Cholecystokinin
Growth hormone releasing factor	Nociceptin/orphanin FQ
Ghrelin	Bombesin
Melanin-concentrating hormone	Glucagon-like peptide-1

Source: Adapted from Campfield et al. (1998), p. 1386, and Rodgers et al. (2002) p. 304.
[a]Some of the receptor subtypes and transporters for these neurotransmitters provide potential therapeutic targets for the treatment of human obesity.

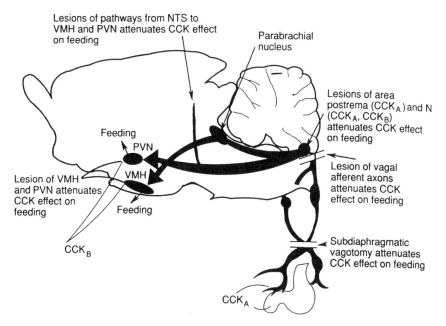

Lesions of pathways from NTS to VMH and PVN attenuates CCK effect on feeding

Parabrachial nucleus

Lesions of area postrema (CCK_A) and N (CCK_A, CCK_B) attenuates CCK effect on feeding

Feeding

PVN

Lesion of VMH and PVN attenuates CCK effect on feeding

VMH

Lesion of vagal afferent axons attenuates CCK effect on feeding

Feeding

CCK_B

Subdiaphragmatic vagotomy attenuates CCK effect on feeding

CCK_A

Figure 7.1 *Many anatomical pathways in the hypothalamus, forebrain, hindbrain, sensory nerves, and gastrointestinal tract contribute to the regulation of feeding behaviors. Diagrammed are some of the pathways mediating the satiety syndrome induced by peripherally administered cholecystokinin.* [Dourish (1992), p. 236.]

and Schwaber, 1984). Afferent axons from the hindbrain project to brain regions mediating feeding behaviors, particularly the hypothalamus (Ritter, 1984; Crawley and Kiss, 1985; Stricker and Verbalis, 1990). The mammalian hypothalamus is a key structure that integrates vagal sensory information from the peripheral gastrointestinal nerves, chemosensory information from circulating hormone levels, and cognitive input from other brain regions, to initiate central processes that lead to feeding and satiation.

The background literature at the end of this chapter includes several excellent books and recent review articles on ingestive behaviors. Most of the literature on rodent ingestive behaviors derives from research with rats. Analogous studies with mice are less complete.

BEHAVIORAL TESTS

Twenty-Four-Hour Consumption

The simplest approach to quantitating feeding and drinking behaviors is to measure 24-hour consumption of standard rodent chow and drinking water in the home cage. As shown in Figure 7.2, commercially available laboratory rodent pellets are weighed on a gram balance, to the nearest 100 mg. Food is weighed before it is placed in the home cage. At the same time the next day, food is again weighed. The difference in weights represents consumption over the 24-hour period (Figure 7.3). Normal home cage feeding can be measured by weighing the food and calculating consumption over any consistent time interval, for example, every 2 weeks (Figure 7.4), every 3 days,

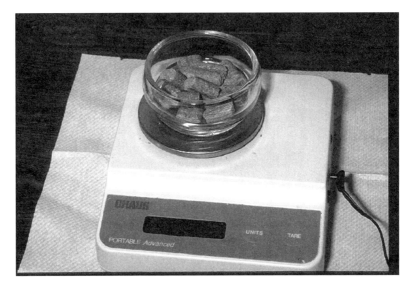

Figure 7.2 *Typical gram balance used to quantitate daily consumption of standard rat chow. Chow is contained in a heavy glass jar with rounded walls, designed to remain upright in the home cage and retard spillage. A standardized amount of food is weighed before placement in the home cage. Remaining food is weighed at the end of the defined consumption period.* (Photograph contributed by the author.)

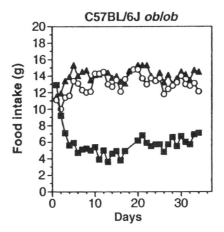

Figure 7.3 *Daily food intake in ob/ob mice treated with the mouse OB protein, leptin. C57BL/6J ob/ob mice are obese animals with a natural mutation in the gene for the leptin protein. Standard rodent chow was available in the home cage. Consumption was measured daily. Triangles indicate typical daily consumption of chow by untreated ob/ob mice. Circles indicate typical daily consumption of chow by ob/ob mice treated with daily intraperitoneal injections of phosphate-buffered saline vehicle. Squares indicate typical daily consumption of chow by ob/ob mice treated with daily intraperitoneal injections of leptin, 5μ g/g/day. Leptin reduced consumption to levels typical of normal C57BL/6J mice, effectively reversing the phenotype of the ob/ob mutation on feeding behavior.* [Adapted from Halaas et al. (1995), p. 544.]

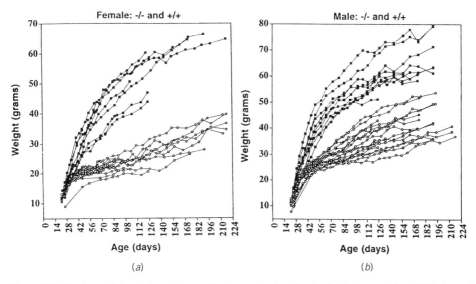

Figure 7.4 *Body weight gain in melanocortin-4 receptor knockout mice between 21 and 220 days of age. Wildtype mice (open circles) of both genders are approximately 10 grams at 21 days, gradually increasing to a maximum of 30 to 50 grams by age 220 days. Mutant mice deficient in the melanocortin-4 receptor (closed circles) begin life at normal body weights but reach 60 to 80 grams by age 220 days.* [From Huszar et al. (1997), p. 134.]

every day, every 8 hours, or at hourly intervals over a 6-hour time period (Reidelberger et al., 1991; Koegler and Ritter, 1998).

Spillage is a critical issue. Rodents tend to pick up food and carry it to another location in the cage, where they nibble the pellets held in their forepaws. Leftover chunks fallen to the cage floor will falsely inflate the measure of consumption. While spillage cannot be totally accounted for, the careful experimenter tries to gather up the larger pieces of spilled food and replace them in the food container for weighing. Rodents tend to turn a lightweight food container upside down. Heavy, rounded glass jars (see Figure 7.2) prevent most rodents from overturning the food container. Powdered chow, or powdered chow mixed with water to make a wet mash, avoids some of the problems of spillage (Halford and Blundell, 1998). Specially designed recessed food cups and perforated food cup lids will minimize spillage of powdered chow. Small pellets delivered individually in response to lever presses in an operant chamber (see Chapter 6) circumvent spillage problems.

Drinking is quantitated by measuring consumption of standard tap water. Liquid diets provide an approach to measure feeding that minimizes spillage and improves measurement accuracy. Sweetened condensed milk, sucrose, glucose, vegetable oil, and specially formulated liquid diets are commercially available. Liquid diets are prepared fresh daily. The simplest method for quantitating drinking of water or a liquid diet is to measure the fluid volume in the drinking bottle before it is placed in the home cage, and again 24 hours later. Measurement is performed by pouring the liquid into a graduated cylinder. Calibrated home cage drinking bottles, similar to graduated cylinders, are commercially available (Wellman and McMahon, 1998). Calibrated drinking tubes, for example, of 30 ml total volume for mice, allow fluid consumption to be observed without disturbing the animal, and recorded at any time interval.

Consumption is monitored at intervals from a distance, without losing the volume that inevitably drips from the drinking spout when the water bottle is removed from the home cage. Methods for quantitating suckling and milk ingestion in pre-weanling rodent pups are available (Young et al., 1996; Fride et al., 2003; Swithers, 2003).

Twenty-four-hour consumption is the least sensitive measure of feeding and drinking behaviors. It takes no account of meal size, meal frequency, nutrient preference, taste preference, or circadian rhythms. Drug treatments with time courses in minutes or hours may be undetectable. Because 24-hour consumption is an average of feeding both during the treatment and afterward, the animal may eat and drink sufficient amounts after the drug effects wear off to compensate for short-term effects of the treatment.

Restricted Daily Access

Short-acting treatment effects can best be detected in tests that restrict daily consumption to a few hours at the same time each day (Moran et al., 1992). Daily intake reaches a stable asymptote after several days of restricted access. Body weight is monitored daily to ensure that consumption is sufficient to maintain body weight in accordance with guidelines for animal care and use. For example, standard rodent chow and drinking water are placed in the home cage for 4 hours per day. Time periods as short as 1 hour may be optimal (Flanagan et al., 1992). Access periods are designed to match the temporal duration of a short-acting treatment. The higher baseline consumption obtained in restricted daily access paradigms yields more sensitive detection of treatments that decrease feeding or drinking.

Specialized Diets and Choice Tests

Special diets are useful to address questions of preference for particular nutrients. For example, studies of genes regulating fat consumption may be optimized by using a diet of rodent pellets that are high in fat, low in carbohydrate, and low in protein, or a liquid diet with a high-fat concentration (Sills and Vaccarino, 1991; Barton et al., 1995; Leibowitz et al., 1998). The special diet can be continuously available, for measurement of 24-hour consumption, or available during restricted access periods (Figure 7.5). Since rodents avoid new foods, any novel diet will require habituation before the start of experimentation, to prevent food neophobia from influencing the experiment.

Pellet and liquid diets that are high in fat, high in carbohydrate, and/or high in protein are commercially available from several commercial sources (Bio-Serve, Frenchtown, NJ 08825; 908-996-2155, *www.bio-serv.com*; Dyets Inc., Bethlehem, PA 18017, 800-275-3938, *www.dyets.com*; Research Diets Inc., New Brunswick, NJ 08901, 732-247-2390, *www.researchdiets.com*). Formulated chow diets are available from animal supply companies such as Purina and Harlan. Specialized powdered chow and liquid diets are formulated to provide nutritionally complete rodent diets. Custom diets are enriched in the particular macronutrient, vitamin, or mineral of interest to the investigation (Corwin et al., 1991).

Our laboratory and others have used highly palatable special diets over short test periods to measure consumption after central administration of short-acting treatments in rats and mice (Figure 7.6). Sweetened condensed milk, sucrose, cookies, chocolate chips, and breakfast cereals are generally preferred by rodents. For example, we

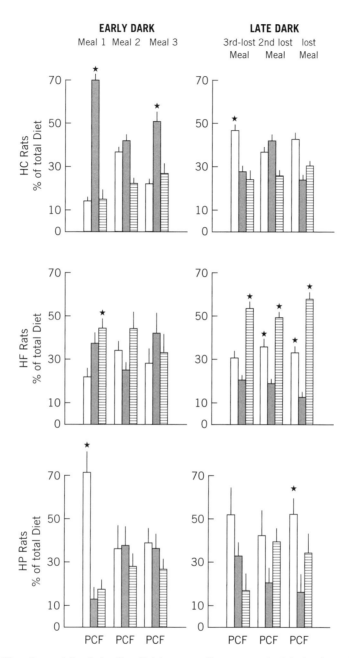

Figure 7.5 *Circadian variation in feeding. Total consumption was greatest during the early dark phase of the feeding cycle. Rats were given a choice of protein (P), carbohydrate (C), and fat (F). Individual preferences for protein (HP = high protein preferring rats), carbohydrate (HC = high carbohydrate preferring rats), and fat (HF = high fat preferring rats) were demonstrated among a group of normal adult male Sprague-Dawley rats.* [From Shor-Posner et al. (1994), p. R1397.]

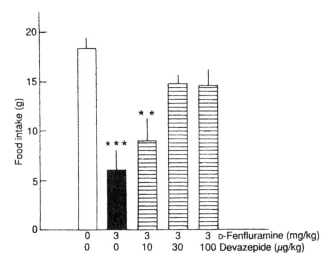

Figure 7.6 *Consumption of a highly palatable food source by rats over a 30-minute test period was in the range of 15 to 20 grams, providing a strong feeding signal against which to test drug treatments for anorectic effects. Fenfluramine, which blocks serotonin activity in the brain, significantly inhibited consumption. Treatment with a cholecystokinin CCK-A receptor antagonist attenuated the anorectic action of fenfluramine, suggesting an interaction between peripheral cholecystokinin and central serotonin systems in the regulation of feeding.* [From Cooper (1992), p. 266.]

employed a wet mash of sugar cookies (Nilla Wafers, Nabisco Brands, Inc., East Hanover, NJ; 14% fat, 79% carbohydrate, 7% protein) to investigate the role of the neuropeptides cholecystokinin and galanin in feeding in rats and mice (Crawley et al., 1986, 1990; Corwin et al., 1993; Crawley and Corwin, 1994). For mice, a mash of 1 cookie/5 ml distilled water is placed in the home cage overnight, several days before the start of the experiments, to ensure that the novel food becomes familiar. On the day of testing, 10 grams of cookie mash in a shallow plastic cup is available in a small, empty cage. Treatments are microinjected into the lateral ventricles, the paraventricular nucleus of the hypothalamus, or intraperitoneally, as appropriate to the experimental questions. The mouse is placed into the test cage with the palatable food source. Duration of the access period is 30 minutes. Mash weight is recorded before and after the 30-minute test period. This short access period for a highly palatable food allows detection of the dramatic increase in consumption of the palatable food induced in rats by central administration of galanin, which has an in vivo half-life under one hour. Galanin treatment produces a significant short-term increase in consumption in normal satiated rats that are maintained in the home cage with free access to lab chow and water (Figure 7.7). Similarly this method allows detection of the robust decrease in food consumption induced by peripheral administration of cholecystokinin, a neuropeptide with a similarly short half-life in vivo, in rats and mice after an overnight fast (Crawley and Corwin, 1994). Trevor Robbins, Barry Everitt, and co-workers at the University of Cambridge employ a choice of foods such as cheddar cheese and chocolate chip cookies dissolved in milk, presented in round, porcelain food bowls in the center of an illuminated open field, over a 10-minute test session to food-restricted rats (Cole et al., 1988). This test allows a comparison of food preferences in a novel, somewhat aversive environment.

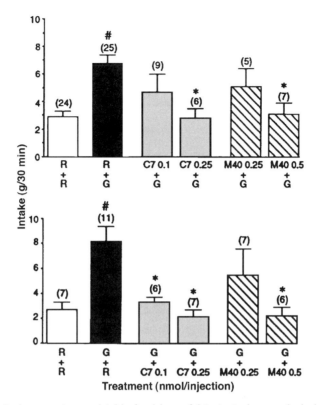

Figure 7.7 *Limited access to a palatable food is useful to test pharmacological responses. The neuropeptide galanin (G), administered into the paraventricular nucleus of the hypothalamus of rats, increased consumption of a high-carbohydrate and high-fat cookie mash, as compared to Ringer's vehicle control microinjections (R). Pretreatment with C7 or M40, two peptidergic galanin receptor antagonists, blocked galanin-induced feeding, demonstrating the in vivo antagonist actions of C7 and M40.* [From Corwin et al. (1993), p. 1530.]

Macronutrient Selection

Rodents have a choice of foods in their natural environment. Choosing the optimal nutrients for health and survival is essential. Mechanisms underlying these choices in rodents may offer insights into the unhealthy food choices often made by humans. Macronutrient selection tests analyze the choice of foods of various nutritional compositions over time. Robin Kanarek at Tufts University presents a step-by-step protocol for quantitating consumption of specialized diets high in carbohydrate, high in fat, and high in protein (Kanarek 2004).

Sarah Leibowitz and co-workers at Rockefeller University employ an ingenious method to measure specific nutrient consumption in rats (Kyrkouli et al., 1990; Shor-Posner et al., 1994; Smith et al., 1997; Leibowitz et al., 1998). Separate sources of pure protein, carbohydrate, and fat, supplemented with vitamins and minerals, are freely available in the home cage. Each single nutrient source is connected to a pan balance (Figure 7.8). Crumbs and spillage generally remain on the pan so that their weight is not included as consumed food (Strohmayer and Greenberg, 1994). Output from the three balances is stored in a computer system that allows continuous quantitation of

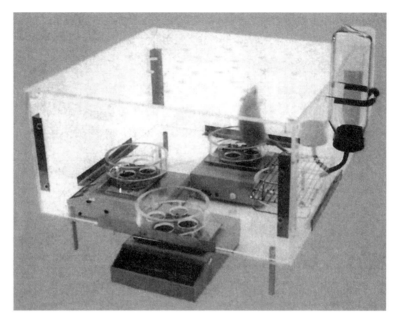

Figure 7.8 *Macronutrient presentation and continuous consumption monitoring apparatus.* (Photography kindly contributed by Columbus Instruments, Columbus, Ohio.)

each macronutrient over time, in the home cage, as the animal continues its normal behaviors throughout the circadian cycle. The animal has continuous opportunity to choose from among the three individual pure nutrient sources. This system quantitates relative preference for protein versus carbohydrate versus fat. Similar equipment can be used to compare consumption of mixed diets that are high in one nutrient. This method is optimal for investigations into mechanisms underlying nutrient selection, meal size, circadian factors, and treatments that target specific food preferences (Thibault and Booth, 1998).

Preferences for specific nutrients can also be quantitated with choice paradigms, using two calibrated water bottles containing liquid diets of differing nutrients. As described below, taste is evaluated by taste preference tests, using different flavors added to water in two or more calibrated water bottles. Food preferences are also measured in choice tests with feeding jars containing powdered food mixed with flavors such as cocoa, cinnamon, thyme, and turmeric, before beginning the social transmission of food preference cognitive task (Bunsey and Eichenbaum, 1995; Wrenn et al., 2003). Valerie Bolivar and Lorraine Flaherty at the Genomics Institute of the Wadsworth Center in Troy, New York, investigated genes mediating novel food preferences to cinnamon and cocoa in the BXD recombinant inbred strain set, finding a significant QTL on chromosome 8 (Bolivar and Flaherty, 2004). Taste aversion tests are discussed below and in the context of learning and memory in Chapter 6.

The *cafeteria diet* is designed to induce obesity over a several week period (Table 7.2). Access is provided to a set of highly palatable and highly caloric foods. Combinations of sweetened condensed milk, sucrose, sausages, peanuts, cheeses, chocolate, and cookies, are offered along with normal chow (Ingle, 1949; Himms-Hagen et al., 1986; Wainwright and Francey, 1987; Trvdik et al., 1997; Pedrazzi et al.,

TABLE 7.2 Evaluation of Food Intake after 1 and 2 Months of Cafeteria Diet[a]

	Body Weight (g)		
	Co	Nov	Ov
1 Month of cafeteria diet			
	202.2 ± 4.7	187.9 ± 5.1	203.8 ± 0.6
Day 1			
Pellet	67.5 ± 2.3	23.5 ± 3.7	28.6 ± 3.0
Sausages		100.2 ± 6.7	101.1 ± 5.5
Day 2			
Pellet	73.1 ± 1.7	36.9 ± 4.1	50.0 ± 3.4
Cakes	—	44.4 ± 5.8	47.1 ± 4.5
Day 3			
Pellet	74.1 ± 1.0	31.1 ± 2.4	33.0 ± 2.8
Condensed milk	—	127.4 ± 8.6	121.4 ± 10.1
2 Months of cafeteria diet			
	442.5 ± 4.7	453.7 ± 1.9	$522.9 \pm 5.9^{*}$
Day 1			
Pellet	72.7 ± 2.8	23.8 ± 2.4	28.6 ± 3.4
Sausages	—	153.0 ± 17.5	35.7 ± 12.9
Day 2			
Pellet	77.6 ± 2.5	49.3 ± 8.0	57.2 ± 4.7
Cakes	—	54.6 ± 5.6	47.9 ± 4.4
Day 3			
Pellet	77.6 ± 1.5	24.1 ± 1.7	29.2 ± 2.8
Condensed milk	—	142.8 ± 15.9	110.5 ± 15.7

Source: From Pedrazzi et al. (1998), p. 159.

[a]The cafeteria diet is designed to induce obesity in rodents. The animal is presented with a choice of several highly palatable foods. Sausages, cakes, and condensed milk were presented along with normal chow. After two months on the cafeteria diet, body weights are significantly higher (*) in a subset of the overfed rats (Ov), as compared to control rats given chow pellets only (Co). Interestingly another subset of rats on the cafeteria diet ate as much but did not gain as much weight (Nov).

1998). This model of human obesity is useful for analyzing mechanisms regulating overeating, and for screening antiobesity drugs. Further, a line of Sprague-Dawley rats that develop cafeteria-diet-induced obesity, and a line of rats that are cafeteria-diet-resistant, have been maintained for several generations (Levin et al., 1998), indicating a genetic component to an underlying predisposition to develop obesity on a high-carbohydrate and high fat diet.

Continuous Monitoring of Microstructural Analysis of Rodent Feeding Behavior

Many interesting behaviors contribute to the ultimate consumption of food. Foraging behaviors in the wild and in laboratory burrows have been elegantly described for mice (Perrigo and Bronson, 1985; Whishaw et al., 1990). Competition for limited food resources in the natural environment requires interactions between appetitive, cognitive, and social behavior processes. Microstructural analysis of the pattern of feeding yields valuable information about meal size, rate of consumption, intermeal interval, and orofacial motor components of ingestion (Burton et al., 1981; Cooper and Francis, 1993; Cooper et al., 1990, 1996b; Davis and Smith, 1988; Houpt and Frankmann, 1996; Rushing et al., 1997). Independent analysis of these components of ingestive behavior can yield important information about larger issues relevant to motivation, appetite, and satiation.

Methods for continuous quantitation of the separate elements of feeding behaviors (Strohmayer and Smith, 1987; Halford and Blundell, 1998 Fox et al., 2001) provide considerable insight into the exact aspect of feeding that is influenced by a single gene mutation. Time-sampling procedures of a videotaped feeding session quantitate parameters including mean local eating rates, mean intake per eating behavior episode, mean eating bout length, and time interval between bouts of feeding. Behavioral codes for scoring the videotape allow the observer to record behavioral categories over time, including biting, gnawing, swallowing, licking the water bottle nozzle, grooming, scratching, sniffing, twitching of vibrissae, locomotion, rearing, and resting periods of inactivity.

Computer-assisted lickometer systems for rats and mice tally licks over time for a liquid diet, or nose-pokes into a food hopper over time for a solid diet (Cooper et al., 1990, 1996b; Houpt and Frankmann, 1996; Rushing et al., 1997; Fox et al., 2001). Meal parameters include quantitation of meal size, meal duration, mean lick rate, intermeal interval, bursts of ingestion, burst size, interburst interval, clusters, cluster size, and intercluster interval. Software calculates the relative percentages of session time composed of interlick intervals, interburst intervals, and intercluster intervals (Houpt and Frankmann, 1996; Rushing et al., 1997).

Sensitivity of the meal pattern apparatus is increased by including a nest box or a partitioned area in which the animal prefers to spend most of its time (Nicolaidis et al., 1979; Ladenheim et al., 2002; Figure 7.9). Since the food hopper is in another area of the cage, the animal emerges from the niche only to eat. This method reduces variability and baseline levels of feeding due to occasional random nibbling (Nicolaidis et al., 1979; Rushing et al., 1997). Electronic lickometers and software that were custom designed for rats (Ackroff and Sclafani, 1998; Moran et al., 1998; Aja et al., 2001) have been redesigned for mice and are commercially available (Coulbourn Instruments, Med Associates; DiLog Instruments, Inc., Tallahassee, Florida).

Ontogeny of Feeding

Methods are available for measuring ingestion at young ages. Consumption of a liquid diet is measured for rat pups at postnatal days 5–18 (Hall and Bryan, 1980; Kowalski et al., 1998). Pups are placed in an incubator, away from the mother, for 4 hours, inducing a mild state of food and water deprivation. The pup is voided of urine and feces and weighed. A solution of half-milk and half-cream or 10% glucose is squirted onto a Kimwipe tissue. The pup is placed on the tissue in a test chamber for 20 minutes. Latency to begin licking is recorded. Total intake is measured as the gain in body weight of the pup, to the nearest 0.01 gram.

Sham Feeding

Sham feeding distinguishes mechanisms triggered by food in the stomach and intestines versus all other mechanisms that influence feeding. Ingested liquid food is prevented

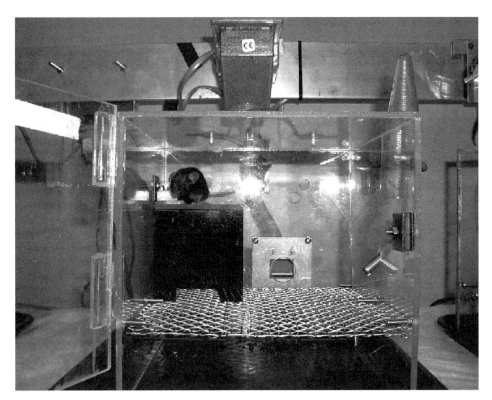

Figure 7.9 *Feeding tests are conducted in home cages fitted with calibrated liquid food dispensers. The liquid diet feeding tube, at right, is distant from a niche, the black partitioned nesting area at left, to minimize nibbling.* [Photograph kindly contributed by Tim Moran, Johns Hopkins University, Baltimore, MD.]

from accumulating in the stomach or emptying into the small intestines by a surgically implanted gastric fistula (Young et al., 1974; Gibbs et al., 1993; Smith 1998a,b). Ingestion increases dramatically when food is not allowed to accumulate in the gastrointestinal tract, due to the lack of peripheral feedback satiety signals to the brain.

Satiety Syndrome

Gerry Smith and Jim Gibbs of Cornell University have greatly enhanced our understanding of meal pattern and satiety (Gibbs et al., 1973; Smith, 1982, 1996, 2000). The postprandial satiety sequence in rodents is the series of behaviors that follows cessation of a meal, including grooming and behavioral sedation (Gibbs et al., 1973; Antin et al., 1975; Smith 1998b). Discrete stages of the satiety syndrome are quantitated, to describe the normal pattern of behavioral changes that accompany the cessation of feeding (Young et al., 1974; Gibbs et al., 1973, 1993; Smith, 1982, 1996, 1998a; Crawley and Schwaber, 1984; Crawley et al., 1986; Dourish, 1992; Moran, 2004). As the rodent stops eating, locomotor activity levels decline, and grooming behaviors rise. Complete behavioral inactivity for several minutes or hours follows, representing postprandial depression. Stages of the satiety syndrome are diagrammed in Figure 7.10.

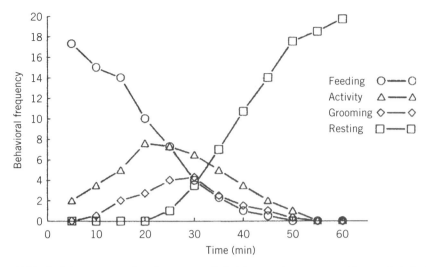

Figure 7.10 *Satiety sequence in rats, as quantitated by an observer recording each behavioral component over 20 observations per 5-minute time bin, for a total of 60 minutes. Four mutually exclusive categories define the satiety sequence: (1) feeding; (2) locomotor activity, which increases and then declines as feeding declines; (3) grooming, which increases and then declines as feeding declines; (4) resting, and the postprandial period of behavioral inactivity after consumption of a meal.* [From Dourish (1992), p. 239.]

Conditioned Taste Preference and Conditioned Taste Aversion

Conditioned taste preference tests measure aspects of taste, feeding, and learning (Booth, 1985; Sclafani, 1995; Ackroff and Sclafani, 1998). Flavors are positively conditioned by temporal pairing with nutrients. Flavor preference learning is accelerated by administering a nutrient solution into the stomach immediately after oral ingestion of a flavored solution. Two bottles are presented, containing distinctly flavored solutions such as saccharine or cherry flavoring. Drinking of solution 1 is paired with intragastric infusion of a liquid nutrient diet. Drinking of solution 2 is paired with intragastric infusion of water. A lickometer system detects number of licks from each bottle. Preference for the nutrient-reinforced flavor solution 1 develops over several daily sessions. Conditioned taste preference test equipment was designed for rats, illustrated in Figure 7.11. Intragastric nutrient infusions for conditioned taste preference in mice has been perfected by Anthony Sclafani at City University of New York (Sclafani and Glendinning, 2003).

Conditioned taste aversions develop rapidly in rats (Garcia et al., 1955; Cappell and LeBlanc, 1977; Dacanay and Riley, 1982; Bernstein et al., 1983; Braverman and Bronstein, 1985; Garcia, 1989; Riley, 1998). Flavors are negatively conditioned by temporal pairing with a drug treatment, which induces interoceptive cues of malaise, abdominal cramps, or sickness. The animals are first water-deprived for 24 hours. Rather than intragastric infusion of a nutrient versus water, the animal is injected with an aversive drug such as lithium chloride versus water. The taste that was paired to the aversive drug produces subsequent suppression of consumption of the paired flavor. Taste aversion conditioned by lithium chloride has been adapted for mice, with greater taste aversion reported in DBA/2J than C57BL/6J (Risinger and Cunningham, 2000).

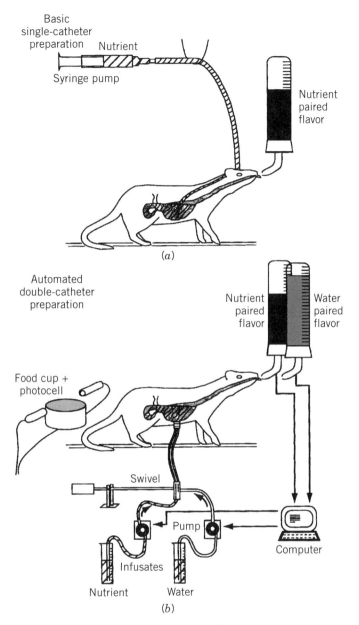

Figure 7.11 *Conditioned taste preference. Rodents quickly learn to prefer one flavor over another flavor in a semiautomated system. Intragastric infusion of the nutrient or water solution is administered through a subcutaneously implanted gastric catheter. The animal is presented with a flavored solution to drink and infused with a nutrient solution through the catheter into the stomach. The animal is similarly presented with a different flavored solution to drink and infused with a non-nutrient water solution through the catheter into the stomach. When given a choice between the two flavors, the animal will drink more of the flavor previously paired with the nutrient, as automatically quantitated by a lick counter connected to each water spout.* [From Ackroff and Sclafani (1998), p. 8.6F.3.]

Conditioned taste aversion in mice has been useful in investigations of drugs of abuse, described in Chapter 11.

TRANSGENICS AND KNOCKOUTS

The first transgenic mice with abnormal body weights were generated by Palmiter and co-workers at the University of Washington in Seattle in 1982 (see the photograph at the beginning of Chapter 2). A rat growth hormone fusion gene was microinjected into C57 fertilized eggs, resulting in offspring with body weights almost twice that of control littermates of the same gender (Palmiter et al., 1982). Divergence of growth rates was detectable as early as 3 weeks of age (Palmiter et al., 1983). Introduction of the growth hormone transgene into dwarf *lit* mice, a model of human hereditary growth hormone deficiency type I, corrected the body weight deficit in the individual mice that expressed the transgene (Hammer et al., 1984).

Leptin was the great discovery of the 1990s in the field of feeding behaviors. Jeffrey Friedman and co-workers at Rockefeller University cloned a gene from adipose tissue that is responsible for the obesity and type II diabetes of the natural mutant *ob/ob* mouse (Zhang et al., 1994). Named after *leptos*, the Greek word for thin (Friedman and Halaas, 1998), the leptin gene is mutated in the *ob/ob* mouse, while the receptor for leptin is mutated in the *db/db* diabetic mouse (Lee et al., 1996a). Both strains of mice are natural mutants that display increased body fat as adults. Body weights in adults reach an average at least 30% higher than wildtype C57BL/6J littermate controls (Pelleymounter et al., 1995).

Expression of the leptin receptor in the arcuate nucleus and paraventricular nucleus of the hypothalamus suggests that leptin secreted from adipose tissue into the circulation reaches hypothalamic cells on the blood side of the blood–brain barrier. The 167 amino acid leptin protein appears to serve as a signaling mechanism, a hormonal "lipostat" (Gura, 1997). Early studies of meal pattern in *ob/ob* mice demonstrated that obese mice ate larger and less frequent meals than lean mice during the dark period of the circadian cycle (Strohmayer and Smith, 1987). The *db/db* natural mutant is deficient in the receptor for leptin and develops obesity and diabetes, displaying an impairment in extinction on conditioned taste aversion (Harris et al. 2001; Ohta et al. 2003). Leptin treatment reduced feeding in the *ob/ob* mouse (Figure 7.3) and in normal mice and rats. Studies measured daily body weights, daily 24-hour consumption of standard chow in the home cage during the 6-day preinjection baseline period, and during a 28-day leptin treatment period (Pelleymounter et al., 1995). The remarkable obesity of leptin-deficient mice contrasts with the normal body weights and food consumption in mice deficient in ghrelin, a peptide released from the stomach that acts on receptors in the arcuate nucleus of the hypothalamus (Wortley et al., 2004).

A subset of obese humans appears to have low levels of circulating leptin (Maffei et al., 1995). Leptin resistance, accompanying high levels of circulating leptin, has been noted in other populations of obese humans (Ioffe et al., 1998). These first reports suggest that mutations of the *ob/ob* gene may underlie some forms of inherited obesity. The *ob/ob* natural mutant mouse, and treatments with leptin or the leptin gene delivered by an adenoassociated viral vector, have revealed the interactions of leptin with hypothalamic neuropeptides that regulate feeding, including neuropeptide Y and ghrelin (Ueno et al., 2004). Mice deficient in both neuropeptide Y and leptin were less obese than mice with the *ob/ob* mutation alone, an interaction that appears to be mediated by leptin acting on neuropeptide Y neurons in the arcuate nucleus (Erickson et al., 1996b; Beck, 2000; Pinto et al., 2004).

Agouti is another natural mouse mutant that exhibits obesity (Yen et al., 1970; Butler and Cone, 2002). *Agouti* null mutants are deficient in the gene for the melanocortin-4 receptor (MC4-R) and the *mahogany* protein (Nagle et al., 1999). The *agouti* obesity syndrome includes body weights that are approximately double those of age-matched adult C57BL/6J wildtype littermates, longer body lengths, approximately 50% greater food intake, hyperinsulinemia and hyperglycemia (Huszar et al., 1997). Melanocortins are peptides derived from proopiomelanocorticotrophin, a precursor protein that act at receptors on melanocyte cells in the hypothalamus to regulate feeding behaviors and body weight (Bray, 1987; Chen et al., 2004). Increased feeding in *agouti* mice and controls was measured as 24-hour consumption of standard chow in the home cage, daily over a two-week period. Melanocortin receptors have been mutated to reveal the receptor subtype mediating obesity. Striking obesity was seen in the melanocortin receptor 4 subtype knockouts (Huszar et al. 1997; Figure 7.4). Unlike wildtype controls and *MC3R* knockouts, *MC4R* null mutants did not increase their physical activity and thermogenesis in response to a high-fat diet (Butler et al., 2001). In addition, MC4R mutants did not reduce their food consumption in response to treatments with anorectic agents including the melanocortin agonist MTII, leptin, corticotropin releasing factor, or urocortin (Marsh et al., 1999). Further, *MC4* null mutants did not increase their feeding in response to the orexigenic actions of neuropeptide Y (NPY) or peptide YY (Marsh et al., 1999). In contrast, mice deficient in the receptor for the melanocortin-5 receptor, which is expressed primarily in peripheral tissues such as sebaceous glands, displayed normal body weights but showed deficits in thermoregulation and in sterol ester lipid content in the hair (Chen et al., 1997b).

Agonists and antagonists of melanocortin were administered in four different models, including the *agouti* mouse, the *ob/ob* mouse, 16-hour fasted C57BL/6J mice, and mice injected with NPY. Consumption of normal chow was measured at hourly intervals (Fan et al., 1997). Dose-related reductions in food consumption were produced by a melanocortin agonist in *agouti, ob/ob*, fasted C57BL/6J mice, and NPY-treated mice (Fan et al., 1997). Conversely, a melanocortin antagonist stimulated hourly consumption of normal chow during both daytime and nocturnal hours in free-feeding C57BL/6J mice (Fan et al., 1997). Thus, use of a variety of feeding paradigms and established pharmacological agents has elucidated the role of melanocortin receptor subtypes in feeding behavior of mutant mice and suggested mechanisms involving metabolic and thermoregulatory dysregulation (Butler and Cone, 2002).

Neuropeptide Y is a 36 amino acid neuropeptide distributed in the central and peripheral nervous system, with high concentrations in the hypothalamus (Chronwall et al., 1985; Sawchenko et al., 1985). Microinjection of NPY into the hypothalamus induces dramatic food consumption in satiated rats (Stanley and Leibowitz, 1985; Colmers and Wahlstedt, 1993; Rowland and Kalra, 1997; Kalra and Kalra, 2002, 2004). However, *NPY* null mutant mice show normal body weights and normal food consumption (Erickson et al., 1996a). The unaffected phenotype of NPY knockout mice has been interpreted as evidence for redundancy of the many hypothalamic mechanisms regulating feeding behaviors. Challenge approaches revealed that NPY knockouts ate less after a 24- or 48-hour fast (Bannon et al., 2000), suggesting a specialized role for NPY in feeding after food deprivation.

In an intriguing follow-up experiment, *NPY* null mutants were bred with *ob/ob* mice (Chapter 7 frontpiece). As mentioned above, the double knockouts showed normal body weights, indicating that lack of NPY prevented the obesity syndrome in *ob/ob* mice (Erickson et al., 1996b). Feeding was measured as 5-day intake at ages 6, 8, 12, and 14 weeks of age. The ability of an *NPY* null mutation to block the obesity effects of a leptin null mutation implicates NPY as a downstream regulator of the actions of leptin.

Neuropeptide Y receptor knockout mice have been analyzed on feeding tests, to understand the receptor subtype that mediates the actions of pharmacologically administered NPY (Lin et al., 2004). Y1 receptor knockout mice display normal body weights in males but higher body weights in females, beginning at about 20 weeks of age (Pedrazzini et al., 1998). Female *NPY-Y1* null mutants had higher amounts of body fat, leptin, and free fatty acids in the plasma. Daily food intake was significantly lower in the knockout mice, particularly after a 24-hour fast. *NPY-Y1* knockout mice continued to show an enhanced feeding response to NPY treatment, albeit not as much as in wildtype controls. Another independent line of *NPY-Y1* knockouts also showed higher body weights in the females, and showed increased weight of white adipose tissue and impaired insulin secretion in response to glucose (Kushi et al., 1998). Mutation of the Y2 receptor using adenoviral vector delivery into the hypothalamus resulted in increased food consumption, concomitant with elevated levels of NPY mRNA (Sainsbury et al., 2002). Other NPY receptor subtypes, for example, NPY-Y5, are hypothesized to contribute a compensatory response to the lack of the NPY-Y1 receptor (Pedrazzini et al., 1998; Heinrichs et al., 1998). Knockout mice deficient in the NPY-Y5 receptor subtype develop mild late-onset obesity, showing increased body weight, higher adiposity in inguinal, retroperitoneal, scapular, and reproductive fat pads, and increased 24-hour consumption of home cage chow (Marsh et al., 1998). Intraventricular microinjection of NPY increased 4-hour food intake in both null mutants and wildtype controls. An NPY-Y1 selective antagonist, 1229U91, blocked the NPY-induced feeding in both null mutants and wildtype controls. These data indicate that NPY-Y1 and NPY-Y5 receptor subtypes both contribute to NPY-induced feeding (Marsh et al., 1998).

Melanin-concentrating hormone (MCH) is another hypothalamic peptide implicated in the regulation of feeding. Mice deficient in the gene for MCH showed reduced body weight, reduced feeding, increased metabolic rate, and reduced expression of proopiomelanocortin (Shimada et al., 1998). Null mutations in the receptor gene *MCHR1* receptor gene conferred resistance to the obesity induced by a high fat diet (Chen et al., 2002).

Opiate drugs, endogenous opiate peptides, and opiate receptors mediate feeding behaviors (Bodnar, 2004; Levine and Billington, 2004). Mu opiate receptor knockout mice did not show the normal increase in running wheel activity in anticipation of food access (Kas et al., 2004).

Cholecystokinin (CCK) is a 33 amino acid peptide distributed in brain and peripheral organs. Systemic administration of the biologically active CCK octapeptide inhibits food consumption in moderately fasted rats, mice, and many other species including humans (Gibbs et al., 1973, 1993; Smith et al., 1985; Crawley and Corwin, 1994; Smith, 1998a). Further CCK appears to induce the full behavioral syndrome of satiety (Antin et al., 1975). Satiety effects of CCK are mediated primarily by the CCK-A receptor subtype at a peripheral location (reviewed in Crawley and Corwin, 1994). Otsuka Long-Evans Tokushima Fatty rats (OLETF) are a natural mutant that is missing the CCK-A receptor gene (Miyasaka et al., 1994). OLETF rats show normal baseline 24-hour food intake, as compared to Long-Evans Tokushima Otsuka (LETO) control rats, but no response to cholecystokinin treatment (Miyasaka et al., 1994), consistent with the absence of the CCK-A receptor. Comparison of liquid diet ingestion by OLETF and LETO rats during 30-minute daily access sessions revealed that OLETF rats had greater intake of the liquid diet due to a slower decline in licking rate over the test session (Moran et al., 1998). In a 24-hour solid food access test, OLETF rats showed increases in meal size, leading to overall hyperphagia and obesity (Moran et al., 1998).

Knockout mice deficient in the CCK-A receptor subtype maintained normal body weight but failed to decrease food intake in response to intraperitoneal CCK administration (Kopin et al., 1999). The difference between rats and mice deficient in CCK-A receptors may relate to the upregulation of neuropeptide Y in the dorsomedial hypothalamus of the OLETF rats but not of the CCK-A null mutant mice (Bi et al., 2004). Knockout mice deficient in the CCK-B receptor subtype, which is localized in the stomach and throughout the brain, diminished food intake by 90% in response to CCK, similar to the reduction in feeding after CCK treatment in the wildtype controls for both knockouts (Kopin et al., 1999). *CCK-B* knockouts showed enhanced gastric emptying and a 10-fold elevation in plasma gastrin concentrations; however, adult body weights appeared to be normal (Langhans et al., 1997; Miyasaka et al. 2004).

Bombesin is a 40 amino acid peptide expressed in the brain and periphery that inhibits feeding behaviors when administered systemically to rats (Kulkosky et al., 1982). Three bombesin receptors have been cloned, one of which is the gastrin-releasing peptide (GRP) receptor (Hampton et al., 1998). *GRP-R* deficient mice had normal body weights but showed no response to administered bombesin on feeding behaviors, as measured by 30-minute glucose consumption (Hampton et al., 1998). Analysis of bombesin receptor knockout mice revealed hyperphagia and obesity only in the BRS-3 receptor subtype (Yamada et al. 2000).

Corticotropin releasing factor (CRF) is a 41 amino acid neuropeptide that is highly expressed in the hypothalamus and regulates the hypothalamic–pituitary–adrenal axis (Vale et al., 1981). CRF is implicated in stress responses (Heinrichs et al., 1992), including stress-induced anorexia (Morley and Levine, 1982; Zorrilla et al., 2003). CRF knockout mice, CRF-R1 receptor knockout mice, and CRF binding protein overexpressing mice all exhibit normal daily food intake and normal body weight gain (Contarino et al., 1999). A higher rate of body weight loss was observed in the *CRF* knockout mice after several days of running wheel activity on a food-restricted diet, as compared to weight loss in wildtype controls under the same conditions. CRF binding protein transgenics displayed a dampened circadian rhythm of feeding. These secondary phenotypes support a role for endogenous CRF in the regulation of feeding under selective conditions (Contarino et al., 1999).

Orexins, also called hypocretins, are a family of peptides expressed in the hypothalamus (Sakurai et al, 1998; Rodgers et al. 2002). Orexin-A is 33 amino acids; orexin-B is 28 amino acids. These peptides were discovered in a search for endogenous ligands of orphan-G-protein-coupled receptors cloned from a hypothalamic library (Sakurai et al., 1998). Central administration of orexins into the lateral hypothalamus of rats stimulated consumption of normal chow, as measured at 1-hour time intervals over a 4-hour test session. Orexin knockout mice eat less but show normal growth curves, while orexin overexpressing transgenic mice do not become obese (Rodgers et al.,

2002). In the *ob/ob* mouse, expression of the prepro-orexin gene was downregulated, in contrast to elevated NPY expression (Yamamoto et al., 1999). These experiments illustrate the application of a mutant mouse model to investigate multiple neurotransmitters regulating different components of ingestion and metabolism.

Serotonin inhibits feeding when microinjected into the hypothalamus (Curzon et al., 1997). Serotonin transporter ligands such as fenfluramine and fluoxetine (Prozac) show anorexic effects in mice (Cooper et al., 1990; McGuirk et al., 1992; Li et al., 1998). Serotonin transporter knockout mice displayed late-onset obesity (Andrew Holmes, personal communication). Null mutation of the serotonin transporter modulated body weight in dopamine transporter and brain derived neurotrophic factor knockout mice (Murphy et al., 2003). Knockout mice deficient in the serotonin 5-HT_{1B} receptor subtype did not show the normal fenfluramine-induced anorexia, implicating this receptor in the regulation of feeding by endogenous serotonin (Lucas et al., 1998). Serotonin receptor subtype 5-HT_{1B} mutant mice had higher body weights and the males drank more water (Bouwknecht et al., 2001). 5-HT_{2C} knockout mice are significantly overweight at older ages (Tecott et al., 1995; Tecott and Abdallah 2003). Partial leptin resistance, insulin resistance, and impaired glucose tolerance in older 5-HT_{2C} knockout mice indicate a predisposition to type II diabetes (Nonogaki et al., 1998).

These several examples highlight a major application of targeted gene mutations to biomedical research. A common scenario is the elaboration of an important physiological or behavioral action of an endogenous transmitter, but a lack of knowledge of the receptor subtype mediating the action. In the absence of pharmacologically selective ligands, the receptor subtype can be revealed through appropriate experiments with receptor subtype knockout mice. Understanding of the critical receptor subtype aids the development of subtype-selective drugs for the treatment of disorders such as obesity.

BACKGROUND LITERATURE

Blundell JE (1977). Is there a role of serotonin (5-hydroxytryptamine) in feeding? *International Journal of Obesity* **1**: 15–42.

Campfield LA, Smith FJ, Burn P (1998). Strategies and potential molecular targets for obesity treatment. *Science* **280**: 1383–1387.

Curzon G, Gibson EL, Oluyomi A (1997). Appetite suppression by commonly used drugs depends on 5-HT receptors but not on 5-HT availability. *Trends in Pharmacological Sciences* **18**: 21–25.

Inui A (1999). Feeding and body-weight regulation by hypothalamic neuropeptides—Mediation of the actions of leptin. *Trends in Neuroscience* **22**: 62–67.

Moran TH (2004). Gut peptides in the control of food intake: 30 years of ideas. *Physiology and Behavior* **82**: 175–180.

Olszewski PK, Levine AS (2004). Minireview: Characterization of influence of central nociceptin/orphanin FQ on consummatory behavior. *Endocrinology* **145**: 2627–2632.

Richter C (1922). A behavioristic study of the activity of the rat. *Comparative Psychology Monograph* **1**: 1–55.

Rodgers RJ, Ishii Y, Halford JCG, Blundell JE (2002). Orexins and appetite regulation. *Neuropeptides* **36**: 303–325.

Schulkin J (1991). *Sodium Hunger: The Search for a Salty Taste*. Cambridge University Press, Cambridge.

Sclafani A (1995). How food preferences are learned—Laboratory animal models. *Proceedings of the Nutrition Society* **54**: 419–427.

Smith GP (1998). *Satiation: From Gut to Brain*. Oxford University Press, New York.

Thibault L, Booth DA (1998). Macronutrient-specific dietary selection in rodents and its neural bases. *Neuroscience & Biobehavioral Reviews* **23**: 457–528.

Wellman PJ, McMahon LR (1998). Basic measures of food intake. In *Current Protocols in Neuroscience*, JN Crawley, CR Gerfen, R McKay, MW Rogawski, DR Sibley, P Skolnick, Eds. Wiley, New York.

Woods SC, Seeley RJ, Porte D, Schwartz MW (1998). Signals that regulate food intake and energy homeostasis. *Science* **280**: 1378–1383.

Woods SC (2004). Gastrointestinal satiety signals. I. An overview of gastrointestinal signals that influence food intake. *American Journal of Physiology-Gastrointestinal and Liver Physiology* **286**: G7–G13.

Special Issues of Journals and Series

Behavior Genetics **2n** (4), July 1997 Special Issue: *The Genetics of Obesity*.

European Journal of Pharmacology **440** (2–3), April 2002 Special Issue: *The Pharmacotherapy of Obesity*.

Current Protocols in Neuroscience (1999). Chapter 8, Unit 6: *Feeding*. Wiley, New York.

Neuropeptides **33** (5), October 1999-Special Issue: *Neuropeptides and Feeding*.

8

Reproductive Behaviors

HISTORY AND HYPOTHESES

Sexual, reproductive, and parental behaviors are highly species-specific. Rodent reproductive behaviors have been investigated with two complementary approaches: studies of natural behaviors in the field and experimental manipulations in the laboratory. Behavioral endocrinology began in the mid-1800s with the insights of A. A. Berthold in Sweden (Berthold, 1849; Becker et al., 1993). One of the first experiments was an investigation of sexual behaviors in roosters. Berthold (1849) removed the testes of a rooster and observed its behaviors. The rooster stopped crowing and ceased its sexual and aggressive behaviors. Reimplantation of one testicle restored crowing, sexual activity, and aggression. The hormone responsible for crowing, aggressive, and sexual behaviors in the rooster was later identified as testosterone.

Courtship, ovulation, sexual receptivity, copulation, pregnancy, parturition, lactation, and parental care of the young in rodents are all under tight hormonal control. The estrus cycle of the female rat, which lasts 4 to 5 days, is characterized by cyclical hormonal fluctuations over the sequence of proestrus, estrus, metestrus, and diestrus (Marler and Hamilton, 1968). The estrus cycle of the mouse follows the same cycle over 4 to 6 days, during which the female is receptive in the late proestrus/early estrus portion of the cycle (Silver, 1995). Testosterone, estrogen, and progesterone act at steroid receptors in the brain, stimulating neural pathways that regulate these behaviors. These steroid hormones and many other peripheral hormones and central neurotransmitters appear to be necessary for the normal development of the gonads and secondary sexual characteristics, adult sexual behaviors, pregnancy, and parental behaviors in mammals (Baulieu, 1989; DeVries, 1990; Becker et al., 1993; Carter and Kirkpatrick, 1997; Blasberg et al., 1998; Cooke et al., 1998; Nelson, 2005).

Gonadal steroid hormones have been extensively characterized in rodents (Becker et al., 1993; Rainbow et al., 1982; Everitt, 1990; De Vries, 1990; Nelson, 2005). The mouse testes produce testosterone and dihydrotestosterone. The mouse ovaries produce estradiol and progesterone. Steroids released from the gonads travel through the bloodstream to the brain. As steroids are lipophilic, they easily cross the blood–brain barrier. High-affinity binding sites for estrogen and progesterone are present in rodent brain, with high concentrations of receptors in the sexually dimorphic nuclei (Pelletier, 2000). Critical sites of action for testosterone, dihydrotestosterone, estradiol, and progesterone in the mouse brain include the preoptic area, arcuate nucleus, bed nucleus of the stria terminalis, ventromedial nucleus, and paraventricular nucleus of the hypothalamus. Two estrogen receptors have been cloned, ERα and ERβ, both of which are distributed in the hypothalamus, limbic structures, and pituitary, where they activate intracellular transcriptional processes (Laflamme et al., 1998; Mitchner et al., 1998; McDonnell et al., 1995; Gustafsson, 2003).

Neurosecretory neurons in the hypothalamus release neuropeptides, including thyrotropin-releasing hormone, gonadotropin-releasing hormone, corticotrophin-releasing factor, and oxytocin. These hypothalamic peptides travel down the hypothalamic–pituitary portal blood vessels to the pituitary gland, where they act at specific high-affinity binding sites. Secretory neurons in the pituitary are selectively stimulated to release specific pituitary hormones, including prolactin, luteinizing hormone, follicle-stimulating hormone, thyroid-stimulating hormone, adrenocorticotropin-releasing factor, vasopressin, and oxytocin. Pituitary hormones then travel through the general circulation to peripheral endocrine glands including the adrenal, thyroid, pineal, testis, and ovaries, where they regulate the physiology of these peripheral target organs.

Sexual behaviors in rodents are mediated through a complex loop of positive and negative feedback mechanisms via these central and peripheral hormones. Testosterone facilitates male sexual behavior. A threshold concentration of circulating testosterone is required to initiate penile erection and sexual activity in male rats (Damassa et al., 1977; Meisel and Sachs, 1994). Sites of action for testosterone are peripheral, at receptors on the surface of the glans penis, and central, including the preoptic area of the hypothalamus (Davidson, 1966). Estrogen facilitates rodent female sexual behaviors. Estrogen and progesterone act on neurons in the ventromedial nucleus of the hypothalamus to stimulate sexual receptivity in female rats (Pfaff and Schwartz-Giblin, 1988). Oxytocin released from hypothalamic neurons acts on the posterior pituitary. Pituitary oxytocin travels through the circulation to the uterus to trigger uterine contractions during labor and delivery, and to the mammary glands to trigger milk ejection during lactation (Lincoln and Paisley, 1982; Ganten and Pfaff, 1986; Wagner et al., 1997). Oxytocin and vasopressin mediate social affiliative behaviors (Carter, 2003), as described in Chapters 9 and 12.

Inbred strains of mice vary greatly on reproductive characteristics. Lee Silver's book, *Mouse Genetics*, provides an excellent summary of strain differences in percentage of productive matings, litter size, and number of litters per mating pair, for 12 strains of mice (Table 4.1 in Silver, 1995). Relative fecundity is highest in the FVB/N strain, which averages over 90% productive matings, 4.8 litters per mating pair, and 9.5 pups per litter. C57BL/6J is among the most prolific strains, with 84% productive matings, 4.0 litters per mating pair, and 7 pups per litter. 129/SvJ is somewhat lower, with 75% productive matings, 4.1 litters per mating, and 5.9 pups per litter. BALB/cJ is lowest, with 47% productive matings, 3.8 litters per mating pair, and 5.2 pups per litter. Anecdotal reports differ concerning breeding success across inbred strains, and in targeted mutations on various strain backgrounds. Season of the year, circadian light cycle, frequency of cage litter changes, noise from ventilated racks or building construction, ultrasonic noise from fluorescent light bulbs in the ceiling, and other unknown environmental variables, appear to affect reproduction. It seems likely that these external stressors may differentially affect inbred mouse strains on their mating, pregnancies, and parenting behaviors.

As described in Silver (1995), the average estrous cycle of these mouse strains is 4 to 6 days in length. Gestational period varies across mouse strains, from 18 to 22 days. Weaning can be as early as 18 days after birth, but survival and general health of the pups is better when weaning occurs between 3 and 4 weeks of age. Onset of puberty is strain-dependent and subject to environmental influences, averaging from 6 to 8 weeks of age. Life span is highly variable across strains, with a range of life spans from 10 months for AKR/J to 27 to 28 months for C57BL/6J. F_1 hybrid mice from the cross of C57BL/6J and DBA/2J survive for as long as 3.5 years.

Parental behaviors in rodents include nest building, retrieving the pups to the nest, nesting with the pups, licking the pups clean, nursing the pups, and defending the pups (Becker et al., 1993; Silver, 1995; Russell and Leng, 1998; Lonstein and Fleming, 2001). Mouse pups are born hairless, with their eyes closed, and with underdeveloped motor skills. Nesting and pup retrieval by the parents is necessary to keep the pups warm in the nest and in close proximity for nursing. Both father and mothers contribute to aspects of mouse parental behaviors. Parental behaviors begin at birth and continue for 3 to 4 weeks. Parenting can influence the subsequent adult behaviors of mouse pups. Emotional reactivity to stressors in adult mice is significantly greater in mice who experienced maternal separation as pups (Meaney, 2001; Francis et al., 2002). The strain of the mother appears to be the determining factor, even when the mother is foster, and even when the fostering begins in utero (Calatayud and Belzung, 2001; Francis et al., 2003). Hormone synthesis during pregnancy and stimuli from the newborn pups are necessary to elicit and maintain maternal behaviors. Poor maternal behaviors sometimes occur with the first pregnancy. Although few or no pups may survive from the first litter; subsequent pregnancies may result in normal maternal behaviors and good pup survival rates.

Nursing requires the synthesis of milk in the mammary glands, parental retrieval of the pups to the nest in sufficiently close proximity to the mother, crouching behavior by the mother, attachment of the pups to the teats, suckling behavior by the pups, and milk ejection (Lonstein and Fleming, 2001; Meaney, 2001). Absence of any of these behaviors or physiological mechanisms will prevent effective lactation, and the pups will die. Milk ejection is a physiological reflex triggered by suckling. The pup's sucking stimulates afferent neurons that travel via the spinal cord and synapse on oxytocin-containing neurons in the supraoptic and paraventricular nuclei of the hypothalamus, which stimulate pituitary cells to release oxytocin into the bloodstream (Lincoln and Wakerly, 1975). Circulating oxytocin reaches the mammary gland, initiating contraction of the myoepithelial cells that surround the alveoli, to trigger milk ejection (Lincoln and Paisley, 1982; Robinson, 1986; Ganten and Pfaff, 1986; Wagner et al., 1997). Oxytocin serves corollary behavioral functions by enhancing sexual receptivity, social pair bonding, and maternal behaviors (Fahrbach et al., 1985; Witt et al., 1992; Caldwell et al., 1986, 1994; Williams et al., 1994; Insel et al., 1995; Insel and Hulihan, 1995; Nishimori et al., 1996, Carter and Kirkpatrick, 1997; Russell and Leng, 1998; Ferguson et al., 2002; Choleris et al., 2004).

Nest building is a natural behavior pattern expressed by both male and female mice (Estep et al., 1975; Bulloch et al., 1982; Batchelder et al., 1982; Manning et al., 1992; Lijam et al., 1997; Lonstein and Fleming, 2001; Garey et al. 2002; Van Loo et al., 2004; Long et al., 2004). Sleeping within the nest serves a heat regulatory function, especially for newborn pups. In the wild, nests offer camouflage from predators for both pups and adults. Some strains of mice are better nest builders than others. C57BL/6J and BALB/cIbg build larger nests than C3H/2Ibg or DBA/Ibg (Batchelder et al., 1982). Genetic components of nesting have been demonstrated in lines of mice selectively bred for the tendency to build large versus small nests (Lynch, 1980; Schneider et al., 1987; Sluyter et al., 1995). Communal nesting may also function to reduce infanticide, perhaps through a shared colony odor, as suggested by evidence that genetic similarity increases kin recognition (Manning et al., 1992). The staggerer and weaver natural mutants with cerebellar dysfunctions showed poor nest building, varying from much slower and less complete nest-building behavior to a complete lack of nest building (Bulloch et al., 1982).

Retrieval behaviors in rodents appear to require extensive feedback from the pups. Mouse pups appear to call to their parents, just as human babies cry to elicit attention from their parents. Infant rodents emit high-frequency vocalizations, 50 to 80 kHz for mouse pups, distinct from the 22 kHz vocalizations emitted by adult rats as alarm calls (Zippelius and Schleidt, 1956; Cohen-Salmon et al., 1985; Hofer and Shair, 1987; Brudzynski and Chiu, 1995; Molewijk et al., 1995; Hofer, 1996; Hahn et al., 1997; Branchi et al., 1998; Hofer et al., 2001). Ultrasonic vocalizations are recorded with a "bat detector" microphone that converts the vocalizations from the ultrasonic range to the human-hearing range, and tallies number of calls over time, or with specialized sonogram equipment. UltraVox, an automated system for recording and analyzing ultrasonic vocalizations, is available from Noldus. Detailed analyses of the calls and of behaviors co-occurring with pup vocalizations have been elegantly described (Liu et al., 2003; Branchi et al., 2004). Vocalizations called "wriggling calls" at frequencies below 10 kHz are emitted while the pup is struggling in the nest or pushing to attach to the teats. Wriggling calls appear to elicit maternal behaviors, especially nest building and licking. Ultrasonic vocalizations in the range of 50 to 80 kHz are emitted specifically when the pup is isolated from the mother, from the nest, and from warmth (Hofer et al., 2001). Although pup vocalizations have been interpreted as distress calls, intentionality by the pup has not been demonstrated, and an explanation based on respiratory reflexes induced by hypothermia has been offered (Hofer and Shair et al., 1993). However, these ultrasonic vocalizations serve an apparent communicatory function, as the calls trigger the parents to search for and retrieve the pups to the nest. Strain differences have been reported for ultrasonic vocalization duration, rates, and frequency characteristics in mouse pups, and strain differences in the ability of the mother to hear the calls (Hahn et al., 1997; D'Amato et al., 2005). Lines of rats with high, moderate, and low pup ultrasonic vocalizations have been generated by selective breeding (Brunelli et al., 2002). Pup retrieval is a very important aspect of parenting in rodents. Since mouse pups have poorly developed thermoregulation during the first week of life, if they remain out of the nest they are likely to die of hypothermia. In the wild, exposure to predators is another cause of death for pups out of the nest.

Olfactory cues are important components of mouse social, sexual, and parental behaviors (Vandenbergh, 1973; Edwards and Burge, 1973; Schoots et al., 1978; Doty, 1986; Coquelin, 1992; Calamandrei et al., 1992; Bartoshuk and Beauchamp, 1994;

Liebenaur and Slotnick, 1996; Bakker et al., 2002a, b). Urine marking by female mice provides an olfactory cue to males, conveying information about sexual receptivity. Olfactory cues from the infant pups contribute to parental retrieval behaviors. Disruption in searching for pups is evident when the sense of smell is disrupted in female mice by intranasal administration of zinc sulfate (Koch and Ehret, 1991). Sensory deficits prevent a deaf or anosmic mother from perceiving ultrasonic vocalizations and olfactory cues emitted by the pups. Lack of recognition and failure to initiate maternal behaviors may result in death of the pups because of parental neglect or cannibalism (Slotnick and Nigrosh, 1975; Cohen-Salmon, 1985; Koch and Ehret, 1991; Calamandrei et al., 1992).

BEHAVIORAL TESTS

Courtship and Copulation

Extensive literature describes tests for sexual behaviors in rats and mice (Baum et al., 1974; Erskine, 1989; Everitt, 1990; Price, 1991; Becker et al., 1993; Sisk and Meek, 1997; Rissman et al., 1997a, b; Wersinger et al., 1997; Ogawa et al., 1997; Baum, 2003; Burns-Cusato et al., 2004). Detailed step-by-step procedures are provided by Cheryl Sisk and co-workers at Michigan State University (Sisk and Meek, 1997). Quantitation of individual components of sexual behaviors is conducted by a well-trained human observer. The best method is to videotape the test session and later score the behaviors from the videotape, to avoid disturbance by the presence of the experimenters during the test session. Digital cameras connected to computers allow remote observation on-screen in real time, and enable the saving of large data files on DVD discs.

The test female is paired with a sexually experienced stimulus male. The test male is paired with a sexually receptive female. In most paradigms to study rodent sexual behaviors, the experimenter regulates circulating estradiol levels rather than relying on the fluctuating estradiol levels over the 4-day estrus cycle of the female rat. Surgical ovariectomy is performed, and exogenous estradiol is then administered via subcutaneous injection of β-estradiol 3-benzoate, or implantation of estradiol in silastic tubing for gradual release over days or weeks (Smith et al., 1977; Becker et al., 1993; Sisk and Meek, 1997).

The female rodent exhibits a characteristic set of behaviors to display sexual receptivity to an adult male. Proceptive behaviors indicating sexual motivation include hopping, darting, and ear wiggling. When the male mounts the receptive female, she arches the back, elevates the head and rump, extends the back feet, and deflects the tail to one side. This posture, known as *lordosis*, is necessary for coital performance in rodents (Figure 8.1). Quantitative analysis of female sexual behaviors in rodents focuses on the frequency of exhibition of the lordosis response. The commonly used lordosis quotient (LQ) is the ratio of the frequency of lordosis to a fixed number of mounts × 100. A completely receptive female will exhibit lordosis each time the male mounts, showing an LQ of 100. A moderately receptive female may exhibit lordosis during 5 out of 10 mounts, showing an LQ of 50.

Male rodents exhibit sexual behaviors that follow a fixed sequence of motor patterns in the presence of a receptive female (Figure 8.2; Pfaus et al., 1990; Everitt, 1990; Weed and Boone, 1992; Becker et al., 1993; Cherry, 1993; Sisk and Meek, 1997; Wersinger et al., 1997). The sequence begins with grooming and anogenital investigation. Mounts then follow. The male mouse mounts the female from the rear

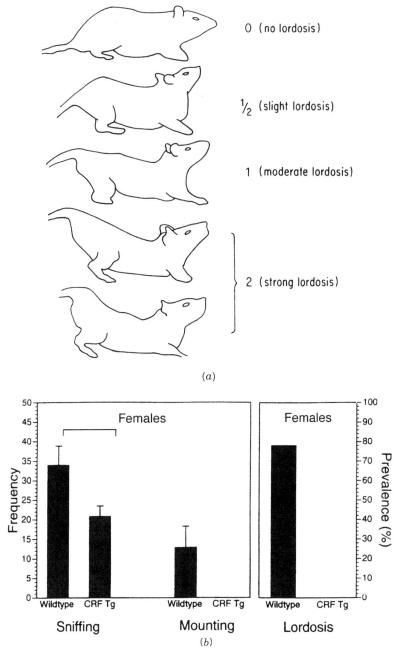

(a)

(b)

Figure 8.1 *Female lordosis behavior. (a) Lordosis posture in a female rodent is scored in the presence of a male.* [Adapted from Brink et al. (1980), p. 73.] *(b) Female transgenic mice overexpressing corticotropin releasing factor display minimal lordosis postures and receive fewer sniffs and mounts by males, as compared to wildtype controls.* [From Heinrichs et al. (1997), p. 219.]

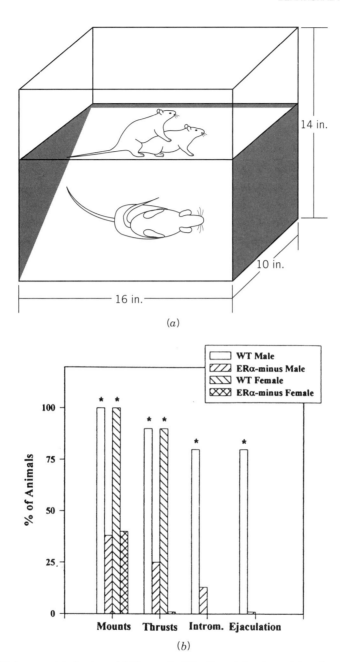

(a)

(b)

Figure 8.2 *Male sexual behaviors. (a) Rissman's "No Secrets" mirrored box for observation of mouse mating behavior. Mount posture is illustrated. (Diagram kindly contributed by Dr. Emilie Rissman, Department of Biology, University of Virginia.) (b) Male knockout mice deficient in the gene for the estrogen receptor ERα displayed fewer mounts, pelvic thrusts, intromissions, and ejaculations, as compared to wildtype controls. Similarly, female knockout mice deficient in the gene for ERα and treated with testosterone showed fewer mounts and pelvic thrusts. [From Wersinger et al. (1997), p. 178.]*

and grasps her flanks with his front feet. The male then begins multiple pelvic thrusts. A sexually inexperienced male may orient incorrectly, resulting in mounts and pelvic thrusts to the side, head, or other region of the female. Intromissions are counted when a pelvic thrust results in vaginal penetration. After a series of rapid intromissions, a longer lasting thrust results in ejaculation of semen. Ejaculation is confirmed by performing a vaginal lavage of the female to determine the presence of sperm in the vagina. Following ejaculation, the male engages in genital grooming and lack of activity toward the female, termed the postejaculatory refractory period. To test sexual behaviors independent of circulating testosterone levels, the gonads are surgically removed and testosterone is administered by subcutaneous injection or in subcutaneously implanted silastic tubing (Meisel and Sachs, 1994; Sisk and Meek, 1997; Wersinger et al., 1997).

Sexual behavior analysis is best conducted during the first 2 to 4 hours of the beginning of the dark period of the circadian cycle, when sexual behaviors normally occur in mice (Sisk and Meek, 1997). The test room is illuminated with a dim red light bulb. A videocamera sensitive enough to record clear images under low red illumination is used to videotape the test sessions for subsequent scoring (Weed and Boone, 1992; Price, 1991; Sisk and Meek, 1997). A mirror is positioned under a glass testing arena, or a clear Plexiglas test cage is placed on a vertical viewing platform, to obtain views of all facets of sexual behaviors (Sisk and Meek, 1997; Wersinger et al., 1997; Figure 8.2). The test mouse is first placed in the testing apparatus for an hour to acclimate to the novel environment. The test session may be of 10 to 15 minutes duration or up to 4 hours duration, with repeated daily test sessions as required for the purposes of the experiment.

Sexual motivation in the male is quantitated in the presence of the female in an adjacent chamber, with visual, auditory, and olfactory cues available but no physical contact (Becker et al., 1993). Latency and number of penile erections are scored. Motivation is sometimes tested by scoring responses to the presence of female olfactory cues only. Indirect measures of sexual motivation are quantitated in operant lever press tasks reinforced by access to a receptive female, exploratory behaviors at a transparent partition with holes, or place preference tests for locations previously shared with a sexual partner (Becker et al., 1993; Matthews et al., 1997; Kudryavtseva, 2003).

Parenting

Nest building (Figure 8.3) is tested after providing clean nesting material such as Nestlet squares (Ancare Corporation, Long Island, New York), Mountain Mist cotton batting, absorbent cotton, facial tissue, or paper strips (Slotnick and Nigrosh, 1975; Estep et al., 1975; Bulloch et al., 1982; Schneider et al., 1987; Sluyter et al., 1995; Thomas and Palmiter, 1997; Lijam et al., 1997; Van de Weerd et al., 1997; Lonstein and Fleming, 2001; Garey et al., 2002; Bond et al. 2002). Step-by-step methods for conducting tests of nesting behavior were provided by Alison Fleming and co-workers at the University of Toronto (Lonstein and Fleming, 2001). Nesting usually begins shortly after the material is placed in the cage. Nests are often completed within the first hour after nesting material is introduced (Bulloch et al., 1982; Lijam et al., 1997). Some procedures involve removing the nest and weighing it, then providing new nesting material every 24 or 48 hours (Batchelder et al., 1982; Lynch, 1980; Schneider et al., 1987). Height or diameter of the nest or weight of nesting material used provide measures to

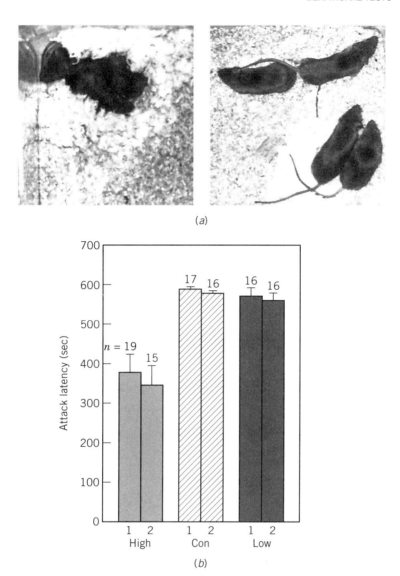

Figure 8.3 *Nest building. (a) Nest built by normal mice (left). Mice with a null mutation in Dv11 failed to build a nest from the Nestlet material supplied (right). [From Lijam et al. (1997), p. 898.] (b) Nest-building behavior correlates with aggression. Replicated lines of mice selected for high nest building (HIGH1, HIGH2) attacked an intruder more quickly than replicated lines of mice selected for low nest building (LOW1, LOW2) or randomly bred controls (CON1, CON2). [Adapted from Sluyter et al. (1995), p. 249.]*

quantitate nesting (Lijam et al., 1997). Quality of the nest is evaluated as (1) no nest, (2) platform-type nest consisting of a pallet on the floor of the cage, (3) bowl- or cup-shaped nest with sides, or (4) bowl- or cup-shaped nest with sides and a cover (Estep et al., 1975; Bulloch et al., 1982). Cup-shaped nests built from Nestlet squares are illustrated in Figure 8.3. Individual scoring of singly housed male and female mice gives gender-independent nesting scores for each mouse (Schneider et al., 1986).

Parental care of the pups is scored for several parameters. Detailed methods are available in Lonstein and Fleming (2001). Time spent within the nest is summed for each parent (Wang et al., 1995). The observer scores frequency of searching for pups and latency to retrieve pups after the pups are removed from the nest (Slotnick and Nigrosh, 1975; Stern and Mackinnon, 1978; Koch and Ehret, 1991; Nishimori et al., 1996; Thomas and Palmiter, 1997; Figure 8.4). Maternal behaviors (Figure 8.5) are scored for number of bouts, time spent licking pups, time spent crouching over pups, specific postures including kyphosis, hunched, and prone, time spent nursing, and pup retrieval (Stern and Mackinnon, 1978; Koch and Ehret, 1991; Wang et al., 1995; Lonstein and Fleming, 2001). Successful lactation, resulting in milk in the stomach of the pups, is indicated by the white coloring of the abdominal region, observed through the thin skin of the newborn pup (Figure 8.6). Conditioned place preference tests and operant tasks that require lever presses to deliver a pup from a chute to a dispenser opening (Figure 8.7), measure parental motivation (Hauser and Gandelman, 1985; Lonstein and Fleming, 2001).

Parental retrieval behaviors are scored during pup separation challenge tests. Ultrasonic vocalizations emitted by the pups are monitored with specialized equipment

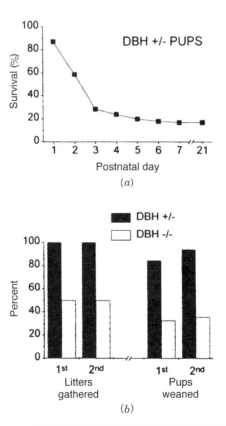

Figure 8.4 *Parents locate a pup that has moved outside the nest and carry the pup back to the nest. Knockout mice deficient in dopamine β-hydroxylase, the synthetic enzyme for norepinephrine, had fewer pups that survived to (a) weaning age due to (b) poor retrieval behaviors.* [From Thomas and Palmiter (1997), p. 584.]

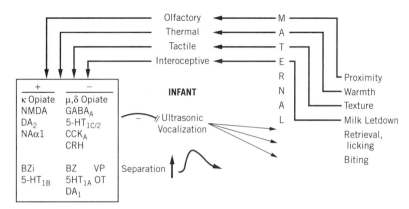

Figure 8.5 *Infant–mother interactions and regulator neurotransmitters. The mother receives sensory stimuli from the pup, including olfactory, tactile, auditory, and visual cues. The pup receives maternal care from the mother, including milk, warmth, licking, retrieval, and protection.* [From Hofer (1996), p. 205.]

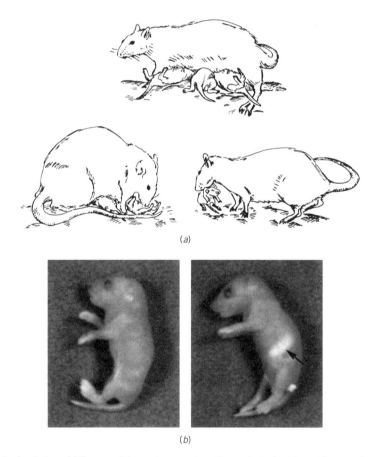

Figure 8.6 *Lactation. (a) Successful nursing requires the rodent dam to retrieve and position the pups.* [Adapted from Alberts and Gubernick, (1990), p. 419.] *(b) Milk is found in the stomachs (arrow) of mouse pups nursed by wildtype mothers (right panel) but not in the stomachs of pups nursed by mothers deficient in the gene for oxytocin (left panel).* (Photograph kindly contributed by Dr. W. Scott Young, National Institute of Mental Health, Bethesda, MD.)

The Pupomat: An operant task for pup reinforcement

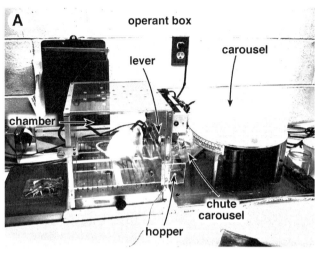

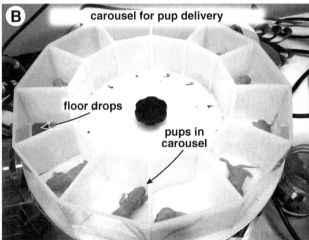

Figure 8.7 *Parental motivation for pup retrieval. An operant chamber is connected to a carousel containing rat pups. The mother presses the lever to deliver a pup from the carousel magazine down a chute. Motivation is measured by varying the reinforcement schedule for lever presses to access pups. Maternal behaviors simultaneously scored include retrieving the pup from the chute into the operant chamber, licking the pup, and crouching over the pup.* [From Lonstein and Fleming (2001), *Current Protocols in Neuroscience*, Figure 8.15.5.]

that detects high frequencies, such as the electronic sonogram recorder (Figure 8.8*a*) and the portable "Bat Detector" (Zippelius and Schleidt, 1956; Cohen-Salmon et al., 1985; Hofer and Shair, 1987; Brudzynski and Chiu, 1995; Hofer, 1996; Hahn et al., 1997; Branchi et al., 1998, Hofer et al., 2001; Winslow and Insel, 2002; Liu et al., 2003; Branchi et al., 2004). Pups vocalizations normally elicit retrieval behaviors by the parents, which are scored as latency and frequency of retrieving pups to the nest (Figure 8.4). As seen in Figure 8.8*b*, pup vocalizations decline when the

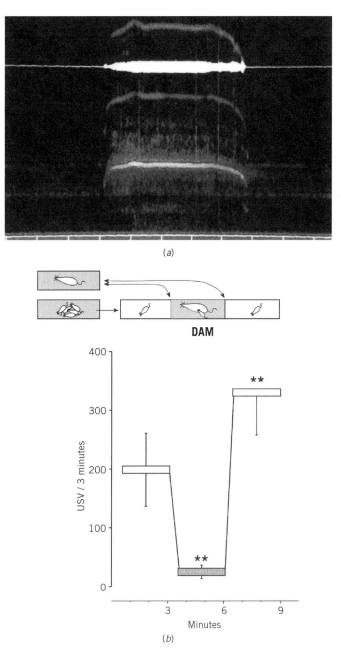

(a)

DAM

(b)

Figure 8.8 *(a) Spectrograms of ultrasonic vocalization of an infant CD-1 mouse.* (Sonogram kindly contributed by Drs. Igor Branchi and Enrico Alleva, Istituto Superiore di Sanita, Rome.) *(b) Maternal potentiation of ultrasonic vocalization in isolated rat pups. The pup is separated from the mother, litter pile, and nest and placed in a novel cage for 9 minutes. During the first 3 minutes after removal from the home cage, the pup emits an average of 200 ultrasonic vocalizations, recorded through a specialized microphone. During the next 3 minutes, the anesthetized mother (dam) is placed in the test cage. Vocalizations decrease to low levels, an effect termed "contact quieting." During the last 3 minutes, the dam is removed from the test cage. Pup ultrasonic vocalizations increase to a level significantly higher than during the first 3 minutes of isolation, an effect termed "maternal potentiation."* [Schematic kindly contributed by Dr. Harry Shair, New York State Psychiatric Institute, New York, and adapted from Shair et al. (2003).]

anesthetized mother is placed near the isolated pup (Shair et al., 2003). When the mother is removed and the pup is re-isolated, vocalizations are higher than during the first isolation period. (Shair et al., 2003). While contact quieting may rely on physiological mechanisms, the maternal potentiation effect suggests the involvement of cognitive components, since physiological components are identical during the first and second isolation periods. Specialized ultrasonic microphones that translate ultrasonic vocalizations into the human hearing range and associated recording equipment and software analysis programs are available from several companies including Noldus and Avisoft Bioacoustics. Noldus Information Technology manufactures equipment for automatically recording and quantitating ultrasonic vocalizations over time in mouse pups.

TRANSGENICS AND KNOCKOUTS

Mutations in the receptors for gonadal steroids provide a new approach to studying the role of these receptor subtypes in facets of reproductive behaviors. Estrogen receptor knockouts have been developed for the ERα and the ERβ (Lubahan et al., 1993; Wersinger et al., 1997; Burns-Cusato et al., 2004), one of the two estrogen receptor subtypes expressed in the rodent brain (Green et al., 1986; Kuiper et al., 1996; Li et al., 1997). *ERα* knockout mice show elevated levels of circulating gonadal hormone levels as compared with wildtype littermate controls (Rissman et al., 1997b; Eddy et al., 1996). Female *ERα* knockout mice did not display sexual receptivity, as compared to wildtype controls, when both genotypes were ovariectomized and treated with estradiol or a combination of estradiol and progesterone (Rissman et al., 1997a).

While the external reproductive organs of the male *ERα* knockouts appeared anatomically normal, spermatogenesis was greatly reduced due to disorganized seminiferous tubules, and the males were infertile (Eddy et al., 1996). Male *ERα* knockout mice displayed reduced sexual performance, which appeared to result primarily from low sexual motivation (Eddy et al., 1996; Rissman et al. 1999). Association preferences measured in male *ERα* knockout mice found that the knockouts spent less time with a receptive female, and did not prefer hormone-primed stimulus females over unreceptive ovariectomized females (Rissman et al., 1997b; Wersinger et al., 1997). However, in another study (Ogawa et al., 1997), male *ERα* knockout mice with intact gonads mounted as frequently and with latencies not significantly different than wildtype controls. Different behavioral methods may account for this discrepancy, for example, conducting the behavioral tests in a large, neutral test arena (Rissman et al., 1997b) versus the home cage (Ogawa et al., 1997).

Male *ERα* knockout mice and their wildtype littermate controls that were castrated and treated with testosterone in silastic implants displayed very low numbers of mounts and pelvic thrusts and virtually no intromissions or ejaculation (Wersinger et al., 1997; Figure 8.2). These results are consistent with the actions of aromatized testosterone at both estrogen receptors and testosterone receptors to activate masculine sexual behaviors in rodents (Meisel and Sachs, 1994; Wersinger et al., 1997).

Estrogen receptor knockouts of the ERβ subtype appear to develop normally and are fertile (Krege et al., 1998). Male *ERβ* knockouts mated and produced offspring, but showed ejaculatory behavior at an older age than wildtype controls (Scordalakes et al., 2002; Temple et al., 2003). At pubertal ages, male *ERβ* knockouts displayed more aggression than wildtype controls, but by adulthood the aggression levels were lower in the mutants (Nomura et al., 2002; Pfaff et al. 2002). Double-knockouts of *ERα* and *ERβ* displayed a striking lack of lordosis in the females, even when stimulated with dopamine agonists (Kudwa and Rissman, 2003)

Progesterone receptor knockout mice (Lydon et al., 1995) displayed a phenotype that differed considerably from the estrogen receptor knockout mice. Null mutants missing the progesterone receptor developed to adulthood and showed normal external genitalia. Male progesterone receptor knockout mice were fertile but exhibited reduced mount frequencies, and heterozygotes showed reduced responses to testosterone (Phelps et al., 1998). Female homozygotes were infertile, consistent with the role of progesterone in early pregnancy. Morphological examination did not indicate any gross abnormalities of the internal reproductive organs. However, hormonal treatments revealed functional abnormalities of the uterus, ovary, mammary gland, and brain. Oocytes were not present in the oviduct or upper uterine horn of the female knockouts. Gonadotropin treatments induced superovulation in the wildtype females but not in the knockouts. Ovariectomized wildtype females treated with estrogen and progesterone showed a normal lordosis response, whereas similarly treated knockouts did not display lordosis. These results from progesterone receptor knockout mice confirm the role of progesterone in ovulation and sexual receptivity in mice.

The progesterone receptor has two naturally occurring receptor isoforms, A and B (McDonnell, 1995). PR_A and PR_B arise from the same gene, but are regulated by separate promoters. The ratio of PR_A to PR_B was altered in mice by insertion of an additional A form as a transgene (Shyamala et al., 1998). PR_A transgenic mice showed highly branched lateral ducts and disorganized basement membranes in the mammary glands, features associated with carcinogenesis.

Aromatase is the enzyme that converts testosterone to estrogen (Bakker et al., 2002). Aromatase knockout mice were investigated for sexual behaviors. Preference for a gonadally intact male, an estrus female, soiled bedding, and volatile odors were conducted in a Y-maze. Female knockouts showed less investigation of conspecifics and sexual odors, and less lordosis behavior in the presence of an intact male, while olfactory and locomotor behaviors were not different from wildtype controls.

Gonadal steroid receptors constitute a superfamily of transcription factors that target nuclear genes. *SRC-1* is a coactivator gene, which functions in transcriptional activation by steroid receptors. Knockout mice in which the *SRC-1* gene was inactivated were viable and fertile and showed normal body weights (Xu et al., 1998). However, estrogen treatment, which produced a 4.3-fold increase in uterine wet weight in wildtype littermate controls, produced a smaller increase of 2.8-fold in the *SRC-1* knockout female mice. Mammary gland development was reduced in the female knockouts, although these mice still produced milk. Testosterone treatment of castrated males produced a greater increase in prostate gland weight in wildtype than in *SRC-1* knockout male mice. This knockout phenotype supports a role for *SRC-1* in efficient gonadal steroid action on reproductive tissues.

Targeted mutation of the gene for oxytocin produced null mutants deficient in this 9 amino acid peptide (Nishimori et al., 1996; Young et al., 1996; Wagner et al., 1997). Oxytocin knockout mice were healthy, with external genitalia that appeared to be normal. Testes weights and histological appearance of testes tissue were normal. Both male and female oxytocin knockouts were fertile. Pregnancies appeared to be normal, with deliveries at 18.5 to 19.5 days. Pregnant homozygous mutant females built typical nests, groomed, retrieved, and nested with the newborn offspring. Pups were observed to latch on to the nipples of the mothers and appeared to suckle. Milk was present in the mammary glands of the homozygous mutant females, but mammary tissue was partially involuted and alveolar cells did not proliferate in the knockout mice. Milk was not observed in the stomachs of the offspring (Figure 8.6*b*). All pups delivered by homozygous mutant female mice died within 24 hours of delivery. If milk was given to the pups by the experimenter, the pups survived. If exogenous oxytocin was given to the mothers through injection by the experimenter, the pups survived. Oxytocin has long been implicated in uterine contractions during labor. Results from oxytocin knockout mice support an interpretation that female oxytocin knockout mice can mate, complete the full term of pregnancy, deliver normal sized litters, engage in maternal behaviors, and synthesize milk, but are unable to release milk for lactation.

Nitric oxide is a gaseous transmitter secreted from mammalian cells that acts as a neuromodulator in the brain and peripheral nervous system (Nathan, 1992; Dawson and Snyder, 1994; Moncada, 1997). Nitric oxide synthase is the synthetic enzyme for nitric oxide (Bredt and Snyder, 1994). Nitric oxide synthase is present in neurons of the pelvic plexus, the dorsal penile nerve, and the cavernous nerve. Nitric oxide synthase inhibitors block stimulated penile erection (Burnett et al., 1992). Mice with targeted disruption of neuronal nitric oxide synthase are viable and fertile (Nelson et al., 1995; Burnett et al., 1992). Western blot analysis showed no neuronal nitric oxide synthase in the penile tissue of the knockouts but greater amounts of endothelial nitric oxide synthase (Burnett et al., 1992). This finding is consistent with an interpretation that the endothelial form of nitric oxide synthase mediates nitric oxide-dependent penile erection.

Neuronal nitric oxide synthase knockout mice were observed for sexual behaviors during normal breeding of homozygous mutant males and homozygous mutant females. Knockout males displayed inappropriate mounting behavior, associated with considerable vocalizations by the females (Nelson et al., 1995). Male nitric oxide synthase knockout mice were tested for sexual behaviors in the presence of ovariectomized, estrogen-primed wildtype stimulus females. Number of mounts, latency to first mount, number of intromissions, number of ejaculations, and penile erection were not significantly different between the male knockouts and wildtype littermate controls (Burnett et al., 1992). These results suggest that nitric oxide synthase knockout mice have the ability to mate, but display inappropriate behavioral interactions between the male and female knockouts that prevent the normal mating sequence. High levels of intermale aggression in these knockouts is described in detail in Chapter 9.

Heinrichs and co-workers provide an interesting test of the adage "If overstressed then undersexed," in an elegant behavioral analysis of corticotropin releasing factor (CRF) transgenic mice (Heinrichs et al., 1997). As described in Chapter 10, over-expression of CRF results in transgenic mice that show persistent activation of the hypothalamic–pituitary–adrenal axis and anxiogenic-like behaviors. Both genders have intact gonads. Sexual and reproductive behaviors were analyzed in these male and female *CRF* transgenics. Sexually mature but inexperienced *CRF* transgenic males were paired with wildtype females who had been hormonally primed with estradiol and progesterone. Normal sniffing, mounting, and latencies to mount were observed in the *CRF* transgenic males, as compared to the wildtype controls. Sexually mature and experienced adult females were treated with estradiol and progesterone and paired with wildtype sexually experienced males. Female *CRF* transgenics showed no lordosis responses, received significantly less anogenital sniffing, and no mounts (Figure 8.1*b*).

Active kicking, rolling over, scurrying, and fending off the male were observed. The authors interpret the dramatic loss of sexual receptivity in the *CRF* transgenic females to the heightened stress and aggression levels in these mice, concomitant with higher levels of plasma corticosterone.

Targeted disruption of the gene for dopamine β-hydroxylase prevents the synthesis of norepinephrine and epinephrine in homozygous mutants (Thomas and Palmiter, 1997). Female *Dbh* −/− knockout mice showed normal fertility and delivered normal size litters. The mammary glands showed normal development. However, all pups died within a few days of birth, unless cross-fostered with wildtype mothers. Pups with knockout mothers were observed scattered within the bedding rather than gathered into the nest, indicating a deficit in maternal retrieval behavior. Further examination of the *Dbh* −/− mothers revealed that they engaged in nest building with the cotton squares provided, and they crouched over pups in the nest, but they did not retrieve pups into the nest (Figure 8.4*b*). Olfaction appeared to be normal in female *Dbh* −/− knockout mice, based on their ability to avoid a water bottle containing an aversive odorant, amyl acetate, in a saccharine solution. Hearing, necessary for parental detection of ultrasonic vocalization in separated pups, was not tested.

Peg3 is a paternally imprinted gene, expressed in the hypothalamus. *Peg3* knockout mice were generated to test the role of *Peg3* in parental behaviors (Li et al., 1999). Paternal transmission of the mutation resulted in heterozygous mutant females that showed longer latencies to retrieve pups, crouch over pups, and build nests. Pups of heterozygous mutant mothers were born with normal body weights but gained less weight during the first 3 weeks of life, as compared to pups of wildtype mothers.

A similar behavioral profile was present in mice with targeted disruption of the immediate early gene *fosB* (Brown et al., 1996). Null mutants of the *fosB* gene were healthy and viable, with normal appearance of the gonads, and normal rate of pregnancy. The majority of pups born to matings of homozygous mutant males and females died within 1 to 2 days of birth, unless cross-fostered to wildtype females. Serum levels of estradiol and progesterone were normal, as were messenger RNA levels and immunostaining for oxytocin, growth hormone, and prolactin, in the *fosB* −/− females. Examination of the home cage revealed pups scattered around the cage and poor pup retrieval by the *fosB* −/− mothers. Further, young nulliparous female and male *fosB* −/− mice showed reduced pup retrieval as compared to wildtype littermate controls. The knockouts were not anosmic, as they avoided an aversive odor, isovaleric acid or pentadecalactone. Performance on the Morris water task was normal, indicating normal learning. The authors conclude that *fosB* knockouts have a defect in a "nurturing" behavior. As hearing was not tested, it remains possible that the mothers were unable to detect the pup ultrasonic vocalizations that normally trigger retrieval to the nest. As described in Chapters 3, 4, and 5, evaluation of hearing, vision, olfaction for social odors, and motor abilities will help to rule out potential false positives and anthropomorphic overinterpretations of a behavioral phenotype.

A fourth case is seen in mice with a null mutation in G_{olf}, the G_{α} subunit of the G protein activated by olfactory receptor ligands (Belluscio et al., 1998), as described in Chapter 5. Two days after birth, 75% of G_{olf} −/− pups died. Mutant mothers did not nest, crouch over their pups, or retrieve pups. No milk was found in the stomachs of the pups. Olfactory receptors appeared normal in the olfactory epithelium of the knockout mice. Olfactory responses were measured electrophysiologically in sensory neurons of the olfactory epithelium, using electro-olfactogram recordings elicited by odors. Seven structurally distinct odors produced extracellular field potentials in wildtype controls. The magnitude of the responses were reduced 70–80% in the homozygous mutants. These results are interpreted by the authors as an anosmic phenotype, characterized by greatly reduced sense of smell in the G_{olf} knockout mice.

These several interesting knockout models address many of the elements of sexual behaviors and parent–pup behavioral interactions in mice. A variety of stimuli are involved in eliciting mating and parental behaviors in rodents, including olfactory, auditory, visual, and tactile cues (Beach and Jaynes, 1956; Doty, 1986; Lonstein and DeVries, 2000). Defects in communication due to deficits in a sensory modality could produce a false positive interpretation that sexual, reproductive, or parenting is abnormal. Ultrasonic vocalizations are particularly important in the parental retrieval response, as described above. It will be important to test knockout mice for the ability of the pups to emit these vocalizations and the ability of the parents to hear in the ranges of 50 to 70 kHz and below 10 kHz. Defects in the vocal system of the pups or the auditory system of the mothers would be an interesting explanation of the observed deficits in parental behaviors, which would preclude an interpretation of a more complex defect in parenting.

BACKGROUND LITERATURE

Baum MJ (2003). Activational and organizational effects of estradiol on male behavioral neuroendocrine function. *Scandinavian Journal of Psychology* **44**: 213–220.

Becker JB, Breedlove SM, Crews D (1993). *Behavioral Endocrinology*. MIT Press, Cambridge.

Burns-Cusato M, Scordalakes EM, Rissman EF (2004). Of mice and missing data: What we know (and need to learn) about male sexual behavior. *Physiology and Behavior* **83**: 217–232.

Carter S (1992). Oxytocin and sexual behavior. *Neuroscience and Biobehavioral Reviews* **16**: 131–144.

Carter CS, Kirkpatrick B, Eds. (1997). *The Integrative Neurobiology of Affiliation*, Vol. 807. Annals of the New York Academy of Sciences, New York.

Devries GJ (1990). Sex differences in neurotransmitter systems. *Journal of Neuroendocrinology* **2**: 1–13.

Erskine MS (1989). Solicitation behavior in the estrous female rat: A review. *Hormones and Behavior* **23**: 473–502.

Estep DQ, Lanier DL, Dewsbury DA (1975). Copulatory behavior and nest-building behavior in the wild house mouse *(Mus musculus). Learning and Behavior* **3**: 329–336.

Everitt BJ (1990). Sexual motivation: A neural and behavioural analysis of the mechanisms underlying appetitive and copulatory responses of male rats. *Neuroscience and Biobehavioral Reviews* **15**: 217–232.

Ganten D, Pfaff D (1986). *Neurobiology of Oxytocin*. Current Topics in Neuroendocrinology, Vol. 6. Springer, New York.

Hofer MA (1996). Multiple regulators of ultrasonic vocalization in the infant rat. *Psychoneuroendocrinology* **21**: 203–217.

Hofer MA, Shair HN, Brunelli SA (2001). Ultrasonic vocalizations in rat and mouse pups. *Current Protocols in Neuroscience*. Wiley, New York, pp. 8.14.1–8.14.6.

Ivell R, Russell J (1995). *Oxytocin*. Plenum, New York.

Lonstein JS, Fleming AS (2001). Parental behaviors in rats and mice. In *Current Protocols in Neuroscience*. Wiley, New York, pp. 8.15.1.–8.15.26.

Marler P, Hamilton WJ (1968). *Mechanisms of Animal Behavior*. Wiley, New York.

Meisel RL, Sachs BD (1994). The physiology of male sexual behavior. In *The Physiology of Reproduction*, 2nd ed., E Knobil, JD Neill, Eds. Raven, New York, pp. 3–105.

Nelson RJ (2005). *An Introduction to Behavioral Endocrinology*. Sinauer, Sunderland, MA.

Pfaff D, Frohlich J, Morgan M (2002). Hormonal and genetic influences on arousal—Sexual and otherwise. *Trends Neuroscience* **25**: 45–50.

Rissman EF, Wersinger SR, Fugger HN, Foster TC (1999). Sex with knockout models: Behavioral studies of estrogen receptor alpha. *Brain Research* **835**: 80–90.

Silver LM (1995). *Mouse Genetics. Concepts and Applications*. Oxford University Press, New York.

Sisk CL, Meek LR (1997). Sexual and reproductive behaviors. In *Current Protocols in Neuroscience*, JN Crawley, CR Gerfen, R McKay, MA Rogawski, DR Sibley, P Skolnick, Eds. Wiley, New York, pp. 8.2.1–8.2.15.

9

Social Behaviors

HISTORY AND HYPOTHESES

In the wild, *Mus musculus* is a social species that establishes group territories (Van Oortmerssen, 1971; Monaghan and Glickman, 1993). The size of the territory is related to the density of available food. Each social group is founded by one male and one or two females. Parents and offspring share a nest, built by the adults, that provides shelter, camouflage from predators, and conservation of body heat (Sluyter et al., 1995). The social group sleeps together in the nest. Both parents retrieve pups into the nest (Van Oortmerssen, 1971). Adult offspring emigrate from the group nest to establish new social groups in new territories. This family organization is in contrast to rat social colonies, in which a larger number of adult males and females may share an extensive underground burrow and tunnel complex (Lore and Flannelly, 1977). Both mouse and rat colonies are defended by a dominant male that attacks intruders that enter its territory (Blanchard and Blanchard, 1977; Kozorovitskiy and Gould, 2004). The present chapter discusses some of the tests commonly used for several types of social behaviors, including exploratory social interactions between two unfamiliar conspecifics, aggressive behaviors, and juvenile play behaviors. Sexual and parental behaviors are extensively discussed in Chapter 8.

In laboratory cage housing, adult mice are prevented from leaving the home environment to colonize new territories. Fighting between adult males in crowded, inescapable, group housing is common. Group-housed male mice often develop a social dominance hierarchy in which the dominant male attacks the others frequently. Overcrowding may lead to severe fighting and subsequent pathologies. Injury or scarring observed in the subordinates often necessitates removal of the dominant mouse from the cage.

A fascinating study of overcrowding in laboratory rats was conducted by John B. Calhoun at the National Institute of Mental Health in Poolesville, Maryland, in the early 1960s (Calhoun, 1962a, b). A 10 × 14 ft room was divided into four pens, with elevated burrows, winding staircases, tunnels, nest boxes, food hoppers, and drinking troughs. The population density increased over time. Then the population crashed, suffering severe mortality. Infant deaths reached 96%. Females did not carry pregnancies to full term or did not survive delivery of the litter. Males engaged in high levels of aggressive encounters, culminating in strong dominance hierarchies. The resulting dominant males patrolled large areas of territory. Subordinates grouped in the middle pen areas, venturing out to the food and water sources only early in the morning and often in pairs. The dominant male established a harem of females within his territory, preventing other males from engaging in sexual activity with these females. Females living within the harem built nests and raised young relatively successfully. Females living in the densely populated middle pen areas gathered fewer strips of nesting paper, did not form cup-shaped nests, did not retrieve pups into the nest, and seldom nursed. Pups abandoned throughout the pens quickly died and were eaten by adults. The abnormal social pressures generated by population densities had profound effects that might model aspects of aberrant social organization in other species.

Similar population dynamics were reported in mice trapped in the wild and contained in a 69 × 24 meter fenced enclosure (Van Oortmerssen and Busser, 1989). After a steady increase in numbers, the population declined from several hundred to 10 mice, then rose rapidly again, and crashed to zero, over a 30 month period. Since food and water were freely available in this experiment, a key factor appeared to be very high levels of fighting.

Early social deprivation in rodents induces behavioral changes that provide another set of models for aberrant social behaviors (Leigh and Hofer, 1973, 1975; Day et al., 1982; Hofer and Shair, 1987). Rat pups were isolated from their littermates at postnatal day 12, by removing all but one pup from the mother. Isolated pups were more reactive to handling, showing greater cardiac acceleration after being picked up and weighed by the experimenter, and lost more body weight after repeated handling. Isolation beginning at postnatal day 21 resulted in aberrant self-manipulation, including frequent chasing and biting of the tail, and higher incidence of shaking and convulsive activities. Rat pups separated from their mothers for short periods of time display marked elevations in stress reactivity and anhedonia at later ages (Meaney, 2001; Matthews and Robbins, 2003). Single housing of young adult male mice, beginning at 2 months of age and maintained for several weeks, induced high levels of aggressive attack behaviors (Valzelli, 1967), as discussed extensively below.

Two unfamiliar mice placed together in a novel environment will explore the new cage and investigate each other. Social interactions in rodents include approaching, following, sniffing, climbing onto and grooming each other (File, 1980, 1985; Winslow, 2003). Social recognition and social preference are measured in choice tasks and habituation/dishabituation sequences (Dantzer, 1998; Ferguson et al., 2002; Winslow, 2003; Bielsky et al., 2004). Strain distributions have been described for aspects of social approach behaviors, reciprocal social interactions, and preference for social novelty (Moy et al., 2004; Nadler et al., 2004; Bolivar and Flaherty, 2003). Allogrooming, wherein one animal grooms the fur of another, appears to be a form of social interaction in mice (Strozik and Festing, 1981; Mondragon et al., 1987). Barbered fur and trimmed whiskers are reportedly more common in females and subordinates (Stozik and Festing, 1981; Militzer and Wecker, 1986; Mondragon et al., 1987; Sarna et al., 2000). Excessive allogrooming can lead to alopecia lesions of the skin (Militzer and Wecker, 1986), and absence of whiskers (Garner et al., 2004; Long et al., 2004), which can impair the health and behavioral abilities of the overgroomed mice but can be reversed by removal of the individual(s) initiating the overgrooming (Sarna et al., 2000; Lijam et al., 1997). Juvenile play behavior has been extensively reported for rodents (Panksepp, 1981; Thor and Holloway, 1984; Siviy and Panksepp, 1987; Siviy et al., 1990; Pellis et al., 1992; Branchi and Ricceri, 2002; Terranova and Laviola, 2006). Rough-and-tumble chasing and wrestling in juvenile rats contains many components of adult fighting behavior, including one animal being pinned down by the other. Aggressive behaviors in the wild serve the functions of defending progeny, mates, and self from attack by a predator, defending the resources of the home territory, gaining access to a sexual partner, and achieving dominant status within the social group (Lorenz, 1966; Blanchard et al., 2003). In the laboratory environment the dominant male monitors and defends the entire cage space, occasionally attacking other males. The resident/intruder test, described below, measures territorial aggression in male rodents (Winslow and Miczek, 1983; Pich et al., 1993; Heinrichs and Koob, 1997; Miczek et al., 2001).

Circulating levels of endogenous testosterone partially determine aggressiveness in male mice (Wilson, 1975; Monaghan and Glickman, 1993; Nelson and Chiavegatto, 2001). Isolation of a male mouse may elevate levels of testosterone and of brain monoamine neurotransmitters, leading to higher levels of attack behavior (Welsh and Welsh, 1969; Eichelman and Thoa, 1973; Monaghan and Glickman, 1993; Monahan and Maxson, 1998). Isolation-induced fighting is a standardized method to compare genetic strains of mice for aggressive tendencies and to investigate neuroanatomical and neurochemical mechanisms underlying aggressive behaviors (Valzelli, 1967; Crawley and Contrera, 1976; Winslow and Miczek, 1983; Guillot and Chapouthier, 1998). Fighting often begins with the dominant male mouse approaching and rapidly "tail rattling," that is, thumping its tail against the floor of the cage in quick staccato bursts. The subordinate male mouse responds by rearing up to face the attacker, and raising its forepaws up in the air, described as the *submissive posture* (Grant and MacIntosh, 1963). Dominant/subordinate postures are seen in many species of rodents and other mammals, as first described by Charles Darwin (Darwin, 1872; Eisenberg, 1963; Marler and Hamilton, 1968).

Genetic components underlying aggressive behaviors have been demonstrated in many studies of inbred strains of rats and mice (Crusio, 1996; Mineur et al., 2003; Feldker et al., 2003; Canastar and Maxson, 2003). Strain distributions of inbred strains of mice on male isolation-induced fighting behavior and standard opponent fighting have been extensively described (Southwick and Clark, 1968; Van Oortmerssen and Bakker, 1981; Schneider et al., 1992; Guillot et al., 1995; Sluyter et al., 1996). Klaus Miczek and co-workers at Tufts University, and Frans Sluyter and co-workers at the Institute of Psychiatry in London, have selectively bred lines of mice with low or high levels of attack behaviors (Weerts et al., 1992; Miczek et al. 2001; Sluyter et al., 2003). Wim Crusio and co-workers cross-bred closely related inbred strains of mice that differed on isolation-induced aggression (Schicknick et al., 1993). Congenic strains of mice were established by backcrossing for 10 generations to select for low levels of isolation-induced aggression (Schneider-Stock and Epplen, 1995). High nest building co-varied with high levels of aggression in the standard opponent test (Lynch, 1980; Sluyter et al., 1995). Stress and neuroendocrine markers of stress correlate to higher levels of attack behaviors in mice (Mineur et al., 2003; Tordjman et al., 2003). Loci on the Y chromosome have been implicated in behavioral genetics studies of mouse aggression (Roubertoux et al., 1994; Maxson, 1996) Results from these studies point to polygenic determinants of aggressive traits.

Novice investigators sometimes confuse aggressive behavior with hyperresponsivity to handling. To all of us who have been bitten by a mouse, I extend my sympathies. However, a mouse biting a human handler is not displaying aggressive behavior, as defined by the natural mouse repertoire. Rather, the mouse is displaying a response to the stress of handling and/or to the fear of a predator, which is more correctly interpreted as a defensive behavior. True measures of aggression are primarily interactions between conspecifics, that is, two mice of the same species.

BEHAVIORAL TESTS

Social Affiliation

Huddling represents the tendency of mice to sleep close together in the home cage. Presence or absence of huddling can be observed while the animals are resting or sleeping, usually during the light phase of the circadian cycle. To avoid disturbing the animals, it is best to videotape the observations for subsequent scoring of frequency of huddling (Lijam et al., 1997). *Nesting* is quantitated in the home cage. As described in Chapter 8, nesting material is placed in the bottom of the cage. The nesting material, such as clean cotton batting (Mountain Mist), Nestlet squares, strips of paper, or scraps of cloth, is freshly provided each day. Materials are placed in the food hopper of the cage lid or on the cage bottom. The mice remove the cotton material and arrange it to form a nest. The cotton remaining in the hopper is weighed daily, to measure use in nest building (Sluyter et al., 1995; Lonstein and Fleming, 2001). Alternatively, the observer scores a videotape of a fixed time period, for example, one hour, from presentation of the nesting material. Latency to begin nest building, time to completion of the nest, total height of the nest, identity of individuals participating in building the nest, and identity of mice subsequently sleeping in the nest, are useful quantitative measures of social nesting behavior (Lijam et al., 1997; Lonstein and Fleming, 2001). Figure 8.3 illustrates typical and atypical huddling and nesting.

Juvenile play behavior in mice and rats is tested by observations of 20- to 31-day-old pups in a 31 × 31 × 32 cm test cage under low red light illumination (Thor and Holloway, 1984; Siviy and Panksepp, 1987; Terranova and Laviola, 2006). Each test animal is placed with a stimulus animal of the same age for a 10-minute observation period, or for a longer 40-minute session. In some experiments the stimulus animal is rendered unresponsive by pretreatment with a sedative. The test animal is scored for play solicitation on three measures. Number of crossovers is used as the measure of approaches to the stimulus animal. Number of play grooms measures the incidence in which the test rat pup pulls at the stimulus animal's fur or tail. Darts represent the test animal's running either toward or away from the stimulus animal. Videotracking with subsequent scoring by event recorder allows scoring of associated measures in mice, including social grooming, social sniff, follow, mutual circle, crawl under/over, and push under. Juvenile play behavior postures in rats are illustrated in Figure 9.1.

Allogrooming and *whisker trimming* in pair housed and group housed mice may be videotaped, and/or the presence or absence of whiskers is scored for individually identified mice in each home cage (Sarna et al., 2000; Garner et al., 2004). Allogrooming

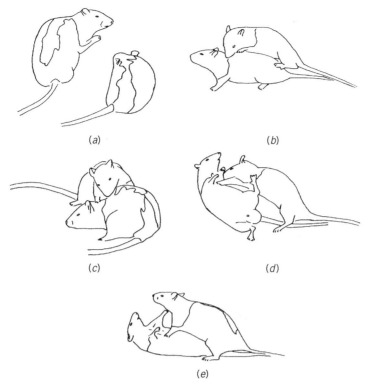

Figure 9.1 *Juvenile rat play behavior is similar to adult fighting, with more defensive responses and more ventral–ventral contact during play fighting. The defender may swerve away from the attacker (a), the attacker may approach the nape of the neck from above (b) or behind (c), and a cephalocaudal rotation (d) may bring the defender to a supine position (e).* [Adapted from Pellis et al. (1992).]

is grooming of the fur of one individual by a conspecific (Mondragon et al., 1987). Barbering or whisker trimming is seen in some mouse populations (Long, 1972; Militzer and Wecker, 1986). Individuals will groom and bite off the whiskers or vibrissae of a cagemate (Strozik and Festing, 1981). Ian Whishaw and co-workers at the University of Lethbridge, Canada, videotaped and scored whisker barbering, during mutual grooming in C57BL6 mice (Sarna et al., 2000). Their publication is entitled "the Dalila effect."* Barbering behavior is initiated by one member of a pair of mice, usually the larger dominant male. The barber uses its incisor teeth to grasp individual whiskers of the cagemate, and plucks out whiskers. The recipient appears to be passive during the whisker barbering, and often follows the barber for additional grooming.

The Jackson Laboratory undertook a study in 1996 and 1997 to determine the factors leading to excessive whisker trimming, body hair loss, and dermatitis that was present in many of their commercially sold production colony strains and stock, particularly C57BL/6J (Dr. David D. Myers, written communication to Jackson Laboratory customers, November 27, 1996). Infectious, immunologic, and genetic factors were ruled out. Noise levels, nonventilated caging, age at weaning, diet, chemical content of the water, and many other factors were examined. Dietary factors and age of weaning

---

*From the Bible story of Delilah's cutting the hair of Samson, who then lost his strength.

were discovered to be major contributors to excessive hair loss in these mouse colonies. Reducing dietary fat from 11 to 6% and delaying weaning from 3 to 4 weeks of age, significantly reduced the incidence of alopecia and dermatitis. Removal of the dominant male or the female(s) that were the primary initiators of the allogrooming, or separating the individual mice showing loss of whiskers and bare patches in the fur, will usually allow recovery and regrowth of the whiskers and fur (Lijam et al., 1997; Sarna et al., 2000). Normal and abnormal appearance of the whiskers is illustrated in Figure 9.2.

Social interaction between two mice who have not met before includes following, sniffing, grooming, and physical contacts that may lead to fighting. Social interactions are scored by an observer from a videotape of a 5- or 10-minute session. Sandra File and co-workers at the University of London perfected a scoring system for rats that shows high reliability between raters and demonstrates aspects of social interaction that are sensitive to anxiolytic drug treatments (File and Hyde, 1978; de Angelis and File, 1979; File, 1980, 1985; File, 1997). As described in Chapter 10, the social interaction test apparatus is a novel or a familiar environment, either a large or a small open

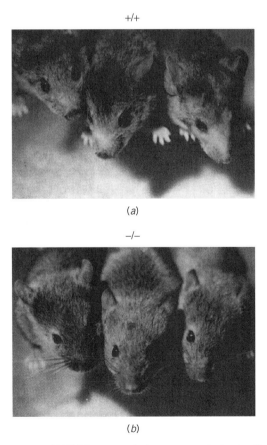

Figure 9.2 *Whisker barbering. (a) Whiskers are missing, apparently due to allogrooming in the home cage. (b) Normal appearance of the vibrissae is seen in the mouse on the left.* [From Lijam et al. (1997), p. 898.]

field, under either low or high illumination. Different levels of social interaction are obtained with each permutation of these environmental variables. Since the quantity of social interaction varies with ambient light level, this task is conceptualized as a anxiety-related test that is responsive to anxiolytic drug treatments in rats, as described in Chapter 10.

Jim Winslow at Emory University has provided step-by-step protocols for measuring social interaction, social recognition, social discrimination, partner preference, and social memory tests in mice (Winslow, 2003). These choice tests employ chambers with two or three compartments (Figure 9.3). In general, mice display good social recognition and social memory, defined by a decline in investigation over repeated encounters between the same mice, and preference for a novel partner (Winslow, 2003; Moy et al., 2004). The subject mouse is scored for olfactory investigation, allogrooming, offensive postures, attack, pursuit, defensive postures, escape, mounting, side-by-side contact, and other social behaviors, as well as for motor abilities. One social choice test protocol involves tethering stimulus mice to the side chambers. The freely moving subject mouse chooses to interact with one or both stimulus partners. Sequential sessions in a habituation–dishabituation design provides a more specific measure of social memory. Test sessions are videotaped for later rating by observers, using data entry software such as the Noldus Observer. Another approach is to evaluate the social tendencies of a subject mouse in the absence of responses from the target mouse. Approaches by the subject to a noninteractive, contained stranger are scored by a human observer in real time or from videotapes, or in an automated apparatus containing photocells that detect entries and time spent in each chamber (Hahn and Schanz, 1996; Moy et al., 2004; Nadler et al., 2004; Brodkin et al., 2005). Strain differences in levels of social approach indicate a genetic basis for sociability (Bolivar and Flaherty, 2003; Moy et al., 2004; Brodkin et al., 2004, 2005). Chapter 12 describes the application of social approach tests to model autism-related behaviors. Figure 12.5 illustrates an automated system for quantitating time spent in a chamber containing a novel conspecific, with controls for time spent exploring a novel object, and general exploratory locomotion in the novel environment.

Aggression

Male mice engage in fighting behavior and develop dominance hierarchies beginning at juvenile ages (Terranova et al., 1998). One method to determine the dominance hierarchy within a group of male mice is the *round-robin* approach (Southwick and Clark, 1968). Each combination of pairs of male mice from the group is tested for dominant versus submissive postures and attacks. Scoring across the test sequence determines the ranking of each male, which defines the dominance hierarchy. Comparisons of inbred strains were feasible with this method when each individual of each strain was used only once (Brain and Poole, 1974; Hahn and Schanz, 1996). The complicated number of comparisons makes the round-robin approach somewhat unwieldy.

The *standard opponent* method is a simpler approach to evaluate aggression and dominance status. One mouse is chosen for its highly replicable behavior as a submissive male in repeated tests with other males. Another mouse is chosen for its highly replicable behavior as a dominant male in repeated tests with other males. These standard test partner opponents are generally chosen from mouse strains known for their high or low levels of fighting behaviors. The chosen mouse is used as a standard

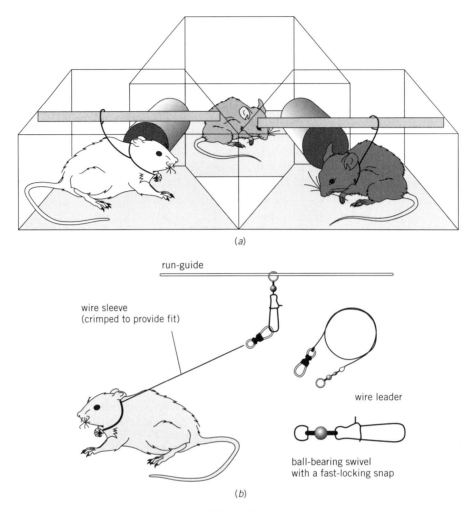

Figure 9.3   *Social preference apparatus. (a) The subject mouse, illustrated by the gray mouse in the central chamber, is given a choice between two stimulus mice, illustrated as the white mouse shown in the left chamber and the black mouse shown in the right chamber. (b) The stimulus animals are tethered, while the subject can freely explore all three chambers. Tethering prevents fighting initiated by the stimulus mouse. Variations on this task include social recognition, where one stimulus mouse is familiar and one stimulus mouse is novel, and social memory, in which a delay of 30 minutes, 120 minutes, 1 day, or several days intervene between the first and second exposure to the familiar stimulus mouse.* [From Winslow (2003), *Current Protocols in Neuroscience*, p. 8.16.10.]

test partner for subsequent pairings with experimental mice. Strain distributions have been defined for aggressive tendencies by pairing a standard opponent with individuals from each genetic strain of interest (Lagerspetz and Lagerspetz, 1974; Brain and Poole, 1974). Antianxiety drugs have been analyzed for reducing aggressive tendencies in mice using a standard opponent method (Valzelli, 1967).

A 5-minute fighting test session is observed from a distance of at least a meter, or through a window in the door of an environmental chamber, or videotaped for subsequent scoring by an investigator. Session lengths as short as 2 minutes or as long as 30 minutes have been used. If attacks and bites become severe, the session is terminated. Observers score the frequency of occurrence of behaviors, including general body sniffs, anogenital sniffs, following, chase, threat, tail rattle, attack, number of bites, location of bites, escape, upright subordinate posture, body contact, and allogrooming (Lagerspetz and Lagerspetz, 1974; Miczek, 1983; Weerts et al., 1992; Sluyter et al., 1995; Eklund, 1996).

Isolation-induced fighting is a modification of the standard opponent test for aggressive traits. Male mice are singly housed in the home cage for one month. Isolation increases the likelihood of attack behaviors (Valzelli et al., 1974; Crawley et al., 1975). Figure 9.4 illustrates typical postures during threat and fighting in mice. Table 9.1 defines some behavioral variables used in aggression tests. Table 9.2 shows some typical data for isolation-induced aggression scores in five strains of mice.

The *resident–intruder* test is a modification of isolation-induced fighting in which the standard opponent test is conducted in the home cage of the test mouse (Winslow and Miczek, 1983; DeBold and Miczek, 1984; File and Guardiola-Lemaitre, 1988; Tornatzky and Miczek, 1993; Pich et al., 1993). The standard opponent is the male intruder, prompting territorial attacks from the male resident test mouse in his home cage. Isolation is not necessary to elicit fighting in the resident–intruder test. Aggression is higher when the resident is living with his family. The test is conducted in a large home cage in which the mother and pups are within the cage during

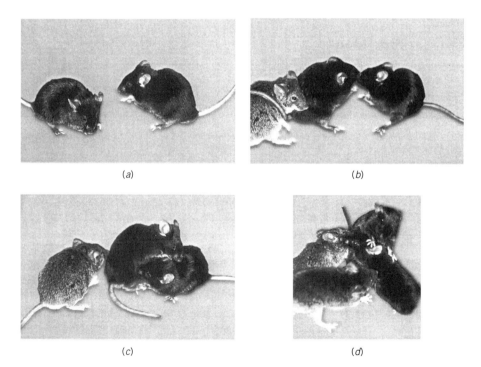

(a) (b)

(c) (d)

Figure 9.4 *Isolation-induced fighting behavior in male mice. Agonistic behaviors generally begin with anogenital contact, tail rattling by the dominant male, defensive upright posture by the subordinate male, attacks and biting by the dominant male on the rump of the subordinate male, and running escape by the subordinate.* [Photograph kindly contributed by Dr. Randy Nelson, Johns Hopkins University.]

TABLE 9.1   Definitions of Variables Used to Score Aggressive Encounters in Mice

Tail rattling: Rapid lateral quivering or thrashing of the tail, usually occurring just prior to or immediately following contact
Attacks: Biting of the opponent mouse
Aggressive grooming: Vigorous grooming of the opponent mouse from lateral position, using teeth and forepaws
Wrestling: Vigorous shoving and sparring when both animals take on an upright posture; it is usually performed by both animals simultaneously
Chasing: A rapid pursuit of the standard opponent by the test male, with or without physical contact
*L*1: Latency time to the first attack (in seconds) from the introduction of the C3H standard opponent male

Source: From Schneider et al. (1992), p. 199.

TABLE 9.2   Representative Data from Five Strains of Mice on Variables to Score Aggressive Encounters

Strain	*n*	Rattling	Attacks	Aggressive Grooming	L1 in (sec)
C57BL/6	25	0.1	0	0.8	600
		(0–2.3)	(0–8.5)	(0–2.7)	(175–600)
C3H	11	0.4	0.9	0.4	303
		(0–2.1)	(0–2.2)	(0.1–1.0)	(140–600)
BALB/c	23	2.4	2.7	0.6	240
		(0–7.5)	(0–8.0)	(0–1.3)	(10–600)
CBA	16	0.2	0.4	0.5	417
		(0–1.6)	(0–4.0)	(0–1.0)	(150–600)
NMRI	30	0.3	1.7	0.5	395
		(0–1.2)	(0–6.2)	(0–1.5)	(33–600)

Source: From Schneider et al. (1992), p. 199.
Medians with ranges are in parentheses.
n = number of animals.

the test but partitioned off from the area used for the fighting test. Females with pups will display aggression toward an intruder. Fighting between females without pups is relatively rare in comparison to male–male fighting in mice. Steve Heinrichs and George Koob at Scripps Research Institute in La Jolla, California, provide a step-by-step protocol for conducting the resident–intruder test (Heinrichs and Koob, 1997).

The *tube test* for social dominance is a simple test to measure aggressive tendencies, without allowing the mice to injure each other in a true fight. Dominant and submissive postures are scored, along with approach/avoidance behavior, for two mice during a brief pairing in a specialized chamber (Messeri et al., 1975; Hahn and Schanz, 1996). The social dominance tube test employs two start boxes, one at each end of a 30-cm long × 3.5-cm diameter clear plastic tube (Hahn and Schanz, 1996; Lijam et al., 1997; Klomberg et al., 2002; Salinger et al., 2003; Long et al., 2004; Spencer et al., 2005). Two mice of the same gender are placed at opposite ends of the tube. Both mice will begin to explore in a forward direction. If one mouse is dominant and the other is subordinate, the dominant mouse will approach while the subordinate backs away. A diagram of a social dominance tube and typical data obtained in the tube test are shown in Figure 9.5.

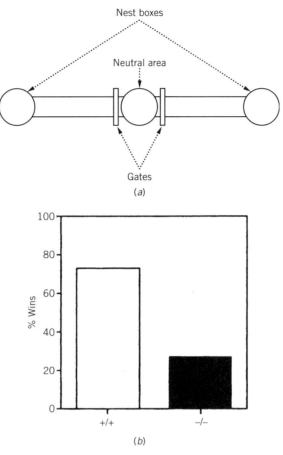

Figure 9.5 *Social dominance is tested in a tube-like apparatus. (a) The test mouse is placed in one nest box and the standard opponent in the other nest box. Gates at the end of each tube allow olfactory but not physical contact. When the gates are removed, the two mice will approach each other. Dominant and subordinate postures are scored. The dominant male generally forces the subordinate male out of the neutral area and back to its original nest box.* [From Hahn and Schanz (1996), p. 467.] *(b) Subordinate behavior is detected as reduced number of wins by dvl-1 knockout mice* [Lijam et al. (1997), p. 899.]

Robert and Caroline Blanchard at the University of Hawaii have elaborated an elegant Mouse Defense Test Battery (Blanchard et al., 1998, 2003a). Observational methods analyze a large number of behaviors including flight, freezing, defensive threat and attack, and risk assessment. Defensive behaviors are an important component of investigations of aggression, distinct from attack behaviors (Blanchard et al., 2003a). As seen in the visible burrow apparatus (Figure 9.6), developed by Drs. Blanchard, a mouse will avoid a rat predator by escaping into a distant chamber.

TRANSGENICS AND KNOCKOUTS

René Hen and co-workers at Columbia University generated one of the first knockout mice that showed a behavioral phenotype on a social behavior (Saudou et al., 1994;

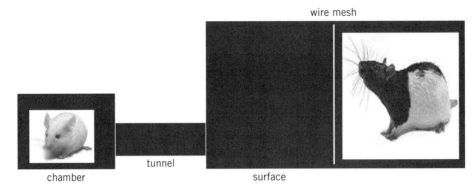

wire mesh

tunnel

chamber   surface

Figure 9.6 *Visible burrow, illustrating a mouse escaping from a rat predator.* [From Blanchard et al. (2005), p. 8.19.8.]

Hen, 1996). Null mutants deficient in the 5-HT$_{1B}$ serotonin receptor subtype were tested for isolation-induced aggression in a 3-minute test session. The test mouse was isolated, that is, first housed alone in the home cage for four weeks. A wildtype mouse that had been group housed was chosen as the intruder. For each test session, an intruder was placed in the home cage of the test mouse (the resident). Tail rattling, latency to attack, intensity of attacks, and number of attacks by the resident were scored. Two sessions, one week apart, were conducted for each resident test mouse. The null mutants showed shorter attack latencies and higher number of attacks, as compared to heterozygote and wildtype littermate controls. The knockouts were normal on baseline locomotion in the open field and showed no signs of any gross physical abnormalities. Feeding and reactivity to aversive stimuli were reported to be normal in the knockouts. The hyperlocomotion response to a 5-HT$_1$ agonist, RU246969, was absent in the knockouts, subsequent to a dose of RU246969 that had prominent hyperlocomotion effects in the wildtype controls, confirming the functional loss of the 5-HT$_{1B}$ receptor. The authors discuss the rodent literature implicating serotonin receptors in aggressive behaviors and the relevance of the observed behavioral phenotype of the *5-HT*$_{1B}$ knockout mice to genes postulated to regulate aggressive behaviors in humans.

Serotonin transmission was altered by targeted mutation of monoamine oxidase (MAO), the enzyme that metabolizes serotonin and catecholamines (Cases et al., 1995). *MAO-A* knockout mice showed elevated brain levels of both serotonin and norepinephrine, and reduced levels of the major serotonin metabolite, 5-hydroxyindoleacetic acid. Resident–intruder testing of 6-month-old male residents with a 2-month-old intruder measured latency to first attack over a 10-minute test session. Knockouts attacked significantly faster than wildtype littermate controls. In a separate test, younger males were isolated for 5 weeks and then tested in a similar resident–intruder pairing for 3 minutes. Latency to attack was again shorter in the knockouts.

Null mutation of the serotonin transporter resulted in lower levels of aggressive behaviors (Holmes et al., 2002). The serotonin transporter (5-HTT) is a presynaptic protein that takes up synaptically released serotonin and recycles the transmitter back into the presynaptic terminal for subsequent reuse. Low levels of serotonin have been implicated in aggression in humans and nonhuman primates (Brown et al., 1982; Heinz 1998; Krakowski, 2003). Male 5-HTT $-/-$ showed longer latencies to attack, and fewer numbers of bite attacks, when tested in two encounters on the resident–intruder task, as compared to $+/+$ littermates, while $+/-$ were not significantly different than $+/+$, indicating that a threshold level of 5-HTT is sufficient to maintain normal levels of aggressive behaviors (Holmes et al., 2002). Reduced home cage activity in the 5-HTT $-/-$, and insensitivity of the 5-HTT $-/-$ to a serotonin agonist that induced hyperlocomotion, suggest that low exploratory locomotion may have contributed to the phenotype of reduced aggression in 5-HTT null mutants (Holmes et al., 2002).

An unpredicted aggressive phenotype was discovered by Randy Nelson and co-workers at Johns Hopkins University in Baltimore for null mutant mice deficient in nitric oxide synthase, the synthetic enzyme for the neurotransmitter nitric oxide (Nelson et al., 1995; Kriegsfeld et al., 1997; Nelson and Young, 1998, Chiavegatto and Nelson, 2003). Cage deaths and intermale fighting were observed in home cages containing grouped male mice, leading to a careful study of aggressive behaviors. Videotapes of 8-hour sessions in the home cage were scored for aggressive and sexual encounters. Persistent fighting and mounting was seen in the male knockouts as compared to wildtype littermate control mice. In a 5-minute resident–intruder test, total number of attacks initiated by the male mutant mice was 3 to 4 times higher than wildtype littermate controls, and latency to attack was significantly faster. Subordinate postures were rarely observed in the knockouts. The high levels of attack behavior in the male mutants was testosterone-dependent (Kriegsfeld et al., 1997).

Adrenergic α_{2C} receptors mediate the release of norepinephrine from locus coeruleus neurons in the rodent brain (Cooper et al., 1996a). Null mutations in the α_{2C} receptor gene generated mice that showed enhanced startle and reduced prepulse inhibition (Sallinen et al., 1998), described in Chapter 10. The test animals were scored for attacks against a BALB/c standard opponent, in a 10-minute test session in a clean, neutral cage. The knockout mice showed faster attack latencies in isolation-induced aggression tests (Sallinen et al., 1998). Number of attacks was not significantly different from wildtype littermate controls.

Adenosine is a neurotransmitter with effects on blood pressure regulation, and adenosine receptors are a target for the psychoactivating effects of caffeine (Cooper et al., 1996a). Adenosine$_{2a}$ receptor knockout mice showed elevated blood pressure and high levels of attack behavior on the resident–intruder test (Ledent et al., 1997). Mice were tested in a 5-minute session, following 4 weeks of isolation, using 8-week-old CD1 male mice as intruders. Latency to attack was faster, number of attacks was highly elevated, and tail rattling episodes were more frequent in the mutants as compared to wildtype littermate controls.

Neural cell adhesion molecule is a glycoprotein that mediates cell recognition events during development (Walsh and Doherty, 1991). Isolated male mice deficient in the *NCAM* gene for neural cell adhesion molecule were tested over a 5-minute session in their home cages in the resident–intruder test with a C57BL/6J male intruder (Stork et al., 1997). Latency to tail rattle and latency to attack were greatly reduced, and number of attacks was greatly elevated in the *NCAM* knockouts, as compared to wildtype littermates. Heterozygotes were intermediate on all three measures. Breeding NCAM null mutants with a line of transgenic mice overexpressing NCAM effectively rescued the aggressive phenotype (Stork et al., 2000).

Substance P is a neuropeptide that transmits pain sensation in small diameter sensory fibers of the spinal cord (Chapter 5). Null mutants deficient in the gene for substance P showed lack of stress-induced analgesia and differential responses to morphine on analgesia tests (De Filipe et al., 1998). Substance P knockout mice displayed much longer latencies to attack, and much lower attack scores, when isolated males were observed over three 5-minute sessions in the resident–intruder test. One interpretation of the reduction in aggressive behaviors could be the lack of stress-induced analgesia found in these knockouts (De Filipe et al., 1998). Assuming that fighting is a stressor for mice, which produces a short-term analgesic effect, then reduced pain perception may allow the individual to continue fighting even if wounded. Since substance P knockouts lack the stress-induced analgesia protection, they may perceive the pain of a wound at a level sufficiently higher than normal to inhibit further fighting.

Vasopressin is a neuropeptide that facilitates aggressive behaviors in hamsters (Ferris, 2000). Vasopressin receptor V1b receptor knockout mice showed reduced aggression on the resident-intruder test (Wersinger et al., 2002, 2004). Male null mutants had significantly longer latencies to attack, number of attacks, and number of bites. Normal scores were obtained on control behaviors for olfaction, vision, motor abilities, growth curves, and time spent sniffing an unfamiliar female (Figure 9.7). Oxytocin is a related neurohormone and neurotransmitter that regulates milk ejection and reproductive behaviors (Carter, 1992, 2003). Phenotypes of oxytocin knockouts on parental behaviors are described in Chapter 8. Reduced aggressive behavior was seen in one line of oxytocin knockout mice (DeVries et al., 1997). In the resident–intruder test, following isolated housing, knockout and wildtype males did not differ on attack latency or number of attacks. In the 5-minute standard opponent test in a neutral arena, the genotypes also did not differ on latency to attack or frequency of attacks. However, on both tests the mutants spent less total time in aggressive encounters, showing significantly lower average durations of each aggressive attack. Another line of oxytocin knockout mice, generated by a different laboratory, demonstrated more aggressive attack behavior by the mutant males in isolation-induced and resident-intruder tests (Winslow et al., 2000). Background strains, DNA constructs, and breeding strategies are postulated differences between these two lines of oxytocin null mutants.

The remarkable role of oxytocin and vasopressin in pair bonding, social recognition, social affiliation, and social memory has been revealed in studies of prarie and montane voles, oxytocin knockout mice, and vasopressin receptor knockout mice (Carter et al., 1997; Winslow and Insel, 2002; Ferguson et al., 2002; Bales and Carter, 2003; Carter, 2003; Bielsky et al., 2003; Insel, 2003; Lin et al., 2004). Social recognition, defined by habituation to a novel conspecific in a social interaction task, was measured by reduced exploration of a novel mouse, and by reduced exploration of a novel mouse as compared to a familiar mouse. Reduced social recognition, termed social amnesia, was seen in oxytocin null mutants (Ferguson et al., 2001), and in vasopressin V1a receptor null mutants (Bielsky et al., 2003), as compared to social memory levels displayed by wildtype controls. Treatment with oxytocin restored social memory in oxytocin null mutants (Ferguson et al., 2000b). Vasopressin V1a null mutants treated with vasopressin displayed excess grooming behaviors (Bielsky et al., 2004). Additional deficits in oxytocin knockouts included reduced levels of ultrasonic vocalizations in separated pups, and less anxiety-like behaviors on the elevated plus-maze (Winslow et al., 2000; Winslow and Insel, 2002). However, olfactory abilities appeared to be normal in carefully control experiments with these knockouts. Impaired avoidance of odors from males infected with parasites, considered a socially relevant discrimination, was reported in female oxytocin mutants (Kavaliers et al., 2003). Spatial abilities

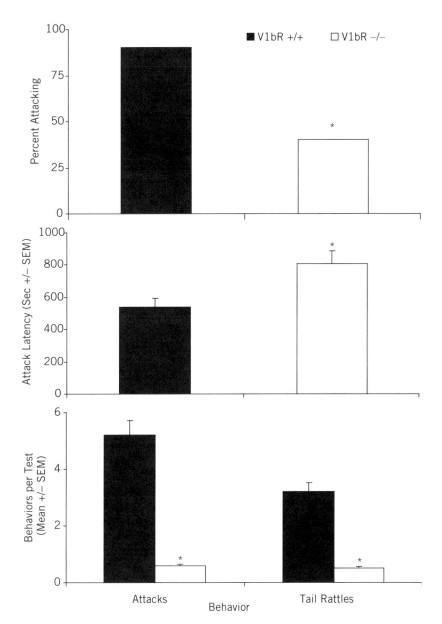

Figure 9.7 *V1bR−/− (N = 10) male mice are less aggressive than V1bR+/+(N = 10) littermates in a resident–intruder paradigm. A significantly lower percentage of V1bR−/− mice displayed aggressive behaviors as compared to wildtype V1bR+/+ mice (*P < 0.05; top panel). The mean latency (seconds ± SEM) to attack was significantly longer in V1bR−/− than V1bR+/+ mice (*P < 0.05; middle panel). The mean number (± SEM) of attacks per test and tail rattles per test was significantly lower in V1bR−/− than V1bR+/+ mice (*P < 0.05; bottom panel).* [From Wersinger et al. (2004), p. 641.]

were also normal in the oxytocin knockouts (Ferguson et al., 2001). Social recognition deficits were seen in both male and female oxytocin mutants (Winslow and Insel, 2004). The neural circuitry mediating the actions of oxytocin on social recognition appear to involve the amygdala, olfactory bulbs, piriform cortex, and dorsolateral septum, with additional activation of the somatosensory cortex, and hippocampus in the oxytocin knockouts (Ferguson et al., 2001; Winslow and Insel, 2004).

Aberrant social behaviors have been detected in mice with null mutations in the genes for other neuropeptide transmitters. Gastrin-releasing peptide receptor knockouts showed increased social investigation during social interaction and directed toward an anesthetized mouse (Yamada et al., 2000b). Greater social investigatory behavior was seen in both males and females, while olfactory ability appeared similar to wildtype controls for locating a buried food (Yamada et al., 2001). Bombesin knockouts displayed abnormal social behaviors along with hyperphagia and obesity (Yamada et al., 2000a). Male mice deficient in pituitary adenylate cyclase-activating polypeptide (PACAP) type 1 receptor (PAC1) displayed unusual social behaviors that differed between sexes (Nicot, 2004). Males engaged in higher levels of sexual mounting toward both females and males, as compared to wildtype controls. Females displayed a delay in affiliative behaviors with novel males.

Dishevelled-1 is a protein in the *wnt* signaling pathway that mediates segment polarity development in *Drosophila* (Klingensmith et al., 1994). Rich Paylor in our laboratory, in collaboration with the molecular genetics laboratory of Tony Wynshaw-Boris at the National Human Genome Research Institute, characterized the behavioral phenotype of null mutant mice deficient in *dvl1* (Lijam et al., 1997), as described in Chapters 5, 8, and 10. Reduced social interactions were seen in the null mutants. Videotapes of home cage behaviors during the light and dark phases of the circadian cycle indicated that the knockouts often slept in separate areas of the home cage, rather than huddled together. Nest building was measured over a 45-minute observation period after delivery of a Nestlet into the home cage. Nest depth was significantly reduced in the *dvl1* mutants, as compared to wildtype littermate controls Figure 8.3(*a*). The percentage of mice in the nest was significantly lower for the knockouts. Body temperature was similar across genotypes, which argues against an interpretation that the altered nesting patterns in the knockouts were caused by altered thermoregulation. The social dominance tube test was conducted between pairs of mutant and wildtype mice. The knockouts backed away from the social encounters, returning to their start area, significantly more often that the wildtype littermates (Figure 9.5*b*). Deficient nesting behaviors and low scores on social dominance in *dvl1* null mutants were replicated in an expanded study conducted in a different laboratory environment at the University of California San Diego (Long et al., 2004).

Many physiological factors can influence social behaviors. Apparent increases in aggressiveness may stem from multiple causes, including illness or hypersensitivity to pain. It is especially important to conduct observations on general health and sensory abilities of the mutant mice to avoid overinterpretations of the role of a gene in social behavior. Cases of abnormal social interactions, in which normal bodily health, sensory abilities, and motor functions are confirmed, become extremely interesting for furthering our understanding of the genes, neurotransmitters, and neural pathways regulating complex social behavior patterns. Mutation of TRP2, a putative ion channel receptor expressed exclusively in the vomeronasal organ, prevented activation of vomeronasal neurons to social odors, inhibited male–male aggression, and resulted in sexual courtship behaviors toward both male and females, in male TRP2 null mutant mice (Stowers et al., 2002). Hormones, neurotransmitters, and their receptors, assayed in knockout mice showing an unusual phenotype on a social interaction, are useful in elucidating direct and compensatory mechanisms in the neuropharmacological and neuroanatomical regulation of social interactions (Bielsky et al., 2003).

BACKGROUND LITERATURE

Blanchard RJ, Blanchard DC (1977). Aggressive behavior in the rat. *Behavioral Biology* **21**: 197–224.

Blanchard RJ, Wall PM, Blanchard DC (2003b). Problems in the study of rodent aggression. *Hormones and Behavior* **44**: 161–170.

Brain PF, Mainardi D, Parmigiani S, Eds. (1989) *House Mouse Aggression: A Model for Understanding the Evolution of Social Behaviour*. Harwood Academic, Chur, Switzerland.

Calhoun JB (1962). Behavioral sink. In *Roots of Behavior*, EL Bliss, Ed. Harper and Row, New York.

Carter CS, Kirkpatrick B, Eds. (1997). *The Integrative Neurobiology of Affiliation*. Annals of the New York Academy of Sciences 807, New York.

Crusio W, Ed. (1996). The neurobehavioral genetics of aggression. Special Issue: *Behavior Genetics* **26**: 459–533.

Ecklund A (1996). The effects of inbreeding on aggression in wild male house mice *(Mus domesticus)*. *Behaviour* **133**: 883–901.

Eichelman BS, Thoa NB (1973). The aggressive monoamines. *Biological Psychiatry* **6**: 143–164.

Gheusi G, Bluthé RM, Goodall G, Dantzer R (1994). Social and individual recognition in rodents: Methodological aspects and neurobiological bases. *Behavioural Processes* **33**: 59–88.

Hahn ME, Schanz N (1996). Issues in the genetics of social behavior: Revisited. *Behavior Genetics* **26**: 463–470.

Heinrichs SC, Koob GF (1997). Application of experimental stressors in laboratory rodents. In *Current Protocols in Neuroscience*. Wiley, New York, pp. 8.4.1–8.4.14.

Hen R (1996). Mean genes. *Neuron* **16**: 17–21.

Lagerspetz KMJ, Lagerspetz KYH (1974). Genetic determination of aggressive behaviour. In *The Genetics of Behaviour*, JHF Van Abeelen, Ed. Elsevier, North Holland, Amsterdam, pp. 321–346.

Lambert KG, Gerlai R (2003). A tribute to Paul McLean: The neurobiological relevance of social behavior. *Physiology and Behavior* Special Issue **70**: 341–547.

Lorenz K (1966). *On Aggression*. Methuen, London.

Marler P, Hamilton WJ (1968). *Mechanisms of Animal Behavior*. Wiley, New York.

Maxson SC (1996). Searching for candidate genes with effects on an agonistic behavior, offense, in mice. *Behavior Genetics* **26**: 471–476.

Murphy DL, Wichems C, Li Q, Heils A (1999). Molecular manipulations as tools for enhancing our understanding of 5-HT neurotransmission. *Trends in Pharmacological Sciences* **20**: 246–252.

Nelson RJ, Chiavegatto S (2001). Molecular basis of aggression. *Trends in Neurosciences* **24**: 713–719.

Panksepp J (1981). The ontogeny of play in rats. *Developmental Psychobiology* **14**: 327–332.

Scott JP (1985). Investigative behavior: Toward a science of sociality. In *Leaders in the Study of Animal Behavior*, DA Dewsbury, Ed. Associated University Presses, Cranbury, NJ, pp. 389–429.

Terranova ML, Laviola G (2005). Scoring of social interactions and play in mice during adolescence. *Current Protocols in Toxicology*. Wiley, New York, pp. 13.10.1–13.10.11.

Thor DH, Holloway WR (1984). Social play in juvenile rats: A decade of methodological and experimental research. *Neuroscience and Biobehavioral Reviews* **9**: 455–464.

Valzelli L (1967). Drugs and aggressiveness. *Advances in Pharmacology* **5**: 79–108.

Van Oortmerssen GA (1971). Biological significance, genetics and evolutionary origin of variability in behaviour within and between inbred strains of mice (*Mus musculus*): A behaviour genetic study. *Behaviour* **38**: 1–92.

Wilson EO (1975) *Sociobiology: The New Synthesis*. Harvard University Press, Cambridge.

Winslow JT (2003). Mouse social recognition and preference. In *Current Protocols in Neuroscience*. Wiley, New York, pp. 8.16.1–8.16.16.

Winslow JT, Insel TR (2002). The social deficits of the oxytocin knockout mouse. *Neuropeptides* **36**: 221–229.

Postdoctoral Pretreatment:

1985   1986   1987

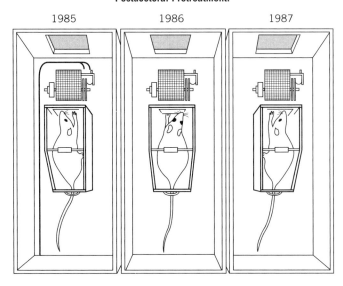

Job-Hunting Shuttle Task:

1988

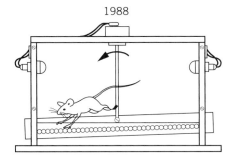

Learned helplessness apparatus *[Drawing kindly contributed by Robert Drugan, University of New Hampshire.]*

10

Emotional Behaviors: Animal Models of Psychiatric Diseases

HISTORY AND HYPOTHESES

Step one: *Don't anthropomorphize*! Emotions are personal, internal, and highly species-specific. There is no way for a human investigator to know whether a mouse is feeling afraid, anxious, or depressed. These are subjective emotional experiences, existing in the mind and body of the individual. Major mental illnesses involve neural circuitry that may be uniquely human. For example, schizophrenia appears to involve the prefrontal cortex, which is highly expanded in humans. Aberrant behaviors symptomatic of human mental illnesses therefore may not occur in a recognizable form in rodents.

What we *can* do is observe the behavioral and physiological responses that a mouse makes to stimuli and events. The anatomy, physiology, and neurochemistry of the mouse is similar to the human in many respects (De Souza and Grigoriadis, 1994; Cooper et al., 1996a; Chrousos, 1998). Most brain structures are homologous throughout mammals, albeit larger, more elaborated, or more extensively connected in some species than in others. While we cannot reproduce a human neuropsychiatric disease in a mouse, individual components of symptoms, causes, and treatment responses, can be mimicked. Single behavioral, anatomical, biochemical, and neurophysiological markers of human diseases have been termed *endophenotypes* (Gould and Gottesman, 2005).

For example, glutamate is the major excitatory neurotransmitter, and γ-amino butyric acid (GABA) is the major inhibitory transmitter, in the mammalian brain. Hormonal and neurotransmitter responses to stress have been explicated, and appear to be similar across mammalian species. The sympathetic nervous system reacts to acute stressors by releasing norepinephrine. The hypothalamic–pituitary–adrenal axis reacts to acute and chronic stressors by releasing corticotropin releasing hormone, adrenocorticotrophin releasing hormone, corticosterones, and epinephrine. Stressors activate the immune system to release cytokines. Benzodiazepine binding sites on the GABA receptor complex bind anxiolytic drugs such as diazepam (Valium) in both rodents and humans. Antidepressant drugs such as fluoxetine (Prozac) inhibit the serotonin transporter in both the rodent and the human brain. Pharmacological mechanisms of psychoactive drug responses are similar across many mammalian species. Typical antipsychotics block dopamine receptors in rodent and human striatum.

Reports of genetic linkages in major mental illnesses (reviewed in Gershon, 1990; Kendler, 1997; Kennedy and Macciardi, 1998; Kennedy et al., 2003; Lipsky and Goldman, 2003; Kelsoe, 2003; Berrettini, 2004) suggest that susceptibility genes contribute to symptoms of depression, manic-depression, obsessive-compulsive disorder, anxiety, and schizophrenia. Identified genetic components of psychiatric diseases raise the possibility of studying the actions of these genes and their products on brain function and behavior in mutant mouse models of psychiatric syndromes.

Animal models of human psychiatric diseases are designed to replicate one or more components of the disease. There are two major goals of animal models: (1) to test hypotheses about the mechanisms of the disease and (2) to predict treatment response in human patients. A good animal model satisfies many of the following criteria:

1. Robust and reproducible
2. Simple enough to be routinely performed in a laboratory environment
3. Quantitative and preferably easily automated
4. Replicates at least one symptom of the human disease (*face validity*)
5. Responds to treatments that are effective in the human disease (*predictive validity*)
6. Is unaffected by treatments that are ineffective in the human disease
7. Conceptual analogy to the cause of the human disease (*construct validity*)
8. Conceptual analogy to multiple components of the human disease:
   a. Behavioral symptoms
   b. Neuroanatomical pathology
   c. Neurochemical abnormalities
   d. Temporal progression
   e. Precipitating event

Recognizing the caveats and limitations of an animal model is fundamental to interpreting the mouse behavioral test results appropriately. Too much hyperbole has appeared in the transgenic and knockout literature, in which a behavioral phenotype is exaggerated as modeling the disease, rather than modeling a component or a symptom of the disease. Mistakes of overinterpretation are not limited to the mutant mouse literature, but have plagued the larger preclinical psychopharmacology literature in general.

Designing a good animal model starts with the set of diagnostic criteria for the human psychiatric disorder, and the symptoms that are unique to that disease. Some mental illnesses are highly comorbid, with symptoms of two or more diseases appearing in the same individual (Baumgartner et al., 1985; Breier, 1989; Swann et al., 1990; Goodwin and Jamison, 1990; Chrousos, 1998; DSM-IV, 1994; Lydiard et al., 1998; Schatzberg et al., 1998; Kendler et al., 1998). Stressful life events may precipitate a variety of illnesses including posttraumatic stress disorder, anorexia, episodes of depression, or episodes of psychosis. Exaggerated startle response and/or deficits in prepulse inhibition are variably symptomatic of schizophrenia, obsessive-compulsive disorder, posttraumatic stress disorder, aging, and Huntington's disease (Grillon et al., 1996; McDowd et al., 1993; Braff and Geyer, 1990; Paylor and Crawley, 1997). Some investigators and review boards suggest that the term "animal model" be avoided entirely (McKinney, 1989). Terms such as "paradigms" and "tasks analogous to a symptom of a human disorder" are more accurate. The present chapter will describe tests that satisfy many of the criteria above. Descriptions focus on rodent tests that have been extensively used to study emotional behaviors in the domains of fear, anxiety, depression, and schizophrenia. Autism, mental retardation, and other neurodevelopmental disorders are described in Chapter 12.

Fear of death in the form of aggressive neighbors, forest fires, predators, injury, starvation, dehydration, and freezing drives a great deal of behavior. Tests of fear-related behaviors in rodents employ stressful stimuli such as footshock, restraint, predators, separation from the nest, and repeated fighting episodes. Stress, defined as a state of threatened homeostasis (Chrousos, 1998), produces adaptive physiological and behavioral responses by the individual. Figure 10.1 diagrams the neurotransmitters, brain

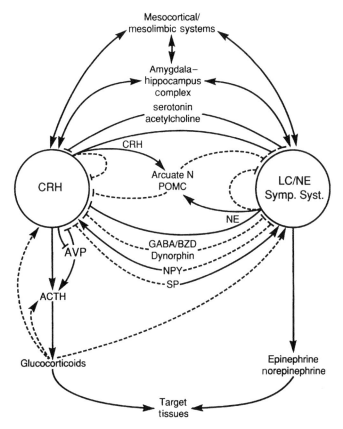

Figure 10.1 Schematic representation of the brain regions and peripheral organs mediating stress responses, indicating their functional connections and neurotransmitters. [From Chrousos (1998), p. 314.]

regions, and peripheral organs that contribute to the components of the stress response. A major brain region mediating fear responses is the amygdala. Rich in afferents, efferents, and neuropeptides, the amygdala is a key structure in the complex neural circuitry linking many forebrain and midbrain nuclei (Miserando et al., 1990; LeDoux, 1992, 1995; Davis et al., 1994; Pape and Stork, 2003; Koob, 2003; Misslin, 2003; Fanselow and Poulos, 2005). The hippocampus appears to mediate some forms of con- ditioned freezing but not fear-potentiated startle (McNish et al., 1997; Sanders et al., 2003). Genetic components underlying conditioned fear have been described in strain distribution studies, quantitative trait loci analyses of inbred strains of mice, and tar- geted gene mutations (Chen et al., 1996a; Wehner et al, 1997; Caldarone et al., 1997; McCaughran et al., 2000; Cook et al., 2002; Talbot et al., 2003; also see Chapter 6 for the learning and memory components of fear conditioning).

Many robust tests of rodent anxiety-related behaviors work well in mice. Most are based on conflicts (File, 1981, 1997a; Crawley, 1989; Finn et al., 2003). This category of anxiety-like behavior is thought to result from the conflict inherent in approach–avoidance situations. For example, you want a good grade in the course to gain admission to a good graduate or medical school, but you don't want to study for the killer final exam. You want to tell the world about your exciting research results, but have fears about the audience's response when you walk up to the podium to give your talk. A mouse may want to explore a new environment to find food, but may fear venturing out into the open, exposed space in the daylight, where it is an easy target for predators. Most rodent anxiety paradigms have analogies to acute anxiety episodes in humans, but are less relevant to severe, chronic anxiety disorders. Table 10.1 lists many of the animal models of anxiety-related behaviors in past and current use.

TABLE 10.1 Animal Models of Anxiety-Related Behaviors[a]

Conditioned Response Tests	Unconditioned Response Tests

1. Conflict tests
    Geller–Seifter conflict
    Pigeon and monkey conflict
    Vogel conflict
	1. Exploration tests
	    Elevated plus-maze
	    Free exploration
	    Light/dark exploration
	    Open field
	    Staircase test
	    Zero-maze

2. Others
    Active/passive avoidance
    Conditioned emotional response (CER)
    Conditioned suppression of drinking (CSD)
    Conditioned ultrasonic vocalization
    Defensive burying
    dPAG-induced flight
    Fear-potentiated startle
    Punished locomotion
	2. Social tests
	    Separation-induced ultrasonic vocalizations
	    Social competition
	    Social interaction
	3. Antipredator tests
	    Anxiety/defense test battery
	    Fear/defense test battery
	    Human threat test (primates)
	    Mouse defense test battery
	4. Others
	    Hyponeophagia
	    Pain-induced ultrasonic vocalizations
	    Startle response (baseline)

Source: Rodgers (1997), p. 478.

[a]Unconditioned response tests are measures of spontaneous, naturalistic behaviors within the innate repertoire of the animal. Conditioned response tests are measures of evoked behaviors in response to experimental conditions.

Rodent tasks relevant to human depression are primarily stress-induced reductions in avoidance or escape, termed *behavioral despair*. Behavioral neuroscientists find it difficult to identify mouse behaviors that display conceptual analogies to the chronic and cyclical components of human depression syndromes (Baumgartner et al., 1985; Breier, 1989; Swann et al., 1990; Goodwin and Jamison, 1990; Chrousos, 1998; DSM-IV, 1994; Lydiard et al., 1998; Schatzberg et al., 1998; Kendler et al., 1998). Related syndromes, including panic disorders, posttraumatic stress disorder, phobias, obsessive-compulsive disorder, anorexia nervosa, and bulemia, have been modeled in mice with moderate success. However, neurochemical abnormalities associated with depression, for example, low levels of serotonin, norepinephrine, and their metabolites, dysfunctions of neurotransmitters and hormones in the hypothalamic–pituitary–adrenal axis, and immune system dysfunctions, may have correlates in mice (Stenzel-Poore et al., 1994; Heninger, 1995; Campbell and Gold, 1996; Greenberg et al., 1998; Capuron and Dantzer, 2003; Watkins and Maier, 2005). Approved treatments for depression, including serotonin reuptake inhibitors such as Prozac and monoamine oxidase inhibitors such as amitriptyline, modulate neurotransmission in mouse brain in a manner that may be analogous to the actions of these drugs in the human brain.

The most widely used rodent models of symptoms of depression are the *Porsolt forced swim* task (Porsolt et al., 1977, 1978; Lucki et al., 2001), *tail suspension* (Cryan et al., 2002; Ripoll et al., 2003), and the Seligman and Maier *learned helplessness* task (Seligman and Maier, 1967; Maier, 1984; Peterson et al., 1995). These tests are based on the assumption that animals will normally try to escape from an aversive stimulus. When the aversive stimulus is inescapable, the animal will eventually stop trying to escape. Early cessation of attempts to escape has predictive validity, and perhaps some face validity, to stress-induced depression. Antidepressant drug treatments restore the escape behavior, providing the pharmacological validation for conceptualizing these tasks as depression-related (Porsolt et al., 1977a, 1977b; Detke et al., 1995; Maier, 1984; Bai et al., 2001; Cryan et al., 2002; Ripoll et al., 2003; Caldarone et al., 2003). In addition anti-anxiety drugs prevent some forms of behavioral despair, and anxiety-producing drugs mimic inescapable aversive stimuli in inducing learned helplessness, indicating an anxiety-like component inherent in some of these tasks (Drugan et al., 1984, 1985). Table 10.2 summarizes some of the mouse models that are sensitive to treatment with antidepressant drugs.

Rodent tests designed to model symptoms of schizophrenia have traditionally involved dopamine agonist-mediated hyperactivity, psychostimulants, stressors, and responses to antipsychotic dopaminergic antagonists (Iversen and Iversen, 1981; Deutch et al., 1985; Segal and Kuczenski, 1997; Laviola et al., 2002; Weiner, 2003). Prepulse inhibition deficits seen in schizophrenic patients have been replicated in rodent paradigms (Geyer et al., 1990; Geyer and Swerdlow, 1998). Symptoms such as auditory hallucinations and delusions are less likely to find an animal correlate than deficits in processing of sensory information. Negative symptoms, cognitive impairments, neurochemical abnormalities, sensorimotor gating deficits, and attentional abnormalities are more amenable to modeling in mice (Weinberger and Berman, 1996; Geyer and Ellenbroek, 2003; Weiner, 2003).

Elevated levels of the dopamine metabolites, dihydroxyphenylacetic acid and homovanillic acid, in the cerebrospinal fluid and urine of schizophrenic patients (Pickar, 1998; Pickar et al., 1990) have been experimentally duplicated in rodents (Deutch et al., 1985; Kalivas and Nemeroff, 1988; Kalivas et al., 1993). Approved treatments

TABLE 10.2 Widely Used Rodent Models Sensitive to the Effects of Antidepressant Agents

Animal Model	Ease of Use	Reliability	Specificity	Applicable to Mice	Comments	Refs
Forced swim test	High	High	High[b]	Yes	Sensitive to acute antidepressant treatments; does not reliably detect SSRIs	[6,7]
Modified forced swim test	High	High	High[b]	?	Sensitive to acute antidepressant treatments; differentiates antidepressants from different classes including SSRIs	[9]
Tail suspension test	High	High	High[b]	Yes	Sensitive to acute antidepressant treatments; certain strains climb their tail	[7,58]
Olfactory bulbectomy	Medium	High	High	Yes	Behavioral effects evident only following chronic treatment; mechanism of action poorly understood	[17,18]
Learned helplessness	Medium	Medium	High	Yes	Sensitive to short-term antidepressant treatments; ethical restrictions in some countries	[7,29,30]
DRL-72	Medium	Medium	Medium	?	Sensitive to short-term antidepressant treatments	[7]
Neonatal clomipramine[c]	Medium	Medium	?	Yes	Only limited testing of antidepressants have been conducted	[29]
Prenatal stress	Medium	?	?	Yes	Only limited testing of antidepressants have been conducted	[59,60]
Chronic mild stress	Low	Low	High	Yes	Reliability has been questioned repeatedly; behavioral effects evident only following chronic treatment	[31,39,61]
Resident intruder	Low	?	Medium	?	Distinguishable behavioral effects only following chronic treatment; requires further validation in other laboratories	[62]
Drug-withdrawal-induced	Low	High	Medium	Yes	Requires further validation: cannot asses baseline strain differences easily	[36,37,40,41]

Source: From Cryan, Markou and Lucki, (2002), p. 239.

for schizophrenia include typical neuroleptics such as haloperidol (Haldol), which are antagonists of the D-2 dopamine receptor, and atypical neuroleptics such as clozapine, risperidone, and olanzapine, which are antagonists of both dopamine and serotonin receptors (Breier, 1996; Kennedy et al., 2001; Meltzer et al., 2003; Lieberman, 2004). Rodent models of symptoms of schizophrenia are often validated by the pharmacological specificity of their responsiveness to these classes of drugs. Further, dopamine and serotonin receptor ligands are used as pharmacological challenge tests to investigate genetic mutations of these receptors in mutant mouse models.

BEHAVIORAL TESTS

Fear-Related Behaviors

The earliest quantitative tests of fear-related behavior in mice measured *freezing* and *defecation* in a brightly lit open field environment (DeFries et al., 1974; Blizard and Bailey, 1979). As described in Chapter 4, the original open fields were large wooden enclosures, marked off into regions or square blocks. The investigator counted the number of squares crossed and number of fecal boli deposited in time intervals ranging from minutes to hours. Photocell-equipped automated open fields are currently used to give a preliminary indication of fear-related behavior, seen when the mouse remains in the corners or near the perimeter of the arena (*thigmotaxis*) rather than moving out in the center of the arena. More time spent in the center of an open field indicates a possible anxiolytic-like (anti-anxiety) effect of a mutation or drug treatment. Less time in the center may reflect fear-related traits. However, center time can be affected by many factors, including locomotor and sensory deficits. A positive finding in an open field test must be considered preliminary, prompting further experiments with more specific fear-related or anxiety-related tests, as described below. Genetic components of open field traits have been demonstrated in strain distributions of inbred strains of mice (reviewed in Crawley et al., 1997a; Bolivar et al., 2000a) and in selected lines of rats such as the Maudsley Reactive strain (Blizzard, 1981). Quantitative trait loci analyses of activity and defecation in an open field (Figure 10.2) mapped loci on mouse chromosomes 1, 12, and 15 for these fear-related behavioral traits (Flint et al., 1995; Henderson et al., 2004).

Freezing is a fear-related behavior common to many species (LeDoux, 1995). When confronted suddenly with a bright light, a painful stimulus, or a predator, the animal may stop whatever it was doing and stand perfectly still, as described in Chapter 6. Rats and mice freeze immediately after a footshock (Kalin et al., 1988; Swiergel et al., 1992; LeDoux, 1992, 1995, 2002; Holmes et al., 1994; Davis et al., 1994; Fanselow and Gale, 2003). Freezing is measured as total immobility, other than respiration. One or more footshocks of 0.3 to 0.5 mA and 1 or 2 seconds duration are administered to the animal. Immediately after the footshock(s), bouts of freezing are scored as present or absent, by observation every 5 or 10 seconds, or more frequently in automated systems, for 3 to 6 minutes.

Contextual and cued fear conditioning is an extension of fear-induced freezing. Rats and mice are conditioned to freeze to environmental cues that were previously associated with the footshock, as extensively described and referenced in Chapter 6. Contextual and cued fear conditioning are measured 24 and 48 hours later. The task assesses memory of the visual, tactile, and olfactory context, previously paired with the

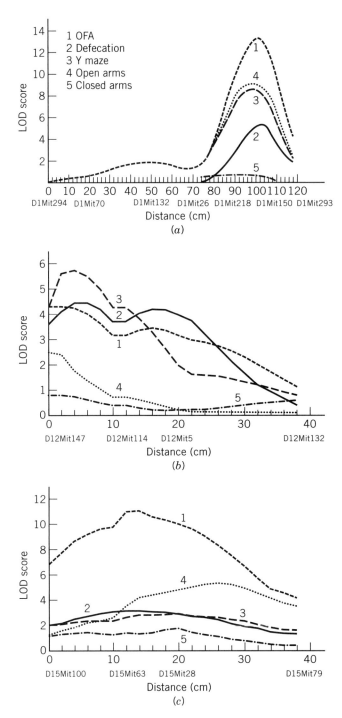

Figure 10.2 *Quantitative trait loci analysis of open field activity (OFA), defecation, activity in a Y-maze, and entries into the open and closed arms of an elevated plus-maze in selected inbred lines of mice. Highly significant LOD scores were detected on chromosomes 1, 12, and 15.* [From Flint et al. (1995), p. 1433.]

footshock, to a single cue, usually the paired auditory tone or white noise (Fanselow and Tighe, 1988; Fanselow 1990; Chen et al., 1996a; Caldarone et al., 1997; Wehner et al., 1997; Fanselow and Poulos, 2005). This task combines learning and memory with an emotional component that depends on an intact amygdala (LeDoux, 1996; Fanselow and Gale, 2003). A commercially available apparatus for measuring cued and contextual conditioning is shown in Figure 6.10.

Fear-potentiated startle is a fear-conditioning test that measures the acoustic startle response after pairing the startle stimulus with a footshock. Rats given a pairing of footshock with an auditory startle cue will show a larger amplitude of whole body flinch to the auditory tone startle stimulus, when it is repeated at a later time without the footshock, as compared to the amplitude of whole body flinch to the same tone by rats with no previous conditioning (Davis et al., 1993, 1994; Gewirtz and Davis, 1998). Bill Falls at the University of Vermont successfully adapted fear-potentiated startle for mice, demonstrating greater fear-potentiated startle in DBA/2J than C57BL/6J (Falls et al., 1997). Bob Hitzemann and co-workers at the State University of New York at Stony Brook found a continuous distribution of fear-potentiated startle responses in F_2 mice from the cross of C57BL/6J and DBA/2J (McCaughran et al., 2000). Step-by-step instructions for conducting the procedures to measure fear-potentiated startle in mice are available in *Current Protocols in Neuroscience* (Falls, 2002). Equipment to induce and measure fear-potentiated startle is shown in Figure 10.3. This task appears to model components of posttraumatic stress disorders in humans (David et al., 1997; Zhao and Davis, 2005).

Ultrasonic vocalizations are emitted by rodents in response to stressors, possibly serving social communication functions. As described in Chapters 8 and 12 and illustrated in Figures 8.5, 8.7, 10.4, and 12.3, infant mouse pups call to their parents in the frequency range of 50 to 70 kHz (Zippelius and Schleidt, 1956; Hofer, 1996; Hofer et al., 2001; Fish et al., 2004). The vocalization appears to function as a distress call in response to separation. The pup vocalizes when isolated from its mother and its nest, and stops when reunited with the mother and retrieved to the nest. Number of calls declines after postnatal day 12 in mice (Hahn et al., 1998; Hofer et al., 2001). Frequency of ultrasonic vocalizations can be quantitated by an observer equipped with headphones and an ultrasonic detector. Noldus offers an automated system, Ultravox, that records, stores, and calculates time events for pup ultrasonic vocalizations. Commercially available equipment is sensitive to dominant frequencies in the range of 2 to 100 kHz (Brunelli et al., 1997). Myron Hofer and co-workers at New York State Psychiatric Institute in New York City demonstrated a genetic component to the trait of high versus low amounts of ultrasonic vocalization in rat pups (Figure 10.4). A line of high vocalizing pups and a line of low vocalizing pups have been maintained for several generations (Brunelli et al., 1997, 2002; Brunelli, 2005). Rat pup ultrasonic vocalizations are inhibited by treatment with anxiolytic drugs such as benzodiazepines (Olivier et al., 1998) and with antidepressant drug treatments (Winslow and Insel, 1990). Klaus Miczek and co-workers at Tufts University in Medford, Massachusetts confirmed the ability of anxiolytics and antidepressants to reduce ultrasonic vocalizations in separated mouse pups (Rowlett et al., 2001; Fish et al., 2004).

Adult rats emit 22-kHz vocalizations in the presence of a predator (Blanchard et al., 1991) and in an aggressive encounter with a dominant male (Vivian and Miczek, 1993). These appear to be alarm calls, conveying information to other conspecifics about the

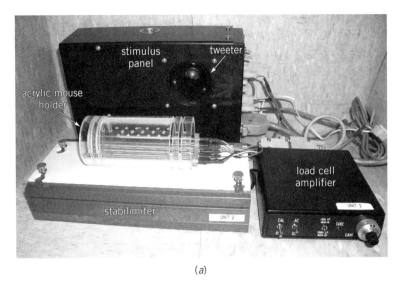

(a)

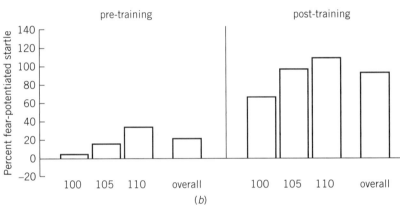

(b)

Figure 10.3 (a) Fear-conditioned startle apparatus. The aversive conditioning grid is inserted in the bottom of the plastic cylinder within an acoustic startle system. (b) Amplitude of whole body flinch to 100, 105, and 110 dB startle tones is greater on day 2, after the aversive stimulus was paired with the startle tones. [From Falls (2003), Current Protocols in Neuroscience, Figures 8.11.B.1 and 8.11.B.5.]

presence of an immediate threat, since more vocalizations to a cat predator were emitted when conspecifics were present than when the subject rat was alone (Blanchard et al., 1991). Ultrasonic vocalizations in adult rats are inhibited by anxiolytic and antipanic drug treatments (Molewijk et al., 1995). Ultrasonic vocalizations in the range of 50 kHz in adult rats were reported to accompany approach behaviors during anticipation of reward (Knutson et al., 2002). Adult mice emit ultrasonic vocalizations in the 70 kHz range during sexual and social interactions (Maggio and Whitney, 1985; Liu et al., 2003). Thus, the context of the vocalization may be related to stress, anxiety, or to other emotional states.

A *mouse defense test battery*, designed by Caroline and Robert Blanchard and coworkers at the University of Hawaii, quantitates fear-related behaviors of mice in stressful naturalistic situations (Griebel et al., 1995, 1996; Blanchard et al., 1998, 2001,

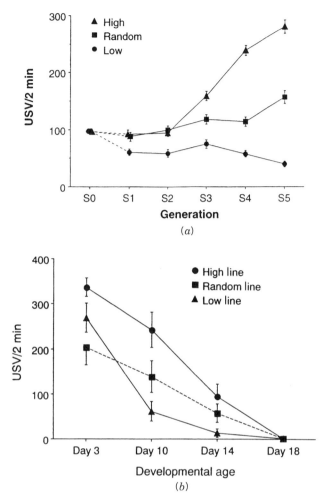

Figure 10.4 *Mouse pups emit ultrasonic vocalizations when they are away from the nest and isolated from their parents. (a) Genetic component to ultrasonic vocalization frequency was demonstrated by selective breeding. After five generations, lines of rat pups were generated that showed high and low levels of ultrasonic vocalizations during isolation, as compared to randomly bred pups. (b) Ultrasonic vocalizations in all lines declined to zero when pups reached age 18 days.* [From Brunelli et al. (1997), pp. 258 and 261.]

2003a). The stressor is a rat, introduced into a runway or straight alley. Mouse behaviors quantitated by the observer include defensive aggression, risk assessment, avoidance distance and frequency, flight speed, immobility/freezing, vocalizations, rearing, wall climbings, and jump escapes (Table 10.3). Elements of these stress responses are inhibited by treatment with anxiolytic and antipanic drugs. This naturalistic paradigm models traits relevant to stress responses in humans and appears to be predictive of therapeutic drug response.

Wildness and *placidity* are terms used to quantitate reactions of mice to handling (Wahlsten et al., 2003). Mice captured in the wild, and/or bred from wild-derived strains, are harder to catch, resist being held, vocalize and bite when captured. Wild

TABLE 10.3 Measures of Defensive Behaviors in Wild Mice and in Six Strains of Laboratory Mice

Sub-Test	Measure	Wild Mice	Lab Mice
Flight/avoidance in oval runway	Avoidance distance (cm)	125.50	64.00
	Avoidance frequency (five trials)	4.90	2.65
	Flight speed (m/s)	0.69	0.57
Risk assessment (RA) in oval runway	Stops	0.30	4.25
	Reversal	1.00	1.45
	Orientation	1.50	2.35
Straight alley	Approach–withdraw	0.20	1.82
	Immobility(s)	34.50	10.40
Response to forced contact	Vocalization	0.00	2.80
	Upright posture	0.20	2.46
	Biting	0.40	1.50
	Jump attack	4.70	0.92

Source: From Blanchard et al. (2001).

strains available through the Mouse Phenome Project at The Jackson Laboratory include MOLF/Ei, CAST/Ei, PERA/Ei, and SPRET/Ei. Along with SWR/J, these wild-derived strains showed high scores on a wildness rating scale and on anxiety-related tests (Wahlsten et al., 2003).

Anxiety-Related Behaviors

The *Geller-Seifter conflict test* was the first pharmacologically validated measure of anxiety-related behavior in rodents (Geller et al., 1962). Rats learn to press a lever for a food reinforcement. A mild shock is delivered through the lever on every tenth or twentieth lever press. This is the "punished" phase of the task. "Unpunished" responding measures number of lever presses for the food reinforcement with no shock contingencies, serving as the control for the procedural components of the task. Anxiolytic drugs increase the number of shocks accepted in the punished condition, without affecting unpunished responding. This task has the advantage of selectivity for anxiolytic drugs, showing no effects of other classes of psychoactive drugs. An additional advantage is that a well-trained rat can be used repeatedly for multiple drug treatments. Disadvantages include the need for long-term food restriction, and a relatively long training time of 1 to 3 weeks, for the rat to reach stable baseline performance on punished responding. Originally designed for rats, the Geller-Seifter protocol has been effectively modified for mice (Kuribara and Asahi 1997; Spooren, 2000; Varty et al., 2005).

The *Vogel conflict test* (Vogel et al., 1971; Johnston and File, 1991) avoids the long dietary restriction and training time by using water reinforcement after 24-hour water deprivation. The mild shock is delivered through the regular drinking water spout, eliminating the need to train the rat to press a lever. Analogous to the Geller-Seifter test, the conflict is the appetitive approach behavior to the drinking spout titrated against the aversive shock. This task has also been well validated for specificity to anxiolytic drugs (Figure 10.5) and has been used extensively by the pharmaceutical industry to discover new anxiolytics. Designed for rats, the Vogel conflict test was adapted for mice by John Glowa at NIMH and Louisiana State University, and has been used for

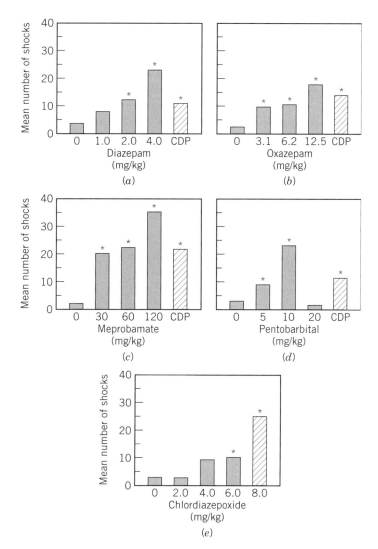

Figure 10.5 *The Vogel conflict test is an operant task in which negative reinforcement accompanies drinking. Anxiolytic drugs such as chlordiazepoxide (Librium), (a) diazepam (Valium), (b) oxazepam, (c) meprobamate, (d) pentobarbital and (e) chlordiazepoxide, are active in the Vogel test. Other classes of drugs, such as amphetamine, pemoline, and scopolamine, that show no anxiolytic activity in humans, are inactive in the Vogel test.* [From Vogel et al. (1971), pp. 3 and 4.]

mice by several laboratories (Crawley et al., 1986; Griebel et al., 1997; van Gaalen and Steckler, 2000).

Sandra File at the University of London utilized the aversive properties of an unfamiliar, brightly lit, open arena to develop a novel anxiolytic test for rodents. *Social interaction* is a naturalistic conflict test that does not employ shock (File and Hyde, 1978; File, 1981, 1996, 1997a; Dunn and File, 1987; File 1997). Occurrences of social interactions between two rats are scored when the rats are in a familiar versus an unfamiliar open field environment under low or high levels of illumination. Sniffing, following, grooming, kicking, mounting, jumping on, wrestling, and other forms of physical contact are quantitated over a 10-minute test period. Videotaping with subsequent scoring is recommended to prevent interfering cues from the human observers. Higher frequencies of social contact occur when lighting conditions are dim and the open field is familiar. Anxiolytic drugs increase the number of social interactions in the high-illumination condition to a level equal to that seen under low illumination (Figure 10.6). This test is also sensitive to an anxiolytic neurokinin receptor antagonist (File, 1997b) and to the anxiogenic effects of corticotropin-releasing factor (Dunn and File, 1987). David Overstreet at the University of North Carolina generated selectively bred lines of rats with high and low sensitivities to the hypothermic effects of a serotonin 5-HT$_{1A}$ receptor agonist, 8-OH-DPAT, which differ on the social interaction test (Gonzalez et al., 1998). While the social interaction task is an excellent discriminator of anxiolytic drug response, the original methods were less useful for mice, due to the higher levels of locomotor activity and lower levels of social interaction seen in mice as compared to rats (unpublished data from the author's laboratory; Sandra File, personal communication). Modified versions of the social interaction task for mice have been developed and successfully applied (File, 1997a; Nakamura and Kurasawa, 2001; Navarro et al., 2004).

Our laboratory combined the concepts behind the brightly lit open field and the social interaction test to create the *light ↔dark mouse exploration test* (Crawley and Goodwin, 1980; Crawley, 1981; Blumstein and Crawley, 1983; Crawley and Davis, 1982; Mathis et al., 1994, 1995). Like the social interaction test, light↔dark exploration is a nonshock, naturalistic conflict, optimized for mice, that titrates the tendency of mice to explore a

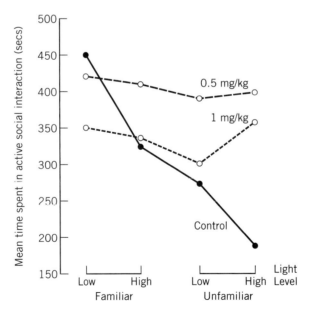

Figure 10.6   *Social interaction test measures several categories of active social exploration between two rats. More time is spent in social interactions in a familiar than in an unfamiliar environment, in a dimly lit than in a brightly lit environment, and after treatment with an anxiolytic drug such as diazepam (Valium). [From File (1980), p. 246.]*

novel environment against the aversive properties of a brightly lit open field. The task was designed to be very simple and easily automated. A standard plastic rat cage is divided by a photocell-equipped panel into two compartments. One compartment comprises two-thirds of the surface area, is transparent, is without a lid, and is illuminated by an overhead lamp. The adjoining compartment comprises one-third of the surface area, is spray-painted black, and is covered with a black Plexiglas lid. The apparatus has also been called the black/white box. An opening in the divider panel allows the mouse to move from one compartment to the other. Four sets of photocells monitor beam breaks across the opening. When the mouse fully exits one compartment and enters the other compartment, the beams are broken and unbroken in the correct sequential order, and the automated data recorder registers a transition from one compartment to the other. A timer, activated by the transition, records total seconds spent in the dark compartment. A 10-minute test session is typically used. This test is sensitive to anxiolytic drugs but not to other categories of psychoactive drugs (Crawley, 1981). Number of transitions is highly correlated to other exploratory behaviors such as rearings (Crawley, 1981). Separate analysis of general locomotion in an open field is necessary, to avoid artifacts that may occur due to nonspecific hyperactivity or sedation. Genetic components underlying the light↔dark exploration task have been demonstrated through strain distribution studies of inbred strains of mice (Crawley and Davis, 1981; Misslin et al., 1989; Mathis et al., 1994; Lepicard et al., 2000; Griebel et al., 2000; Bouwknecht and Paylor, 2002; Holmes et al., 2003c). Quantitative trait loci analyses in AXB and BXD recombinant inbred mouse strains revealed several chromosomal loci linked to light↔dark transitions, time in the dark, and response to diazepam (Mathis et al., 1995; Gershenfeld and Paul, 1997; Turri et al., 2001; Finn et al., 2003). A standard light↔dark apparatus and representative data are shown in Figure 10.7.

The *elevated plus-maze* rests on the same naturalistic conflict between the tendency of mice to explore a novel environment and the aversive properties of a brightly lit, open area (Handley and Mithani, 1984; Pellow et al., 1985; Lister et al., 1987b; File, 1997a). This test builds on the light↔dark exploration conflict with two new components, the openness and the height of the runways, which are raised as much as a meter above the floor. The mouse is placed in the central intersection, from which it can walk down any of four narrow runways. Two are open, such that the mouse can see the cliff. The two alternating arms are enclosed with walls. Mice prefer the closed arms but will venture out into the open arms. Walls made of painted wood, black Plexigas, and clear Plexiglas appear to yield similar results, suggesting that visualization of the cliff elevation is not essential. Instead, this test may be measuring thigmotaxis, the tendency of mice to stay in physical contact with wall surfaces, as detected by the whiskers and fur.

The investigator scores the number of entries into each arm and the time spent in each arm, on a series of successive trials. Risk assessment behaviors can also be scored, including stretch-attend postures and head dips over the edges of the open sides. Behaviors are scored from a videotape and/or by an observer over a 5-minute test session. Automated versions are available from several companies including Lafayette, Coulbourn, Noldus, Hamilton-Kinder, and Clever Sys, Inc. Anxiolytic drugs specifically increase the number of entries into the open arms and the time spent in the open arms. Anxiogenic as well as anxiolytic effects are detectable. A major advantage is that the total number of entries into all arms provides a built-in control measure for

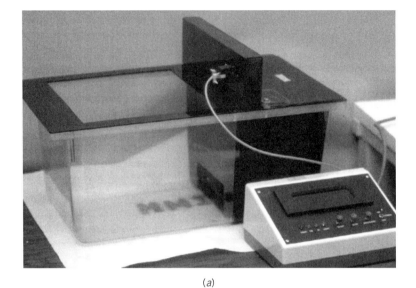

(*a*)

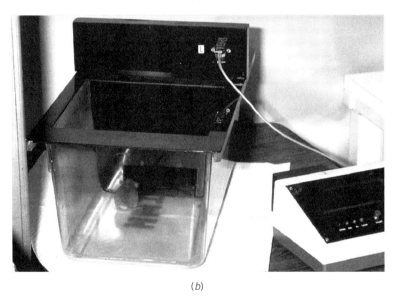

(*b*)

Figure 10.7 *(a,b) Light↔dark exploration test measures the conflict between the natural tendencies of mice to explore a novel environment and the aversive properties of a brightly lit open field. Photograph contributed by the author. (c) Genetic component is seen in the number of transitions between the two compartments in two inbred strains of mice. [From Mathis et al. (1994), p. 174.]*

general hyperactivity or sedation. Therefore a separate test for general locomotion in an open field is unusually not necessary to verify a positive finding.

The elevated plus-maze is presently the most widely used model of anxiety-like behaviors for drug discovery in pharmaceutical companies (Dawson and Tricklebank, 1995). Genetic components underlying performance on the elevated plus-maze have been demonstrated in strain distribution studies of inbred strains of mice (Trullas and

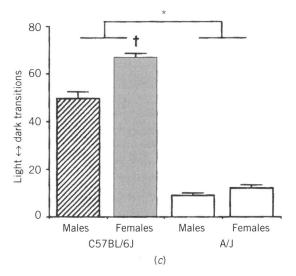

Figure 10.7 *(Continued)*

Skolnick, 1993; Garrett et al., 1998; Griebel et al., 2000; Lepicard et al., 2000; Rodgers et al., 2002; Finn et al., 2003). Chromosomal loci linked to elevated plus-maze performance have been identified by quantitative trait loci analysis (Flint et al., 1995; Cohen et al., 2001; Henderson et al., 2004). A typical elevated plus-maze apparatus and representative data are shown in Figure 10.8. Rat and mouse versions of the elevated plus-maze are essentially identical, except for the smaller dimensions of the mouse version.

A clever modification of the elevated plus-maze is the *elevated zero maze*. Brightly lit, open areas alternate with dark, covered areas, comprising the annulus of an elevated circular runway (Shepherd et al., 1994; Heisler et al., 1998; Cook et al., 2001; Zorner et al., 2003; Mombereau et al., 2004). Entries into the open areas and time spent in the open areas are quantitated by the investigator over a 5-minute test session, by direct observation, later scoring of a videotape, or automation. The advantage of the elevated zero maze over the elevated plus-maze lies in the continuous nature of the runway. In the plus-maze, mice often stay in the start area, the central square box where the four arms meet. Mice often quickly return to the start box square rather than fully entering an arm. Start box behaviors introduce ambiguity in scoring an arm entry and increase the variability of the data. The zero maze has no start box, and the session is continuous, obviating these problems.

The *emergence test* includes components of the light↔dark, open field center time, and elevated plus-maze tests. A small enclosed dark chamber is placed within a standard open field. The mouse is placed in the enclosed box at the start of the test session. Latency to emerge from the enclosed chamber into the main open field, and time spent in the enclosure versus the open area, are scored (Rodriguez de Fonseca et al., 1996; Dulawa et al., 1999; Tang et al., 2002; Holmes et al., 2003b, e; Zomer et al., 2003).

The *mirrored chamber test* includes components of the light↔dark and emergence tests. A cube made of reflective mirrored sides is placed within an open field. Latency to enter and time spent in the cube are scored (Toubas et al., 1990; Cao et al., 1993; Seale et al., 1996; Bowers et al., 2000; Kliethermes et al., 2003; Salas et al., 2003; Lang and

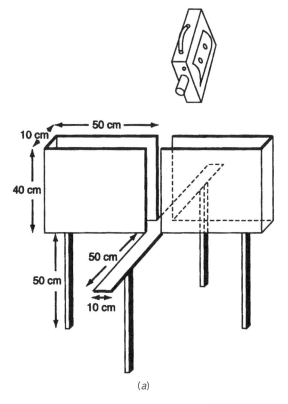

(a)

	Test 1	Test 2	Test 3	Test 4
% Entries into open arms	42.1±2.6	40.9±2.9 0.57	41.9±2.9 0.46	43.3±2.6 0.41
% Time on open arms	33.9±2.4	30.0±2.7 0.64	28.8±2.6 0.55	31.7±3.1 0.29
Total entries	18.9±0.8	21.3±1.1 0.44	22.8±1.3 0.45	24.9±1.1 0.34

(b)

Figure 10.8 *Elevated plus-maze measures the conflict between the natural tendency of mice to explore a novel environment versus the natural tendency of mice to avoid the aversive properties of an open, narrow, elevated surface. (a) Schematic drawing of an elevated plus-maze. [From File (1997a), p. 8.3.7.] (b) Representative data from NIH Swiss mice over a 5-minute test session, and over subsequent 5-minute test sessions, each separated by 48 hours. [From Lister (1987b), p. 181.]*

de Angelis, 2003). The mirrored chamber test appears to measure the response of the mouse to the reflective surface, rather than the mirrored image of another mouse, since replacement of the mirrors with white or gray reflective squares produced similar results (Lamberty, 1998).

The *staircase test* (Figure 4.9) measures the tendency of mice to walk up a set of five steps. This task is sensitive to anxiolytic drug treatments (Simiand et al., 1984; Steru et al., 1987; Quock et al., 1992; Pick et al., 1996; Weizman et al., 1999; Mombereau et al., 2004). Number of steps climbed and rearing are the variables measured.

These several naturalistic anxiety-related conflict tests have been extensively validated with a variety of anxiolytic drugs, including benzodiazepines, ethanol, nicotine, neurosteroids, and ligands for receptor subtypes of neuropeptide Y, galanin, corticotropin releasing factor, and serotonin. Both mice and rats have been successfully used in most of these tasks. In general, these various exploratory conflicts appear to measure similar anxiety-related traits. However, mutant mice may show an anxiety-related phenotype on one test but not another (Holmes et al., 2003c). Factor analyses, by correlational and principal component analysis of behavioral parameters of the elevated plus-maze, light↔dark transitions, open field and others, suggest that different fundamental behavioral traits, brain regions, and chromosomal loci may mediate underlying components of each task (Wall and Messier, 2000; Carola et al., 2002; Yilmazer-Hanke et al., 2003; Holmes et al., 2003b; Henderson et al., 2004). At present, the elevated plus-maze is the most widely used, perhaps because it incorporates a simultaneous control for locomotor activity, the entries into the closed arms. To evaluate the robustness of an anxiety-related trait, it is best to run two or three of these anxiety-related tasks in a cohort of mutant mice. The elevated plus-maze, light↔dark transitions, and Vogel conflict tests are three that have been well validated pharmacologically, and measure somewhat different forms of anxiety-related behaviors. Conducted in that order, these three tasks do not seem to influence each other, so a batch of mice can be tested sequentially on the elevated plus-maze, light↔dark transitions, and Vogel test, with a few days intervening between each task. Similar results on several related tasks will provide a strong corroboration of the behavioral phenotype.

Depression-Related Behaviors

The *Porsolt forced swim test* measures the time spent swimming versus the time spent floating in a tall cylinder filled with water (Porsolt et al., 1977, 1978, 1998; Drugan et al., 1989; Detke et al., 1995; Hansen et al., 1997; Redrobe and Bourin, 1998; Lucki et al., 2001; Crowley et al., 2004). Rats and mice will generally swim in the water, apparently seeking an escape route. After some time the animal may stop swimming and instead will float on the surface of the water, appearing to have "given up the search." Fatigue is a factor but does not appear to explain the cessation of swimming, since a minor disturbance will reactivate swimming. Also termed "behavioral despair" and "learned helplessness," this task is sensitive to acute and chronic treatment with antidepressant drugs (Willner, 1984; Lucki et al., 2001). A genetic component underlying the forced swim test was originally demonstrated in rats by comparison of the Maudsley Reactive and Nonreactive strains (Overstreet et al., 1992). Inbred strains of mice vary in both baseline immobility and in response to antidepressant drugs in the forced swim test (Lucki et al., 2001; Bai et al., 2001).

For mice, the water is filled to a depth of at least 10 cm, which exceeds the distance to which the tail can extend, so the mouse cannot balance on its tail at the bottom of the cylinder. The top of the cylinder is at least 15 cm above the upper surface of the water, such that the mouse cannot climb out of the cylinder. Our laboratory uses a cylinder that is 20 cm in diameter and a waterline that is 15 cm above the bottom of the cylinder. The water is maintained at room temperature or warmer, up to 35°C. The mouse is placed in the water and undisturbed for the test session. Session durations between 4 and 20 minutes have been used for mice. A preexposure period of 2 to 5 minutes before a short test session has been used, with scoring beginning immediately thereafter. Rogert Porsolt, at Phoenix International Pharmacology Company in Le Kremlin-Bicêtre, France, employs a 15 minute habituation session, one day before a 5-minute test session (Porsolt et al., 1998). Floating is defined as lack of swimming, and can include minimal movement of one leg, sufficient to keep the head above water. The investigator records the total number of seconds spent floating over the course of the test session. Irwin Lucki at the University of Pennsylvania recommends a 6-minute session, with the last 4 minutes of the test scored from videotapes (Lucki et al., 2001). An automated version of the forced swim test has been developed by Lucki and co-workers using a videotracking system (Crowley et al., 2004), and by Jim Meyerhoff at the Walter Reed Army Institute of Research using photocell beams (Hebert et al., 1998). Commercially available systems are manufactured by Coulbourn Instruments, ViewPoint in Montreal, Canada, and others. Stoelting offers a videotracking system, ANY-maze, that detects floating when viewed from above or from the side of the cylinder; the same videotracking system is used for the Stoelting elevated plus-maze, radial maze, and open field. A water-filled cylinder used in the Porsolt swim test for mice, views of mice in the cylinder, and representative data, are shown in Figure 10.9.

Treatment with antidepressants reduces the total time spent floating in the Porsolt forced swim test. Although this task does not have obvious construct validity in terms of modeling a specific symptom of human depression, it is sensitive to antidepressant drug treatments, including selective serotonin reuptake inhibitors (Borsini, 1995; Redrobe and Bourin, 1998; Cryan et al., 2002). However, other categories of psychoactive drugs can also produce positive responses on forced swim (Willner, 1984). Novel approaches to treating depression, such as corticotropin releasing factor receptor CRF1 antagonists and agonists, produce antidepressant-like actions, meaning reducing immobility, on the

| (a) | (b) | (c) |

Figure 10.9 *(a) Overhead view of mouse in a Porsolt swim task cylinder filled with water. (b) Side view of mouse swimming in the Porsolt forced swim test. (c) Side view of mouse floating with head above water in the Porsolt forced swim test. [Photograph kindly contributed by Dr. Andrew Holmes, National Institute on Alcoholism and Alcohol Abuse, Rockville, MD USA.] (d) Stressors affect performance on the Porsolt forced swim task. Mice previously defeated in a series of resident–intruder encounters showed higher levels of immobility, as measured in an automated swim test, as compared to mice that were undefeated or had no prior resident/intruder encounters. [From Hebert et al. (1998), p. 262.]*

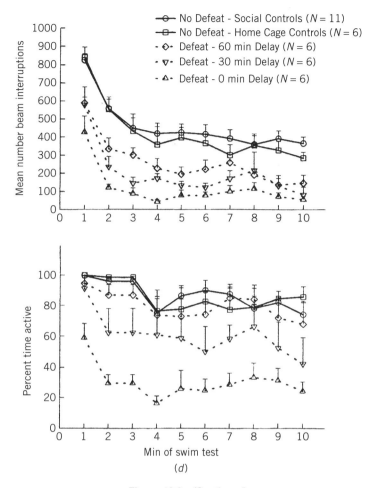

Figure 10.9 *(Continued)*

forced swim test (Nielsen et al., 2004; Tezval et al., 2004). Neuropeptide Y similarly reduced immobility time, indicating an antidepressant-like action of this neuropeptide transmitter (Redrobe et al., 2002).

Stressors such as immobilization and footshock increase immobility in the Porsolt forced swim test (Weiss et al., 1981; Zebrowska-Lupina et al., 1990). Social defeat produced depression-like effects on the Porsolt swim test, as shown in Figure 10.9*b* (Hebert et al., 1998). DBA/2 male mice were paired with highly aggressive C57BL/6 male mice in a series of four encounters. Defeat induced reductions in swim time in the DBA/2 mice. Swim time versus immobility was quantitated in this study using a photocell array located under the transparent bottom of the cylinder, which detects beam breaks when the animal moved on the surface of the water (Hebert et al., 1998). Automated systems appear to provide a useful technical advance for more accurate quantitation and higher throughput screening of drugs in the Porsolt swim test.

Learned helplessness is another test that measures lack of trying to escape an aversive stimulus. Developed by Steve Maier at the University of Colorado, a standard method for rats employs three parallel wheel-turn boxes (Weiss, 1971; Maier, 1984; Drugan et al., 1984, 1985, 1989, 1997; Maier and Watkins, 2005). A rat is placed in each of three identical small Plexiglas boxes, in which a grooved Plexiglas wheel is located in front of the rat. The apparatus is diagrammed in the cartoon at the beginning of this chapter. In the escapable shock condition, tailshocks are administered. The wheel is connected to the tailshock control mechanism. Turning the wheel switches off the shock generator. Rats quickly learn to turn the wheel to avoid the shock. In the inescapable shock condition, the first wheel is yoked to a second wheel, which is in front of the rat in the inescapable shock condition. The rat in the inescapable condition receives the same number of shocks as the rat in the escapable condition but has no personal control over the delivery of the aversive stimulus. The rat in the baseline condition has the electrode connected to its tail, but the electrode is not connected to the shock generator. The baseline rat thus experiences no contingencies and receives no tailshock. At the end of a 1-hour session, the rats are returned to their home cages. Twenty-four hours later each animal is placed in a new apparatus, a shuttlebox. The task on the second day is a standard active avoidance. The animal moves from the left chamber to the right chamber of a two-chambered shuttle box, and back again, to escape or avoid footshock that is administered to the grid floor of one chamber at a time. Rats previously in the baseline and escapable condition perform well on the active avoidance task, receiving very few shocks. Rats previously in the inescapable condition perform poorly on the active avoidance task. Approximately half the rats trained in the inescapable condition never learn the active avoidance response. The failure to learn a simple escape task, 24 hours after exposure to an inescapable stressor, has been interpreted as behavioral despair. The rat appears to have learned that it has no control over the aversive stimuli and is helpless to avoid stressors. Hymie Anisman and co-workers at Carleton University in Ottawa have adapted the inescapable shock paradigm for mice (Anisman et al., 1979; Shanks and Anisman 1989, 1993), as shown in the data of Figure 10.10.

The *tail suspension* test has conceptual similarities to the Porsolt forced swim test and has been successfully used for mice in many laboratories (Stéru et al., 1985; Van der Heyden et al., 1987; Trullas et al., 1989; Vaugeois et al., 1996; Cryan et al., 2002; Ripoll et al., 2003; Caldarone et al., 2003; Crowley et al., 2004; Strekalova et al., 2004; Holmes et al., 2004; Cryan and Mombereau, 2004; Cryan et al., 2005). The mouse is suspended by the tail, such that the body dangles in the air, facing downward (Figure 10.11). A clip or adhesive tape suspends the mouse for 6 minutes or more, after one or more training sessions, over one or more repeated tests (Vaugeois et al., 1996; Ukai et al., 1998). Mice tend to struggle to face upward and reach for a solid surface. After several minutes mice generally stop struggling and hang immobile. Cumulative immobility time is recorded by the observer or by measurement through a hook connected to a strain gauge (Vaugeois et al., 1996). The strain gauge transmits movements of the mouse to an automated unit that sums the total seconds spent immobile. Commercially available tail suspension equipment is sold by Med Associates (Figure 10.11), Hamilton Kinder and Lafayette Instruments. Treatment with antidepressant drugs reduces immobility time (Stéru et al., 1985; Van der Heyden et al., 1987; Vaugeois et al., 1996; Crowley et al., 2004; Cryan et al., 2005). A genetic component underlying differences in immobility time was demonstrated in strain distribution studies of inbred strains of mice (Van der Heyden et al., 1987; Trullas et al., 1989; Liu and Gershenfeld, 2003; Ripoll et al., 2003). Selective breeding of high and low

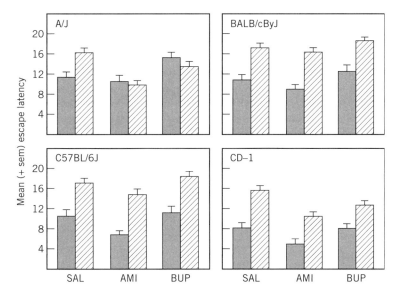

Figure 10.10 *Four strains of mice were tested after vehicle treatment (SAL) or 14 days of antide-pressant drug treatments (amitriptyline, AMI, or bupropion, BUP), on escape deficits following an inescapable stressor. Mice receiving no footshock (black bars) showed faster escape in a shuttlebox escape task 15 days later, as compared to mice previously receiving inescapable footshock (striped bars). Antidepressant drug treatments significantly improved escape behavior in one strain (A/J) but not in the other strains (BALB/cByJ, C57BL/6J, CD-1).* [From Shanks and Anisman (1989), p. 124.]

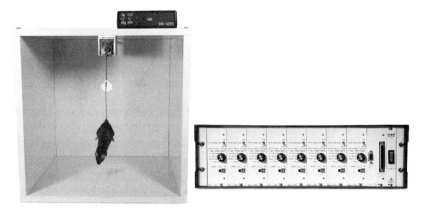

Figure 10.11 *Automated tail suspension test for mice. The mouse hangs by its tail. Time spent struggling versus time spent in immobility is recorded through a load cell amplifier.* [Photograph kindly contributed by Christina St.Denis, Med Associates.]

immobility scoring mice, shown in Figure 10.12, further indicates a genetic compo-nent to performance on the tail suspension task (Vaugeois et al., 1996; El Yacoubi et al., 2003).

Newer mouse models of depression are under development that better predict antide-pressant efficacy in humans, as summarized in Table 10.2 by John Cryan at Novartis in Basel, Switzerland, Athena Markou at Scripps Research Institute in La Jolla, CA, and Irwin Lucki at the University of Pennsylvania (Cryan et al., 2002; Cryan and Mombereau,

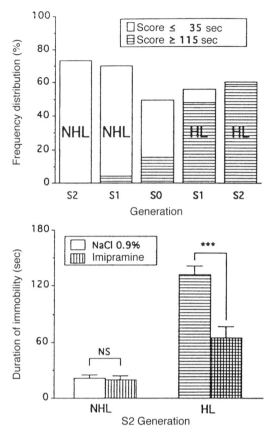

Figure 10.12   *Immobility in the tail suspension test was selectively bred from founder mice showing high versus low numbers of seconds of behavioral immobility. Treatment with an antidepressant drug, imipramine, reduced immobility time in the high-scoring line of mice. [From Vaugeois et al. (1996), p. R2.]*

2004). *Anhedonia* is a major symptom of depression, described as the loss of interest or pleasure in daily activities. Anhedonia is conceptualized in rodents as the loss in responsiveness to reward (Moreau, 1997). In mice, tail suspension and resident–intruder stressors decrease preference for sucrose reinforcement (El Yacoubi et al., 2003; Strekalova et al., 2004). This lack of intake of a rewarding food in mice is similar to that seen in rats after similar stressors (Moreau 1997; Konkle et al., 2003). Athena Markou and George Koob at Scripps Research Institute in La Jolla, California, discovered that cocaine withdrawal precipitated an anhedonic-like state in rats, characterized by elevated thresholds for reward by intracranial self-stimulation (Markou and Koob, 1991). Withdrawal from cocaine and amphetamine increased immobility scores in the forced swim and tail suspension tests in rats and in mice (Barr et al., 2002; Cryan et al., 2003). *Olfactory bulbectomy* is considered a depression model in rats. Removal of the olfactory bulbs affects exploratory activity, cognition, neurotransmitters, hormones, and immune responses, and antidepressants reverse some of these effects (Kelly et al., 1997; Primeaux and Holmes, 2000). The olfactory bulbectomy model of depression is beginning to be applied to mice (Cryan et al., 2002).

Mania and bipolar disorder (manic-depression) have been difficult to model in rodents (Einat et al., 2003a,b). Drug-induced approaches include administration of ouabain, an inhibitor of the sodium pump through which lithium acts, and biphasic locomotor activation by the dopaminergic agonist quinpirole (Shaldubina et al., 2002; El-Mallakh et al., 2003).

Behaviors Related to Symptoms of Schizophrenia

Schizophrenia is the psychiatric disease that has been the hardest to model in animals. The prefrontal cortex, considered the structure most relevant to the pathophysiology of schizophrenia (Goldman-Rakic and Selemon 1997; Weinberger et al., 2001), is minimally developed in the rodent brain. Therefore schizophrenia may be a uniquely human disease. Psychotic symptoms such as hallucinations and delusions appear to be impossible to intuit in a mouse. Negative symptoms, such as flat affect and lack of sociality, along with underlying biochemical and genetic abnormalities, may be more amendable to modeling in mice. Guidelines for evaluating an animal model of schizophrenic-like symptoms for responses to antipsychotic drugs are shown in Table 10.4. Tasks designed for rats have been adapted for mice in some cases.

The majority of early tests for behaviors related to symptoms of schizophrenia were based on the antipsychotic activity of dopamine D-2 receptor antagonists (Ellenbroek and Cools, 1990; Higgins, 1998). Rodent behaviors induced by dopaminergic D-2 receptor agonists acting at the mesocorticolimbic dopamine pathway include *locomotor hyperactivity* and *stereotyped sniffing/grooming*. Dopamine, apomorphine, amphetamine, the D-2 agonist quinpirole, and related drugs are administered systemically or through bilateral indwelling cannulae previously implanted into the nucleus accumbens. Locomotor activity in an automated open field is recorded over a 10- to 60-minute session. Higher doses of these drugs are administered in a separate experiment to quantitate stereotypy. Incidence of repetitive, stereotyped grooming, sniffing, and head movements are recorded every 15 seconds by an observer using a standardized stereotypy scoring system (Creese and Iversen, 1973). Drugs that block dopamine agonist-induced hyperactivity and stereotypy and are antagonists at the D-2 receptor represent good candidates for antipsychotics with a dopaminergic mechanism of action (reviewed in Ellenbroek and Cools, 1990; Higgins, 1998). This approach was useful in generating new D-2 antagonists that were biologically and behaviorally active. However, D-2 antagonists often exacerbate the negative symptoms of schizophrenia, and long-term treatment with D-2 antagonists result in the development of tardive dyskinesias in some patients (Breier, 1996). Current paradigms seek to expand beyond the dopamine hypothesis of schizophrenia.

TABLE 10.4 Criteria for Assessing Predictive Validity of Animal Models for Schizophrenia

1. Neuroleptics of various chemical classes should be effective.
2. No false negatives should occur.
3. No false positives should occur.
4. Anticholinergic drugs should not reduce the effects of neuroleptic drugs.
5. Chronic treatment should not reduce the effects of neuroleptic drugs.
6. There should be a relationship between the clinical potency of neuroleptic drugs and their potency in the model.

Source: Ellenbrock and Cools (1990), p. 470.

Sensitization is a consequence of repeated administration of psychostimulants, such as amphetamine and cocaine, in rats and mice (Iversen and Iversen, 1981; Robinson and Becker, 1986; Kuczenski and Segal, 1989; Cooper et al., 1996a; Segal and Kuczenski, 1997; Kalivas et al., 1993; Hooks et al., 1994a; Kalivas et al., 2004). Enhanced hyperlocomotion and increased neuronal release of dopamine appear after the second or third psychostimulant administration, indicating an underlying sensitization process. Since psychotic symptoms are seen in some people after repeated amphetamine and cocaine use, this animal analogue has heuristic appeal (Robinson and Becker, 1986). Amphetamine or cocaine is administered in daily repeated or escalating doses. Another approach is escalating intravenous self-administration of psychostimulant drugs. *Behavioral sensitization* is measured by repeated testing for hyperlocomotion in an automated open field. Normally locomotor scores for untreated and vehicle-treated rats and mice decline over repeated exposures to the open field, as the animal habituates to the novel environment and stops exploring. Conversely, locomotor scores for rats and mice treated repeatedly with cocaine or amphetamine will increase over repeated testing in the open field (Park et al., 2000; Cornish and Kalivas, 2001; Dong et al., 2004).

Behavioral stressors activate the mesocorticolimbic dopamine pathway, increasing dopamine release, as measured by in vivo microdialysis and tissue levels of dopaminergic metabolites (Thierry et al., 1976; Deutch et al., 1985; Abercrombie et al., 1989; Ellsworth et al., 2001). Repeated intermittent tailshock or footshock induces a robust release of dopamine in rats. The percentage increase in dopamine release is considerably higher in the mesocortical pathway, from the ventral tegmental neurons projecting to the prefrontal cortex, than in the mesolimbic pathway, from ventral tegmental neurons projecting to the nucleus accumbens (Deutch et al., 1985). Since stress may precipitate a psychotic episode in schizophrenic patients (Breier 1996), these stress paradigms represent an approach to modeling an etiological factor of schizophrenia.

Barbara Lipska, Daniel Weinberger, and co-workers at the National Institute of Mental Health in Bethesda, Maryland, discovered that hippocampal lesions in neonatal rats produce a behavioral syndrome in the adults that mimics some of the symptoms of schizophrenia (Lipska et al., 1993, 1995; Lipska, 2004). Increased amphetamine-induced hyperlocomotion, deficits in prepulse inhibition of the acoustic startle response, impaired choice accuracy on a radial maze task, and reduced levels of social interaction are seen in juvenile rats at postnatal days 35–65, following ibotenic acid lesion of the ventral hippocampal formation at postnatal day 7 (Lipska et al., 1993; Chambers et al., 1996; Sams-Dodd et al., 1997; Lipska 2004). This lesion appears to produce changes in rat behaviors that have analogues to both positive and negative symptoms of schizophrenia. Genetic variation in vulnerability to the behavioral changes following this lesion have been reported (Lipska et al., 1995). However, the neonatal hippocampal lesion has not yet been extensively analyzed in mice.

Attentional abnormalities represent a symptom of schizophrenia that may have direct analogs in rodents. Deficits in sensory processing of excessive stimuli and distracting incoming information may contribute to the cognitive impairments and auditory hallucinations that characterize many schizophrenic patients (Swerdlow et al., 1994). Two attentional tasks have been developed that can be used in both humans and rodents. *Latent inhibition* is the impaired performance on a learning task when conditioning includes a pre-exposed stimulus that is presented without any reinforcement contingencies (Lubow and Gewirtz, 1995; Weiner 2003). As parodied in Figure 10.13 and shown in Figure 10.14, (1) tones are presented to a rat in an operant chamber, without pairing the tone to any other stimulus. (2) The tone is subsequently paired with foot-shock. (3) The rat is then given a water bottle in the chamber and allowed to drink. (4) The tone is then introduced. Suppression of drinking is recorded during the next 5 minutes while the tone remains on. Another group of rats is not pre-exposed to the tone alone, eliminating step 1. The tone is paired with a footshock, and the rat is then given the water bottle. Drinking is suppressed to a greater extent by the tone in the second group, whose experience with the tone was only when the tone was paired with footshock. The first group shows less suppression of drinking to the tone, presumably because the animal had initially correctly learned to ignore the tone. Ina Weiner, Joram Feldon, and co-workers at Tel Aviv University discovered that rats treated with amphetamine show disruptions in latent inhibition similar to those seen in schizophrenic patients. Clozapine, an atypical neuroleptic, reverses amphetamine-induced deficits in latent inhibition in rats (Figure 10.14; Weiner et al., 1996). Latent inhibition tasks have been successfully adapted for mice (Gould and Wehner, 1999; Miyakawa et al., 2003; Wang et al., 2004). Jeanne Wehner at the University of Colorado Institute for Behavioral Genetics conducted a strain distribution that demonstrated robust latent inhibition in 129/SvEv, C57BL/6, BALB/cByJ, AKR, an DBA/2 mice, whereas 129/SvJ, CBA, A, and C3H inbred strains of mice did not develop latent inhibition (Gould and Wehner, 1999).

Prepulse inhibition of the startle reflex is a neurophysiological and behavioral measure of sensorimotor gating (Braff and Geyer, 1990; Geyer et al., 1990; Swerdlow et al., 1994; Geyer and Ellenbroek, 2003). A weak stimulus, such as a 90 dB tone, inhibits the subsequent response to a strong stimulus, the very loud 120 dB tone, if presented within 100 msec. Similarly, in tactile prepulse inhibition, a puff of compressed air directed at the head or body inhibits the startle response to a second puff of air or to a 120 dB tone. In humans, eyeblink is the startle response that is measured. In rodents, the whole body flinch is quantitated by an electrostatic sensor measuring amplitude of movement of the animal within a cylindrical holder (Figure 5.2). Abnormal sensory inhibition may reflect a deficit in processing and prioritizing incoming sensory information. Mark Geyer, David Braff, and co-workers at the University of California San Diego discovered that schizophrenic patients show deficits in prepulse inhibition (Braff et al., 1978; Geyer et al., 1990; Braff and Light, 2004). This team then found that rats treated with apomorphine, amphetamine, phencyclidine, and glutamate receptor antagonists show deficits in prepulse inhibition (Geyer et al., 1990, 2001). A genetic component underlying prepulse inhibition was suggested in strain distribution analyses of inbred strains of mice (Paylor and Crawley, 1997; Logue et al., 1997b; Patel et al., 1998; Stevens et al., 1998; Geyer et al., 2002). Treatment with antipsychotic drugs improved prepulse inhibition in normal rats (Curzon et al., 1994) and in strains of mice showing deficits on this task (Patel et al., 1998; Geyer et al. 2002; Geyer and Ellenbroek, 2003).

Of the several rodent behaviors described above as relevant to symptoms of schizophrenia, the prepulse inhibition test has been most effectively optimized for mice. An appealing advantage of the prepulse inhibition task is that almost identical methods are used in mice, rats, and humans. In addition, the sensorimotor gating component of this task appears to be independent of performance on acoustic startle (Paylor and Crawley, 1997). A related task, reduced habituation to acoustic startle, also appears to represent a trait marker in schizophrenia (Meincke et al. 2004), and is impaired by psychomimetic drug treatments in mice (Dirks et al., 2002; Dulawa and

Figure 10.13 *Latent inhibition.* [Drawings kindly contributed by Uri Shalev and Ina Weiner, Tel Aviv University, Tel Aviv, Israel.] *See legend to Figure 10.14.*

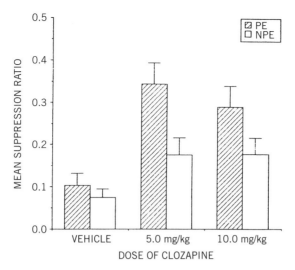

Figure 10.14 *Latent inhibition. (a) Rats are either preexposed (PE) or nonpreexposed (NPE) to a tone. The tone is then paired with a mild footshock (conditioning). (b) The next day, responses to the tone are measured. Preexposure to the tone, before pairing of the tone with a shock, reduces the response to the tone, indicating that the unpaired preexposure interferes with the subsequent association. (c) Suppression ratios for rats exposed to the prior stimulus (PE, preexposed, striped bars) versus rats nonpreexposed (NPE, white bars), on the latent inhibition task. Clozapine, an atypical antipsychotic drug, significantly improved performance on the latent inhibition task, as compared to vehicle controls. [From Weiner et al. (1996), p. 838.]*

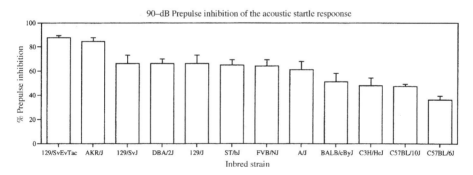

Figure 10.15 *Prepulse inhibition strain distribution of 13 inbred strains of mice. Representative data for a 90 dB auditory prepulse, administered 100 msec before the onset of an acoustic startle stimulus, consisting of a 120 dB burst of sound. Data are shown as percentage inhibition of the startle response. [From Paylor and Crawley (1997), p. 172.]*

Geyer, 2000). As described below, the prepulse inhibition task has been extensively used in transgenic and knockout mice. One caveat is that the prepulse response is sensitive to hearing loss, such as the age-related decline in high-frequency hearing in C57BL/6J (McCaughran et al., 1999). Commercially available equipment to quantitate acoustic startle and prepulse inhibition, manufactured by San Diego Instruments, is illustrated in Figure 5.2*a*. Representative data are shown in Figures 2.6 and 10.15.

TRANSGENICS AND KNOCKOUTS

Mutations in genes for neurotransmitters and receptors relevant to anxiety disorders, depression, and schizophrenia have been been extensively generated in mice. The behavioral tasks described above proved useful in elucidating the outcome of the mutation. This large body of literature has been summarized in several excellent review papers (Holmes, 2001; Finn et al., 2003; Cryan and Mombereau, 2004). Several particularly insightful examples are described below.

The most common and effective treatment for anxiety disorders are benzodiazepines such as diazepam (Valium). Benzodiazepines bind to an allosteric modulatory site on the receptor for gamma amino butyric acid (GABA), the major inhibitory transmitter in the mammalian brain (Tallman et al., 1980; Whiting, 2003). Knockout mice deficient in genes for the individual subunits of the GABA-A receptor are elucidating the subunit compositions that distinguish the anxiolytic actions of benzodiazepines from the sedative, hypnotic, and muscle relaxant actions of benzodiazepines (Rudolph and Mohler, 2004). Gerry Dawson and co-workers at Merck Neuroscience Research Laboratories in Terlings Park, England, discovered a point mutation in the α_1 subunit of the GABA receptor that eliminated the sedative but not the anxiolytic actions of benzodiazepines, as measured by the elevated plus-maze and light↔dark transitions tests (McKernan et al., 2000). Florence Crestani and co-workers at the University of Zurich, Switzerland, found that mice with a point mutation in the α_2 subunit of the GABA receptor failed to display an anxiolytic-like action of diazepam on the elevated plus-maze and light↔dark transitions tests (Löw et al., 2000). In contrast, mice deficient in the α_5 subunit of the GABA receptor showed normal anxiety-related behaviors on the elevated plus-maze but enhanced performance in spatial learning and memory on the Morris water maze (Collinson et al., 2002). Development of anxiolytic drugs with reduced sedative properties may emerge from behavioral assays of these GABA subunit mutant lines of mice (Greibel et al., 2003; Atack, 2003; Rudolph and Mohler, 2004).

Serotonin is a monoamine neurotransmitter that has been implicated in many mental illnesses, including mood and anxiety disorders, schizophrenia autism, and impulse control disorders. Mutant mice deficient in some serotonin receptors subtypes, and in the serotonin transporter (5-HTT), the transmitter reuptake site of action of antidepressants such as fluoxetine (Prozac), have been analyzed on anxiety-related behavioral tasks to understand the role of each element of serotonergic neurotransmission. Andrew Holmes at the National Institute of Mental Health investigated the full behavioral phenotype of knockout mice with null mutations in the serotonin transporter, generated and bred into two different inbred strain backgrounds by Dennis Murphy and co-workers at NIMH. Null mutants lacking the serotonin transporter were normal on measures of general health but displayed less aggression toward intruder mice in the resident–intruder test (Holmes et al., 2002a). Reduced immobility in the tail suspension test was detected when the 5-HTT mutation was bred onto the 129SvEvTac background (Holmes et al., 2002d). 5-HTT null mutants on a C57BL/6J genetic background displayed increased anxiety-like behaviors on the elevated plus-maze, light↔dark exploration, center time in an open field, and emergence into an open field (Holmes et al., 2003e). However, mice with a null mutation of the serotonin transporter on a 129SvEvTac genetic background were not significantly different than wildtype littermates on components of the elevated plus-maze and light↔dark exploration, indicating that background genes such as the 5-HT$_{1A}$ receptor may compensate for reduced serotonin transporter expression (Holmes et al., 2003c, d, e). René Hen at Columbia University and Miklos Toth at Cornell University in New York found increased anxiety-like behaviors in 5-HT_{1A} knockout mice on open field center time and elevated plus-maze (Parks et al., 1998; Ramboz et al., 1998), phenotypes that were rescued by conditional expression of the 5-HT_{1A} gene in the hippocampus and cortex of 5-HT_{1A} knockout mice (Gross et al., 2002). In contrast, 5-HT_{1B} knockout mice displayed increased exploration of an open field, reduced startle reactivity, and greater aggression (Malleret et al., 1999; Zhuang et al., 1999; Dirks et al., 2001). 5-HT_{1A} knockout mice showed significantly lower immobility scores on the tail suspension test, indicating an antidepressant-like phenotype, while 5-HT_{1B} knockout mice were normal on tail suspension immobility (Mayorga et al., 2001).

Transgenic mice overexpressing corticotropin-releasing factor (CRF) have endocrine abnormalities of the hypothalamic–pituitary–adrenal axis that are characteristic of excess CRF (Stenzel-Poore et al., 1994; Contarino et al., 1999b). These mice show anxiogenic-like behavior, as measured on the elevated plus-maze (Figure 10.16). Percentage of time on the open arms of the elevated plus-maze is approximately one-third that of the wildtype littermate controls. An anxiogenic-like profile for CRF overexpressing transgenic mice was confirmed on the light↔dark exploration conflict test (Heinrichs et al., 1996). In addition deficits on cognitive tasks were detected (Heinrichs et al., 1997). Similarly, mice with excess CRF due to mutation of the CRF binding protein showed anxiogenic-like behavior on the elevated plus-maze (Karolyi et al., 1999). Further CRF overexpressing transgenics showed greater neuroendocrine responses to stressors (Contarino et al., 1999b), while CRF deficient knockout showed reduced neuroendocrine responses to some stressors (Muglia et al., 2001), These findings are consistent with a large neuroendocrine literature on CRF as a key regulator of stress responses in the hypothalamic-pituitary axis, and with the pharmacological literature demonstrating anxiogenic behaviors induced by central administration of CRF, and anxiolytic-like actions of CRF receptor antagonists in rats (Heinrichs et al., 1992; Contarino et al., 1999b).

Elegant work by Lisa Gold and co-workers at Scripps Research Institute in La Jolla, California, and by a team at the Max Planck Institute in Munich, Germany, compared the behavioral phenotypes of mice deficient in each of the two CRF receptor subtypes. Null mutation of the CRF-R1 receptor subtype resulted in the opposite behavioral phenotype to CRF peptide transgenics (Smith et al., 1998b; Contarino et al., 1999b; Timpl et al., 1998). On the elevated plus-maze, number of open arm entries and time on the open arms was higher in the *CRF-R1* null mutants than in the wildtype controls (Figure 10.17). Closed arm activity, validated as a control measure of general activity, was normal in the knockout mice, supporting the interpretation of the increased open arm entries as a true anxiolytic-like phenotype. In additional tests for anxiety-related behaviors, the *CRF-R1* null mutants spent more time in the lighted, open area in both the light↔dark test, and in a dark-light emergence test which is similar to the light↔dark exploration test, again indicating low levels of anxiety. Further, these knockouts showed reduced neuroendocrine responses to restraint stress, as measured by significantly less stress-induced increases in plasma levels of adrenocorticotrophin-releasing hormone and corticosterone. In contrast, two independent lines of CRF-R2 receptor subtype knockout mice showed an anxiogenic-like phenotype in the same tests (Contarino et al, 1999a; Bale et al., 2000; Kishimoto et al., 2000). These findings support an interpretation that CRF-R1 is the receptor subtype that mediates

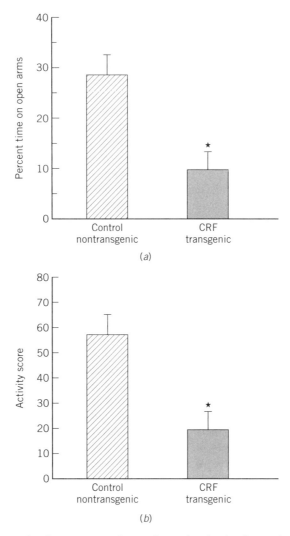

Figure 10.16 *Transgenic mice overexpressing corticotropin-releasing factor show more anxiety-like behavior on the elevated plus-maze, as compared to wildtype controls.* [From Stenzel-Poore et al. (1994), p. 2581.]

CRF-induced stress-related responses. Taken together, the anxiogenic-like phenotype of CRF overexpressing transgenic mice and the anxiolytic-like phenotype of *CRF-R1* knockout mice provide a proof of principle for the usefulness of targeted gene mutations that further our understanding of the behavioral actions of brain neurotransmitters mediating anxiety. These investigations have led to the development of CRF and urocortin receptor ligands as potential treatments for anxiety and stress-related disorders (Keck et al., 2004; Bale and Vale, 2004).

Several other neuropeptide transmitters have been implicated in mood and anxiety disorders. Antagonists of substance P, a member of the neurokinin peptide family, produced antidepressant-like and anxiolytic-like effects in some animal models, leading to an interest in neurokinin receptor antagonists for the clinical treatment

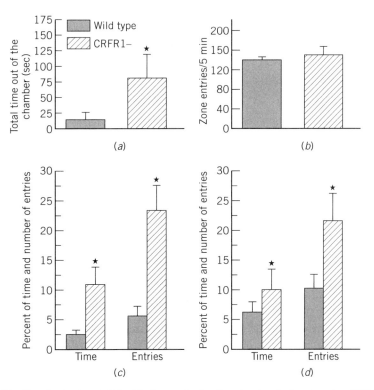

Figure 10.17 *Knockout mice deficient in the corticotropin-releasing factor receptor, CRF-R1, show lower levels of anxiety-like behavior on the elevated plus-maze, as compared to wildtype controls.* [From Smith et al. (1998), p. 1098.]

of depression (Kramer et al., 1998). Neurokinin receptor NK1R knockout mice displayed reduced anxiety-related behaviors and stress responses (Santarelli et al., 2002; Gadd et al., 2003). Cholecystokinin tetrapeptide, acting at the CCK-2 receptor subtype, induces panic-like attacks in normal humans and exacerbates symptoms in patients suffering from panic disorders (Bradwejn and Koszycki, 2001). CCK-2 receptor antagonists show anxiolytic-like actions on the elevated plus-maze and light↔dark tasks (Hughes et al., 1990; Costall et al., 1991). Cholecystokinin receptor CCK-2 knockout mice displayed open field hyperactivity that was reversed by the opiate antagonist naloxone (Dauge et al., 2001; Noble and Roques, 2002). Another line of CCK-2 knockouts displayed anxiolytic-like behaviors on the elevated plus-maze and light↔dark exploration tests (Raud et al., 2003; Hourinouchi et al., 2004). Galanin is a neuropeptide that coexists with norepinephrine in the locus coeruleus (Hökfelt et al., 1999). Galanin overexpressing transgenic mice were normal at baseline on the elevated plus-maze, light↔dark transitions, open field center time, and tail suspension tests (Holmes et al., 2004) but were less responsive to the anxiogenic actions of the noradrenergic activator yohimbine (Holmes et al., 2002c). Galanin receptor GAL-R1 knockout mice displayed anxiety-like behaviors specifically on the elevated plus-maze (Holmes et al., 2003b). Neuropeptide Y also coexists with norepinephrine in a smaller population of locus coeruleus neurons (Hökfelt et al., 1999). While neuropeptide Y knockout mice were normal on the elevated plus-maze, they displayed lower center time in an open field test, and increased response to acoustic startle (Bannon et al., 2000). An NPY overexpressing transgenic rat displayed normal elevated plus-maze behaviors but an anxiolytic-like phenotype in the more stressful Vogel conflict test (Thorsell et al., 2000; Thorsell and Heilig, 2002). Knockout mice deficient in the NPY receptor subtype NPY-Y2 showed anxiolytic-like increases in time spent on the open arms of the elevated plus-maze and light compartment of the light↔dark task, and reduced immobility on the forced swim test, suggesting that NPY-Y2 antagonists may have anxiolytic and/or antidepressant actions (Redrobe et al., 2003; Tschenett et al., 2003).

Knockout mice deficient in the biogenic amine metabolic enzyme monoamine oxidase A were tested for floating time on the Porsolt swim task (Cases et al., 1995). These mice have elevated brain levels of serotonin at 1 to 90 days of age. In a 4-minute test session, immobility time was significantly lower for the mutant mice as compared to wildtype controls. This antidepressant-like phenotype is consistent with the clinical use of serotonin-elevating drugs such as Prozac. Serotonin receptor subtype and serotonin transporter knockout mice have been characterized on models of anxiety and depression and on responsiveness to anxiolytic and antidepressant drug treatments (Murphy et al., 1999). Null mutants deficient in 5-HT_{1A} receptors showed fewer entries into the open arms and less time on the open arms of the elevated plus-maze, indicating higher baseline anxiety-like behaviors (Ramboz et al., 1998; Heisler et al., 1998). The 5-HT_{1A} knockout mice also showed less immobility on the Porsolt swim test, and on a tail suspension test, indicating less depression-like behavior (Ramboz et al., 1998; Heisler et al., 1998).

Fyn tyrosine kinase deficient mice showed unusual performance on several behavioral tests relevant to fear and anxiety (Miyakawa et al., 1994; Niki et al., 1996). Homozygous *fyn* knockout mice showed fewer transitions and spent less time in the lighted chamber of the light↔dark box, as compared to heterozygous littermate controls. The mutants also showed fewer open arm entries and spent less time on the open arms of the elevated plus-maze. No differences between genotypes were detected on a radial maze task, indicating normal cognitive function. The authors suggest that the *fyn* knockout mouse could have some abnormality in the amygdala, which would explain their elevated "fearfulness," as well as their increased incidence of audiogenic seizures.

Prepulse inhibition deficits, analogous to those seen in schizophrenia, have been reported in several knockout mice. Similarly, the latent inhibition task to model attentional abnormalities in schizophrenia has proved useful in characterizing knockout mice with mutations in genes for neurotransmitter receptors. Dopamine receptor subtype D1, D2, D3, D4, and D5 knockout mice appeared normal on baseline prepulse inhibition, but D2 knockouts were more sensitive to the disruptive effects of dopaminergic agonists (Ralph et al., 1999; Holmes et al., 2001; Ralph-Williams et al., 2002). Dopamine transporter knockout mice displayed reduced prepulse inhibition (Ralph et al., 2001, Figure 10.18), an effect that was reversed by treatment with a serotonin receptor 5-HT_{2A} antagonist (Barr et al., 2004). Serotonergic receptor subtype 5-HT_{1B} knockout mice demonstrated reduced prepulse inhibition, reduced startle, and differential responses to serotonergic drugs (Dulawa et al., 1997). Latent inhibition was improved, while prepulse inhibition remained normal, in serotonin receptor 5-HT3 overexpressing transgenic mice (Harrell and Allan, 2003). Mice deficient in metabotropic glutamate receptors mGluR1 and mGluR5 displayed reduced prepulse inhibition of the acoustic

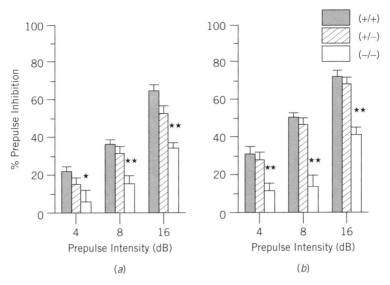

Figure 10.18 *Dopamine transporter null mutant mice (white bars) displayed significantly less prepulse inhibition at three prepulse intensities, as compared to wildtype controls (gray bars) and heterozygotes (striped bars), in both females (a) and males (b). [From Ralph et al. (2001), p. 307.]*

startle response (Brody et al., 2003; Kinney et al., 2003a; Brody and Geyer, 2004). Prepulse inhibition deficits were seen in adrenergic α_{2C} receptor knockout mice (Sallinen et al., 1998). These mice also showed an enhanced startle reflex and shortened attack latency on isolation-induced aggression, discussed in Chapter 9. Latent inhibition was improved in cholinergic muscarinic m5 receptor knockout mice (Wang et al., 2004). Corticotropin-releasing factor overexpressing transgenic mice lacked habituation to acoustic startle, as well as reduced acoustic startle reactivity and increased motor activity during startle testing (Dirks et al., 2002). Reduced prepulse inhibition and habituation to acoustic startle was seen in adenosine A2A receptor knockouts (Wang et al., 2003). Forebrain-specific calcineurin knockouts displayed impaired prepulse inhibition and latent inhibition, along with increased locomotion and reduced social interaction (Miyakawa et al., 2003). Null mutants deficient in *NCAM-180*, the gene for neural cell adhesion molecule, had greatly reduced prepulse inhibition, concomitant with enlarged lateral ventricles (Wood et al., 1998). The growing use of prepulse inhibition in mutant lines of mice emphasizes the usefulness of this fully automated task for understanding the neurotransmitters critical to the circuitry for sensorimotor gating.

RECOMMENDATIONS

Tread softly when approaching a mouse model of a human psychiatric disease. Investigators have no insight into whether a mouse feels "anxious" or "depressed." At most we can develop a behavioral test for mice that has some validity for a symptom of the human condition, or an assay that is sensitive to therapeutic drug responses in humans. Members of our laboratory are taught from day 1 to use cautious terminology such as "anxiety-like," "depression-related," and "relevant to schizophrenia." The credibility of our field depends on avoiding the impression that it is possible to develop a comprehensive mouse model of a human mental illness. Rather, a mutant mouse model that displays one or two symptoms with face validity (analogous behavioral traits), construct validity (common biological mechanisms), or predictive validity (therapeutic drug effects) will be appreciated for testing hypotheses and developing better treatments for psychiatric syndromes.

This chapter described many mouse behavioral tests relevant to anxiety disorders, two relevant to depression-like behavior, and a few relevant to symptoms of schizophrenia. The elevated plus-maze is presently the "gold standard" for anxiety-related behaviors, in that it is simple to construct and conduct, and has built-in controls for general locomotion by scoring closed arm time and entries. Since the elevated plus-maze is highly sensitive to previous experience (File 1990; Holmes and Rodgers 1999), this task is best run as the first in the series of behavioral phenotyping experiments. A second and third test then follows, to confirm an anxiety-related phenotype. The light↔dark exploration task and the Vogel conflict test are good choices, with enough similarities to the elevated plus-maze to provide strong corroboration of a positive finding, and enough differences to confirm generalization of the behavioral trait (see Chapter 13). For depression-related behaviors, both the forced swim test and the tail suspension test have predictive validity for antidepressant drug responses, but only tenuous face validity for analogies to the symptoms of human depression. The tail suspension task appears to be more useful across many transgenic and knockout mouse models, perhaps because the forced swim test is more sensitive to artifacts due to motor abnormalities (Holmes et al., 2002d, 2003d). Prepulse inhibition is a relatively simple, automated, high-throughput task that models a specific symptom of schizophrenia. Prepulse inhibition is currently the "gold standard" for schizophrenia-related behaviors in mice. However, deficits in prepulse inhibition are seen in other neuropsychiatric diseases, and it is not certain that this particular symptom is central to the psychopathology of schizophrenia.

Thus the transgenic and knockout technology spotlights the need for more and better mouse behavioral tests relevant to human psychiatric symptoms. The National Institute of Mental Health in Bethesda, Maryland, and many other research institutes and organizations around the world, are encouraging investigators to develop new and improved behavioral tasks that will more effectively model symptoms of major mental illnesses (Nestler et al., 2002; Lederhendler, 2003). Rigorous paradigm development by our behavioral neuroscience colleagues, and recommendations on the optimal constellation of mouse tests, will greatly advance the translational and preclinical strategies for developing more effective treatments for psychiatric disorders.

BACKGROUND LITERATURE

Blanchard DC, Blanchard RJ (1988). Ethoexperimental approaches to the biology of emotion. *Annual Review of Psychology* **39**: 43–68.

Blanchard DC, Griebel G, Blanchard RJ (2003a). The mouse defense test battery: Pharmacological and behavioral assays for anxiety and panic. *European Journal of Pharmacology* **463**: 97–116.

Blanchard RJ, Hebert MA, Ferrari P, Palanza P, Figueira R, Blanchard DC, Parmigiani S (1998). Defensive behaviors in wild and laboratory (Swiss) mice: The mouse defense test battery. *Physiology and Behavior* **65**: 201–209.

Capuron L, Dantzer R (2003). Cytokines and depression: The need for a new paradigm. *Brain, Behavior, and Immunity*. **17** (Suppl 1): S119–S124.

Clement Y, Chapouthier G (1998). Biological bases of anxiety. *Neuroscience and Biobehavioral Reviews* **5**: 623–633.

Contarino A, Heinrichs SC, Gold LH (1999b). Understanding CRF neurobiology: Contributions from mutant mice. *Neuropeptides* **33**: 1–12.

Cooper JR, Bloom FE, Roth RH (1996). *The Biochemical Basis of Neuropharmacology*, 7th ed. Oxford University Press, New York.

Crawley JN (1989). Animal models of anxiety. *Current Science* **2**: 773–776.

Cryan JF, Mombereau C (2004). In search of a depressed mouse: utility of models for studying depression-related behavior in genetically modified mice. *Molecular Psychiatry* **9**: 326–357.

Cryan JF, Mombereau C, Vassout A (2005). The tail suspension test as a model for assessing antidepressant activity: Review of pharmacological and genetic studies in mice. *Neuroscience and Biobehavioral Reviews* **29**: 571–625.

Csernansky JG, Grace AA, Eds. (1998). New models of the pathophysiology of schizophrenia. *Schizophrenia Bulletin*, Special Issue **24**: 185–317.

Current Protocols in Neuroscience (1997). Behavioral Neuroscience. Wiley, New York, Chapter 8.

Davis M (1986). Pharmacological and anatomical analysis of fear conditioning using the fear-potentiated startle paradigm. *Behavioral Neuroscience* **100**: 814–824.

Dawson GR, Tricklebank MD (1995). Use of the elevated plus-maze in the search for novel anxiolytic agents. *Trends in Pharmacological Sciences* **16**: 33–36.

Dowling JE (1998). *Creating Mind: How the Brain Works*. Norton, New York.

Egan MF, Weinberger DR (1997). Neurobiology of schizophrenia. *Current Opinions in Neurobiology* **7**: 701–707.

Falls WA (2002). Fear-potentiated startle in mice. *Current Protocols in Neuroscience*, pp. 8. 11B.1–8.11B.16.

Fanselow MS (1994). Neural organization of the defensive behavior system responsible for fear. *Psychonomic Bulletin and Review* **1**: 429–438.

File SE (1996). Recent developments in anxiety, stress, and depression. *Pharmacology Biochemistry and Behavior* **54**: 3–12.

File SE (1997a). Animal tests of anxiety. In *Current Protocols in Neuroscience*. Wiley, New York, pp. 8.3.1–8.3.15.

Finn DA, Rutledge-Gorman MT, Crabbe JC (2003). Genetic animal models of anxiety. *Neurogenetics* **4**: 109–135.

Fox MW (1968). *Abnormal Behavior in Animals*. Saunders, Philadelphia.

Geyer MA, Krebs-Thomson K, Braff DL, Swerdlow NR (2001). Pharmacological studies of prepulse inhibition models of sensorimotor gating deficits in schizophrenia: A decade in review. *Psychopharmacology* **156**: 117–154.

Geyer MA, Swerdlow NR, Mansbach RS, Braff DL (1990). Startle response models of sensorimotor gating and habituation deficits in schizophrenia. *Brain Research Bulletin* **25**: 485–498.

Gould TD, Gottesman II (2006). Psychiatric endophenotypes and the development of valid animal models. *Genes, Brain and Behavior* **5**:113–119.

Gray JA (1982). *The Neuropsychology of Anxiety: An Inquiry into the Functions of the Septohippocampal System*. Oxford University Press, New York.

Greenberg BD, McMahon FJ, Murphy DL (1998). Serotonin transporter candidate gene studies in affective disorders and personality: Promises and potential pitfalls. *Molecular Psychiatry* **3**: 186–189.

Griebel G, Blanchard DC, Jung A, Lee JC, Masuda CK, Blanchard RJ (1995). Further evidence that the Mouse Defense Test Battery is useful for screening anxiolytic and panicolytic drugs: Effects of acute and chronic treatment with alprazolam. *Neuropharmacology* **34**: 1625–1633.

Higgins GA (1998). From rodents to recovery: Development of animal models of schizophrenia. *CNS Drugs* **9**: 59–68.

Hofer MA (1996). Multiple regulators of ultrasonic vocalization in the rat. *Psychoneuroendocrinology* **21**: 203–217.

Holmes A (2001). Targeted gene mutation approaches to the study of anxiety-like behavior in mice. *Neuroscience and Biobehavioral Reviews* **25**: 261–273.

Holmes A, Heilig M, Rupniak NMJ, Steckler T, Griebel G (2003). Neuropeptide systems as novel therapeutic targets for depression and anxiety disorders. *Trends in Pharmacological Sciences* **24**: 580–588.

Ingle DJ, Shein HM (1975). *Model Systems in Biological Psychiatry*. MIT Press, Cambridge.

Iversen SD, Iversen LL (1981). *Behavioral Pharmacology*. Oxford University Press, New York.

Jacobs BL, Ed. (1999). Serotonin 50th Anniversary. *Neuropsychopharmacology*, Special Issue **21**: 51.

Kalivas PW, Nemeroff CB (1988). *The Mesolimbic Dopamine System*. Annals of the New York Academy of Sciences, **537**. New York Academy of Sciences, New York.

Koob GF, Ehlers CL, Kupfer DJ (1989). *Animal Models of Depression*. Birkhauser, Boston.

LeDoux JE (1995). Emotion: Clues from the brain. *Annual Review of Psychology* **46**: 209–235.

Lucki I, Dalvi A, Mayorga AJ (2001). Sensitivity to the effects of pharmacologically selective antidepressants in different strains of mice. *Psychopharmacology* **155**: 315–322.

Maier SF, Watkins LR (2005). Stressor controllability and learned helplessness: the roles of the dorsal raphe nucleus, serotonin, and corticotropin-releasing factor. *Neuroscience and Biobehavioral Reviews* **29**: 829–841.

Maier SF (1984). Learned helplessness and animal models of depression. *Progress in Neuropsychopharmacology and Biological Psychiatry* **8**: 435–446.

Martin P (1998). Animal models sensitive to anti-anxiety agents. *Acta Psychiatrica Scandanavica* **98** (Suppl 393): 74–80.

Menard J, Treit D (1999). Effects of centrally administered anxiolytic compounds in animal models of anxiety. *Neuroscience and Biobehavioral Reviews* **23**: 591–613.

Murphy DL, Wichems C, Li Q, Heils A (1999). Molecular manipulations as tools for enhancing our understanding of 5-HT neurotransmission. *Trends in Pharmacological Sciences* **20**: 246–252.

Panksepp J (1998). *Affective Neuroscience: The Foundations of Human and Animal Emotions*. Oxford University Press, New York.

Plomin R (1995). Molecular genetics and psychology. *Current Directions in Psychological Science* **4**: 114–117.

Porsolt RD, Brossard G, Roux S (1998). Models of affective illness: Behavioral despair test in rodents. *Current Protocols in Pharmacology*, pp. 5.8.1–5.8.7.

Ramboz S, Oosting R, Amara DA, Kung HF, Blier P, Mendelsohn M, Mann JJ, Brunner D, Hen R (1998). Serotonin receptor 1A knockout: An animal model of anxiety-related disorder. *Proceedings of the National Academy of Sciences USA* **95**: 14476–14481.

Rodgers RJ (1997). Animal models of "anxiety": Where next? *Behavioural Pharmacology* **8**: 477–496.

Rosen JB, Schulkin J (1998). From normal fear to pathological anxiety. *Psychological Review* **105**: 325–350.

Rosenberg RN, Morris A, Zale W, Prusiner S, DiMauro S, Barchi RL, Nestler EJ (2003). *Molecular and Genetic Basis of Neurologic and Psychiatric Disease*. Butterworth Heinemann, Stoneham, MA.

Weiner I (2003). The "two-headed" latent inhibition model of schizophrenia: Modeling positive and negative symptoms and their treatment. *Psychopharmacology* **169**: 257–297.

Weiss JM, Kilts CK (1995). Animal models of depression and schizophrenia. In *The American Psychiatric Press Textbook of Psychopharmacology*, Schatzberg AF, Nemeroff CB, Eds. American Psychiatric Press, Washington, DC, pp. 81–123.

Special Issues:

Serotonin's 50th Anniversary Symposium, *Neuropsychopharmacology* **21**, (Supplement 1) August, 1999.

Psychopharmacology of Prepulse Inhibition: Basic and Clinical Studies, *Psychopharmacology* **156** (2–3), July, 2001.

Serotonin: Pharmacology, Biochemistry and Behavior, *Pharmacology Biochemistry and Behavior* **71**, April, 2002.

Don't Stop You Win! Good Cool All Right! Excellent Beautiful Again Perfect True Yes Okay More Correct Nice Work! Yummy

DA→

**NUCLEUS
ACCUMBENS**

11

Reward

HISTORY AND HYPOTHESES

Why would anyone, man or mouse, repeatedly self-administer a substance that wrecks their career, obliterates their social life, destroys their body, or kills them? Because drugs of abuse have rewarding properties, of course (don't tell your kids). Many thoughtful insights have been offered on the subject of rewarded behaviors and substance abuse (Stellar and Stellar, 1985; Koob et al., 1987; Wise and Bozarth, 1987; Swerdlow et al., 1986; Kalivas and Nemeroff, 1988; Liebman and Cooper, 1989; Wise, 1989; Crabbe et al., 1994b; Blum et al., 1996; Robbins and Everitt, 1996; Wise, 1996a, 1996b; Koob and Nestler, 1997; Kreek, 1997; Domino, 1998; Di Chiara, 1998; Piazza and Le Moal, 1998; Koob et al., 1998; Kalivas and Nakamura, 1999; Enoch and Goldman, 2001: Crabbe 2002; Koob, 2003; Kreek et al., 2004). Gene candidates that define a predisposition to develop alcoholism and drug addictions in human populations have been proposed, although replication and interpretation of some of these linkage analyses remain controversial (reviewed in Noble, 1998; Reich et al., 1998; Kranzler et al., 1998; Goate and Edenberg, 1998; Kennedy and Macciardi, 1998; Heinz et al., 1998; Buck et al., 2002; Bergeson et al., 2003; Tabakoff et al., 2003; Nestler 2004; Albertson et al., 2004; Mohn et al., 2004; Kreek et al., 2004). A mutation in the dopamine D4 receptor was reported to be linked to novelty seeking personality traits in two out of three quantitative trait loci (QTL) studies (Benjamin et al., 1996; Ebstein et al., 1996; Pogue-Geile et al., 1998). A microsatellite polymorphism at the dopamine D5 receptor gene, the 148 base pair allele 9, was highly correlated with substance abuse and with a temperament/personality measure of novelty seeking behavior in one report (Vanyukov et al., 1998). Polymorphisms in the serotonin transporter gene have been linked to sensitivity to alcohol in primates and humans (Heinz et al., 2000; Barr et al., 2003; Sen et al., 2004).

What's Wrong With My Mouse?   By Jacqueline N. Crawley
Copyright © 2007 John Wiley & Sons, Inc.

Rodent studies have focused on some of the critical elements of the human condition. Tolerance, sensitization, dependence, withdrawal effects, and psychomotor activation have been documented for rats and mice given repeated doses of morphine, heroin, amphetamine, methamphetamine, cocaine, alcohol, cannabis, and nicotine (Griffiths et al., 1975; Yokel and Wise, 1975; Koob et al., 1987; Corrigall and Coen, 1991; Lister, 1992; Hooks et al., 1994a; Domino, 1998; Kalivas et al., 1998; Epping-Jordan et al., 1998; Crabbe et al., 2003; Crabbe and Phillips, 2004; Rustay et al., 2003; McClung and Nestler, 2003; Tanda and Goldberg, 2003; Hall et al., 2004b; Thiele et al., 2004; Valverde et al., 2004; Kreibich and Blendy, 2004). The rewarding properties of these drugs in humans have led to the study of brain mechanisms underlying many different classes of rewarded behaviors.

Humans can describe their pleasurable experiences to an investigator. Mice cannot. The term *reward* therefore is best left to human studies. In an incisive discussion by Wise (1989), the term *reinforcement* is delineated from *reward*. Simply stated, *reinforcers* elicit approach behaviors and cause the responses that preceded them to be repeated. Natural behaviors that appear to be reinforcing, in which rodents spontaneously engage, include feeding, drinking, exploration, sex, nest building, parenting, fighting, and escape from a predator. Experimentally elicited self-sustained behaviors in rodents include intracranial self-stimulation, self-administration of morphine and cocaine, lever pressing for a food or water reinforcement, spatial maze learning for a food or water reinforcement, escape from a footshock, escape from a brightly lit open environment, and escape from a pool of deep water.

Brain mechanisms underlying several forms of reinforcing behavior in rodents were revealed with two major techniques. (1) Intracranial self-stimulation is reinforcing in rats, eliciting lever pressing behavior to deliver repeated pulses of electrical current at several identified brain sites. The discovery of the intracranial self-stimulation phenomenon in the hypothalamus (Olds, 1956, 1962) led to a search for the brain pathways mediating reward. In the rat, anatomical sites at which intracranial self-stimulation is elicited include hypothalamic nuclei, septal nuclei, the medial forebrain bundle, and the mesolimbic dopaminergic pathway (Porrino et al., 1984; Phillips and Fibiger, 1989; Fiorino et al., 1993). (2) Neurotransmitter release, particularly dopamine release in the nucleus accumbens, accompanies the performance of reinforced behaviors. In vitro postmortem assays of monoamine metabolites, in vivo microdialysis, and in vivo electrochemistry have been applied to study the neuropharmacology of reinforced behaviors.

The mesolimbic dopamine pathway, from the ventral tegmentum to the nucleus accumbens, is strongly implicated in reinforced behaviors and drug self-administration (reviewed in Koob, 1992; Salamone, 1994; Cannon and Bseikri, 2004; DiChiara et al., 2004). Dopamine pathways in the rat brain are diagrammed in Figure 11.1. The shell of the nucleus accumbens appears to play a central role in rewarded behaviors and in responses to drugs of abuse (Deutch and Cameron, 1992; Pontieri et al., 1995; King et al., 1997a; Sokolowski et al., 1998). In vivo microdialysis and in vivo electrochemistry studies revealed that dopamine is released from the nucleus accumbens in response to many reinforcing stimuli, including self-stimulation, feeding, sexual activity, exercise, and treatment with morphine, amphetamine, cocaine, ethanol, Δ^9-tetrahydrocannabinol, and nicotine (Zetterstrom et al., 1983; Di Chiara and Imperato, 1988; Hernandez and Hoebel, 1988; Bradberry and Roth, 1989; Kuczenski and Segal, 1989; Chen et al., 1990; Kalivas and Duffy, 1990; Robinson and Camp, 1990;

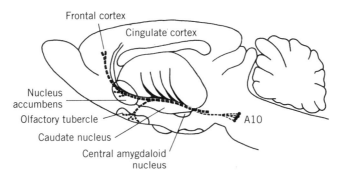

Figure 11.1 *Dopaminergic pathways in the rat brain. The mesocorticolimbic pathway originates in the A10 ventral tegmental cell bodies. A10 projections terminate in the nucleus accumbens, olfactory tubercles, and medial frontal cortex.* [Adapted from Cooper et al. (1996), p. 295.]

Pfaus and Phillips, 1991; Weiss et al., 1992; Damsma et al., 1992; Nakahara et al., 1992; Deutch and Cameron, 1992; Fiorino et al., 1993, 1997; Castner et al., 1993; Westerink et al., 1994; Salamone et al., 1994; Pontieri et al., 1995; Smith, 1995; Tanda et al., 1997; Di Ciano et al., 1998; Marinelli et al., 1998; Rada et al., 1998; Sokolowski et al., 1998; Martellotta et al., 1998; Donny et al., 1998; Rocha et al., 2002; Kelley, 2004). Antipsychotic drug treatments increase dopamine release from the nucleus accumbens, striatum, and prefrontal cortex (Moghaddam and Bunney, 1990; Marcus et al., 1996). Neurophysiological recordings from a subset of neurons in the shell of the nucleus accumbens demonstrated specific responses during consumption of a palatable food or water, during spontaneous nose-poke behavior on a holeboard in a novel environment, and during cocaine self-administration (Lee et al., 1998; Carelli, 2004).

Dopaminergic mesocorticolimbic pathway activation is not restricted to positive reinforcers. Stressors also induce dopamine release in the nucleus accumbens and prefrontal cortex (Herman et al., 1982; Deutch et al., 1985, 1990; Roth et al., 1988; Dunn, 1988; Abercrombie et al., 1989; Deutch, 1993; King et al., 1997a). The mesocortical dopamine pathway from the ventral tegmentum to the prefrontal cortex is particularly responsive to stressors (Roth, 1988; Abercrombie et al., 1989; Deutch, 1993). Physical and psychological stressors appear to facilitate drug self-administration (Table 11.1), through adrenal glucocorticoid actions on mesencephalic dopamine neurons (Piazza and Le Moal, 1998; Spencer et al., 2004).

In addition to the mesocorticolimbic dopamine pathway, several other neurotransmitter systems mediate rewarded behaviors and responses to drugs of abuse. The endogenous enkephalins, which are peptide neurotransmitters acting at opiate receptors in the brain, contribute to the neurobiology of drug self-administration and reinforcement (Di Chiara and North, 1992; Hayward et al., 2002; Zhang et al., 2003). GABA-A receptors in the midbrain—forebrain—extrapyramidal circuit act on endogenous opiate and dopamine neurons (Koob, 1992; Zhang et al., 2003). Serotonin systems contribute to the behavioral responses to psychostimulants (Kuczenski and Segal, 1989; Murphy et al., 1999; Rothman and Baumann, 2003; Auclair et al., 2004). The noradrenergic locus coeruleus is dramatically activated during opiate withdrawal, an effect that can be blocked by clonidine, an α2-adrenergic receptor ligand, the neuropeptide galanin, and by NMDA and AMPA receptor compounds acting on glutamate transmission (Rasmussen, 1995; Zachariou et al., 2003).

TABLE 11.1   Stressful Environmental Experiences Contribute to Self-Administration of Drugs of Abuse in Rodents

Types of Stress	Approaches to Study of Drug Self-Administration			
	Acquisition	Dose–Response	Progressive Ratio	Reinstatement
Food restriction	↑Psychostimulants ↑Opiates ↑Alcohol	↑Psychostimulants ↑Opiates ↑Alcohol		
Tail pinch	↑Amphetamine			
Foot shocks	↑Cocaine		↑Heroin	↑Heroin ↑Cocaine
Restraint	↑Morphine			
Social aggression	↑Cocaine	↑Cocaine		
Social competition	↑Amphetamine			
Social isolation	↑Opiates ↑Alcohol	↑* Cocaine ↑Heroin		
Witnessing stress	↑Cocaine			
Prenatal stress	↑Amphetamine			

Source: From Piazza and Le Moal (1998), p. 68.
Note: ↑Depending on the type of self-administration used: facilitation of acquisition, upward shift of the dose-response curve, higher breaking point, induction of responding on the device previously associated with the infusion of the drug. *For social isolation also slightly higher, equal, and lower sensitivities to the reinforcing effects of cocaine have been reported.

Genetic substrates for drug self-administration have been demonstrated in rodent studies, using inbred strains of rats and mice (Crabbe et al., 1994a, b; Deroche et al., 1997; Kuzmin and Johansson, 2000; Figure 11.2). C57BL/6J mice show a high propensity to consume ethanol (Crabbe et al., 1983; Phillips and Crabbe, 1991; Belknap et al., 1997) and show greater responses to ethanol on measures of ataxia, open field activity, balance beam walking, hypothermia, plasma corticosterone levels, and β-endorphin release (Crabbe et al., 1983; 1994a; De Waele et al., 1992; Roberts et al., 1995; Meliska et al., 1995; Crawley et al., 1997a; Belknap et al., 1997; Crabbe et al., 2003a, b). C57BL/6 mice consumed more ethanol, amphetamine, and nicotine in a two-bottle choice paradigm (Meliska et al., 1995). Selected lines of ethanol-induced "long-sleep" and "short-sleep" time mice have been successfully bred for many generations (Kakihana et al., 1966; Erwin et al., 1976; DeFries et al., 1989; Belknap et al., 1997).

Cocaine self-administration was greater in C57BL/6J and C57BL/6 × SJL than in DBA/2J or BALB/cByJ mice (George and Goldberg, 1989; Grahame and Cunningham, 1995; Deroche et al., 1997; Kuzmin and Johansson 2000). Conditioned place preference for cocaine was higher in C57BL/6J than in 129/SvJ mice, with the F_1 offspring of this cross resembling the C57BL/6J parental phenotype (Miner, 1997). Caffeine-induced seizures were greater in CBA than SWR inbred strains of mice, with caffeine-induced lethality of the F_1 generation from this cross resembling the CBA parental phenotype (Seale et al., 1985). Comparison of A/J, C57BL/6J and the AXB and BXA recombinant inbred lines of mice revealed higher levels of dopamine transporter protein and FosB in the C57BL/6J than in the A/J mice (Brodkin et al., 1998). Rat strains, including Sprague-Dawley, Wistar, Lewis, and Fisher 344, show individual differences in dopamine-mediated behaviors, including amphetamine-induced

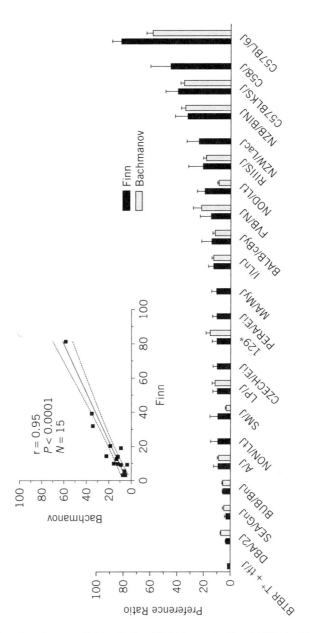

Figure 11.2 *Ethanol preference drinking [ratio of fluid from a bottle of 10% v/v ethanol in tap water to total fluid (ethanol tube + water tube)] was measured in a panel of inbred strains in two separate studies. Depicted are the strain means ± SEM for 8 (Finn) or 12 (Bachmanov) male mice per strain. Data for the 15 strains in common were significantly, positively correlated (see inset). The goodness of fit of the regression line ± 95% confidence intervals was 90.5%, confirming the reliability of measures of ethanol preference across laboratories. *Data for 129 reflects values in 2 different substrains, as 129P3/J were tested in the Bachmanov study and 129S1/SvImJ were tested in the Finn study. [Results adapted from Wahlsten et al. (2006), and kindly contributed by Drs. Deborah Finn and John Crabbe, Oregon Health Sciences University, Portland, OR.]*

hyperlocomotion, sugar consumption, and dopamine release (Hooks et al., 1994b; Sills et al., 1998; Rouge-Pont et al., 1998; Brodkin et al., 1998).

Quantitative trait loci analysis of selectively bred offspring of a BXD cross in the two-bottle choice preference test for 10% ethanol preference identified significant linkages at nine chromosomal loci (Belknap et al., 1997). Genes mediating ethanol withdrawal in C57BL/6J and DBA/2J mice were identified with DNA microarrays (Daniels and Buck, 2002). Quantitative trait loci analysis revealed several chromosomal loci linked to morphine preference in a two-bottle choice test in BXD F_2 mice (Berrettini et al., 1994; Ferraro et al., 2005). Quantitative trait loci analyses detected several chromosomal markers significantly linked to open field hyperlocomotion induced by amphetamine, phencyclidine, ethanol, and cocaine, and methamphetamine-induced stereotypy (Alexander et al., 1996; Grisel et al., 1997; Palmer et al., 2002; Downing et al., 2003; Gill and Boyle, 2003).

These several lines of evidence in rodents support the notion that genetic loci are linked to the reinforcing properties of drug self-administration in humans. Transgenic and knockout mice, with mutations in genes relevant to dopaminergic and enkephalinergic neurotransmission, or in genes identified by QTL and DNA microarray analysis, provide excellent tools to pursue hypotheses about the genetics of drug abuse.

BEHAVIORAL TESTS

Intravenous and Intracranial Self-Administration

Elegant procedures for self-administration of drugs into the peripheral circulation or directly into the brain have been described for rats. Delivery systems allow the rat to receive a bolus dose of drug through an implanted cannula into a vein or into a brain site. Intravenous self-administration requires the rat to press a lever in an operant chamber to receive a dose of drug through an indwelling jugular vein catheter. The catheter is connected to a drug reservoir, usually a syringe, with a computer-controlled delivery system, data collection, and data analysis capabilities. Self-administration into the brain is accomplished by drug administration through an implanted cannula placed in the lateral ventricles or into a discrete brain site such as the nucleus accumbens (Bozarth and Wise, 1981; Hoebel et al., 1983; Goeders and Smith, 1987; Rodd-Henricks et al., 2002). This approach is probably the best measure of the reinforcing properties of a drug in the brain. Rats will self-administer cocaine, amphetamine, morphine, heroin, alcohol, nicotine, and a cannabinoid agonist (Pickens and Thompson, 1968; Yokel and Wise, 1975; Koob et al., 1984; Collins et al., 1984; Pettit et al., 1984; Vaccarino et al., 1985; Corrigall and Coen, 1989; Donny et al., 1998; Martellotta et al., 1998; Carelli 2004; Self et al., 2004, Paterson and Markou, 2004). Drug preloading, in which the experimenter delivers priming doses of the drug to the animal, is generally required before administration becomes self-initiating by a rodent. Once the animal begins to self-administer, drug intake is gradually escalated over successive days of access. High rates of responding for cocaine are reached with one hour access per daily session, over several days (Ahmed and Koob, 1998). The dose–response curve for each individual animal is used to define the *hedonic set point* (Koob, 1992; Ahmed and Koob, 1998).

The smaller size of the mouse, and the greater variability in response to anesthetics in mice, make the procedures for self-administration more challenging in this species. Remarkably many outstanding laboratories have succeeded in conducting drug self-administration in mice (Carney et al., 1991; Grahame and Cunningham, 1995; Martellota et al., 1995; Deroche et al., 1997; Picciotto et al., 1998; Rocha et al., 1998a; Caine et al., 2002; Griffin and Middaugh, 2003; Fink-Jensen et al., 2004; Tessari et al., 2004). Nick Grahame and Chris Cunningham at the Oregon Health Sciences University, Portland, Oregon were among the first to develop an intravenous self-administration system for mice, to investigate genetic differences between C57BL/6J and DBA mice on cocaine self-administration (Grahame and Cunningham, 1995). Anesthesia is a cocktail of ketamine, acepromazine, and xylazine. Silastic tubing is eased into the jugular vein and threaded down to the right atrium. The cannula is fastened to the vein with surgical silk thread, and attached to 22 gauge hypodermic stainless steel tubing. The hypodermic tubing emerges subcutaneously through a small incision in the upper back. Tygon tubing for drug delivery is attached to this external connector. The connector is mounted on a swivel on the lid of the apparatus and connected to a syringe pump containing saline or cocaine. To improve survival during and after surgery, mice are kept warm and body temperature is monitored with a rectal probe. Home cages are warmed with heating pads during the 24-hour period after surgery. Catheters are flushed daily with a heparin solution to prevent blood clots. Grahame's self-administration equipment is illustrated in Figure 11.3.

Rodents generally require exposure to a reinforcing drug before they will begin to self-administer the drug. However, Grahame and Cunningham (1995) found that C57BL/6J mice began cocaine self-administration without priming. Stable self-administration of cocaine, 0.5 mg/kg i.v., 5 µl per reinforcement, delivered over less

Mouse Intravenous Self-Administration Apparatus

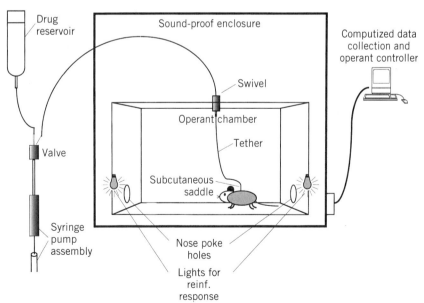

Figure 11.3 *Operant system for intravenous self-administration of drugs in mice.* [Drawing kindly contributed by Nick Grahame and Chris Cunningham, Oregon Health Sciences University.]

than one second, is reached by most C57BL/6J mice in 3 to 13 days of training. Rather than pressing a lever, the mouse operant chamber requires the mouse to poke its nose into a photocell-equipped opening, a behavior that appears to be more natural and easier for mice to acquire. A fixed ratio-3 schedule of partial reinforcement is used; that is, every third response is rewarded. The concentration of the dose is gradually increased over the course of training, up to the final delivery of 0.5 mg/kg, to avoid accidental overdose. Number of nose-pokes at the reinforced hole versus number of nose-pokes at a second, unreinforced hole at the opposite end of the operant chamber is the measure of choice accuracy. Number of nose-pokes for cocaine versus nose-pokes for saline is the quantitative measure of cocaine self-administration behavior. This study reported that both C57BL/6J and DBA/2 mice acquired stable cocaine self-administration. C57BL/6J mice reached a significantly higher asymptote for total number of reinforcers. The authors carefully discuss the issues of drug preference, differential sensitivity to the rate-decreasing effects of cocaine, drug toxicity, and drug pharmacokinetics in their interpretation of genetic strain differences. For example, faster clearance or metabolism of cocaine by liver enzymes, rather than a difference in the preference for cocaine, could explain different levels of cocaine self-administration between inbred strains of mice.

Marina Picciotto, Emilio Merlo-Pich, and co-workers measured the reinforcing properties of nicotine in mice (Picciotto et al., 1998). Similar surgery was employed with a silastic catheter implanted into the jugular vein under halothane anesthesia. Nose-poke detectors were used in the operant chambers. Training was initiated with a cocaine reinforcer, 0.8 mg/kg i.v. After stable baselines were reached with cocaine self-administration, the reinforcer was switched to nicotine, 0.03 mg/kg i.v. Normal mice performed an average of about 90 nose-pokes per hour for cocaine and 20 nose-pokes per hour for nicotine, as compared to about 5 nose-pokes per hour for saline.

Oral Self-Administration

Some drugs are self-administered by rodents when presented in the drinking water. Ethanol is the best example (Crabbe and Harris, 1991; De Waele et al., 1992; Meliska et al., 1995; Belknap et al., 1997; Crabbe and Phillips, 2003). Mice are presented with two bottles in their home cage, one containing tap water and one containing ethanol diluted in tap water. Varying concentrations of ethanol have been used, from 3% to 22%. Concentrations begin low and are gradually increased over the course of habituation and training. A 10% maximal ethanol concentration is commonly used for C57BL/6J mice. The water bottles for mice are 25 ml, with graduations marked on the side for easy observation without disturbing the mouse. Volume of water consumed and volume of ethanol solution consumed are measured to the nearest 0.1 ml, every 24 hours. The two-bottle choice test has been extensively used to characterize ethanol preference. This method has consistently shown that the C57BL/6J inbred strain is alcohol-preferring (Belknap et al., 1997). The two-bottle choice test has proved useful to search for genes linked to ethanol preference, with the recombinant inbred strains approach for analysis of BXD (Crabble et al., 1983; Crabbe et al., 1994), as discussed in Chapter 14. The two-bottle choice procedure is also used in studies of morphine (Figure 11.4), nicotine, and amphetamine self-administration (Berrettini et al., 1994; Meliska et al., 1995). The oral self-administration task is technically less demanding

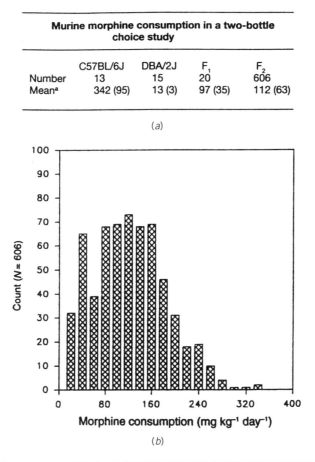

Murine morphine consumption in a two-bottle choice study

	C57BL/6J	DBA/2J	F_1	F_2
Number	13	15	20	606
Mean[a]	342 (95)	13 (3)	97 (35)	112 (63)

(a)

(b)

Effect of genotype on phenotype for morphine consumption

Genotype[a]	D6MIT13	D6MIT15	D10MIT28	D10MIT3
B6	171 (51)[b]	148 (61)	188 (45)	181 (38)
D2	35 (43)	45 (57)	36 (40)	34 (38)
B6 + D2	108 (80)	110 (88)	84 (69)	98 (80)

(c)

Figure 11.4 *Two-bottle choice test for oral self-administration of drugs. Oral consumption of morphine was very high in C57BL/6J mice and very low in DBA/2J mice. Quantitative trait loci analysis of the F_2 offspring of the cross between these two extreme strains revealed genes linked to morphine self-administration. [From Berrettini et al. (1994), pp. 55 and 56.]*

than intravenous self-administration for both the mouse and the investigator. Limiting issues include the stability of the drug solution at room temperature, the caloric value of the drug, and the taste of the drug. Ethanol has a high caloric value, requiring appropriate controls for its nutritive value. Morphine has a bitter taste, requiring masking with saccharin to encourage self-administration, and comparison to a bitter-tasting control solution such as quinine (Berrettini et al., 1994).

Drugs as Discriminant Interoceptive Stimuli

The two-bottle choice test is also used to assess drug psychopharmacology. In these protocols, the choice is between a sucrose solution and water. Glucose and fructose are also used (Flaherty and Mitchell, 1999). To avoid the confound of nutrient value, the choice may be between a non-caloric but similarly sweet solution of saccharine and water. This approach is based on an innate preference for the taste of a sweet solution. After the mouse has established baseline consumption of the sucrose or saccharine solution, drug treatment is given before the choice test session. If the drug acts on reward pathways, the reinforcing efficacy of the sweet solution will be altered. For example, dopamine D1 and D2 receptor antagonists reduced the preference for a saccharine solution, and also attenuated the preference for a fructose solution, as compared to water in a two-bottle choice test (Baker et al., 2003).

Rodents can detect many tastes, pleasant and unpleasant (see Chapter 5, Sensory Abilities, and Chapter 6, Learning and Memory), and can apparently detect more subtle internal physiological cues produced by a drug treatment. Drug-induced internal stimuli, such as lithium-induced gastrointestinal distress or the hallucinogenic effects of some psychostimulants, produce an interoceptive sensation that becomes a biologically meaningful cue (Appel et al., 1982; Barrett et al., 1994). Rats or mice are trained to press one lever to obtain a food or water reward only when the drug cue is present. That lever does not lead to reinforcement in the absence of the drug. Thus the drug becomes a *discriminative interoceptive stimulus*. Rodents will generalize their responses to closely related drugs that produce similar Interoceptive cues. For example, rats trained with cathinone, a psychoactive amphetamine-like alkaloid, generalized their responses to methamphetamine and cocaine (Schechter and Glennon, 1985). Rats were able to discriminate the interoceptive cues from phencyclidine (PCP) versus lysergic acid diethylamide (LSD), and respond differentially to glutamatergic and serotonergic drugs that appeared to mimic or block the effects of PCD and LSD (Winter et al., 2004). Drug discrimination procedures in rats and monkeys are being used to discover drugs that block the rewarding properties of cocaine, toward developing a therapeutically effective cocaine antagonist (Mello and Negus 2000; Mori et al., 2002; Kozikowski et al., 2003). It is intriguing to imagine that drug discrimination procedures could be expanded to allow us to ask mice about their internal feelings, induced by a variety of physiological and behavioral challenges.

Conditioned Place Preference

The conditioned place preference test is an application of the concept of drugs as discriminative interoceptive stimuli. Conditioned place preference is based on the association of the putative rewarding internal state produced by a drug with a set of neutral environmental cues. When given a choice, rats and mice will spend more time in an environment in which they were previously under the influence of a reinforcing drug, as compared to time spent in an environment in which they previously had no drug on board (Figure 11.5). Conditioned place preference has been demonstrated in rats for cocaine, amphetamine, apomorphine, methylphenidate, morphine, heroin, β-endorphin, enkephalin, cannabinoids, neurotensin, gamma-hydroxybutyric acid, and ethanol (Black et al., 1973; Sherman et al., 1980; van der Kooy et al., 1983; Phillips et al., 1983; Spyraki et al., 1983; Glimcher et al., 1984; Swerdlow and Koob, 1984; Morency and Beninger, 1986; Leone and Di Chiara, 1987; Almaric et al., 1987;

Conditioned Place Preference

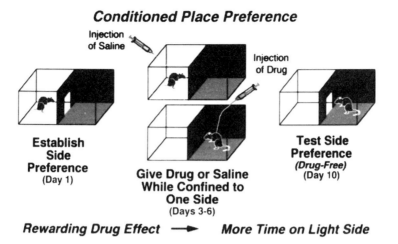

Rewarding Drug Effect ⟶ **More Time on Light Side**

Figure 11.5 *Conditioned place preference apparatus for systemically administered drugs.* [Drawing kindly contributed by George Uhl, Intramural Research Program, National Institute on Drug Abuse, Baltimore, MD.]

Valverde et al., 1996; Tzschentke, 1998; Fattore et al., 2000; Maldonado 2002; Tirelli et al., 2003; Cunningham et al., 2003; Ettenberg, 2004). Conceptually, the association between the internal cues of the drug and the external cues of the environment may be relevant to the tendency of human drug addicts who have successfully completed a drug rehabilitation program to relapse into addiction when they return to their home environment (*Addiction*, 1996).

The conditioned place preference apparatus is a two-chambered box, with a door at the opening between the two compartments (Figure 11.5). The two chambers differ on several sensory environmental cues. One chamber may be highly illuminated, with clear Plexiglas walls and ceiling, while the other chamber has black Plexiglas walls and ceiling and remains dark. One side may be larger than the other. Shapes of the two chambers may differ, one being square or rectangular, while the other is triangular or semicircular. Patterns may differ on the walls of the two chambers, for example, horizontal versus vertical stripes, or no stripes, or checkerboard versus white. The floor texture may differ between the two chambers, including wire mesh, metal bars, smooth Plexiglas, or bedding litter. Olfactory cues may be introduced by painting solutions on the walls or placing a container of an odor extract in one chamber, for example, almond extract, vanilla extract, peppermint extract, lemon juice, cedar chips, or vinegar. Two or three of these different sensory cues are generally used. The specific cues and specific modalities do not appear to be critical in the conditioned place preference test. Rather, the distinctiveness of the differences between the two chambers must be sufficiently obvious for easy recognition by the animal.

The test begins with a *preexposure session* in which the mouse is placed in the two chambered apparatus and given an opportunity to explore both chambers for 10 or 15 minutes. If a significant preference is shown for one of the chambers in the preexposure, the investigator may wish to change the properties of the two chambers to eliminate an a priori side preference or position bias in naïve animals. The mouse is then injected with the drug and confined to one compartment. The time period when the mouse is placed in the compartment must coincide with the time period when the

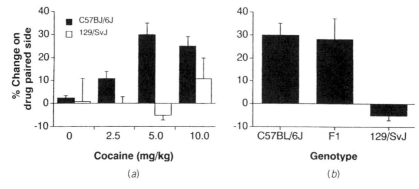

Figure 11.6 *C57BL/6J mice show a strong conditioned place preference for cocaine. 129/SvJ mice do not show a conditioned place preference for cocaine. The F₁ offspring of C57BL/6J × 129/SvJ resemble the C57BL/6J phenotype.* [From Miner (1997), p. 28.]

drug is present in the brain at biologically meaningful concentrations. A simple *drug pairing* exposure may be sufficient. Several drug pairings on successive days are often used, to increase the magnitude of the conditioned place preference. The *preference test* is conducted after the last drug pairing exposure, often 24 hours later. The preference test may be repeated at subsequent time points, to evaluate reinstatement or extinction. During the preference test, the door between the two chambers is open. Time spent in the chamber where the drug was previously paired, as compared to time spent in the other chamber, represents the place preference. Preference to drug pairing is compared with preference to saline pairing, as a control for the injection procedure.

Mice demonstrate good conditioned place preferences for cocaine, amphetamine, morphine, testosterone, dihydrocodeine, toluene, neurosteroids, and ethanol (Matthes et al., 1996; Miner, 1997; Takahashi et al., 1997; Maldonado et al., 1997; Murtra et al., 2000; Beauchamp et al. 2000; Mas-Nieto et al., 2001; Maldonado, 2002; Arnedo et al., 2002; Funada et al., 2002; Zhang et al., 2002; Kamei et al., 2003; Ribeiro Do Couto et al., 2003; Cunningham et al., 2003; Fink-Jensen et al., 2003; Hall et al., 2004b; Crabbe et al., 2005). Like drug discrimination, conditioned place preference has been extensively used to test drug interactions, toward developing therapeutics agonists and antagonists, as well as the effects of lesions and gene mutations, described below. To investigate place preference for cocaine in mice, Cindy Miner of the National Institute on Drug Abuse in Baltimore, Maryland, used cues including a wire mesh floor versus a corncob bedding floor, 4 sessions of drug pairings, and a 24-hour interval between the last drug pairing and the preference test (Figure 11.6). C57BL/6J mice developed a conditioned place preference for cocaine, while 129/SvJ mice did not, indicating a genetic component to the propensity to self-administer cocaine (Miner, 1997). Representative data for conditioned place preference to morphine in mice, reported by Maldonado and co-workers (1997) in Paris, is shown in Figure 11.7.

Hyperlocomotion

Several behavioral and physiological measures yield valuable information about the effects of abused drugs in rodents. *Motor activation* is a major indicator of central activation by psychostimulants, including cocaine, amphetamine, methamphetamine

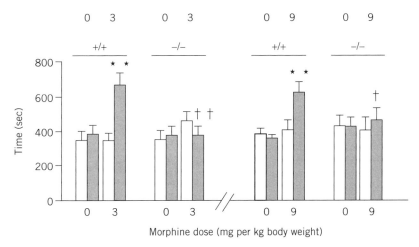

Figure 11.7 *Knockout mice deficient in dopamine D2 receptors (−/−) did not show conditioned place preference to morphine (3 or 9 mg/kg subcutaneously administered), as compared to wildtype control mice (+/+), in the choice phase (black bars). White bars represent the time spent during the preconditioning phase in the chamber in which the drug was injected. "O" represents vehicle control injections.* [From Maldonado et al. (1997), p. 587.]

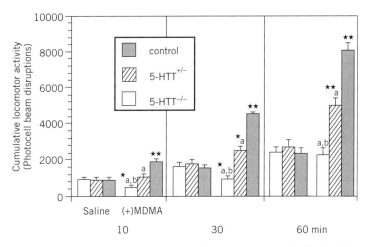

Figure 11.8 *Psychomotor activation as a measure of response to psychostimulant drugs of abuse. Knockout mice deficient in the serotonin transporter (white bars) were not significantly different from heterozygotes (striped bars) or wildtype controls (black bars) on baseline locomotion (saline). Knockouts did not show the normal hyperlocomotion response to treatment with the psychostimulant 3,4-methylenedioxymethamphetamine ([+]MDMA, "ecstacy").* [From Bengel et al. (1998), p. 653.]

(Figure 11.8), and phencyclidine (Swerdlow et al., 1986; Alexander et al., 1996; Grisel et al., 1997; Spielewoy et al., 2001; Todtenkoft et al., 2002; Nishio et al., 2003; Trinh et al., 2003). Acute treatment with these drugs induces hyperlocomotion in rats and mice, which is easily measured in an open field apparatus, described in Chapter 4. *Sensitization* to the motor activating properties of psychostimulants is seen with repeated acute administrations of cocaine, amphetamine, or enkephalin (Post et al., 1988; Kalivas et al., 1988; Hooks et al., 1994a; Segal and Kuczenski, 1997; Spielewoy

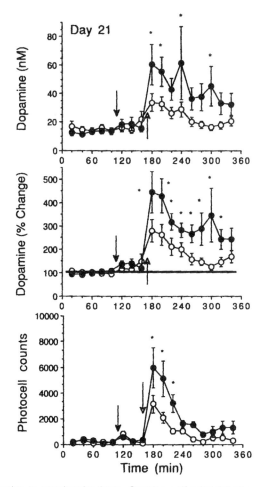

Figure 11.9 *Sensitization to psychostimulants. Cocaine self-administration results in an enhanced response to subsequent doses of cocaine, as long as 21 days after discontinuing self-administration. Saline (black arrow) did not significantly increase dopamine release or increase locomotion in rats that previously self-administered cocaine (black circles) nor in rats that received saline yoked to the self-administrations (white circles). Cocaine (white arrow, 15 mg/kg intraperitoneal injection) increased dopamine release and locomotion more in the rats that previously self-administered cocaine (black circles) than in the yoked saline controls (white circles).* [From Hooks et al. (1994a), p. 268.]

et al., 2001; Itzhak et al., 2002; Kikusui et al., 2005; Orsini et al., 2004) as described in Chapter 10. Dopaminergic mechanisms in the rat brain appear to mediate the increased behavioral responsivity to repeated administration of psychostimulants over time (Figure 11.9), a mechanism that may underlie aspects of addictive behaviors in humans.

Opiate Tolerance and Withdrawal

Analgesia is the primary medical use of opiates. However, *tolerance* develops to chronic treatment with morphine. Patients with neuropathic pain or pain due to cancer require increasing doses of morphine to achieve pain reduction. Continuous treatment with high doses of opiates will result in withdrawal symptoms if drug treatment is abruptly terminated. Tests for analgesia are described in Chapter 5. The two primary tests for analgesia in mice are the tail-flick test and the hot-plate test. *Tail flick* measures the latency for the rodent to move its tail out of the path of a narrowly focused beam of light, which delivers a mildly painful radiant heat. *Hot-plate* analgesia measures the latency for the rodent to lift and/or lick a hindpaw from the floor of the apparatus, which is heated to a temperature of 52–55°C. Tolerance to repeated analgesic drug administration is measured by repeated testing on the tail-flick and hot-plate tests.

Withdrawal symptoms in rodents comprise a set of behaviors including wet dog shakes, excessive grooming, sniffing, tremor, teeth chattering, ptosis, and diarrhea (Boulton et al., 1992). Withdrawal symptoms are evident when rats and mice are treated with morphine twice a day for a week and then treated with naloxone, an opiate receptor antagonist, several hours after the last morphine treatment (Matthes et al., 1996; Maldonado et al., 1996, 1997, 2004; Akbarian et al., 2001; Mas-Nieto et al., 2001; Basile et al., 2002; Zachariou et al., 2003). Scoring of withdrawal symptoms follows a standardized checklist procedure. The human observer is "blind" to the treatment condition, that is, does not know which animals received drug and which received saline as controls, to avoid experimenter bias during the subjective scoring procedure. Behaviors scored include wet dog shakes, tremor, jumping, teeth chatter, eye ptosis, diarrhea, and weight loss (Figure 11.10).

TRANSGENICS AND KNOCKOUTS

Opiate receptor knockout mice provide an excellent validation of the knockout technology. Two independent laboratories, directed by Brigitte Kieffer at the Centre National de la Recherche Scientifique 9050 in Strasbourg, France, and by George Uhl at the National Institute of Drug Abuse in Baltimore, Maryland, developed null mutants for the μ-opioid receptor gene (Matthes et al., 1996; Sora et al., 1997; Kieffer, 1999; Kieffer and Gaveriaux-Ruff, 2002). Self-administration of morphine was absent in the μ-opioid receptor null mutants as compared to wildtype controls (Becker et al., 2000). Heterozygotes showed reduced morphine self-administration (Sora et al., 2001a). Both knockouts showed a deficit in morphine-induced analgesia on the hot-plate and tail-flick tests, demonstrating that mutation of the μ-opiate receptor results in lack of normal behavioral effects of the agonist. Heroin analgesia was also absent, as tested on the hot-plate and tail-flick tests (Kitanaka et al., 1998). In one experiment, untreated knockout mice had greater nociceptive responses on the tail-flick and hot-plate tests than wildtype controls (Sora et al., 1997). Conditioned place preference for morphine was completely lacking in the knockouts (Matthes et al., 1996). Congenic μ-opioid receptor knockouts, bred into a C57BL/6 background, were similar to the mutants on a mixed C57B/129Sv background on baseline hyperalgesia, lack of morphine place preference, and absence of morphine-induced analgesia (Hall et al., 2003), supporting an interpretation that the behavioral phenotype resulted from the μ-opioid receptor mutation and was not differentially modified by other genes in these two genetic backgrounds. Conditioned place preference to nicotine was also absent in the μ-opioid receptor knockouts (Berrendero et al., 2002). Ethanol self-administration and conditioned place preference were similarly lower in the mutants (Roberts et al., 2000; Hall et al., 2001; Becker et al., 2002). In addition, naloxone-precipitated morphine

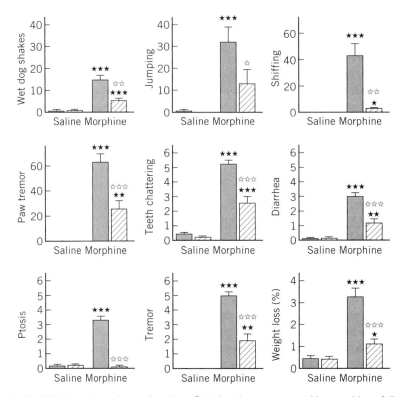

Figure 11.10 *Withdrawal syndrome in mice. Subchronic treatment with morphine, followed by naloxone-precipitated withdrawal, induces typical withdrawal symptoms in rodents, including wet dog shakes, jumping, sniffing, tremor, teeth chattering, eye ptosis, and diarrhea. Saline vehicle treatment + naloxone do not produce any of these behaviors. Knockout mice (striped bars) deficient in the transcription factor gene, cyclic AMP response element-binding protein (CREB), showed reduced withdrawal symptoms on these behaviors, as compared to wildtype controls (gray bars).* [From Maldonado et al. (1996), p. 658.]

withdrawal behaviors were absent in the μ-opioid receptor knockouts (Matthes et al., 1996). Morphine-induced immunosuppression was absent in the knockouts, supporting the hypothesized role of endogenous opiates in lymph, thymus, and spleen immune responses (Gavériaux-Furr et al., 1998). Quantitative mapping of the three opiate receptor subtypes in the brains of these knockouts revealed the absence of μ-opioid receptors, as expected, with no change in anatomical localization or concentrations of the two other opioid receptor subtypes, δ and κ, indicating that compensation by related genes did not occur in this case (Kitchen et al., 1997). A slight diminution in δ receptor agonist-induced analgesia was detected in the μ-opioid receptor deficient mice, suggesting functional overlap or interactions between these two opiate receptor subtypes (Matthes et al., 1998).

Micro-opioid receptor (OPRM1) knockout mice showed less conditioned place preference to cocaine, although cocaine-induced hyperlocomotion was normal and sensitization was enhanced (Hall et al., 2004a). Behavioral sensitization to nicotine was absent in OPRM1 null mutants (Yoo et al., 2004). Heroin-induced conditioned place preference and locomotor activation were similarly absent in OPRM1 knockouts

(Contarino et al., 2002). Infant OPRM1 null mutants did not show the normal preference for their mother's cues, emitted fewer ultrasonic vocalizations when separated from their mothers, and did not show the normal potentiation of separation vocalizations after a brief preexposure to their mother, suggesting reduced attachment behaviors (Moles et al., 2004).

Upregulation of the transcription factor cyclic AMP response element-binding protein (CREB) is implicated in the development of tolerance and drug dependence (Chao and Nestler, 2004). Reduction in symptoms of morphine abstinence was seen in mice missing the gene for CREB (Maldonado et al., 1996). *CREB* knockout mice were normal on morphine-induced analgesia but showed almost no withdrawal behaviors during naloxone-precipitated morphine withdrawal (Figure 11.10), suggesting that CREB plays an important role in opiate dependence. Conditional CREB mutation restricted to the central nervous system, and a CREB alpha/delta hypomorph mutant line, displayed normal motivation for morphine, food, and cocaine, but attenuation of withdrawal symptoms during abstinence (Valverde et al., 2004). Swim stress-induced reinstatement of conditioned place preference to cocaine was absent in mice deficient in the alpha and delta isoforms of CREB (Kreibich and Blendy, 2004). In an elegant study using an alternate technique, microinjection of the CREB gene in a herpes simplex viral vector directly into the rat nucleus accumbens decreased the rewarding effects of cocaine as measured by conditioned place preference (Carlezon et al., 1998).

Mutations of elements of the dopamine system test the long-standing hypothesis that the mesolimbic dopamine pathway mediates the rewarding properties of self-administered drugs. D1 dopamine receptor subtype knockout mice displayed normal conditioned place preference to cocaine (Miner et al., 1995), but lower operant responding for sucrose reinforcement (El-Ghundi et al., 2003). D2 dopamine receptor knockout mice did not show increased responding for intravenous self-administration of morphine on a fixed ratio schedule, in comparison to wildtype controls (Elmer et al., 2002). Null mutants for the dopamine D2 receptor subtype gene showed normal exploratory locomotion, normal conditioned place preference to a food reinforcer, and normal hyperlocomotion in response to morphine (Maldonado et al., 1997). However, these D2 receptor knockouts showed a complete lack of conditioned place preference to morphine (Figure. 11.6), supporting the interpretation that the D2 receptor is necessary for the motivational component of drug addiction (Maldonado et al., 1997). D2 knockouts did not show the reduction in firing rate of nucleus accumbens neurons during the anticipatory phase of reinforcement (Tran et al., 2002). Conditioned place preference for ethanol was absent in D2 knockouts (Cunningham et al., 2000). However, greater self-administration of cocaine was observed in the D2 knockouts as compared to wildtype controls (Caine et al., 2002). D3 receptor knockout and knockdown mice showed greater rewarding effects of morphine and higher morphine-induced hyperlocomotion, implicating this dopamine receptor subtype in a negative modulation of dopaminergic pathways mediating reward (Narita et al., 2003). However, D3 knockouts showed normal conditioned place preference to ethanol (Boyce-Rustay and Risinger, 2003). D4 dopamine receptor subtype knockout mice displayed greater discriminative stimulus effects of cocaine, and greater cocaine-induced hyperlocomotion (Katz et al., 2003). D5 null mutants showed normal discrimination of 10 mg/kg cocaine, but higher levels of cocaine-induced hyperactivity (Elliott et al., 2003), although these mice were normal on baseline measures of locomotion, anxiety-like behaviors, and fear conditioning

(Holmes et al., 2001). From the current body of literature, therefore, the D2 receptor subtype appears to be essential for the reinforcing properties of many drugs of abuse.

Serotonin receptor subtypes are implicated in many behavioral traits, include drug and alcohol abuse (Gingrich et al., 2003). Serotonin-1B receptor knockout mice, discussed in Chapter 10, were tested for responses to cocaine (Rocha et al., 1998b). Self-administration of cocaine was enhanced. Hyperlocomotion was greater, while stereotypy was less, in response to cocaine in the knockouts as compared to wildtype controls. The 5-HT1B receptor subtype appears to be a major site of action for the stimulant and rewarding actions of cocaine, in concert with the dopamine systems described above. Serotonin 1C receptor subtype knockout mice also displayed higher cocaine-induced hyperlocomotion and elevated lever pressing for cocaine (Rocha et al., 2002). These reports suggest that some serotonin receptors may normally act as inhibitory modulators of the effects of cocaine, such that their inactivation enhances the behavioral actions of cocaine.

Dopamine and adenosine $3',5'$-monophosphate-regulated phosphoprotein 32 kilodaltons (DARPP-32) is a protein that is phosphorylated when dopamine binds to D1 receptors. DARPP-32 deficient mice showed attenuated responses to 10 mg/kg cocaine on open-field activity, although a normal response to 20 mg/kg cocaine was seen (Fienberg et al., 1998). DARPP-32 knockout mice displayed a higher rate of locomotor activity sensitization to repeated cocaine treatments, as compared to wildtype controls (Hiroi et al., 1999). Conditioned place preference to cocaine was reduced in DARPP-32 knockout mice (Zachariou et al., 2002). DARPP-32 knockout mice similarly showed reduced conditioned place preference for ethanol, and a lack of concentration-dependent self-administration of ethanol (Risinger et al., 2001).

The dopamine transporter (DAT) is the presynaptic protein that actively takes released dopamine back up into the nerve termin, thereby reducing synaptic concentrations of extracellular dopamine and terminating dopaminergic transmission. Many lines of DAT mutant mice have been generated (Uhl and Lin, 2003). DAT knockouts showed stronger conditioned place preference to morphine (Spielewoy et al., 2000), although conditioned place preference to cocaine remained detectable (Sora et al., 1998). Disruption of the dopamine transporter gene resulted in high levels of spontaneous hyperlocomotion (Giros et al., 1996). The dopamine transporter knockout mice failed to show the normal hyperlocomotion responses to cocaine and amphetamine (Giros et al., 1996; Jones et al., 1998). High doses of amphetamine, cocaine, and methylphenidate, which increased locomotion in the wildtype controls, produced a paradoxical decrease in locomotion in the knockouts (Gainetolinov et al., 1999), suggesting that the dopamine transporter knockout mouse may model the paradoxical effects of Ritalin, a dopamine transporter inhibitor, in attention deficit hyperactivity disorder. Dopamine release in response to amphetamine was completely absent in the knockout mice, as measured in the striatum by in vivo microdialysis. Treatments that increase synaptic levels of serotonin, including fluoxetine, quipazine, 5-hydroxytryptophan, and l-tryptophan, produced large decreases in horizontal locomotor activity in dopamine transporter knockout mice, as compared to minimal changes in wildtype controls, suggesting that a serotonergic mechanism may underlie the paradoxical actions of dopamine transporter antagonists (Gainetdinov et al., 1999). Self-administration of cocaine persists in DAT mutant mice, implicating other neurotransmitter systems such as serotonin (Rocha et al., 1998a).

The serotonin transporter (SERT, 5-HTT) performs a similar synaptic reuptake function for the neurotransmitter serotonin. Clinical studies have implicated polymorphisms in the serotonin transporter gene in affective disorders (Greenberg et al., 1998; and see Chapter 11). Disruption of the *SERT* gene yielded mice with normal baseline locomotion, higher levels of anxiety-like behaviors and aggression, and higher stress responses (Holmes et al., 2003) but lack of responsiveness to methamphetamine (Figure 11.7; Bengel et al., 1998; Murphy et al., 1999). Conditioned place preference to cocaine was retained in serotonin transporter knockout mice (Sora et al., 1998). Double-knockouts with null mutations in both the dopamine and serotonin transporters failed to display conditioned place preference for cocaine, indicating that both transporters contribute to the reinforcing actions of psychostimulants (Sora et al., 2001b).

The vesicular monoamine transporter is a less specific protein that transports dopamine, norepinephrine, serotonin, and histamine from the cytoplasm into presynaptic secretory vesicles. Disruption of the vesicular monoamine transporter-2 gene resulted in early postnatal death (Wang et al., 1997; Takahashi et al., 1997). *VMAT-2* heterozygotes were supersensitive to the locomotor actions of cocaine, amphetamine, apomorphine, and ethanol, and did not show sensitization to repeated cocaine treatments (Wang et al., 1997). Further, *VMAT2* heterozygote mice showed less conditioned place preference to amphetamine than did wildtype controls (Takahashi et al., 1997). These several lines of evidence from mutant mice are consistent with current hypotheses of specific roles of each type of monoamine reuptake mechanism in the neuropharmacology, psychostimulant, and reinforcing properties of abused substances (Hall et al., 2002).

Behavioral responses to ethanol have been investigated in several mutant mouse lines. Ethanol self-administration was decreased in transgenic mice overexpressing the serotonin 5-HT$_3$ receptor (Engel et al., 1998). In a two-bottle choice test, daily ethanol consumption was significantly less in mice that expressed this serotonin receptor subtype selectively in the forebrain when a calcium/calmodulin-dependent protein kinase IIα promoter was used. Decreased ethanol-induced sedation, hypothermia, and tolerance was seen in γ-protein kinase C null mutant mice (Bowers et al., 1999). Ethanol-induced hypothermia, loss of righting reflex, and locomotor activation were enhanced in nicotinic receptor subtype α7 knockout mice (Bowers et al., 2005). Ethanol-induced footslips on the balance beam were not as prevalent in 5-HT(1B) null mutant mice as in their wildtype controls, although ethanol-induced deficits on the rotarod and righting reflex tests did not differ between genotypes (Boehm et al., 2000). Ethanol hypersensitivity was demonstrated in a null mutation of *fyn*, the gene for fyn-kinase, a nonreceptor-type tyrosine kinase that phosphorylates NMDA and GABA receptors (Miyakawa et al., 1997). *Fyn* knockout mice showed longer durations of loss of righting reflex in response to ethanol, while showing normal durations of loss of righting reflex in response to flurazepam, a benzodiazepine sedative-hypnotic, indicating a selective role for this gene in the behavioral actions of ethanol.

Ethanol consumption in a two-bottle choice test was elevated in knockout mice deficient in neuropeptide Y (NPY) (Thiele et al., 1998). The authors assayed taste preference for sucrose and for quinine in control experiments, reporting no genotype differences, while carefully ruling out a change in taste sensitivity to ethanol in the *NPY* knockouts. Todd Thiele and co-workers further ruled out an explanation based on anxiety, since the *NPY* knockout mice were normal on fear conditioning and elevated plus-maze tasks. Knockout mice lacking the neuropeptide Y receptor subtype

Y1 showed increased consumption of ethanol solutions, while consumption of sucrose and quinine were not significantly different than wildtype controls (Thiele et al., 2002). Conversely, mutant mice deficient in the Y2 receptor subtype drank less ethanol than wildtype controls when the mutation was on a mixed 129/SvJ × Balb/cJ background, but drank normal amounts of ethanol solutions when the Y1 mutation was backcrossed into a Balb/cJ background (Thiele et al., 2004a). These findings support a growing literature on neuropeptide transmitters mediating alcohol consumption (Froehlich and Li, 1994; Menzaghi et al., 1994; Thiele et al., 2004b; Lewis et al., 2005).

The reinforcing properties of nicotine underlie human addiction to cigarettes (reviewed in Domino, 1998). Nicotine acts at nicotinic acetylcholine receptors in the brain. Nicotinic receptors are composed of subunits. Several different combinations of 10 different subunits have been identified in various tissues and tested for responses to nicotine (Picciotto et al., 2000; Champtiaux and Changeux, 2004). Mutant mice deficient in the $\alpha 4$ subunit of the nicotinic receptor had higher basal levels of dopamine in the striatum, and displayed a longer hyperlocomotion response to cocaine, but failed to show nicotine-induced dopamine release (Marubio et al., 2003). Targeted mutation of the $\beta 2$ subunit of the nicotinic acetylcholine receptor generated null mutant mice lacking high-affinity binding for nicotine in the brain (Picciotto et al., 1998). Knockouts were bred onto a C57BL/6J background. $\beta 2$ knockout mice failed to show nicotine-elicited dopamine release. Behaviorally, the null mutants showed normal baseline locomotion in a novel environment but reduced baseline locomotion in a familiar environment. Most interestingly, these knockouts showed much lower levels of intravenous drug self-administration. $\beta 2$ knockout mice performed significantly fewer nose-pokes for both nicotine and cocaine, while performing normal numbers of nose-pokes for saline, indicating that the $\beta 2$ receptor subunit is required for nicotine and cocaine self-administration behavior. These findings support the interpretation that the nicotinic acetylcholine receptor subunit composition including $\beta 2$ mediates the reinforcing properties of nicotine (Picciotto et al., 1998).

Two receptors in the mammalian brain specifically bind cannabinoids, including Δ^9-tetrahydrocannabinol, the psychoactive component of marijuana, and anandamide, the endogenous ligand of the cannabinoid receptors (Ledent et al., 1999; Mechoulam and Parker, 2003). Mice deficient in the cannabinoid CB_1 receptor were generated to investigate the role of this receptor subtype in the biological actions of cannabinoids (Tanda and Goldberg, 2003). Reduced sensitivity to the reinforcing properties of sucrose and saccharine was detected in CB_1 null mutants (Sanchis-Segura et al., 2004). CB_1 receptor knockout mice failed to show the normal responses to cannabinoid agonists, including Δ^9-tetrahydrocannabinol, WIN55,212-2, and the endogenous ligand, anandamide, on hot-plate jumping, locomotion, body temperature, blood pressure, and heart rate (Ledent et al., 1999). Self-administration of morphine was completely absent, and withdrawal behaviors following chronic morphine administration and naloxone treatment were greatly reduced in the CB_1 knockout mice, as compared to wildtype controls (Cossu et al., 2001). Conditioned place preference to morphine was absent in one line of CB_1 knockout mice in one report (Rice et al., 2002) but present in another (Martin et al, 2000). Conditioned place preference to nicotine was lacking in CB_1 knockout mice (Castane et al., 2002). These findings in knockout mice add to the pharmacological literature that implicates anandamide and the CB_1 cannabinoid receptor as mediators of the analgesic and rewarding effects of marijuana, opiates, nicotine, alcohol, and other addictive substances (Mechoulam and Parker, 2003).

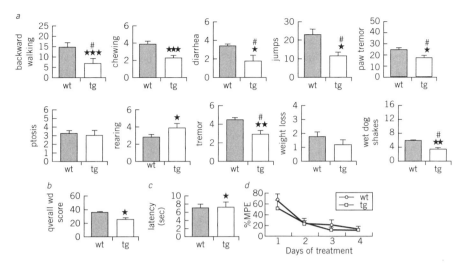

Figure 11.11 *Galanin overexpression under control of the DβH promoter can attenuate some signs of morphine withdrawal. (a) Transgenic mice overexpressing galanin under control of the DβH promoter* $(N = 10)$ *and their wildtype siblings* $(N = 9)$ *were administered increasing doses of morphine every 8 hours for 3 days (20, 40, 60, 80, 100, 100, and 100 mg/kg s.c.), and withdrawal was precipitated 2 hours after the last morphine dose by administering naloxone (1 mg/kg s.c.). After naloxone treatment, withdrawal signs were scored for 30 minutes by an observer blind to treatment. One-way ANOVA was used to compare differences in withdrawal signs between genotype. (b) The overall withdrawal score was significantly reduced in Gal-tg mice as compared with their wild-type siblings* $(P < 0.01)$. *(c) No difference in pain threshold was seen between Gal-tg and their wildtype siblings as measured on the 52°C hotplate. (d) No differences in initial analgesic response to morphine or the development of morphine tolerance was seen between Gal-tg and wild-type mice. tg, Gal-tg mice; wt, wildtype siblings;* *, $P < 0.05$; **, $P < 0.01$; ***, $P < 0.001$. [From Zachariou et al. (2003), p. 9031.]

Interest is growing in the role of neuropeptides in reward processes. As described in Chapter 5, substance P and galanin are neuropeptides that mediate pain transmission in the spinal cord. Mutant mice deficient in the substance P receptor NK1 performed normally on an operant task for food or cocaine reinforcement, but failed to self-administer morphine (Murtra et al., 2000; Ripley et al., 2002). Opiate withdrawal symptoms were also reduced in *NK1* knockout mice (Murtra et al., 2000). Ventral tegmental microinjections of CART, cocaine and amphetamine-regulated transcript peptide, produced place preferences in rats, and *CART* knockout mice displayed excessive consumption of a high caloric diet (Kuhar et al., 2002). Galanin overexpressing transgenic mice showed decreased symptoms of morphine withdrawal, while galanin deficient knockout mice displayed elevated withdrawal (Figure 11.11, Zachariou et al., 2003). When administered into the third ventricle of rats, galanin increased ethanol consumption (Lewis et al., 2004).

The knockout mouse technology appears to be a good tool for dissecting the neuropharmacological receptor subtypes mediating reward and drug addiction. Mutant mouse models such as the dopamine and serotonin transporter knockout mice, knockouts of selected receptor subtypes for opiates, monoamines, cannabinoids, and neuropeptides, galanin overexpressing transgenic mice, and *CREB* knockout mice may assist in identifying treatments to prevent withdrawal symptoms and speed recovery from addiction.

BACKGROUND LITERATURE

Boulton AA, Baker GB, Wu PW, Eds. (1992). *Animal Models of Drug Addiction.* Humana, Totowa, NJ.

Crabbe JC, Harris RA (1991). *The Genetic Basis of Alcohol and Drug Actions.* Plenum, New York.

Crabbe JC, Belknap JK, Buck KJ (1994). Genetic animal models of alcohol and drug abuse. *Science* **264**: 1715–1723.

Crabbe JC, Metten P, Cameron AJ, Wahlsten D (2005). An analysis of the genetics of alcohol intoxication in inbred mice. *Neuroscience and Biobehavioral Reviews* **28**: 785–802.

Crabbe JC, Phillips TJ (2003). Pharmacogenetic studies of alcohol self-administration and withdrawal. *Psychopharmacology* **174**: 539–560.

Crabbe JC, Phillips TJ, Buck, KJ, Cunningham CL, Belknap JK (1999). Identifying genes for alcohol and drug sensitivity: Recent progress and future directions. *Trends in Pharmacological Sciences* **22**: 174–179.

Deutch AY, Cameron DS (1992). Pharmacological characterization of dopamine systems in the nucleus accumbens core and shell. *Neuroscience* **46**: 49–56.

Di Chiara G, North RA (1992). Neurobiology of opiate abuse. *Trends in Pharmacological Sciences* **13**: 185–193.

Drago J, Padungchaichot P, Accili D, Fuchs S (1998). Dopamine receptors and dopamine transporter in brain function and addictive behaviors: Insights from targeted mouse mutants. *Developmental Neuroscience* **20**: 188–203.

Engel J, Oreland L (1987). *Brain Reward Systems and Abuse.* Raven, New York.

Hall FS, Sora I, Drgonova J, Li XF, Goeb M, Uhl GR (2004b). Molecular mechanisms underlying the rewarding effects of cocaine. *Annals of the New York Academy of Sciences* **1025**: 47–56.

Kalivas PW, Nakamura M (1999). Neural systems for behavioral activation and reward. *Current Opinion in Neurobiology* **9**: 223–227.

Kalivas PW, Nemeroff CB (1988). *The Mesolimbic Dopamine System.* Annals of the New York Academy of Sciences **537**. New York Academy of Sciences, New York.

Kieffer BL (1999). Opioids: First lessons from knockout mice. *Trends in Pharmacological Sciences* **20**: 19–26.

Kieffer BL, Gaveriaux-Ruff C (2002). Exploring the opioid system by gene knockout. *Progress in Neurobiology* **66**: 285–306.

Koob GF (1992). Drugs of abuse: Anatomy, pharmacology and function of reward pathways. *Trends in Pharmacological Sciences* **13**: 177–184.

Koob GF, Ahmed SH, Boutrel B, Chen SA, Kenny PJ, Markou A, O'Dell LE, Parsons LH, Sanna PP (2004). Neurobiological mechanisms in the transition from drug use to drug dependence. *Neuroscience and Biobehavioral Reviews* **27**: 739–749.

Koob GF, Nestler EJ (1997). The neurobiology of drug addiction. *Journal of Neuropsychiatry and Clinical Neuroscience* **9**: 482–497.

Kuhar MJ, Adams S, Dominguez G, Jaworski J, Balkan B (2002). CART peptides. *Neuropeptides* **36**: 1–8.

Leshner AI, Koob GF (1999). Drugs of abuse and the brain. *Proceedings of the Association of American Physicians* **111**: 99–108.

Liebman JM, Cooper SJ (1989). *The Neuropharmacological Basis of Reward.* Oxford University Press, New York.

Mammalian Genome (1998). Special Section: Genetics of Alcohol-Related Traits. **9**(12).

Piazza PV, Le Moal M (1998). The role of stress in drug self-administration. *Trends in Pharmacological Sciences* **222**: 67–74.

Pich EM, Epping-Jordan MP (1998). Transgenic mice in drug dependence research. *Annals of Medicine* **30**: 390–396.

Rasmussen K (1995). The role of the locus coeruleus and *N*-methyl-D-aspartic acid (NMDA) and AMPA receptors in opiate withdrawal. *Neuropsychopharmacology* **13**: 295–300.

Salamone JD (1994). The involvement of nucleus accumbens dopamine in appetitive and aversive motivation. *Behavioral Brain Research* **61**: 117–133.

Stellar JR, Stellar E (1985). *The Neurobiology of Motivation and Reward*. Springer, New York.

Thiele TE, Sparta DR, Hayes DM, Fee JR (2004b). A role for neuropeptide Y in neurobiological responses to ethanol and drugs of abuse. *Neuropeptides* **38**: 235–243.

Tzschentke TM (1998). Measuring reward with the conditioned place preference paradigm: A comprehensive review of drug effects, recent progress and new issues. *Progress in Neurobiology* **56**: 613–672.

Wise RA (1996). Addictive drugs and brain stimulation reward. *Annual Review of Neuroscience* **19**: 319–340.

Cartoon of the New Year's Baby with Father Time

12

Neurodevelopment and Neurodegeneration

HISTORY AND HYPOTHESES

Behaviors wax and wane. You can't do everything at once. Hopefully, you don't behave the same way now as you did when you were six years old. At the two ends of the ontogeny spectrum are development and aging. Many diseases appear to be linked to genetic dysfunctions that occur at a critical period of development, or to dysregulations that occur primarily at older ages. Specialized behavioral tasks and equipment are tailored for infant and juvenile mouse pups. Special attention is paid to the design of behavioral tasks addressing functional declines during the aging process. This chapter first describes behavioral tasks for mouse pups and juveniles, to evaluate early behaviors relevant to neurodevelopmental disorders such as mental retardation, muscular dystrophy, and autism. This chapter then addresses behavioral measures of the outcomes of aging, and the consequences of neurodegenerative diseases that usually begin at older ages, such as Alzheimer's and Parkinson's diseases.

Neurodegenerative and neurodevelopmental diseases are the focus of highly active research fields. Proximal causes are being discovered, and treatments are available in some cases. Alzheimer's disease is conclusively diagnosed by high levels of plaques and tangles in the postmortem brain (Selkoe, 2001). Hypothesized causes, including mutations in the amyloid precursor protein, incorrect processing of amyloid protein into toxic peptide fragments; mutations in presenilins 1 and 2 and β- and γ-secretase mutations causing the incorrect processing of amyloid peptides; mutations in genes for the microfilament protein tau; APOE4 polymorphisms; loss of cholinergic neurons in the basal forebrain nucleus basalis of Meynert; widespread degeneration of many neurons throughout the brain, reductions in many neurotransmitters and overexpression of the neuropeptide galanin; oxidative stress causing neuronal death, and inflammation are being investigated with many techniques including mutant

What's Wrong With My Mouse? By Jacqueline N. Crawley
Copyright © 2007 John Wiley & Sons, Inc.

mouse models (Schenk et al., 2001, 2004; Dewachter and Van Leuven 2002; Buttini et al., 2002; Crawley et al., 2002; Higgins and Jacobsen, 2003; Yan et al., 2003; Das et al., 2003; Dodart et al., 2003; Gasparini et al., 2004; Cotman et al., 2005; Rosi et al., 2005; Lee et al., 2005). The defining features of Parkinson's disease are the loss of nigrostriatal dopamine neurons leading to tremors, rigidity, bradykinesia, and postural instability (McInerney-Leo et al., 2005). Symptoms of Parkinson's disease are fairly well controlled by treatment with dopaminergic replacement drugs (Kostrzewa et al., 2005). Transplants of mesencephalic dopamine neurons, cell lines containing genes for dopamine and neurotrophic factors, and embryonic stem cells are potentially effective treatments that are being evaluated in mouse and rat models of Parkinson's disease (Rohrer et al., 1996; Veng et al., 2002; Nishismura et al., 2003). In some cases, the gene(s) responsible for the human disease has been identified, but its function is not immediately apparent. Huntington's disease appears to be caused by a single gene mutation (Gusella et al., 1983; Wexler et al., 1991). However, because the precise functions of the *huntintin* protein remain elusive, animal models are essential to understanding the etiology of Huntington's disease (Kirik and Bjorklund, 2003; Beal and Ferrante, 2004; Levine et al., 2004). Phenylketonurea is a form of mental retardation caused by a lack of the enzyme phenylalanine hydroxylase or its tetrahydrobiopterin cofactor (Cederbaum, 2002). While lifetime dietary restriction of phenylalanine consumption is a relatively successful treatment, gene replacement therapies are being attempted, beginning with mouse models of phenylketonurea (McDonald et al., 2002; Ding et al., 2004). Fragile X, the most common inherited form of mental retardation, is caused by a mutation in the *FMR1* gene on the X chromosome (Verkerk et al., 1991). Like *huntingtin, FMR1* mutation results in large expansions of DNA triplet repeats (Warren and Nelson, 1993). However, FMR1 is an RNA binding protein that influences the expression of hundreds of downstream genes (Brown et al., 2001). Mouse models are proving essential to working out the relevant functions of these downstream genes, and the behavioral consequences of their aberrant expression patterns (Kooy, 2003). Some neurodevelopmental and neurodegenerative diseases appear to be highly heritable but not linked to a single gene mutation. For example, linkage studies of autism indicate a minimum of ten genes contributing significantly to the variance (Muhle et al., 2004). Mice with mutations in each of the candidate genes may reveal behavioral phenotypes relevant to autism (Crawley, 2004).

Mouse models provide important tools to increase our understanding of mechanisms underlying core symptoms, and to discover efficacious therapeutics for developmental, aging, and degenerative syndromes. Good behavioral test batteries are becoming available for measuring components of mouse behavior that are unique to postnatal days 1–60 (Silver, 1995; Heyser et al., 2003; Branchi et al., 2003, 2004). The fields of behavioral toxicology and teratogenics have contributed some of the best mouse observational test batteries for this early stage of development (Moser et al., 1995; Bignami 1996). Methods for modifying standard rodent behavioral tasks to separate the symptoms of normal aging from the symptoms of neurodegenerative disorders are being elaborated (Sarter, 1987; Schoenbaum et al., 2002; King and Arendash, 2002; Higgins and Jacobsen, 2003). Tests that can be repeated in the same mice when they are young, adult, and aged are useful for following the progression of symptoms. Incorporating control measures for sensory and motor abilities will avoid artifactual overinterpretations in behavioral domains such as learning and memory in mutant mouse models of neurodegenerative disorders such as Alzheimer's disease. Most important, robust behavioral symptoms in a mouse model can offer surrogate markers for evaluating efficacy of novel therapeutics. Valid mouse models are especially crucial for the many neurodevelopmental and neurodegenerative diseases that are currently without effective treatments.

BEHAVIORAL TESTS

Birth to Six Weeks

Developmental disease models are among the best applications of the conventional knockout technology. Paralleling the human disease condition, the mutation is present in the mouse genome beginning at the blastula stage, and is expressed at time points when the fetus is influenced by the uterine environment, during the birth process, while the infant interacts closely with the parents, during stages of juvenile play and puberty, and throughout adult life. Longitudinal behavioral tests can be designed to evaluate the phenotypic outcome of the mutation as early as the first day after birth.

Many neurodevelopmental disorders display motor dysfunctions as primary symptoms. Tay-Sachs, muscular dystrophy, spina bifida, and cerebral palsy are examples of early childhood diseases in which neuromuscular abilities are severely impaired. Mouse models are phenotyped with standard motor tasks when the mutants are adult, including open field, rotarod, balance beam, hanging wire, grip tests, and pole climb, described in Chapter 4. Evaluations of mouse models at infant ages is problematic, partly because the motor test equipment is not small enough for mouse pups and partly because mice do not engage in these standard behaviors until they are older. Similarly, mice with targeted mutations in genes for congenital blindness, deafness, and other sensory abnormalities can be tested as adults using measures described in Chapter 5. However, sensory tasks for mouse pups are problematic because the eyes and ears of mice do not open until about two weeks after birth, and standard responses to olfactory stimuli may be difficult to measure until the pups attain motor coordination and motility. Sensitive assays of sensory and motor development need to be generated, to characterize mouse models during infancy. Particularly when the goal of the mouse model is to evaluate therapeutics that can be administered to young children, robust measures of developmental milestones are essential.

Chapters 3 describes several excellent observational test batteries for young mice (Fox, 1965b; Irwin, 1968; Moser et al., 1995; Rogers et al., 1997). Many of these include behaviors specific to the earliest stages of life, such as suckling, opening of the eyes and ear canals, and coordination of movements. Charles Heyser at Franklin and Marshall College in Lancaster, Pennsylvania published a step-by-step protocol for assessing developmental milestones in rodents (CPNS Unit 8.18, Heyser, 2003). As shown in Table 12.1, the sequence includes quantitation of physical landmarks, neurological reflexes, and locomotor behaviors. The pup is evaluated in a clear plastic holding cage with woodchip bedding. The holding cage is placed on a heating pad to maintain the cage temperature at 22–24°C, to avoid hypothermia after the pup is removed from its nest. Identification of individual pups employs a tattoo or repeated applications with an indelible nontoxic marking pen. Daily scoresheets are used to record the day when the eyes are open, when the pinnae of the ears detach completely from the cranium, when the upper and lower incisor teeth erupt, when dorsal and ventral fur appear, and daily body weight. The surface righting reflex involves placing the pup onto its back and recording the number of seconds for the pup to turn over onto its belly. Over the first weeks of development, latency to right the body decreases from a few seconds to instantaneously. A 60-second cutoff time indicates complete inability to right the body. Negative geotaxis involves placing the pup on an inclined plane with the head facing down the plane. Time to change orientation to face up the incline is recorded, again with a 60-second cutoff. The rooting reflex, associated with suckling, is detectable when the face region is gently stimulated with two cotton swabs and the pup pushes its head forward. The acoustic startle response in mouse pups is a slight jerk, kicking, and/or squirming, immediately after a clicker noise or automated tone. Tactile startle responses include exaggerated jumping and running in response to an air puff, such as a small jet of compressed air or the experimenter's breath. The grasp reflex is the ability of the pup to grasp a thin toothpick or blunt metal dissecting rod that is stroked against the paw. Level and vertical screen grasp tests measure the ability of the pup to hold onto a 16-mesh metal screen when dragged horizontally across the screen by the tail, and when the screen is then rotated to a vertical position. Development of coordinated quadruped walking emerges from crawling motions, paddling movements of the paws, and slipping or dragging of the hindlimbs. Table 12.1 indicates the age range

TABLE 12.1 Typical Ages for Appearance of Developmental Milestones

Measure	Average Age for Response (days)	Range (days)
Physical landmarks		
Pinnae detachment	15	10–20
Eye opening	13	7–17
Incisor eruption	7	5–10
Fur development	11	3–15
Reflexes		
Surface righting	5	1–10
Air righting	18	16–21
Negative geotaxis	7	3–15
Cliff avoidance	8	2–12
Visual placing	15	11–18
Forelimb/hindlimb placing[a]	5	1–10
Vibrissa placing response	9	5–15
Auditory startle	15	11–21
Tactile startle	15	3–20
Crossed extensor reflex[a]	3	1–10
Rooting reflex[a]	2	1–15
Grasp reflex[a]	7	3–15
Bar holding	14	10–21
Level screen test	8	5–15
Vertical screen test	19	15–21
Locomotor behavior		
Elevation of the head	12	9–21
Elevation of the forelimbs and shoulders	7	5–15
Pivoting[a]	7	2–17
Crawling[a]	11	7–16
Walking	16	12–21

Source: From Heyser p. 8.18.2 (2003), *Current Protocols in Neuroscience*.
[a]This behavior either disappears or reduces to ∼0 in frequency.

for appearance of these developmental milestones in mice. For example, the rooting response is detectable between postnatal days 1 to 15, the eyes open at about 13 days, surface righting emerges around day 5, crawling appears about day 11 and disappears as walking emerges around day 16, and startle responses are detectable at about day 15.

Many developmental milestones in Table 12.1 are simply observed to be present or absent at the age of testing. For these landmarks, the data are recorded in a dichotomous fashion (yes or no). These include pinnae detachment, eye opening, incisor eruption, fur development, and body weight. Such tests are noninvasive and require very little time to complete. Table 12.1 presents average ages at which each physical landmark is observed.

Dierssen and co-workers at the Medical and Molecular Genetics Center in Barcelona, Spain, employed a set of developmental landmarks and neurobehavioral tests to compare two hybrid inbred lines, 129Sv × C57BL/6 and C57BL/6 × SJL (Dierssen et al., 2002). Developmental screening of 118 pups beginning on postnatal day 2 included daily inspection of pinna detachment, incisor eruption, eye opening, surface righting response, forepaw/hindpaw grasping, cliff drop aversion, forelimb/hindlimb placing, disappearance of rooting response, disappearance of crossed extensor, negative geotaxis, vibrissae placement, tactile orientation, vertical climbing, Preyer reflex/startle response, suspension test, visual placing response, air puff response, reaching response, pivot locomotion, walking test, homing test to reach the home litter, and body weights. The B6129 hybrids showed a delay in neurobehavioral development on many of these parameters. In B6129F2 pups, forepaw grasping appeared later, the rooting response disappeared later, and the Preyer startle response was delayed.

Robertoux and co-workers at CNRS in Orléans, France, employed a comprehensive neurodevelopmental battery to pursue quantitative trait loci mapping of the C57BL/6By × NZB/BINJ cross (Roubertoux et al., 1985; Le Roy et al., 1999). Lod scores reached 3.13 for loci linked to cliff drop aversion, geotaxia, vertical climbing, bar holding, age of eye opening, visual placing, and acoustic startle response at 10 to 40 kHz frequencies. These approaches may be useful in identifying genes mediating neurodevelopmental delay components in mouse models of developmental diseases.

Igor Branchi and Laura Ricceri at the Instituto Superiori di Sanita in Rome summarize the many behavioral tests that can be conducted in mice at postnatal days 2–35 (Branchi and Ricceri, 2002). As shown in the timelines in Figure 12.1, sensory tasks based on olfaction, somatosensory, vestibular, and gustatory responses can be conducted from birth, while acoustic and visual tasks begin at postnatal day 12–13. Motor abilities measurable from birth include suckling, crawling, and face washing, while true grooming, walking, and wall-climbing begin at postnatal days 7–8. Ultrasonic vocalization and odor aversion learning are feasible as early as three days after birth. Branchi and Ricceri offer step-by-step procedures for conducting learning and memory tasks in young mice (Branchi and Ricceri, 2005). Three-week-old mice are tested in a smaller Morris water maze, with lower walls and warmer water temperatures. Novel object exploration in an open field is feasible at about four weeks of age (Figure 12.2).

Schizophrenia, depression, drug abuse, and attention deficit hyperactivity disorder are among the psychiatric diseases that may begin at early ages. Susceptibility genes for these diseases are hypothesized to mediate the development of brain structures needed for higher cognitive abilities, including the ability to cope with stress (Charney and Manji, 2004). Traumatic stress may impair brain development, as well as triggering psychotic and depressive episodes in adults (Nemeroff, 2004). Rodent models of these

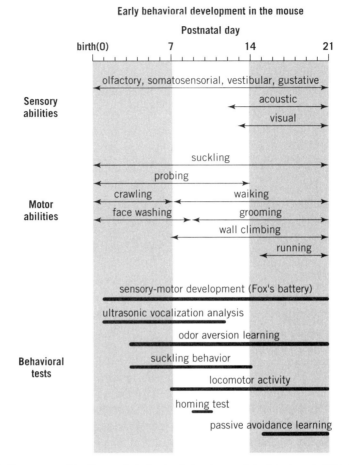

Figure 12.1 *Sequence of testing early behavioral development in the mouse. Thin lines indicate approximate age phases relative to the appearance of different mouse pup sensory and motor capabilities. Thick lines indicate age ranges during which selected behavioral tests can be performed.* [From Branchi and Ricceri (2002), p. 138.]

diseases evaluate adult animals in standard tasks described in Chapters 10 and 11. Infant and juvenile mice are physically and conceptually unsuitable for most of these tests.

One option for measuring a stress response in infant mice is the ultrasonic vocalization "distress call." As described in Chapters 8 and 10, ultrasonic vocalizations are emitted by mouse pups when they are removed from the nest and separated from their parents. Detailed procedures for quantitating ultrasonic vocalizations are described for rat pups by Myron Hofer and co-workers (in *Current Protocols in Neuroscience* Unit 8.14, Hofer et al., 2001), and for mouse pups by Branchi and Ricceri (*in Current Protocols in Toxicology, Unit 13.11, 2006*). The question of whether pup vocalizations are intentional distress calls remains controversial (Hofer et al., 1993; Shair and Jasper, 2003). However, it is clear that their outcome serves a communication function by alerting the parents to the location of the pup. The parental response is to search and retrieve the pup to the nest, thereby rescuing the offspring from hypothermia,

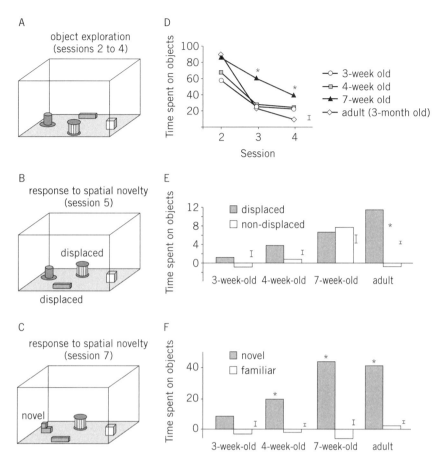

Figure 12.2 *(a, d) Mice learn an object recognition memory task beginning at age 4 weeks. (a,d) Four novel objects are placed in the arena. Time spent exploring the four objects declines across repeated 5-minute test sessions, indicating habituation as the objects become familiar. (b,e) During sessions 5 and 6, two of the objects are displaced to a new location. (c, f) During session 7, one of the objects is replaced by a novel object.* [From Branchi and Ricceri (2005), *Current Protocols in Toxicology* (2006), p. 13.11.7.]

starvation, and predators. Branchi and co-workers describe behaviors that occur in conjunction with ultrasonic vocalizations in 7-day-old CD-1 Swiss mouse pups (Branchi et al., 2004). Their ethogram shows head rising behavior occurring immediately before ultrasonic emissions, while locomotion increases during and after vocalizations. These associated behaviors may provide corroborating parameters to score during a separation challenge, in evaluating developmental milestones in infant mice.

Reflexive versus cognitive components of mouse pup vocalizations have been extensively discussed (Hofer and Shair, 1987, 1993; Hofer et al., 2001). Ultrasonic vocalizations in infant mice separated from the nest may be a reflexive response of the larynx to cold. Contact quieting occurs when a rat pup is placed back with its mother, or even with an anesthetized dam (Figure 8.8*b*). Similarly, if the pup is placed back with an anesthetized adult male rat, vocalizations are inhibited (Figure 12.3). As described in Chapter 8, if the separation and replacement with the dam is repeated, pups emit more

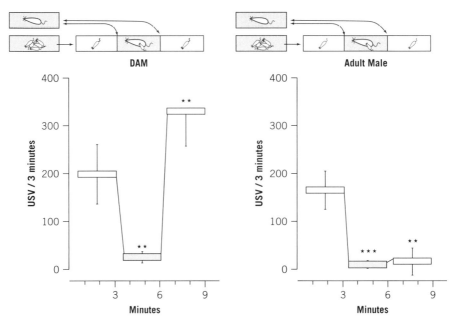

Figure 12.3 *Separation of a rat pup from its parents, siblings, and nest elicits vocalizations at ultrasonic frequencies. Contact quieting occurs when the anesthetized mother (dam) or an anesthetized adult male (sire) is placed with the pup in the isolation cage. If the mother is then removed for a second 3-minute isolation period, the number of vocalizations increases to a level significantly greater than during the first 3-minute isolation period. Since the physiological conditions are identical during the first and second isolation periods, and the potentiation effect occurs when the mother is replaced but not when the father is replaced, maternal potentiation indicates cognitive components of the pup's vocalization response to separation.* [From Hofer et al. (2001), *Current Protocols in Neuroscience*, Figure 8.14.2.]

vocalizations during the second separation than during the first separation (Figure 8.8*b*). This phenomenon is termed *material potentiation* of ultrasonic vocalizations (Hofer et al., 2001). Maternal potentiation may be a useful approach to modeling attachment behaviors in infants. Representative data are illustrated in Figure 12.3.

In addition to scoring pup behaviors, it may be useful to score parental behaviors directed toward the pups. As described in Chapter 8, parental care includes physical contact with pups in the nest, maternal licking, crouching, and nursing, and parental retrieval of pups to the nest. Maternal care influences development and affects subsequent adult behaviors in mice and rats (Carlier et al., 1983; Denenberg, 1999; Meaney, 2001). Michael Meaney and co-workers at McGill University in Montreal demonstrated the critical role of maternal licking and grooming on the development of endocrine and behavioral responses to stress in rats (Champagne et al., 2003). Higher maternal levels of licking and grooming of rat pups during their first postnatal week increased the survival of hippocampal neurons and hippocampal synaptic density, resulting in better spatial learning and memory (Liu et al., 2000; Bredy et al., 2003). To fully analyze stress responsivity, developmental milestones, and adult behaviors in a mutant line of mice, it is important to identify any unusual parental care caused by the mutation that could indirectly influence the development of the offspring, as opposed to direct effects of the mutation on the phenotype of the offspring.

Mental retardation has a genetic basis in neurodevelopmental diseases, including Down's, Fragile X, Smith-Lemli-Opitz, Tay-Sachs, Sandhoff, Prader-Willi, Angelman, Williams, and Rett syndromes. Behavioral phenotypes of mice with mutations in genes linked to these diseases are described in the last section of this chapter. In many cases standard learning and memory tasks described in Chapter 6 are appropriate for testing cognitive abilities in mouse models of these diseases. However, adaptations of learning and memory tasks that can be applied to infant and juvenile mice would provide measures for evaluating therapeutics, potentially including gene therapies, for reversing symptoms in young children. Most of the standard cognitive tasks are beyond the physical abilities of mouse pups. However, modifications and new tasks show promise. Mouse pups learn the spatial pattern of odor cues to find the nest at 12 days of age (Wiedenmayer et al., 2000). Odor aversion learning is detectable in CD-1 Swiss mice at postnatal days 7–10, from pairings of lemon or mint scented wood shavings with a treatment of lithium chloride (Alleva and Calamendrei, 1986), as previously demonstrated for neonatal rats (Rudy and Cheatle, 1977). Taste aversion learning has been demonstrated in neonatal rats, by pairing lithium chloride with sweet or salty substances during suckling in five day old pups (Kehoe and Blass, 1986), or pairing sucrose with ethanol in 10 to 20-day-old pups (Hunt et al., 1991), although no publications have appeared on taste aversions in infant mice. Passive avoidance learning has been demonstrated in 17 to 18-day-old mice, including the use of a specialized step-down passive avoidance apparatus (Burkhalter and Balster, 1979; Ricceri et al., 1996; Ricceri and Berger-Sweeney, 1998).

Autism is a neurodevelopmental disorder that is diagnosed by three core symptoms: aberrant reciprocal social interactions, communication deficits, and stereotyped, repetitive, ritualistic behaviors with narrow restricted interests (Lord et al., 2000). Associated symptoms may include mental retardation, delayed or absent language, minimal eye contact and directed gaze, poor interpretation of facial expressions and of language nuances, lack of empathy and intuition (Theory of Mind), impaired attentional disengagement, anxiety, seizures, sleep disturbances, and motor clumsiness (Piven 2001; Volkmar and Pauls, 2003). Monozygotic twins show 60–90% concordance for autism spectrum disorders. The male to female ratio is as high as 4 to 1. Multiple chromosomal linkage sites and candidate genes have been detected through linkage analyses and association studies (Muhle et al., 2004; Polleux and Lauder, 2004; Wassink et al., 2004). As mouse models with mutations in these candidate genes are generated, mouse behavioral tasks relevant to the symptoms of autism will be increasingly in demand (Crawley, 2004). Standard rat and mouse social interaction tasks (File, 1980, 1997a; Lijam et al., 1997; Bolivar and Flaherty, 2003; Brodkin et al., 2004), social affiliation (Carter et al., 1997), and juvenile play behaviors (Panksepp, 1981; Terranova et al., 1998), described in Chapter 9, are being applied to model the deficits in reciprocal social interaction that define the first diagnostic criterion of autism. Tom Insel, Jim Winslow, and Larry Young at Emory University developed an elegant method to measure social recognition and social memory in mice (Insel et al., 1995; Ferguson et al., 2001; Winslow, 2003), using interactions between the subject and a tethered stranger mouse (Figure 9.3). To more specifically model the deficit in social approach behaviors seen in autism, our laboratory designed an automated three-chambered apparatus to quantitate sociability in mice (Moy et al., 2004a, b; Nadler et al., 2004; Crawley 2004; Figure 12.4). Similar equipment for measuring social approach has been described by Ted Brodkin at the University of Pennsylvania (Brodkin et al., 2004, 2005). Conceptual analogues of the deficits in social

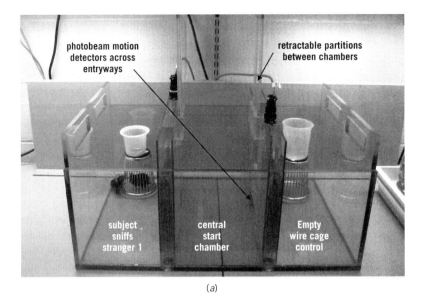

(a)

(b)

Figure 12.4 *Sociability apparatus. (a) The subject mouse is given a choice between exploring a habituated central start chamber, a side chamber containing an enclosed unfamiliar stranger mouse, or a side chamber containing a novel object with no social valence, an empty wire cup. Photocells across the openings between the chambers are connected to timers and software that automatically records time spent in each chamber and entries into each chamber.* [Equipment described in Moy et al. (2004) and Nadler et al. (2004), built by George Dold and co-workers, National Institute of Mental Health Research Services Branch. Photograph by Selen Tolu, contributed by the author.] *(b) Social sniffing initiated by the subject toward the stranger is quantitated by an observer with a stopwatch.* (Photograph by Selen Tolu, contributed by the author.) *(c) Representative data from 3 standard inbred strains, illustrating significant sociability.* [Data adapted from Moy et al. (2004).] *(d) Representative data from 9 strains of mice, illustrating sociability in six inbred strains of mice that spent more time with the stranger mouse than with the novel object or in the habituated center, and three inbred strains of mice that did not spend significantly more time in the chamber with the stranger mouse.* [Data adapted from Moy et al. (in press).]

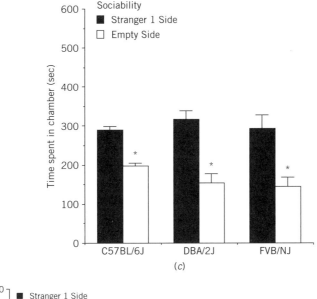

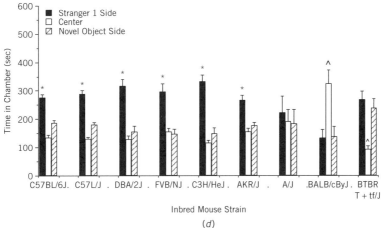

Figure 12.4 (Continued)

communication in mouse models of autism could include reduced vocalizations, lack of response to vocalizations by others, impairments in social transmission of food preference, and inappropriate or absent emission and exploration of social odors (Winslow et al., 2000; Hofer et al., 2001; Nevison et al., 2003; Bolivar and Flaherty, 2003; Moles et al., 2004; Crawley, 2004; Petrulis et al., 2005). Conceptual analogues of stereotypies, repetitive behaviors, narrow interests, and resistance to change in routine in a mouse model of autism may include spontaneous overgrooming, jumping, repeated sniffing of a single location, restricted exploration of only one of a variety of novel objects, reduced extinction of fear conditioning, and poor reversal learning on spatial tasks (Bolivar and Flaherty, 2003; Lewis, 2004; Crawley, 2004). Associated symptoms of mental retardation, anxiety, and seizures, for example, can be modeled with standard tasks available in the literature and described in earlier chapters. Adapting many of these behavioral tests for use in young mice will be an interesting challenge.

Twelve Months and Older

Designing mouse behavioral tasks to investigate neurodegenerative diseases is complicated by the need to dissociate the phenotype caused by the mutation from the phenotype caused by general aging. Some neurodegeneration occurs during the normal aging process. Some inbred strains display age-related pathologies such as deafness and blindness. As described in Chapter 5 and in The Jackson Laboratory strain database (*http://jaxmice.jax.org/jaxmicedb/html/inbred.shtml*), age-related hearing loss occurs in C57BL/6J, DBA/2J, and BALB/c, strains commonly used in behavioral genetics. As described in Chapter 2 and the JAX strain database, progressive visual impairments occur in FVB/NJ, C3H/HeJ, and SJL/J, strains commonly used in breeding knockouts. Motor abilities, such as swim speed in the Morris water maze, running speed in the radial arm maze, and coordination and balance on the rotarod, appear to decline as mice age (Lamberty and Gower, 1991; Ammassari-Teule et al., 1994; King and Arendash, 2002; Figure 12.5). Background genes that are increasingly expressed during the aging process, such as those inducing retinal degeneration, deafness, motor dysfunction, obesity, decline in olfactory sensitivity, cardiovascular disease, and tumors, can strongly influence the phenotype obtained from a targeted gene mutations. In some cases this concordance is advantageous. For example, the combination of targeted amyloid overexpression with normal aging processes in mouse models is consistent with the situation in most Alzheimer's patients (Chen et al., 2000; King and Arendash, 2002; Higgins and Jacobsen, 2003). Martin Sarter at Ohio State University emphasizes that motivational status and emotional reactivity may contribute to impaired cognition in senescent rodents, as is often observed in human geriatric studies (Sarter, 1987).

Problems arise when background genes are variably expressed during aging. By chance, poor vision may occur in more mice in the transgenic group than in the wild-type group. For example, a Morris water maze deficit in a mouse model of Alzheimer's disease may be an artifact of some mice not seeing the visual cues. Three solutions can avoid artifactual overinterpretations: (1) Within each day's experiments, include all genotypes, bred as littermates. Sufficient total Ns of each genotype should allow detection of the phenotype due to the mutation, over and above any phenotypes due to general sensory or motor loss during aging in all genotypes. (2) Conduct the control measurements for general health, neurological reflexes, sensory abilities, and motor functions at approximately the same age as the behavioral tests addressing the specific hypothesis are conducted. This often requires repeating the control tests at each age of interest. Repeating a sensory or motor task at time points that are several months apart usually does not produce a carryover effect. (3) Choose behavioral tasks that circumvent any sensory or motor dysfunction that is detected. For example, if motor abilities are declining in a colony of 20-month-old mice with targeted mutations relevant to learning and memory, test the aged mice on cued and contextual fear conditioning, taste aversion, and/or operant tasks. These learning and memory tasks take place within a small arena or operant chamber, thus requiring minimal motor activity, as compared to the swimming demands of the Morris water maze or the locomotor demands of a radial arm maze. If age-related deafness is an issue, use a bright light cue, or an olfactory scent, instead of the auditory cue in the contextual and cued fear conditioning task. Choose a strain that does not develop age-related obesity when breeding a targeted gene mutation relevant to appetite. Use C57BL/6J, which has a longer life span than most strains (Silver, 1995), to investigate the beneficial effects of caloric restriction on longevity. Measure fat pads in the feet of aged mice to control for insulation against

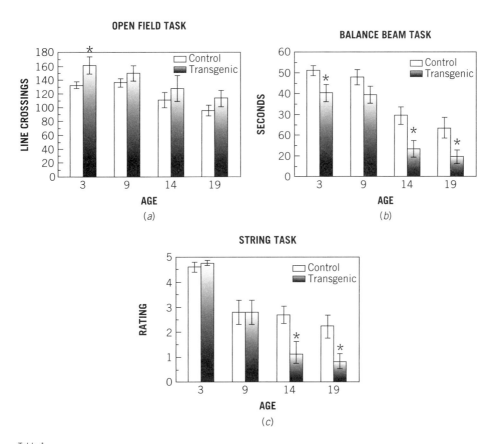

OPEN FIELD TASK

BALANCE BEAM TASK

STRING TASK

Table 1
Summary of significant differences by age and genotype

| Task | Age | | Genotype | | | | |
	Tg⁻	Tg⁺	All	3 Months	9 Months	14 Months	19 Months
Open field	3, 9 > 14, 19	3, 9 > 19	T > N	T > N			
Balance beam	3, 9 > 14, 19	3, 9 > 14, 19	N > T	N > T		N > T	N > T
String agility	3 > 9,14, 19	3 > 9, 14, 19, 9 > 14,19	N > T			N > T	N > T
Y-maze arm entries	3, 9,14 > 19	3, 9, 14 > 19	T > N				
Y-maze alternation			N > T	N > T			N > T
Water maze acquisition							
Water maze retention							
Circular platform errors	14 > 3, 19						
Water maze visible platform	3, 9, 14 < 19	3,9 < 19	N > T		N > T	N > T	N > T
Passive avoidance	3, 9, 19 > 14		T > N		T > N	T > N	
Active avoidance	3 > 9, 19						

(d)

Figure 12.5 *Motor functions declined with age in Tg2576 transgenic mice with the Swedish double mutation K670N/M671L in the amyloid precursor protein gene (T), and also in wildtype littermates (N). (a) Locomotion in an open field, (b) motor equilibrium on a balance beam task, and (c) grip strength on a string suspension test, were progressively impaired between ages 9 and 19 months. (d) Performance on the Morris water maze, Y-maze, and passive avoidance was progressively impaired by age 19 months. However, both genotypes displayed similar rates and magnitudes of decline on both motor and cognitive tasks.* [From King and Arendash (2002).]

the footshock, in evaluating nociceptive thresholds or analgesic drug responses in mice with mutations relevant to pain transmission. Finally, alternate approaches are available to test the role of a gene in a neurodegenerative disorder only in aged mice. Techniques include adeno-associated viral vector delivery of genes into the brains of aged mice, and inducible promoters to allow activation of the mutation only at specific time points, such as 20 months of age.

TRANSGENICS AND KNOCKOUTS

Neurodevelopment

Kei Watase and Huda Zoghbi at Baylor College of Medicine in Houston, Texas, review the large number of mutant mouse models of neurodevelopmental and neurodegenerative diseases (Watase and Zoghbi, 2003). Syndromes caused by mutations in genes mediating brain development often involve more than one gene, complex etiology, and variations in symptoms across patients. Mouse models are available for mental retardation syndromes including Fragile X, phenylketonuria, Rett, Down, and Angelman syndromes (Dierssen et al., 2001; Branchi et al., 2003; Cabib et al., 2003; Watase and Zoghbi, 2003; Moretti et al., 2005).

Down syndrome in humans is the result of an extra copy of chromosome 21. Complete aneuploidy in human trisomy 21 appears as three instead of two of each gene on human chromosome 21 (Reeve et al., 2001). Human trisomy 21 is a major cause of mental retardation, although not all Down syndrome individuals are mentally retarded (Reeves et al., 1995; Holtzman et al., 1996; Demas et al., 1996; Coussons-Read and Crnic, 1996; Escorihuela et al., 1998 Sago et al., 1998). Correlating the functions of the 225 genes on human chromosome 21 to the causes of the symptoms of Down's syndrome has been aided by mutant mouse models (Altafaj et al., 2001; Branchi et al., 2003; Berger-Sweeney, 2003). Chromosome 21 in humans is syntactic to chromosome 16 in mice. Trisomy 16 is an embryonic lethal mutation in mice. Partial trisomy 16 mice have triplicates of some, but not all, of the genes on mouse chromosome 16. Extensive behavioral phenotyping has been conducted on the Ts65Dn mouse with a partial trisomy of the distal portion of chromosome 16 (Gearhart et al., 1986; Davisson et al., 1993). Neurodevelopmental milestones were delayed in Ts65Dn mice, including cliff aversion, swift righting, screen climbing, vibrissa placing, ear twitch response, and ultrasonic vocalization (Holtzman et al., 1996). Ts65Dn mice display spontaneous stereotypies in the home cage, including repetitive jumping and cage top twirling (Turner et al., 2001), and were more active in the open field and elevated plus-maze (Coussons-Read and Crnic, 1996). On learning and memory tasks, Ts65Dn were found to be normal on passive avoidance (Holtzman et al., 1996). Performance on the Morris water maze showed longer latencies to reach both the visible and the hidden platforms, as compared to wildtype controls (Reeves et al., 1995; Holtzman et al., 1996; Escorihuela et al., 1998), indicating procedural deficits beyond the memory components of the task. Radial maze performance was similarly impaired in Ts65Dn mice, in both procedural and memory components of the task (Demas et al., 1996). Incremental repeated acquisition of behavioral chains in an operant learning task was normal in Ts65Dn conducting the simpler components of the task, but impaired in the more difficult sequences (Wenger et al., 2004). Deficits in sustained attention (Driscoll et al., 2004) and faster walking speeds (Hampton et al., 2004) may underlie some of the procedural deficits in Ts65Dn mice. The Ts65Dn model has been applied to evaluate potential treatments for cognitive dysfunctions in Down syndrome. Cholinergic enhancement with the putative nootropic drug piracetam improved Morris water maze performance in littermate controls, but did not improve performance in Ts65Dn mice (Moran et al., 2002). Estrogen treatment for 60 days, by a subcutaneously implanted pellet, improved learning and reversal learning of a T-maze task (Granholm et al., 2002). Trisomy in other segments of mouse chromosome 16 include Dyrk1A, Ts1Cje, and Ms1Ts65 models (Dierssen et al., 2001; Branchi et al., 2003). Dyrk1A mice display higher open field locomotion at both young and adult ages, more entries and time in the open arms in the elevated plus-maze, shorter latencies to fall in the wire hang, and slower learning of the hidden platform and reversal tasks in the Morris water maze (Altafai et al., 2001). Smaller deficits in cognitive performance were detected in the Ts1Cje, and Ms1Ts65 models of Down syndrome (Sago et al., 1998, 2000).

Fragile X syndrome is the most common inherited form of mental retardation, caused by a mutation in the *FMR1* gene that results in CGG nucleotide repeats (Verkerk et al., 1991). FMR1 is an RNA binding protein that influences many downstream genes. Mouse models with *Fmr1* mutations provide a tool to investigate the outcome of mutations in the FMR protein, and of mutations in some of the downstream genes (Kooy, 2003). An early knockout mouse model of fragile X syndrome was developed by D'Hooge and co-workers (1997). Mice with mutations in *Fmr1* display larger numbers of immature dendritic spines, indicating pruning deficits (Comery et al., 1997; Irwin et al., 2002), along with increased cerebral glucose metabolism (Qin et al., 2002). Mice mutant for the *Fmr1* gene showed normal acquisition of the Morris water task but impaired reversal learning. Swim speed was significantly lower for the knockouts in the visible platform task. The mild impairment in water maze performance and possible motor deficits may be relevant to the impaired spatial abilities and visual-motor coordination impairments seen in some fragile X patients. Behavioral phenotypes differed when the *Fmr1* mutation was bred onto B6, 129, or FVB backgrounds (Paradee et al., 1999; Lauder et al., 2004). As compared to wildtype littermates, *Frm1* knockout mice on a B6 background displayed small deficits in acquisition and reversal on a plus-shaped water maze, along with normal fear conditioned freezing and normal conditioned emotional response suppression (Van Dam et al., 2000). Morris water maze and fear conditioning were impaired in an *Frm1* knockout bred onto FVB, and on an F_1 hybrid of B6 and 129 lines, but not when the mutation was bred into a pure B6 background (Paradee et al., 1999; Dobkin et al., 2000; Peier et al., 2000). Radial arm maze learning was impaired in *Fmr1* knockouts (Mineur et al., 2002). Elevated plus-maze behaviors did not differ between *Fmr1* and wildtype controls, on either the C57BL/6J or FVB/NJ backgrounds, along with elevated acoustic startle at some decibel levels (Nielson et al., 2002; Mineur et al., 2002; Frankland et al., 2004b). Prepulse inhibition of acoustic startle was found to be better in *Fmr1* knockout mice, although impaired prepulse inhibition was reported in Fragile X patients (Chen and Toth, 2001; Frankland et al., 2004). Rescue of behavioral deficits in *Fmr1* mice is being attempted, with evidence that insertion of an intact copy of the *Fmr1* gene into the DNA construct reverses some aspects of the phenotype (Gantois et al., 2001).

Alcino Silva and colleagues at the University of California Los Angeles and Cold Spring Harbor Laboratory conducted elegant investigations into learning and memory deficits in a mouse model of neurofibromatosis (Silva et al., 1997; Costa et al., 2002;

Costa and Silva, 2003). Neurofibromatosis type 1 is a single gene, autosomal dominant developmental disease of abnormal cell growth, with symptoms including severe learning deficits, particularly in visual-spatial tasks, as well as neurofibromas and malignant tumors (Costa and Silva, 2003). The *NF1* gene activates Ras GTPase (Costa and Silva, 2003). *Nf1* knockouts are deficient on the Morris water maze, particularly in the probe trial, and on the development of long-term potentiation (Silva et al., 1997; Costa et al., 2002). Learning and memory deficits were rescued by crossing *Nf1* heterozygotes with heterozygote mice mutant for the gene *Ras* (Costa et al., 2002), suggesting a mechanism for the development of treatments of neurofibromatosis type 1.

Rett syndrome, another form of mental retardation, is caused by a mutation in the gene *MECP2*, an X-linked gene that regulates DNA methylation (Amir et al., 1999). *Mecp2* deficient mice, like Rett patients, appear normal until postweaning ages, when reductions in brain volume begin (Guy et al., 2001; Chen et al., 2001; Berger-Sweeney 2003b). Ataxia, myoclonic seizures, tremor, and deficits on wire hang, pole climb, open field activity, tube test social dominance, and resident–intruder social dominance were detected in *Mecp2* knockouts, while performance on the Morris water maze was not affected by the mutation (Shahbazian et al., 2002). Nest building and nest use were lower in *Mecp2* mice than wildtype littermate controls, indicating a social deficit (Moretti et al., 2005).

Autism is a unique neurodevelopmental disorder that is diagnosed by three core symptoms: impaired reciprocal social interaction, deficient social communication, and components of stereotypies, repetitive behaviors, rituals, resistance to change in routine, and narrow restricted interests (Lord et al., 2000). The 4 to 1 male to female ratio, and the 60% or greater concordance between monozygotic twins, indicate a strong genetic component that has been investigated in several multicenter linkage analyses and whole genome scans (Muhle et al., 2004; Polleux and Lauder, 2004). As candidate genes for autism emerge, hypotheses about the causes of autism are being explored in mouse models (Crawley 2004; Moy et al., 2006). Mice deficient in oxytocin, a hypothalamic neuropeptide mediating social attachments behaviors, are impaired in social recognition and social memory, while normal on other forms of memory and on olfactory abilities (Ferguson et al., 2001; Winslow and Insel, 2002). Similar impairments in social recognition are reported in mice deficient in the V1a receptor for another hypothalamic neuropeptide, vasopressin (Lim et al., 2005). Mice lacking the μ-opioid receptor do not show the normal maternal potentiation of ultrasonic vocalizations when separated from their mothers after a brief maternal exposure, indicating less responsiveness to maternal social cues (Moles et al., 2004). Male PACAP receptor 1 knockout mice, deficient in the neuropeptide pituitary adenylate cyclase-activating polypeptide type 1 receptor, exhibited less investigation of social olfactory cues over time, equal mounting of males and females, reduced aggression, and increased grooming toward intruder males (Nicot et al., 2004). Since autism is co-morbid in a large percentage of patients with other neurodevelopmental disorders, including Fragile X, Rett, Angelman's, and Smith-Lemli-Opitz syndromes, mice with mutations relevant to these diseases are being tested on behavioral tasks relevant to the symptoms of autism (Crawley, 2004; Moy et al., 2006). Fragile X knockouts on an FVB/129 background, but not on a C57BL/6 background, fail to show the normal preference for a socially novel stranger mouse in the three chambered sociability test (Lauder et al., 2004). Several inbred strains of mice display reduced sociability for a novel stranger on this task (Figure 12.4*d*) and similar social tasks (Bolivar and Flaherty, 2003; Moy et al., 2004, in press; Brodkin et al.,

2004). *Mecp2* mutant mouse models of Rett syndrome displayed fewer approaches to unfamiliar mice, spent less time in proximity and engaged in less social interaction than wildtype controls (Moretti et al., 2005). Increasing interest in the genetics of autism is likely to produce many new lines of mice with mutations in genes relevant to autism, and many new behavioral tasks designed to model the symptoms of autism (Insel, 2001; Crawley, 2004).

Heyser and co-workers at Scripps Research Institute in La Jolla, California, conducted an elegant evaluation of the Coloboma mutant line, deficient in the gene for the synaptic vesicle docking fusion protein SNAP-25 (Heyser et al., 1995), as described in Chapters 3 and 4. The expression pattern of two SNAP-25 isoforms switches during the first week of postnatal development. Coloboma pups were delayed by 5 days in developing the righting reflex, and by 7 days in the development of bar holding. These developmental delays were independent of lifetime effects of the mutation, such as decreased body weights that began at 6 days old and continued through adulthood.

Growth-associated protein-43 (GAP-43) is expressed in presynaptic terminals of neurons and appears to regulate axonal growth during brain development (Donovan et al., 2002). GAP-43 null mutant mice were thoroughly characterized on a comprehensive set of motor and sensory tasks (Metz and Schwab, 2004). Deficits in wire hang, grip strength, footprint stride length, visual contact placing, and hyperactivity demonstrated the importance of the GAP-43 axonal growth protein in early development. Although the GAP-43 homozygous mutants are small and usually die at a young age, heterozygotes survive and show a milder motor phenotype, as well as deficits on contextual fear conditioning (Metz and Schwab, 2004; Rekart et al., 2005), indicating that half the gene dose is insufficient for normal brain development.

Many other neurotrophic factors appear to be essential for brain development. Brain-derived neurotrophic factor (BDNF) and serotonin (5-HT) have been implicated in the development of the cerebral cortex, hippocampus, hypothalamus, and additional brain regions in mice and humans (Ren-Patterson et al., 2005). Multiple behavioral abnormalities in BDNF heterozygote mutant mice and in mice with mutations in the serotonin transporter and receptor subtypes (Yamada et al., 2002; Einet et al., 2003; Murphy et al., 2003) suggest that deficiencies in critical developmental genes may underlie some human neurodevelopmental disorders (Lam et al., 2006.

Neurodegenerative

Alzheimer's disease is characterized by loss of neurons and loss of memory, usually at late stages of life. While the true cause of Alzheimer's disease remains mysterious, neuropathological amyloid plaques and tau filament tangles in the brain are consistent diagnostic symptoms. The great majority of cases are late onset, beginning after 60 years of age. An early onset form of Alzheimer's, beginning at less than 60 years of age, occurs in less than 5% of cases, and shows strong genetic linkages to mutations in the amyloid precursor protein gene in some familie pedigrees (Tanzi and Bertram, 2005). Mouse models of components of Alzheimer's disease are mostly conventional transgenics, present at all ages (Buttini et al., 2002; Higgins and Jacobsen, 2003; Dodart et al., 2003; Schenk et al., 2004; Kobayashi and Chen, 2005; Bloom et al., 2005; van Dooren et al., 2005; Lee et al., 2005). Many of these mouse models focus on the incorrect processing of amyloid precursor protein. Instead of cleavage to the usual Aβ

1–40 peptide sequence, amyloid protein is cleaved into the Aβ 1–42 peptide sequence, which has neurotoxic effects. Amyloid plaque load was found in some, but not all, of the transgenic mice overexpressing the amyloid precursor protein gene APP, often developing at later ages (Buttini et al., 2002).

Mutant mouse models of Alzheimer's memory loss test the functional role of specific genes that have been implicated in the etiology or symptoms of the disease (Price et al., 1998; Seabrook and Rosahl, 1999). Genes linked to Alzheimer's disease include amyloid protein processing genes, the ApoE4 polymorphism susceptibility factor, and presenilins (Higgins et al., 1997; Selkoe, 1998; Kovacs and Tanzi, 1998; Price and Sisoda, 1998; Seabrook and Rosahl, 1999; St George-Hyslop and Petit, 2005). Mutations in some of these genes result in incorrect processing of β-amyloid precursor protein (APP) in mouse brain, leading to the production of β-amyloid peptides that aggregate to form amyloid-containing neuritic plaques, similar to human Alzheimer's disease. Alternative splicing of the *APP* gene, and alternate cleavages of the β-amyloid precursor protein, produces varying isoforms of the secreted Aβ amyloid peptides in mouse brain. Direct neurotoxic and memory-impairing actions of some Aβ peptides have been reported, while lower concentrations of amyloid peptides appear to enhance performance on memory tasks (Sweeney et al., 1997; Meziane et al., 1998). In addition, genetic models of aging have been identified in mice. For example, senile amyloidosis is seen in a spontaneous genetic model, the senescence-accelerated strain of mice, which are prone to early onset aging and deterioration of performance on memory tasks (Kawamata et al., 1997; Markowska et al., 1998).

Transgenic mice overexpressing amyloid protein have been generated in several laboratories. Mutations at different sites in the amyloid precursor protein gene were bred onto different genetic backgrounds. Varying degrees of overexpression of Aβ deposits, amyloid plaques, neurodegeneration, and memory impairments have been reported. Karen Hsiao at the University of Minnesota and co-workers (Hsiao et al., 1995, 1996) inserted human *APP* containing a mutation from a Swedish pedigree with early-onset Alzheimer's disease into the genome of CH3, C57B6/SJL, or FVB/N mice. The transgene was lethal in a large proportion of FVB/N and C57BL/6J mice (Hsiao et al., 1995, 1996; Carlson et al., 1997). Survival was improved by crossing C57B6 with a C57/B6/SJL background. F_1 C57B6 × C57B6/SJL transgenics displayed extracellular amyloid deposits in multiple plaques in the cerebral cortex and subiculum. These transgenic mice showed slower acquisition of the Morris water task, longer latencies to find the hidden platform, and deficits on selective search on the probe trial. Performance deficits were more severe in older mice, over the range of 3, 6, and 9 months old, compared to age-matched littermate control mice. Number of total platform crossings was normal on the visible platform task, indicating that swimming was normal in the aged transgenics. Interestingly, the same transgene inserted into FVB/N mice did not result in plaque formation in the surviving mice (Hsiao et al., 1995), pointing to background genes as critical susceptibility and protective factors in the *APP* phenotype.

Moran and co-workers reported similar findings in another mutant mouse model of Alzheimer's disease (Moran et al., 1995). Female transgenic mice homozygous for the transgene of human β-APP_{751} cDNA, a splice variant of the β-*APP* gene, were generated. Twelve-month-old transgenic mice showed poor performance on the Morris water task, with increased latencies to find both the hidden and visible platform and lack of selective search in the probe trial, as compared to wildtype controls (Figure 6.7). Six-month-old transgenics showed comparatively minor deficits. Transgenics at both ages were normal on the rotarod, the elevated plus-maze, and a string test of muscle strength. However, the transgenics showed less home cage locomotor activity in the dark phase of the circadian cycle and lower spontaneous alternation in a Y-maze, suggesting reductions in spontaneous exploratory activity that may have affected water maze performance. Spatial learning deficits on the Morris water task were confirmed by another laboratory working with a 751 amino acid isoform of human *APP*, under the control of a neuron-specific enolase promoter (D'Hooge et al., 1996). This line of transgenics displayed longer path length in swimming trials and also reduced open field activity in an independent test of exploratory activity. Passive avoidance was normal in these amyloid overexpressing transgenics. Taken together, these data indicate a potential motor confound that prevents a clear interpretation of a memory deficit.

Transgenic mice expressing the human mutant *APP* with residue 717 substituted from valine to phenylalanine showed high levels of Aβ deposition in the brain, with plaque density increasing with age (Games et al., 1995). Hippocampal levels of Aβ at 18 months of age were over 500-fold higher than the levels of Aβ at 4 months of age (Johnson-Wood et al., 1997). Age-dependent deficits were detected on Morris water maze reversal tasks and object recognition memory, a phenotype that correlated with density of β-amyloid plaque burden (Chen et al., 2000; Morris 2001; Kelly et al., 2003b; Figure 12.6). Various types of cognitive tasks are impaired in transgenic mice that overexpress the amyloid precursor protein and display high levels of amyloid deposition, including radial arm water maze (Leighty et al., 2004), Y-maze (King and Arendesh, 2002), contextual fear conditioning (Corcoran et al., 2002), and eyeblink conditioning (Weiss et al., 2002). It is interesting to note that there are several reports of normal performance on the Morris water maze, active avoidance, and other memory tasks in some lines of amyloid overexpressing transgenics that display amyloid plaques (King and Arendesh, 2002; Savonenko et al., 2003; Leighty et al., 2004). Similarly intriguing are cognitive impairments found in amyloid precursor protein transgenics on a B6 background that do not display plaque deposition (Lee et al., 2004). Careful analysis of sensorimotor functions revealed no genotype differences, indicating true cognitive declines in these mouse models of amyloid overexpression (Savonenko et al., 2003; Leighty et al., 2004). Thus plaque burden does not always predict cognitive decline in *APP* transgenic mouse models of Alzheimer's disease.

Transgenic mice expressing the carboxy terminal 100 amino acid sequence of human Alzheimer's precursor protein have been generated (Nalbantoglu et al., 1997). These mice showed high levels of β-amyloid immunoreactivity, extensive gliosis at older ages, and reduced magnitude of long-term potentiation. Poor performance on the Morris water task was seen, on measures of escape latency, number of trials to locate the hidden platform, and selective quadrant search on the probe trial. Poor spatial learning in mice transgenic for the carboxyl terminus of the amyloid precursor protein was confirmed in a second, independent report (Berger-Sweeney et al., 1999).

Null mutation of the β-*APP* gene was employed to investigate the normal physiological function of amyloid precursor protein (Dawson et al., 1999). Amyloid precursor protein null mutants displayed significantly longer latencies to locate the hidden platform and reduced selective search on the probe trial in the Morris water task. Scores on neurological, swimming, and rotarod tests were normal in the *APP* knockout mice. These mice also showed impaired long-term potentiation and marked reactive gliosis in the cortex and hippocampus.

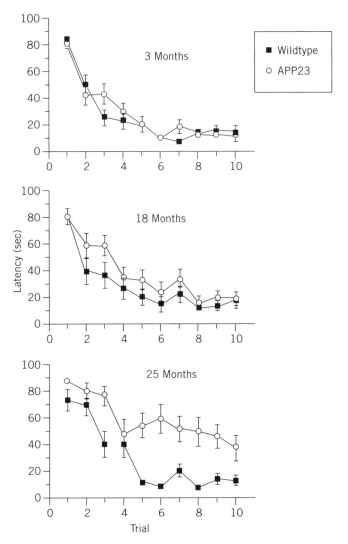

Figure 12.6 *Transgenic APP23 mice expressing human APP* $_{751}$ *with the K670N/M671L mutation displayed (a) an age-related deficit on latency to find the hidden platform in a small Morris water maze.* [From Kelly et al. (2003b), p. 369.]

Knockout mice deficient in amyloid precursor-like protein 2 (APLP2) showed normal body weight, forelimb grip strength, balance, postural reflexes, spinal reflexes, and nociceptive reflexes (Von Koch et al., 1997). *APLP2* knockouts were reported as normal on the Morris water task and conditioned avoidance tasks. Von Koch and co-workers then crossed the *APLP2* knockouts with *APP* knockouts, resulting in a double-knockout mouse, *APLO2* $^{-/-}$/*APP* $^{-/-}$ (Von Koch et al., 1997). Eighty percent of the double-knockouts died within the first week of birth. The surviving mice developed severe motor abnormalities, including ataxia, spinning, head tilt, and difficulty in righting, as well as reduced body weight. These findings raise the possibility that APP and APLP2, which are highly homologous proteins and perform similar functions in vivo, can compensate for each other. However, no compensatory upregulation of APP was found in *APLP2* knockouts. Similarly no compensatory upregulation of another related amyloid protein, APLP1, was seen in the *APLP2* knockouts. Basal expression levels of APP and APLP2 may be sufficient to substitute for each other in this case, and/or other mechanisms contribute to the severe behavioral phenotype observed in the double-knockout mice.

Presenilin 1 and presenilin 2 are components of the gamma secretase enzyme that cleaves amyloid precursor protein (Beglopoulos and Shen, 2006). Presenilin 1 gene mutations were lethal in mice, although heterozygotes survived (Davis et al., 1998). Presenilin 2 mutation N141I from a familial Alzheimer's disease in a German pedigree was inserted into mutant PS2 transgenic mice, resulting in increases in Aβ levels over 2 to 8 months of age (Oyama et al., 1998). Double mutants carrying both the transgene for mutant amyloid precursor protein and presenilin 1 display large numbers of fibrillar Aβ deposits in the cortex and hippocampus (Holcomb et al., 1998). Reduced spontaneous alternation behavior was impaired in the double mutants at earlier ages than the appearance of substantial Aβ deposition. Performance on the Morris water task indicated normal acquisition of the hidden platform task and selective quadrant search on the probe trial across genotypes at 6 to 9 months of age (Holcomb et al., 1999). More standard memory tests at older ages of this double transgenic will be needed to test for specific deficits in learning and memory to evaluate this interesting double transgenic model of Alzheimer's symptoms. Conditional presenilin 1 and 2 mutants with partial loss of function in the forebrain only were normal on memory tasks at 2 months of age, but displayed deficits on contextual fear conditioning and Morris water maze acquisition at 6 months of age (Saura et al., 2004).

Double mutants and downstream genes are proving useful to understand the mechanisms underlying the deleterious effects of amyloid overexpression. APP/PS1 mice, transgenic for amyloid precursor protein and mutant in presenilin-1, have severe deficits in amyloid processing that favor the neurotoxic peptide Aβ 1-42 over the normal Aβ 1–40 peptide (Janowsky et al., 2004). AAPP + PS1 transgenics show age-related plaque deposition correlated with gradually developing memory deficits and with age-related reductions in genes for synaptic proteins, including the metabotropic glutamate receptor GluR1, the neurotrophin BDNF, and alpha CaM Kinase II (Dickey et al., 2004; Saura et al., 2004). Deficits in short-term working memory appeared at age 3 months, along with reduced long-term potentiation and appearance of plaques, while long-term memory deficits began at 6 months, when amyloid plaque burden was considerably higher (Trinchese et al., 2004).

APP mice with the Swedish double mutation APP23, showing high amyloid plaque load, displayed elevated levels of galanin in the hippocampus and loss of cholinergic and catecholamine fibers (Diez et al., 2003). Galanin is a neuropeptide transmitter that is highly overexpressed in the basal forebrain in Alzheimer's disease, where galanin immunoreactive fibers and terminals hyperinnervate the remaining cholinergic nucleus basalis of Meynert neurons (Counts et al., 2003). Galanin overexpressing transgenic mice display similar elevations in galanin levels in the forebrain, hippocampus, and cortex, along with reductions in a cholinergic marker (Steiner et al., 2001; Crawley et al., 2002). Galanin transgenics displayed deficits on the more difficult components of several learning and memory tasks, including the Morris water maze probe trial, cued trace fear conditioning, and social transmission of food preference (Steiner et al., 2001; Kinney et al., 2002; Wrenn et al., 2003). In these studies, cognitive deficits were detected at ages from 4 months to 24 months, consistent with the conventional lifetime overexpression of galanin (Figure 12.7). Control measures of general health, body weight, neurological reflexes, open field, rotarod motor coordination, acoustic startle, hot-plate and tail-flick nociceptive responses, buried food and habituation/dishabituation olfactory abilities, attention on the five-choice serial reaction time task, and vision and swimming abilities on the Morris visible platform test were all normal, supporting the interpretation of a selective memory deficit in galanin transgenics (Holmes et al., 2002d; Kinney et al., 2002; Rustay et al., 2005; Wrenn et al., 2006).

Mice deficient in apolipoprotein E (*ApoE*) appeared normal on memory tasks in most reports (Anderson et al., 1998). Dramatic phenotypes were revealed after other types of *ApoE* mutations. The combination of APP overexpression and absence of ApoE produced more severe behavioral deficits than APP overexpression alone (Dodart et al., 2000). *ApoE* transgenic mice expressing the carboxyl-terminal-cleaved product, apoE4(Delta272-299), displayed neurodegeneration in the cortex and hippocampus, showed abnormally phosphorylated tau protein resembling preneurofibrillary tangles, and impaired learning and memory in individuals that survived to age 6–7 months (Harris et al., 2003). Transgenic mice expressing the human *ApoE4* isoform displayed deficits on object recognition and Y-maze avoidance (Grootendorst et al., 2005).

Mice deficient in the neurotrophin receptor p75, localized on the cholinergic cell bodies that degenerate in Alzheimer's disease, showed significantly fewer cholinergic neurons in the basal forebrain and significant performance deficits on several behavioral tasks (Peterson et al., 1999). Acquisition and retention of inhibitory avoidance, and acquisition of the hidden platform spatial navigation component of the Morris water task, were significantly impaired in the *p*75 null mutants. However, latency to locate the visible platform was also impaired, and activity and habituation in an open field were reduced as well. The authors interpret the behavioral phenotype with appropriate care, suggesting either a cognitive dysfunction or a more global sensorimotor impairment.

Alzheimer's disease and other progressive neurodegenerative diseases are associated with inflammation and the expression of interleukin 6 (IL-6), a cytokine that regulates inflammatory responses (McGeer and McGeer, 2001). Heyser and co-workers developed a transgenic mouse model of interleukin 6 overexpression in the brain (Heyser et al., 1997). Interleukin 6 transgenic mice showed poor performance on shock avoidance in an aversively motivated Y-maze task. Performance declined over the test ages of 3, 6, and 12 months. The deficits did not appear to result from nonassociative or motivational factors since escape latencies were normal for the transgenics.

Neurofibrillary tangles seen in the brain in Alzheimer's and Parkinson's disease have been replicated in mutant mouse models. P301L mice overexpressing tau, the micro tubule protein, show neurofibrillary tangles in cortex, brainstem, and spinal cord (Gotz et al., 2001). R406W tau mutation mice showed tau inclusions in the hippocampus and striatum of aged mice (Tatebayashi et al., 2002). Behavioral analyses of tau mutant mice are in progress. Deficits in contextual fear conditioning were detected in 19-month-old R406W mice, while control measures of body weight, open field activity, rotarod, and acoustic startle were not significantly different between genotypes in these aged mice (Tatebayashi et al., 2002).

Mutant mouse models with high amyloid burden and/or neurofibrillary tangles are ideal research tools to evaluate treatments designed to reduce amyloid plaques in the brain. Promising findings emerged from immunotherapy treatment of APP

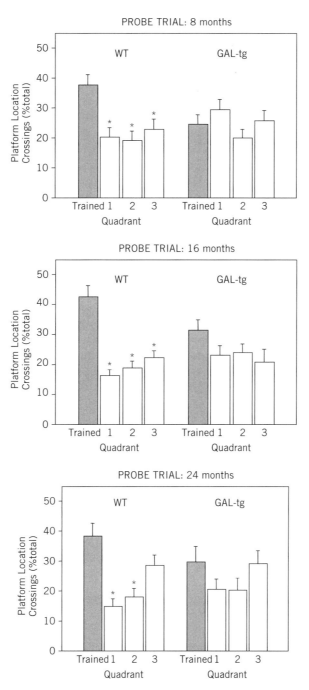

Figure 12.7 *Galanin overexpressing transgenic mice, modeling the galanin overexpression seen in the basal forebrain in Alzheimer's disease, failed to acquire the Morris water maze hidden platform task using distal spatial cues. Wildtype littermates showed selective quadrant search on the trained quadrant, while GAL-tg swam equally in the trained and untrained quadrants. The learning deficit was apparent at all ages tested, consistent with the lifetime overexpression of galanin in these conventional transgenics.* [From Steiner et al. (2001), p. 4187.]

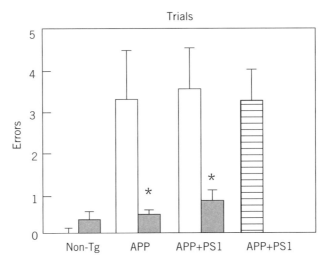

Figure 12.8 *Amyloid overexpressing transgenic mice (white bars) displayed more errors than littermate controls (Non-Tg) on a radial arm maze task. APP transgenics vaccinated with Aβ antigen displayed a significant reduction in errors (black bars). APP transgenics vaccinated with a control vehicle continued to display high numbers of errors.* [From Morgan et al. (2000).]

overexpressing transgenic mice. (Figure 12.8) Direct immunization with an Aβ antigen, or passive immunization with anti-Aβ antibodies, attenuated amyloid deposition and reduced the number of plaques in the brain in APP transgenic mice (Morgan et al., 2000; Janus et al., 2000; Higgins and Jacobsen, 2003; Schenk et al., 2004). Monthly injections of Aβ 1–42 in APP + PS1 mice, beginning at age 2 months, produced only modest reductions in amyloid plaques but significant protection from performance deficits on the Morris water maze and radial arm maze (Jensen et al., 2005). This study incorporated many relevant control measures, including open field, balance beam, and anxiety-related behaviors, at ages $4\frac{1}{2}$ to 6 months and 15 to $16\frac{1}{2}$ months of age (Jensen et al., 2005). The exciting results from mouse models encouraged clinical trials of immunotherapy for Alzheimer's disease. Unfortunately, lethal and sublethal meningoencephalitis and brain inflammation occurred in several patients, halting the clinical protocol and prompting a search for safer immunization approaches (Janus, 2003; Schenk et al., 2004).

Parkinson's and Huntington's disease both involve neurodegeneration of the basal ganglia. Neuropathology and behavioral concomitants are diagrammed in Figure 12.9. Parkinson's is characterized by progressive loss of substantia nigra dopamine neurons innervating the striatum, leading to symptoms of tremor and rigidity (Calne et al., 1992). Dopamine replacement therapy with the precursor L-DOPA and related drugs is the standard treatment, effective but not optimal (Kostrzewa et al., 2005). Rat models of Parkinson's disease focused on replicating the dopaminergic lesion with the neurotoxin 6-hydroxydopamine (6-OHDA). Unilateral nigrostriatal lesions with 6-OHDA produced agonist-induced contralateral circling (Ungerstedt, 1971), quantitated with the rotameter equipment illustrated in Figure 4.10. Mouse models of Parkinson's replicated the lesion with the neurotoxins 6-OHDA and 1-methyl-4-phenyl-1,2,3,6-tetrahydropyridine (MPTP; Schober, 2004). Reduced rotational behavior in unilateral 6-OHDA lesioned mice was improved by mouse embryonic stem cells

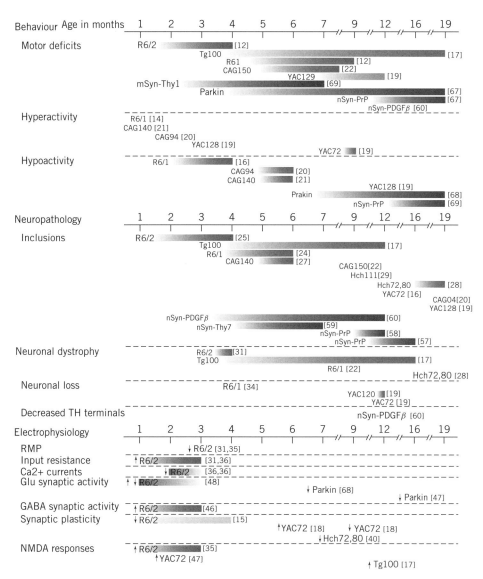

Figure 12.9 *Genetic mouse models of Huntington's disease (HD) and Parkinson's disease (PD). Shaded lines represent the ages at which symptoms appear.* [From Levine et al. (2004), p. 692.]

microinjected into the lesioned striatum (Nishimura et al., 2003), demonstrating the potential benefits of stem cell treatment of Parkinson's disease. Open field locomotion is reduced after MPTP lesions, and also in knockout mice deficient in *Fas*, an apoptosis gene elevated in the substantia nigra and striatum in Parkinson's disease (Hayley et al., 2004). While Parkinson's disease is generally not inherited, one form of familial Parkinson's disease is caused by accumulation of α-synuclein, a protein found in Lewy body inclusions in the cytoplasm of midbrain dopaminergic neurons (Trojanowski and Lee, 2003). Transgenic mice overexpressing α-synuclein on a tyrosine hydroxylase promoter display mild motor impairments (Masliah et al., 2000). Delivery of α-synuclein by adeno-associated viral vector into the rat substantia nigra produced a behavioral phenotype including rotation and reduced paw reaching (Kirik and Björklund, 2003). *Parkin* is a gene mutated in familial early-onset Parkinson's disease (Levine et al., 2004). Mice with mutations in *Parkin* show tremor and reduced open field activity; however, they do not show the predicted loss in dopaminergic neurons or reduced striatal dopamine (Levine et al., 2004).

Huntington's disease models include mice with mutations in the *huntingtin* gene. Expanded polyglutamine repeats in the *huntingtin* gene vary from 12 to over 150 CAG triplets (Kirik and Björklund, 2003; Levine et al., 2004; Morton et al., 2005). R6/2 Huntington's disease transgenic mice carry a 141–157 CAG repeat that results in reduced open field activity beginning at age 4 weeks, and less freezing on the fear conditioning memory task (Bolivar et al., 2003). Rebecca Carter, Steve Dunnett, and Jenny Morton at the University of Cambridge, contributed a full behavioral characterization of the motor deficits in R6/2 mice, using methods described in Chapter 4 and *Current Protocols in Neuroscience*, Unit 8.12. At 5 weeks old, Huntington's mice displayed impairments in swimming, rotarod at 33 to 44 rpm, and traversing the narrowest square (5 mm) balance beam (Carter et al., 1999). Deficits on fore- and hindpaw footprinting (Figure 4.5) became apparent at 8 weeks, and performance on all motor tasks declined between 8 and 15 weeks, progressively until death (Carter et al., 1999). The circadian cycle was disrupted, with daytime activity increasing and nighttime activity decreasing (Morton et al., 2005b). Cognitive deficits included less freezing on fear conditioning (Bolivar et al., 2003), and impaired performance on the Morris water maze (Murphy et al., 2000). Reduced spontaneous activity and reduced reactivity to handling accompanied lower levels of anxiety-like behaviors (File et al., 1998). Two tests for anxiety-related behaviors were conducted by Sandra File and co-workers. On the elevated plus-maze, the transgenic mice spent more time on the open arms and showed more open arm entries than wildtype controls, indicating lower levels of anxiety. However, total number of entries into the closed arms was reduced, indicating lower levels of general exploratory activity. On the holeboard test, total number of head dips was lower in the transgenic mice, which could indicate greater anxiety. However, beam breaks due to general motor activity was lower in the transgenic mice, representative of reduced activity. The apparent anxiolytic effect of the mutation and the reduction in general locomotion began at 8 weeks of age and exceeded in magnitude the differences between two inbred strains of mice, C57 and CBA, on these tasks.

A knock-in with chimeric mouse/human exon 1 containing 140 CAG repeats inserted into the mouse *huntingtin* gene, generated by Marie-Francoise Chesselet at the University of California Los Angeles, displayed higher open field horizontal and vertical activity, and smaller stride lengths in the footprint test (Hickey and Chesselet, 2003). The YAC129 mouse model of Huntington's disease displayed deficits on motor learning on the rotarod, swimming to a platform in a water T-maze, and habituation to acoustic startle (Van Raamsdonk et al., 2005). Drug therapies that boost brain levels of acetylcholine and monoamines have improved the cognitive deficits, but not the motor dysfunctions, in mouse models of Huntington's disease (Morton et al., 2005a). Paroxetine, an antidepressant, was reported to attenuate motor dysfunctions when administered either before or after the start of symptoms in *huntingtin* mice (Duan et al., 2004). An adenosine receptor A(2A) agonist was reported to slow the deterioration of motor performance in R6/2 mice (Chou et al., 2005). Along with neuroanatomical and neurochemical measures, the motor phenotype of R6/2 mice provides a robust parameter for evaluating potential treatments for Huntington's disease.

Bcl-2 transgenic mice contain a human transgene that may regulate apoptosis, naturally occurring cell death (Rondi-Reig et al., 1997). This gene is expressed at high levels in the amygdala. Null mutants spent more time on the open arms of the elevated plus-maze, entered a novel environment more frequently, and displayed slower acquisition of the hidden platform task than wildtype littermate controls (Rondi-Reig et al., 1997, 2001a). The authors suggest that reduced anxiety, neophobia and/or fearfulness, and cognition are characteristics of the *bcl-2* transgenic genotype.

Amyotrophic lateral sclerosis (ALS) is an adult-onset disease characterized by motor neuron degeneration and loss of voluntary movement (Rothstein, 2003). The majority of cases are sporadic, with unknown etiology. In the small number of ALS cases with apparent familial inheritance, about 25% appear to be caused by a mutation in the gene for superoxide dismutase 1 (SOD1), leading to glutamate excitotoxicity, oxidative stress, and cell death (Rowland and Shneider, 2001). Knockout mice with mutations in SOD1 show clear evidence of motor neuron degeneration and neuromuscular weakness (Lalonde and Strazielle, 2003). Significant delays in the righting reflex and in hindpaw grasping responses were seen in SOD1G85R mice during the first postnatal week (Amendola et al., 2004). Longitudinal analysis of footprint stride length, grip strength, and rotarod balance showed progressive declines from 3 to 8 months (Puttaparthi et al., 2003; Derave et al., 2003). SOD1 mutant mice have been used to test hypotheses about the cause of ALS, including neurotoxins in cycad seeds (Cox and Sachs, 2002; Shaw and Wilson, 2003). Efficacy in slowing the progression of the motor dysfunctions has been tested for potential treatments including *N*-acetyl-L-cysteine, creatine, minocycline, the glutamate antagonist riluzole, the calcium channel blocker nimodipine, viral vector delivery of insulin-like growth factor 1 and glial cell line derived neurotrophic factor (Gurney et al., 1998; Andreassen et al., 2000; Derave et al., 2003; Kriz et al., 2003; Zhang et al., 2003; Kaspar et al., 2003; Boillíe and Cleveland, 2004). Jeff Rothstein at Johns Hopkins University in Baltimore notes that some drugs that slowed the disease symptoms significantly in SOD1 mice did not have a significant effect in ALS patients (Rothstein, 2003). Dose and route of administration differentially affect steady state drug levels in mice versus humans, and drugs that can be administered before the start of symptoms in the mouse model can only be tested only after diagnosis in patients. Most important, the SOD1 mouse models a gene mutation that is present in only 1–2% of ALS cases.

The SOD1 mouse represents a research tool that addresses a genetic hypothesis about a human disease, which adds incrementally to our knowledge. As with so many neurodevelopmental and neurodegenerative diseases, the research process is iterative. Discovery of gene candidates leads to mutant mouse models. Phenotypes of the mutant mice serve to test hypotheses about the candidate genes. Results inform the further investigation of genes linked to the disease, and downstream mediators. Further hypotheses are then tested in mice. Potential treatments are evaluated for reversal of the phenotype in mice, including multidrug cocktails (Figure 12.10). Efficacious therapeutics may then move into human clinical trials. While results from mouse models predict therapeutic drugs only sometimes, the mouse assays offer relatively simple preclinical evaluations to rule out the least likely candidates among the large numbers of FDA-approved compounds under consideration for ALS (Rothstein, 2003).

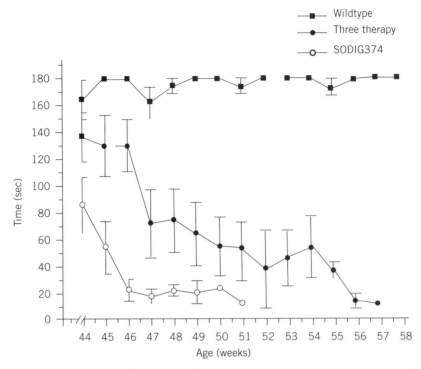

Figure 12.10   *A mouse model of amyotrophic lateral sclerosis (SOD1, open circles) displayed muscle weakness on a wire grip test, as compared to wildtype controls (black squares). Treatment with a three-drug cocktail (black circles) improved muscle strength in the SOD1 knockout mice.* [From Kriz et al. (2003), p. 432.]

RECOMMENDATIONS

Longitudinal studies are highly appropriate for mouse models of neurodevelopmental and neurodegenerative diseases. Repeated tests at time points relevant to the progression of the human disease may reveal the exact temporal correlation between symptoms and pathology, such as increasing amyloid load and memory loss in mouse models of Alzheimer's disease or lack of juvenile social play beginning at weaning in mouse models of autism. Some mouse behavioral tasks can be repeated in the same animal, while some cannot. The main issue is carryover effects; that is, experience in the first test session influences behavior in the second test session of a behavioral task. Measures of developmental milestones and neurological reflexes can usually be repeated at frequent intervals. A mouse can usually be tested more than once on most sensory and motor tasks, with intervals of several days between experiences without carryover effects. The Morris water task involves procedural learning that will influence subsequent training, but mice can be re-trained with the platform in a new location. Fear conditioning is a very long-lasting phenomenon, for which pairing of a shock with environmental cues will affect future associations, an advantage when studying extinction but a complication when initiating repeated fear conditioning at a later age. Anxiety-like behaviors on the elevated plus-maze are dramatically influenced by earlier experiences on the plus-maze, although the approach-avoidance conflict in the light↔dark test appears stable across two or three repeated tests (Blumstein and Crawley, 1983; Holmes and Rodgers, 1999). Social approach behaviors appear to be consistent for a mouse tested in two test sessions separated by days or weeks (Moy et al., 2004). Thus the repeatability of a given task needs to be evaluated before it is used for multiple evaluations of the same mice across the ages relevant to a progressive disease.

Controls for general health, sensory abilities, and motor functions are best run at ages near those relevant to the expected phenotype in a progressive disease. For example, mice need to be tested for vision, hearing, and open field activity at older ages to avoid artifacts and overinterpretations of a cognitive phenotype at advanced ages in mouse models of Alzheimer's disease. The principle of conducting appropriate control measures at the same ages as the hypothesis-driven behavioral tests is equally applicable to targeted gene mutations, lesions, drug treatments, and environmental manipulations.

BACKGROUND LITERATURE

Beglopoulos V, Shen J (2006). Regulation of CRE-dependent transcription by presenilins: Prospects for therapy of Alzheimer's disease. *Trends in Pharmacological Sciences* **27**: 33–40.

Bignami G (1996). Economical test methods for developmental neurobehavioral toxicity. *Environmental Health Perspectives* **104** (Suppl 2): 285–298.

Branchi I, Bichler Z, Berger-Sweeney J, Ricceri L (2003) Animal models of mental retardation: From gene to cognitive function. *Neuroscience and Biobehavioral Reviews* **27**: 141–153.

Branchi I, Ricceri L (2002) Transgenic and knock-out mouse pups: The growing need for behavioral analysis. *Genes, Brain and Behavior* **1**: 136–141.

Carter RJ, Lione LA, Humby T, Mangiarini L, Mahal A, Bates GP, Morton AJ, Dunnett SB (1999). Characterization of progressive motor deficits in mice transgenic for the human Huntington's disease mutation. *Journal of Neuroscience* **19**: 3248–3257.

Champagne FA, Francis DD, Mar A, Meaney MJ (2003) Variations in maternal care in the rat as a mediating influence for the effects of environment on development. *Physiology and Behavior* **79**: 359–371.

Crawley JN (2004). Designing mouse behavioral tasks relevant to autistic-like behaviors. *Mental Retardation and Developmental Disabilities Research Reviews* **10**: 248–258.

Dodart JC, Mathis C, Bales KR, Paul SM (2002) Does my mouse have Alzheimer's disease? *Genes, Brain and Behavior* **1**: 142–155.

Heyser CJ (2003). Assessment of developmental milestones in rodents. *Current Protocols in Neuroscience*, Unit 8.18.

Hickey MA, Chesselet MF (2003) The use of transgenic and knock-in mice to study Huntington's disease. *Cytogenet Genome Research* **100**: 276–286.

Higgins GA, Jabobsen H (2003) Transgenic mouse models of Alzheimer's disease: Phenotype and application. *Behavioural Pharmacology* **14**: 419–438.

King DL, Arendash GW (2002) Behavioral characterization of the Tg2576 transgenic model of Alzheimer's disease through 19 months. *Physiology & Behavior* **75**: 627–642.

Kirik D, Björklund A (2003) Modeling CNS neurodegeneration by overexpression of disease-causing proteins using viral vectors. *Trends in Neuroscience* **26**: 386–392.

Kooy RF (2003) Of mice and the fragile X syndrome. *Trends in Genetics* **19**: 148–154.

Le Roy I, Perez-Diaz F, Cherfouh A, Roubertoux PL (1999) Preweaning sensorial and motor development in laboratory mice: Quantitative trait loci mapping. *Developmental Psychobiology* **34**: 139–158.

Menalled LB, Sison JD, Dragatsis I, Zeitlin S, Chesselet MF (2003) Time course of early motor and neuropathological anomalies in a knock-in mouse model of Huntington's disease with 140 CAG repeats. *Journal of Comparative Neurology* **465**: 11–26.

Rothstein JD (2003) Of mice and men: Reconciling preclinical ALS mouse studies and human clinical trials. *Annals of Neurology* **53**: 423–426.

Sarter M (1987). Measurement of cognitive abilities in senescent animals. *International Journal of Neuroscience* **32**: 765–774.

Watase K, Zoghbi HY (2003) Modelling brain diseases in mice: The challenges of design and analysis. *Nature Reviews Genetics* **4**: 296–307.

Zoghbi HY (2003). Postnatal neurodevelopmental disorders: Meeting at the synapse? *Science* **302**: 826–830.

Behavioural Brain Research. Special Issue on Animal Models of Autism, in press.

Neuroscience and Biobehavioral Reviews (2003). Special Issue on Brain Development, Sex Differences and Stress: Implications for Psychopathology **27**: 1–188.

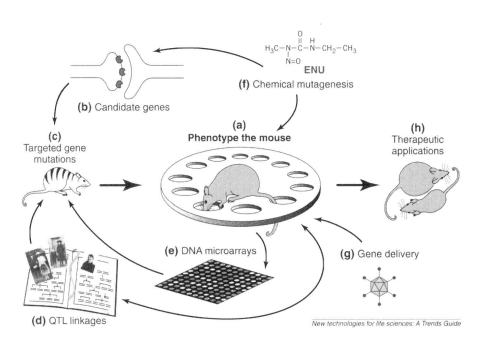

Robust, replicable, accurate phenotyping is key to the success of diverse genetic approaches. [*From Crawley (1996)*.]

13

Putting It All Together

Swedish smörgåsbords, Spanish tapas bars, and American buffets offer a huge selection of small dishes, from which you assemble your ideal meal. The preceding chapters offered a huge selection of mouse behavioral tasks. How do you choose the ideal number, assortment, and sequence of tests for analyzing your mouse? This chapter offers strategies and advice on putting together an optimal constellation of behavioral tests to address specific hypotheses. Numbers of mice, background strains for breeding, general health issues, preliminary observations, critical control measures, choice of relevant tests within each behavioral domain, multiple uses of each mouse, order of testing, and replication requirements are some of the important considerations for your experimental design. While the field of mouse behavioral phenotyping continues to evolve, the guiding principles of this chapter will help in selecting a reasonable set of behavioral tests to answer your specific questions about what's wrong with your mouse.

WHEN TO START

The first chimera with germ-line transmission prompts a big lab celebration. Traditionally, this success marks the completion of the major molecular genetics investment in the project, and the beginning of the major phenotyping investment. Recently, repositories of transgenic and knockout mice have been created to offer existing mutant lines for sale to investigators. The Jackson Laboratory JAX Mice Database (*http://jaxmice.jax.org/jaxmice-cgi/jaxmicedb.cgi*) includes a searchable database for purchasing hundreds of lines of mice that have been donated by academic investigators. JAX repositories also offer storage and purchase of frozen embryos or frozen sperm from mutant lines (Anagnostopoulos et al., 2001). Cryopreserved embryos and sperm can be purchased directly, or the investigator may pay for recovery of frozen embryos into breeding pairs of mutant mice. Bay Genomics (*http://baygenomics.ucsf.edu*), partially supported by the National Institutes of Health in the United States, offers hundreds of embryonic stem cell lines with single gene mutations, generated by gene trap vectors. NIH is funding a Knockout Mouse Project to produce a comprehensive set of embryonic stem cell lines with mutations for each gene in the mouse genome (Austin et al., 2004). Companies including Deltagen (*http://www.deltagen.com*) and Lexicon Genetics Inc. (*http://www.lexicon-genetics.com*) sell thousands of privately developed embryonic stem cell lines and mice with single gene mutations. The European Program on Mouse Phenotyping, EUMORPHIA (*http://www.eumorphia.org*), develops and validates methods for phenotyping mice, including standardized behavioral screens, and offers training courses for applying these methods to mutant lines of mice, including a growing consortium of mouse mutagenesis centers in Europe and Asia. Repositories of mutagenesis lines and targeted gene mutation mouse models, available to the international research community, are speeding the discovery of gene functions and the development of treatments for genetic diseases. Science policy makers such as the NIH Institute Directors are considering ways to make the available lines of mutant mice more affordable and more accessible to the scientific community. In the future, phenotyping experiments may begin immediately, bypassing the need to generate a new targeted gene mutation in-house. The Trans-NIH Mouse Initiative lists major resources for locating available mutant mice at repositories, universities, and research institutes around the world (*http://www.nih.gov/science/models/mouse/resources*).

Behavioral phenotyping begins when the mutation is successfully bred into large numbers of offspring. Basic information gained during breeding includes number of pregnancies, frequency of pregnancies, number of litters born, number of offspring per litter, survival of pups, survival of homozygous mutants, and growth rates. These simple biological observations are made while keeping the breeding records. As pups are genotyped and sexed, sufficient numbers of each genotype and sex are raised to reach the age at which the mice can conduct the procedures required for the behavioral tests that address the scientific hypothesis.

MINIMUM NUMBERS OF MICE (*N*s)

Numbers required for behavioral experiments may seem shockingly high to molecular geneticists. Some behavioral neuroscience laboratories begin with 10 littermates of each genotype for the first pilot study. Specific tasks may require an N of 15, 20, or more mice per treatment group. Other tasks may succeed with a N of 5. The precise number depends on a statistical power analysis (Figure 13.1; Wahlsten, 1999, 2000). Mixed genetic backgrounds often contribute to the variability of the data, as described below. Larger numbers of each genotype may be needed to detect a genotype difference above high "background noise."

Recommended Ns indicate the number of mice for each genotype and for each sex. For example, $N = 15$ for a conventional knockout of a recessive gene translates into $N = 15$ for each of the genotypes +/+, +/−, and −/−, meaning $3 \times 15 = 45$ mice. Comparison of data by sex is conducted to detect any significant differences between males and females. If a sex difference is detected, $N = 15$ males and $N = 15$ females of each genotype translates into 15 mice $\times$ 3 genotypes $\times$ 2 sexes $= 90$ mice for the first round of testing.

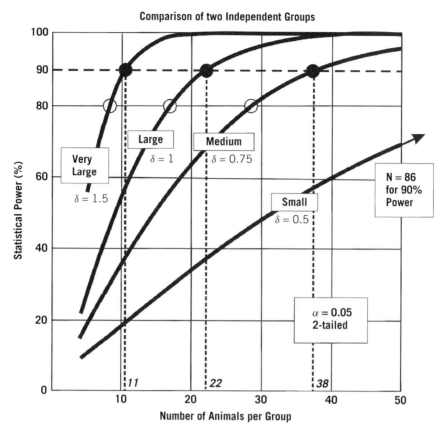

Figure 13.1 *Power versus sample size for a 2-tailed t test of the difference between two independent groups (e.g., wildtype and mutant) when the null hypothesis is that there is no true difference (∗ = 0). The sample size needed to yield a power of 90% is shown as an italicized numeral along the X axis.* (Kindly contributed by Professor Douglas Wahlsten, University of Windsor, Windsor, Ontario, Canada.)

As in every area of scientific research, a positive finding needs to be replicated. Replication experiments will require at least one more set of mice, usually of similar Ns. Independent cohorts of mice must be used to repeat the experiment. To confirm the robustness of the phenotype, the second and third rounds of mice are obtained from different parents, at different times of the year, and the tests are conducted by different investigators. Mice bred for behavioral phenotyping should not be used for other purposes before or during the behavioral testing. Therefore a commitment to breeding and housing these large Ns, dedicated for use in the behavioral experiments, is the first step in the phenotyping project.

Not everyone can afford to maintain enough breeding and housing cages to generate 90 mice simultaneously. Practical approaches to accumulating large Ns include pooling mice and pooling data. If the pilot experiments show no effect of sex, then Ns for subsequent experiments can be the pooled Ns of males and females. Mice reach sexual maturity between 6 and 8 weeks of age (Silver, 1995). Mice are usually considered aged at 20 months, with the aging process probably underway after 12 months. Therefore mice can be considered as normal adults when they are between ages 3 and 10 months old. For most behavioral tests, individuals within this age range can be combined within one treatment group to reach the required Ns. However, strains of mice demonstrate varying longevity, with AKR/J living to an average age of 10 months while C57BL/6J lives to an average age of 27 to 28 months (Silver, 1995). Definitions of normal adult ages may need to be adjusted for very short-lived and very long-lived strains.

Data from the wildtype controls of different cohorts are compared. If no statistically significant differences are found between sets of wildtypes, then pooling of data is justified. Pooling data from younger and older adults will require consideration of potential artifacts arising during the ontogenic progression, such as deafness, loss of vision, and motor deterioration, as described in Chapter 12. Choice of the background strain for breeding, and choice of breeding strategy, are extensively discussed in Chapter 2. In cases where the background strain expresses age-dependent deficits relevant to the behavioral tasks of interest, such as learning and memory, control measures are conducted to assess general health, sensory abilities, and motor functions, at the age close to when the hypotheses-driven behavioral tasks are conducted, as described in Chapter 12. This strategy avoids age-related artifacts and misinterpretations, such as deafness causing poor scores on auditory cued fear conditioned learning. The need to compare littermates rather than mice separately bred by genotype is discussed in Chapter 2. The most important consideration in pooling mice from many litters to achieve large Ns is that each genotype be approximately equally represented within each experiment and on each testing day.

HOUSING ENVIRONMENT

Because environmental conditions have direct effects on behavior, it is crucial to maintain consistency within the breeding colony throughout the course of the behavioral experiments. Variables such as temperature, noise, lighting, circadian cycle, humidity, cage cleaning, type of cage, number of mice per cage, and age at weaning have been reported to affect behaviors in mouse pups and adults (Crabbe et al., 1999b; Wahlsten et al., 2003a). Parental care, especially separation of pups from their mothers, can profoundly change adult behaviors (Meaney, 2001). Dominance hierarchies develop among cagemates if one individual is highly aggressive, resulting in wounding and stress-related syndromes in the subordinates. There is no foolproof way to avoid the occasional disaster, except to watch for malfunctions, keep records, and institute corrections as quickly as possible. For example, removing the aggressive dominant male from the cage usually allows the remaining males to recuperate sufficiently for use in behavioral tests after a week or two. The aggressive male may no longer be usable within the experimental design because of possible behavioral effects of isolation stress.

The safest way to control for the effects of environmental perturbations is to always compare mutant mice to their wildtype littermates. Even intrauterine environment during gestation can influence behavioral results (Francis et al., 2003). Littermates are the proper control group for behavioral phenotyping. For example, if the air conditioning breaks down on a hot summer day, all of the genotypes will suffer from the excessive heat. It is possible that an environmental factor has a selective interaction with only one genotype, For example, failure of the circadian timer would affect wildtype littermate controls more than homozygous mutants for the retinal degeneration gene *rd/rd*.

Even across different environments, or after a mild environmental perturbation, a very robust phenotype in null mutants usually remains significant in comparison to wildtype littermates. The issue of environmental standardization for mouse behavioral testing has been extensively discussed (Wurbel, 2002). Impressive standardization of environmental variables across laboratories was conducted by John Crabbe in Portland, Oregon, Doug Wahlsten in Edmonton, Alberta, Canada, and Bruce Dudek in Albany, New York, to evaluate the consistency of behavioral phenotypes of inbred strains and of a serotonin receptor subtype 5-HT$_{1B}$ knockout. Results published in *Science* in 1999 showed similar phenotypes across laboratories on some, but not all, behavioral tasks (Crabbe et al., 1999). In-depth analyses in a subsequent comprehensive publication of the dataset (Wahlsten et al., 2003c) revealed highly consistent behavioral results across laboratories on the robust differences between inbred strains and in the 5-HT$_{1B}$ mutants on the Morris water maze and on ethanol preference. A few tests, including the elevated plus-maze, which measures anxiety-related behaviors, detected significant differences in results across the three laboratories (Wahlsten et al., 2003c). The authors conclude that laboratory environments cannot be made fully identical, to guarantee exact replications. However, the similarity of results obtained by different laboratories testing the same lines of mutant mice appears to be approximately the same in behavioral phenotyping as in other biological assays.

Breeding strategies are extensively discussed in Chapter 2. Breeding the mutation onto a single genetic background using a congenic breeding scheme may dramatically minimize the variability in behavioral data. Choice of background strain is one of the fundamental decisions preceding the start of breeding. Breeding the mutation onto two or more different backgrounds, to compare the outcome of the mutation in two independent lines of mice on differing genetic backgrounds, addresses fascinating hypotheses about gene interactions (Dobkin et al., 2000; Holmes et al., 2003d; Bailey et al., 2006). If the phenotype of the mutation is significantly different when bred onto another background, protective or susceptibility genes are implicated. The search for modifier genes is one of the fastest growing areas of mutant mouse phenotyping and genetic medicine (Egan et al., 2001; Grice et al., 2002; Buchner et al., 2003; St George-Hyslop and Petit, 2005).

Not everyone can afford to breed congenic lines on multiple backgrounds. Many labs begin with the mutation on a mixed background. The mix often includes one of the 129 strains from the embryonic stem cells, one of the C57BL/6 strains from the blastula donor, and a third strain for the breeder females. Can useful pilot data be obtained by comparing genotypes on a mixed background? Maybe. If the phenotype is subtle, its expression may not be statistically significant, because the variability contributed by the background strains will mask the small effect of the mutation. However, if the mutation produces a very large phenotypic effect, its overall expression may be sufficiently large to override the variability contributed by the background genes. For example, the *FGFR-3* knockout mice described in Chapter 3 display serious bone malformations that lead to a severe loss of motor function by age 13 weeks (McDonald et al., 2001). Rotarod impairments were easily detected on a mixed background from 129 embryonic stem cells and B6 blastocysts. Sometimes investigators pursue pilot experiments from the first mice on the mixed background, looking for a significant phenotypic difference between null mutants and wildtypes to include in their initial grant application. After funding is obtained, the lengthy and expensive process of backcrossing into a single genetic background can be undertaken, toward more thorough behavioral phenotyping.

TESTING RESOURCES

Space and Equipment

Planning your series of phenotyping experiments involves many practical considerations. A quiet testing room is usually needed for conducting a behavioral task. Running behavioral experiments within a busy laboratory will introduce so many distractors that the mice may not be able to perform tasks normally or consistently. It is best to identify a dedicated procedure room that is close to the mouse housing facility. Transportation from housing room to test room across different buildings, floors, or long hallways is stressful to mice. Generally an hour in the new location is needed for the mice to settle down if the transportation is minimal. If cages are taken on a long ride on a rattling cart across bumpy floors and up a noisy elevator, the adjustment time will be longer. The habituation period after transportation needs to be built into the daily experiments. A separate alcove, shelf, or cart, outside the test room, is the best place to keep the cages during this habituation time at the beginning of the experimental day.

Behavioral test equipment can be large or small. After deciding on the constellation of behavioral tasks to address your hypotheses, you will want to peruse catalogs for equipment to purchase (see the partial list of commercial suppliers at the end of Chapter 2). Issues to consider include quality, cost, and size. Large pieces of equipment with computer controls may need a dedicated location in an appropriate behavioral test room. Several small pieces of equipment can be kept in the same room and used as needed. Behavioral neuroscientists generally schedule the use of the room to maximize time-sharing of equipment use.

In general, only one behavioral test can be conducted at a given time within a test room. This is because mouse behaviors are influenced by the smells, sounds, and sights of other animals engaged in other tasks. Therefore many small test rooms are more useful than one large test room. Equipment is distributed in the available test rooms according to relative size. For example, larger rooms are needed to accommodate the Morris water maze, or a set of 8 Accuscan open field boxes, or 16 lickometers, with their associated computer controls. Small rooms can be used to run the rotarod, the hotplate test, or an olfactory task. The small pieces of equipment are stored on shelves in the room. The equipment for that day's specific experiment is set up on a table in the small room. At the end of the test day, the equipment is cleaned with soap and water, wiped dry, sprayed with ethanol or another antiseptic, and returned to a shelf. An example of a functional arrangement of a suite of behavioral test rooms is shown in Figure 13.2.

Personnel

Great research starts with great researchers. Choose your team carefully. If you are not experienced in mouse behavioral tasks, find an expert collaborator. Remember that all collaborators are not created equal. Throughout this book, prominent behavioral neuroscientists are mentioned, along with their areas of expertise. The references in

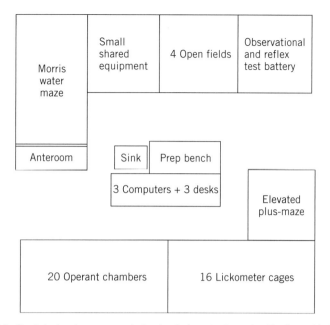

Figure 13.2 *Idealized design for a mouse behavioral phenotyping suite. Dedicated behavioral proce- dure rooms are arranged around the perimeter. Investigator stations are located in the central space. Open entrance/exit areas are located at the left and right sides of the suite. [From Crawley (2003), p. 143.]*

this book were chosen for quality. Referenced publications provide further ideas for good potential collaborators. PubMed searches are another method to identify well- published experts. Try to attend a local seminar or a conference talk by a potential collaborator, to judge the quality of their research and the range of behavioral tasks ongoing in their laboratory. Introduce yourself after the talk, in person or by email or phone call later. Sending copies of your relevant publications, grant application, or a summary of findings from your mutant mouse, will give the potential collaborator some insight into your scientific questions and the possible scope of the collaborative project. Follow up with a fair proposal. Offering resources toward the collaboration, such as funding for a student or postdoctoral fellow, purchasing a piece of equipment, providing rack space, cage costs, or behavioral test rooms, may increase the chances that the behavioral neuroscientist will agree to the collaboration. Expecting a collaborator to run your experiments for free may be unrealistic and insulting. When positive pilot data are obtained, the collaboration can be cemented by writing a grant application together, with the behavioral component sharing Co-Principal Investigator status and proportional funding. Negotiate authorships on potential publications, particularly who will be first and last author. Discuss future availability of the line of mice for the behavioral collaborator to pursue additional independent behavioral studies. Usually it is best to reach these agreements at the beginning, to avoid later disputes and to ensure the development of a good collaborative relationship.

Perhaps you have some experience with mouse behavioral tasks, from a previous collaboration or from taking one of the training courses listed in Chapter 1. It may be possible to set up two or three specific tasks that you have learned and train a technician to run them in a routine fashion. In this case it is advisable to develop a working relationship with one of the behavioral neuroscience experts in these tasks, to help you as a consultant on the project. A consultant willing to look over your experimental methods and assist in interpreting the results can prevent some of the rookie mistakes that we all make in setting up new lab assays. Another approach is to hire a young research assistant professor or core facility manager to run the behavioral component of your research program. It may be hard to find an outstanding behavioral neuroscientist with comprehensive expertise in mouse behavioral tasks who will accept a non-tenure-track appointment. Many university departments offer combination positions, for example, a half-time appointment as core manager along with a half-time faculty position doing independent research. In some cases a tenure-track position is offered with part-time responsibilities for managing the core facility.

Perhaps you are a behavioral neuroscientist experienced in mouse anxiety tasks and wanting to introduce schizophrenia-related tasks to your lab. Or you have run rat learning and memory tasks for decades and now want to switch to mice. A short visit to the laboratory of a colleague with the appropriate expertise may be sufficient. Recruiting a well-trained graduate student from an outstanding behavioral neuroscience laboratory to become a postdoctoral fellow in your laboratory is another excellent way to bring a specific expertise into your lab. Visiting the vendor booths at conferences, such as the Federation of European Neuroscience Societies and Society for Neuroscience meetings, may provide a helpful opportunity to see the advantages and disadvantages of the different equipment that you are considering for setting up the new tasks.

GENERAL HEALTH AND NEUROLOGICAL REFLEXES

The first step in any careful phenotyping battery is confirming that the mutant line is generally healthy (Chapter 3). Sick mice are likely to fail most behavioral tests. Observing the appearance of the mice, their responses to handling, and their home cage behaviors will usually be sufficient to detect a major health problem. Severe sickness or malaise will preclude any further behavioral testing. The second step involves testing basic neurological reflexes, including the righting reflex, acoustic startle, and visual forepaw reaching (Chapter 3). Gross physical deficits will prevent the use of a mutant line in many behavioral tasks. Investigators will save themselves a great deal of time and effort by finding out early that their mice are not healthy enough to proceed with behavioral testing, rather than hoping for the best and later having to discard useless data.

Evaluating general health, observing home cage behaviors, and checking neurological reflexes such as acoustic startle, visual forepaw reaching, and the righting reflex, are usually completed before the start of any more complex behavioral testing (Figures 13.3 and 13.4). Tests for general health and neurological reflexes can be repeated on the same animal (Table 13.1). Carryover effects have not been reported for these simple, quick observational measures. When there is reason to believe that a mutation or a genetic background will express a different phenotype at a later age, tests for health and reflexes can be conducted again at the later age (Chapter 12). These brief tests, each taking a few seconds, can often be performed sequentially on the full Ns in a day or two.

Figure 13.3 *Observations of general health. One strategy for comprehensive behavioral phenotyping of a new transgenic or knockout mouse is to begin with observations of general health (Chapter 3). The investigator notes the appearance of the fur, a measure of good grooming. Matted, discolored fur, bald patches, missing whiskers, scarring, or wounds could indicate poor health, insufficient grooming, overgrooming, and fighting. Body weight is often measured at weekly intervals, to plot growth rates. Unusual responses to handling include extreme jumping, vocalizations, and attempts to bite the human handler. Extreme levels of these responses to handling may indicate poor health. Repeated observations of home cage behaviors are conducted by an investigator standing at a distance in the mouse housing facility, or by videotaping the cages in the housing facility. Observational periods of 15 minutes or longer are conducted at morning, afternoon, early evening, and late night time points, since mice show different behavior patterns across the circadian cycle. Repeated observations at different times of day may be conducted on the same day, on consecutive days, or over several weeks, by a human observer scoring in real time or scoring videotapes. Nesting patterns, sleeping in a huddle, allogrooming, sexual, parental, and aggressive behaviors are measures of social interactions that are easily noticed in the home cage environment. Unusually high or unusually low levels of home cage activity may also be obvious. Abnormalities in these measures may be revealing and lead to more intensive, quantitative testing in that behavioral domain, such as feeding tests (Chapter 7) or social challenges (Chapter 9). Serious health abnormalities, such as inactivity in the home cage and lack of responsiveness to handling, may indicate illness or physical defects that will preclude further behavioral testing.*

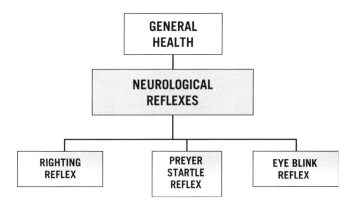

Figure 13.4 *Measures of neurological reflexes. Many quick tests are available to assess simple neurological reflexes (Chapter 3). The righting reflex is measured by turning the mouse onto its back and timing the latency for the mouse to turn back over onto all four paws. Most mice immediately right themselves, and/or are not able to be turned onto their backs at all. Serious motor dysfunctions such as spinal cord injury or motor neuron degeneration are easily detectable by the righting reflex (Chapter 3). The Preyer startle reflex measures the flinch response to a sudden loud tone, such as a clicker or buzzer close to the ear of the mouse. Deaf mice fail the Preyer test (Chapter 5). Eyeblink is elicited by approaching the eye of the mouse with a cotton swab or similar soft object, and noting the eyelid blink response. Blind mice fail the eyeblink test (Chapter 5). If neurological reflexes are severely impaired, it is likely that further behavioral testing will be compromised by fundamental physical dysfunctions.*

TABLE 13.1   Measures of General Health, Home Cage Behaviors, and Neurological Reflexes in a Cohort of Galanin Overexpressing Transgenic Mice and Their Wildtype Littermate Controls

	Females		Males	
	WT (8)	GAL-tg (21)	WT (11)	GAL-tg (23)
General health				
Body weight (g)	19.77 ± 4.2	21.22 ± 3.4	26.99 ± 3.7	25.1 ± 3.1
Poor coat condition	13	18	0	0
Missing whiskers	14	0	0	4
Piloerection	0	0	0	0
Unusual body tone	14	5	10	0
Unusual limb tone	0	0	10	8
Home cage behaviors				
Solitary sleeping	0	0	0	0
Fighting and aggression	0	0	0	4
Motoric, muscular abilities				
Positional passivity	0	0	0	8
Trunk curl	100	100	100	100
Forepaw reaching	100	100	100	100
Righting reflex	100	100	100	100
Wire hang (seconds to fall)	56.4 ± 3.2	58.4 ± 4.4	49.3 ± 3.8	57.8 ± 3.8
Reflexes				
Eye blink	100	100	100	100
Ear twitch	100	100	100	100
Whisker response	86	95	100	100
Toe pinch response	100	100	100	100
Reactivity				
Moving away on petting	100	100	100	100
Struggle on restraint	100	100	100	100
Vocalization on restraint	57	65	50	58
Dowel biting (3 point scale)	1.8	2.0	2.0	1.9
Empty cage behaviors				
Freezing on transfer	0	0	0	0
Wild running	0	0	0	0
Stereotypies	0	0	0	0
Cage exploration (3 point scale)	2.4	2.2	2.1	2.2
Grooming (3 point scale)	1.6	1.5	1.8	1.7

Measures for body weight and wire hang are stated as mean $\pm$ SEM. Measures of general health, home cage behavior, and neurological reflexes were not significantly different between genotypes, 3 point scale: 1 = deficient, 2 = moderate, 3 = excessive.

Source: From Kinney et al. (2002), p. 181.

Note: Numbers of females and males of each genotype are shown in parentheses. Data for most measures are expressed as the percentage of mice displaying the phenotype. Numerical scales are described in parentheses. No genotype differences were detected on any measure, supporting an interpretation that all mice were generally healthy.

HANDLING

Most behavioral neuroscience laboratories follow a routine of handling their mice or rats before the beginning of behavioral testing. The goal of handling is to habituate the mouse to the stress of removal from the cage, physical contact with humans, and other standard handling manipulations within experiments. The idea is to reduce stress levels due to human handling before getting started on actual experiments. Handling procedures may include picking up the mouse by the tail, resting it on a gloved hand, stroking the fur for a few seconds, letting the mouse walk freely on a sleeve for a few seconds, and replacing the mouse in its home cage. Repeated handling for a few minutes on two or three days is generally sufficient. Handling occurs naturally during the observations for general health, and while conducting tests for neurological reflexes. Routine handling also occurs naturally during cage changes by the animal caretakers, and during identification procedures by researchers and caretakers involved in maintaining the breeding colony.

SENSORY AND MOTOR TASKS

Choice of sensory and motor tasks is based on the hypothesized phenotype resulting from the mutation. Consider the following options:

A. The gene of interest is hypothesized to regulate a motor function.
B. The gene of interest is hypothesized to regulate a sensory ability.
C. The gene of interest is hypothesized to regulate behavior in another domain, such as learning and memory or feeding. Behavioral tests for the hypothesized domain require normal motor functions and sensory abilities.

Option A

The hypothesis driving many targeted gene mutations concerns a motor function. As described in Chapters 4 and 12, many mutant mouse models target human genes responsible for motor dysfunctions, such as Parkinson's disease, ataxias, and amyotrophic lateral sclerosis. A battery of motor tasks is designed to model the human traits. Ataxias are analyzed with gait-sensitive tests such as footprint path and balance beam walking. Motor coordination and balance deficits are analyzed with the rotarod, balance beam, and pole climb. Outcomes of spinal cord injury are detected in the open field test, rotarod, and balance beam. Neuromuscular defects are measured in the wire hang and grip tests. Many motor tasks can usually be conducted repeatedly on the same mice. Most of these tasks are relatively quick, from 60 seconds to 15 minutes. It is possible to test a large cohort of mice on one task within the same day. To our knowledge, the set of mice can be tested the next day on a different motor task without carryover effects (Paylor et al., 2006). This flexibility allows analyses of time courses, such as the rate of recovery after a lesion, or the progression of neurodegeneration. Therefore several relevant motor tasks can be conducted to confirm a motor dysfunction or improvement. The recommendation for a constellation of behavioral phenotyping tests for evaluation of motor function would be (1) general health battery, (2) neurological reflexes, and (3) motor tasks including two or more corroborating tests (Figure 13.5).

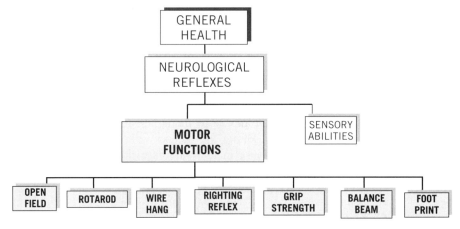

Figure 13.5 *Tests for motor functions. Automated and observer-scored tests for motor functions are described in detail in Chapter 4. Wire hang and grip strength are corroborative measures that detect neuromuscular dysfunctions in mouse models of human diseases such as amyotrophic lateral sclerosis. Balance beam and footprint tests detect ataxias in mouse models of human diseases such as Huntington's, Parkinson's, and Ataxia telangiectasia. Rotarod measures of motor coordination and balance are sensitive to mutations of genes in the cerebellum. Open field locomotion measures many components of motor functions and exploratory activity. Spinal cord injury, sedative drugs, and psychostimulants such as amphetamine produce abnormal scores on open field tests. If general health and neurological reflexes are normal and a motor dysfunction is detected, motor deficits may optimally define the consequences of the mutation in a gene relevant to motor functions.*

Option B

The hypothesis driving the targeted gene mutations concerns a sensory ability. If the goal of the research program is to understand genes mediating a sensory ability, then the most sensitive tasks to measure function in that sensory domain are chosen (Figure 13.6). Two basic approaches are standard in the literature. As described in Chapter 5, neurophysiological recording from the sensory cortex during presentations of sensory stimuli is the most precise technique with the greatest sensitivity. Operant behavioral tasks, based on reinforced responses to selected sensory stimuli, provide detailed information about sensory perception, as described in Chapters 5 and 6. Both techniques require a major investment in dedicated equipment, and highly trained researchers. The advantage of these sensitive approaches is the detection of graded responses. Quantitative assessments of amount of hearing loss, pain thresholds, just noticeable differences in visual acuity, discrimination between similar tastes, and detection of odor gradients, require these sophisticated technologies. Laboratories investigating deafness genes, retinal development, and olfactory receptors, or discovering better analgesics, may focus their efforts on just one intensive neurophysiological or operant method for all of their research questions.

Option C

The hypothesis concerns a more complex behavioral domain. When the goal of the research program is to investigate genes mediating complex traits, such as anxiety, learning and memory, or social behaviors, the sensory and motor tasks are required

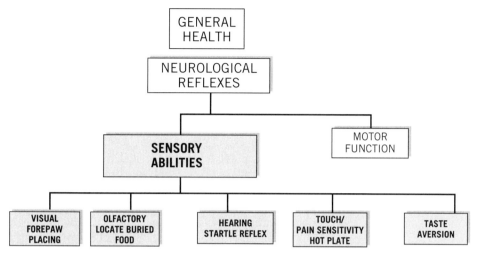

Figure 13.6 *Tests for sensory abilities. Quick yes-or-no measures of sensory abilities, and more intensive tests for sensory acuity, are described in detail in Chapter 5. Forepaw reaching when lowered to a table surface is a simple measure of vision. Latency to find a small piece of food buried in the litter is a measure of olfaction. Acoustic startle to a sudden loud tone is a measure of hearing. Paw licking, jumping, or vocalization to the touch of a 55°C hot plate is a measure of pain sensitivity. Avoiding food or drink with a noxious flavor is a measure of taste. If general health and neurological reflexes are normal and a sensory dysfunction is detected, the sensory defect may optimally define the consequences of the mutation in a gene relevant to blindness, anosmia, deafness, analgesia, or taste perception. Investigations focused on a sensory domain will benefit from more intensive tests for threshold responses and acuity, including neurophysiological recording from the sensory nerve or sensory cortex during presentation of sensory stimuli, or operant behavioral responses to sensory stimuli presentation.*

primarily as controls. Simple measures may be sufficient to ensure that the mouse is physically able to conduct the procedures, such as being able to hear a tone or smell its pups. As described in Chapter 5, and shown in Figure 13.6, there are some reasonably good, quick-and-dirty, yes-or-no, low-tech tests for sensory abilities. These only reveal that a mouse is blind, deaf, or anosmic (cannot smell). Quick visual tests include forepaw reaching, moving from a lighted to a dark area, swimming to a highly visible platform, and a visual cliff test. A quick hearing test is the Preyer startle reflex, using a clicker, buzzer, hand clap, or jangling a set of keys. Simple smell tests involve sniffing novel scents and locating buried food. Quick tests for pain perception are the hot plate and tail flick. As described in Chapter 4, there are several quick tests that are sufficient to detect severe motor abnormalities. A 5-minute open field test may be sufficient to quantitate deficits in locomotor activity.

COMPLEX BEHAVIORAL DOMAINS

Learning and memory, feeding and drinking, social and reproductive behaviors, and mouse analogs of human neuropsychiatric diseases are examples of complex traits. Each single behavioral task in these domains has its own limitations and potential pitfalls (Chapters 6–11). Tasks within a specific behavioral domain may address different hypotheses. For example, in the feeding domain, a macronutrient selection task can reveal preference for a high fat diet, while microstructural analysis of circadian feeding patterns can distinguish reductions in meal size from reductions in frequency of snacking (Tempel and Leibowitz, 1990; Currie, 1993; Rodgers et al., 2000; Kanarek, 2004; Goodson et al., 2004). Body weight measurements document longitudinal outcome over days, weeks, and months. Spatial learning tasks may test hypotheses about genes regulating hippocampal development, while cued fear conditioning may test hypotheses about genes expressed in the amygdala. Ideally we conduct three or more different tests within the behavioral domain of interest. When similar results are obtained in all three tests, strong conclusions can be drawn about the underlying generality of the behavioral phenotype. When different results are obtained from the three tests, refined hypotheses can be further pursued. For example, contextual fear conditioning is sensitive to hippocampal lesions, while auditory cued fear conditioning is sensitive to amygdala lesions, as described in Chapter 6. Deficits on only the cued component will prompt researchers to focus future experiments on various functions of amygdala nuclei.

If your mutant line is normal on measures of general health, neurological reflexes, sensory abilities, and motor functions, then you can proceed to conduct the more complex tasks with confidence that physical procedural abilities will not produce artifactual false positives. If one sensory or motor ability is impaired, it is usually possible to design around the deficit. For example, olfactory learning tasks can be substituted for visually cued learning and memory tasks, and vice versa (Chapter 6).

Recommendations for specific groups of tasks within a domain are based on insights from the behavioral tests described in the preceding chapters. Learning and memory may best be evaluated in a new line of mice by choosing three very different types of tasks described in Chapter 6. The Morris water maze measures spatial learning, contextual and cued fear conditioning measures emotional memory, social transmission of food preference measures olfactory memory, and rotarod performance measures motor learning (Figure 13.7). A deficit in all of these tasks, coupled with normal measures of general health, neurological reflexes, motor functions, and sensory abilities, would be interpreted as a truly fundamental cognitive dysfunction. A deficit in just one of these tasks would be followed up with a second test within that cognitive subdomain. For example, a Morris water maze impairment may be pursued by testing the mice on the Barnes maze, radial maze, or T-maze, to confirm a spatial learning deficit. Abnormal contextual or cued freezing may be further explored with passive avoidance and fear-conditioned acoustic startle, to confirm a fear conditioning deficit. Social transmission of food preference impairments may be confirmed with a social recognition task and an operant task dependent on olfactory cues, to confirm an olfactory memory deficit. Feeding and drinking abnormalities may best be analyzed with several of the measures described in Chapter 7. For example, daily body weights, daily chow consumption, and microstructural analysis of feeding with a lickometer, would measure patterns of consumption and total ingestion. Similarly, daily water consumption and a two-bottle choice test would measure different components of drinking. Sexual and parental behaviors described in Chapter 8 are scored by standardized observational methods involving many parameters, including anogenital investigation, lordosis posture, mounts, intromissions, and ejaculations, nest building, crouching, lactation, and pup retrieval to the nest. Social behaviors detailed in Chapter 9 are combined to address specific hypotheses, such as tube dominance, isolation-induced fighting, and

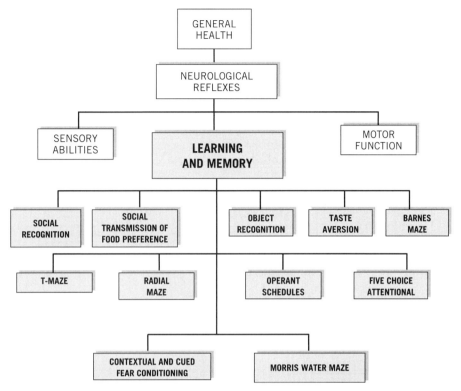

Figure 13.7 *Learning and memory tasks. Many learning and memory tasks are available for testing cognition in mice, as described in detail in Chapter 6. To evaluate a new line of mice with a mutation in a gene hypothesized as relevant to learning and memory, it is useful to conduct several tasks to evaluate the global nature of a cognitive deficit. For greatest scientific rigor, two or three tests are conducted within each category of learning and memory, for example, spatial learning, fear-induced emotional memory, olfactory tasks, and schedule-induced operant chamber tasks. Attentional tasks such as the five choice serial reaction time task will confirm or rule out an attentional confound. The same mice can be tested in a several tasks, preferably beginning with the least stressful and ending with the most stressful. Tasks with very similar properties, such as passive avoidance and contextual fear conditioning, are best conducted in separate sets of mice to avoid carryover effects. If a sensory or motor deficit was detected in earlier testing, the choice of learning and memory tasks are designed to avoid reliance on that sensory modality or motor function. As examples, mutations in genes that cause blindness preclude cognitive tasks that rely on visible spatial cues, such as the Morris water maze or operant chambers with light cues. Mutations in genes that cause ataxia preclude cognitive tasks that depend on intact locomotion, such as the radial arm maze.*

resident–intruder attack to characterize aggression; or group nesting, social recognition, and social approach scoring to characterize social affiliation. In the domain of anxiety-related behaviors, discussed in Chapter 10, complimentary tasks include the elevated plus-maze, light↔dark transitions, and open field emergence (Figure 13.8). If the mutation or background strain has motor deficits that could confound the interpretation of anxiety-like reductions in activity on these tasks, the Vogel thirsty lick conflict test, fear-conditioned freezing, and fear-conditioned startle are anxiety-related tasks that do not depend on extensive locomotion. Two complimentary tasks sensitive to antidepressants are the tail suspension and Porsolt forced swim tests, described in Chapter 10.

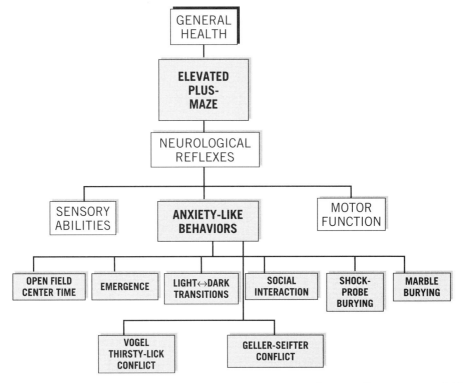

Figure 13.8 *Anxiety-related behavioral tests. Many approach–avoidance conflict tests are available for evaluating anxiety-like behaviors in mice, as described in detail in Chapter 10. For greatest scientific rigor, several different tests are conducted to corroborate an anxiety-related phenotype, such as elevated plus-maze, light↔dark transitions, marble burying, and the Vogel thirsty-lick conflict test. The same mice can be used for many tasks. The elevated plus-maze is conducted first because it is most sensitive to prior experience. If a sensory or motor deficit was detected in earlier testing, the choice of anxiety-related tasks must avoid reliance on that sensory modality or motor function. For example, a mutation that reduces locomotor activity may best be tested for an anxiety-like phenotype on the Vogel task, which requires minimal locomotion. The light↔dark task would be inappropriate for a blind background strain of mice or a mutation that causes blindness.*

Well-validated measures for investigating drugs of abuse are the conditioned place preference task and self-administration by intravenous or two-bottle choice methods, illustrated in Chapter 11. Most of these tests can be administered sequentially to the same set of mice. Exceptions and caveats are discussed below. Corroborative phenotypic results obtained in two, three, or four tests within a behavioral domain, in the absence of procedural artifacts, are the strongest proof that behavioral neuroscience can offer to confirm the behavioral outcome of a targeted gene mutation.

Order of Testing

One cohort of mice can be tested on a large number of behavioral tasks. Many laboratories conduct a comprehensive set of behavioral tasks using the same set of mice (Bannon et al., 2000; Ferguson et al., 2000; Paylor et al., 2001; Kinney et al., 2002; Salinger et al., 2003; Holmes et al., 2003b; Brooks et al., 2004; Wrenn et al., 2004;

Moretti et al., 2005). Large-scale mutagenesis projects (Chapters 3 and 14) often use a standardized set of simple behaviors (Rogers et al., 1997), administered sequentially to all mice as a first screen. High-throughput screening for drug discovery may similarly begin with a sequence of simple tasks, or automated equipment that measures many behavioral parameters simultaneously (Brunner et al., 2002). Repeated use of the same animals lowers the total costs, saves on breeding time, and is consistent with good animal care and use guidelines for minimizing the numbers of animals used in biomedical research. However, multiple tests can produce carryover effects. One task could directly influence the results obtained in a subsequent task. Even more problematically, there could be a summation of small effects from multiple sequential tasks, that cumulatively impacts the data obtained in the later tests.

In general, it is usually possible to use the same group of mice for multiple tasks, both within a behavioral domain and across behavioral domains. Figures 13.3 through 13.8 illustrate a potential sequence for consecutive testing. A common design is to allow one week between tests, to minimize carryover effects. The one week interval is more a rule-of-thumb than a comprehensively verified recommendation. Richard Paylor and co-workers at Baylor School of Medicine discovered that more frequent testing is feasible (Paylor et al., 2005). One- or two-day test intervals between the neurological exam, open field activity, light↔dark exploration, rotarod, prepulse inhibition, and startle habituation tests yielded results very similar to those obtained when the same test battery was conducted with one-week intervals between tests (Paylor et al., 2005).

Systematic analysis of carryover effects has been conducted only infrequently. Most behavioral phenotyping experts make educated guesses when choosing the sequence of testing and intervals between tests. The general principle is to conduct the most stressful tasks last. Within the learning and memory domain, for example, the less stressful social transmission of food preference and object recognition tests are conducted first. Somewhat stressful learning and memory tasks are conducted next, including passive avoidance, which involves a single footshock, or maintenance on a food restricted diet for radial maze and operant chamber tasks. Learning and memory tasks that are considered most stressful include the multiple pairings of footshock and tone in trace fear conditioning, and repeated swimming in the Morris water maze. A sensible recommendation is to run the mice in the fear conditioning and Morris water maze tasks as the last experiments in the series. For example, in the galanin research project in our laboratory, we start with the elevated plus-maze, then conduct the measures of general health, home cage observations, neurological reflexes, motor tasks, sensory measures, and then continue through other hypothesis-driven learning and memory tasks, saving fear conditioning and the Morris water maze for the end of the sequence. This decision is based on the assumption that a highly stressful task produces the most deleterious and longest lasting carryover effects on other tasks. However, the ability of a human researcher to intuit a mouse's perception of stress is limited at best. One obvious limitation is the human sensory range. Mice and rats excrete urinary pheromones and vocalize in response to stressors (Vivian and Miczek, 1993; Cocke et al., 1993; Griebel et al., 2002; Kiyokawa et al., 2005), emitting ultrasonic vocalizations in a frequency range above human perception. It seems likely that mice awaiting testing may detect stress-related smells and sounds from other mice engaged in a task in the same room or nearby. Mice in the waiting area may be influenced by hearing these distress calls, in ways that change their performance when it is their turn to conduct the task. Researchers may therefore undervalue the perceptions of the subject mice, and misinterpret the level of stress associated with a behavioral test. More evidence on carryover effects between tasks would be helpful in determining the levels of stress inherent in behavioral tasks, and any long-term consequences on subsequent tasks.

An obvious but sometimes overlooked detail is where to place a mouse after it is tested. The right answer is to take the mouse out of the equipment when it finishes the test, and place it into a holding cage. Not back into its home cage. Odors, vocalizations, changes in behavior, and other cues emitted by the subject mouse as it exits the behavioral equipment will be immediately transmitted to the untested mice awaiting their turn. An empty mouse cage located far from the behavioral testing room is a good repository for the subject mice as they finish up the task. One holding cage accommodates all of the mice from an original housing cage. At your leisure, you later return the group to its original housing cage and bring them back to the vivarium.

One excellent study on order of testing was conducted by Richard Paylor and co-workers at Baylor College of Medicine (McIlwain et al., 2001). Nine behavioral tasks are routinely conducted by Dr. Paylor's laboratory in phenotyping a new transgenic or knockout mouse. The test battery consists of (1) a neurological screen, (2) open field activity, (3) anxiety-related light↔dark exploration, (4) rotarod motor coordination and balance, (5) acoustic startle and prepulse inhibition, (6) habituation of acoustic startle, (7) contextual and cued fear conditioning, (8) Morris water maze, and (9) hot plate. Experiment 1 compared mice tested on the entire battery to mice tested on only one of the tasks. Compared to naïve mice, the battery mice showed less open field activity, less anxiety-related behavior, better rotarod scores, greater pain sensitivity on the hot-plate test, and worse selective quadrant search in the Morris probe trial. Startle, habituation to startle, prepulse inhibition, and fear conditioning showed no effect of prior testing. Experiment 2 compared mice tested on 4 of the battery tasks (open field, light↔dark, acoustic startle/prepulse inhibition, and fear conditioning), administered at weekly intervals. The order of the tasks was systematically varied. Two different strains of adult male mice were evaluated, C57BL/6J (B6) and 129/SvEvTac (129S6). Naïve mice showed higher exploratory activity in open field and light↔dark tasks than mice who went through a previous battery of nine assays. When only four behavioral tests were administered in various sequences, order of testing did not affect open field activity, prepulse inhibition of acoustic startle, or contextual or cued fear conditioning for either strain. Light↔dark latencies and maximum startle amplitudes varied with testing order. When the light↔dark task was conducted first, B6 mice showed faster latencies to enter the dark chamber, as compared to B6 groups that first experienced open field and startle testing. In contrast, 129S6 mice showed slower latencies to enter the dark chamber when compared to 129S6 mice that first experienced open field testing, prepulse inhibition, and fear conditioning. Higher maximum startle amplitudes were found in B6 mice tested for startle as the third task in their sequence, as compared to B6 mice receiving the startle task first or last in the sequence. Order of testing did not affect startle amplitude in 129S6 mice. The authors conclude that there can be test order effects, but that many behaviors are resistant to test order and previous testing experience. Dr. Paylor's laboratory now conducts the neurological screen last, with no adverse outcomes (Richard Paylor, personal communication).

Vootele Voikar and co-workers at the University of Helsinki in Finland investigated training history on behavioral tasks administered to C57BL/6JOlaHsd and 129S2/SvHsd mice (Voikar et al., 2004). Naïve mice tested on only one task were compared to mice that went through a battery of tasks. The battery sequence consisted of handling, open field, elevated plus-maze, light↔dark transitions, Y-maze spontaneous alternation, beam walking, wire hang, rotarod, hot-plate, fear conditioning, water maze, forced swim, and conditioned taste aversion. No difference in performance was detected between naïve and battery mice on most components of the open field activity, light↔dark parameters, % spontaneous alternation, beam walking, wire hang, rotarod, fear conditioning, and water maze learning. Elevated plus-maze open arm time and % open arm entries were lower in experienced battery mice than naïve mice. Hot-plate latencies were shorter in battery than naïve mice. The forced swim test displayed the biggest effect of test history, with significantly more immobility in battery than naïve mice of both strains. Conditioned taste aversion was significantly stronger in naïve than battery mice in the 129 strain.

Order of testing may be most critical for the elevated plus-maze test of anxiety-related behaviors. Anxiolytic drug effects are less when the rat or mouse has previous experience on the elevated plus-maze (File et al., 1992; Holmes and Rodgers, 1999). Prior handling and prior experience on the elevated plus-maze influences baseline performance on the elevated plus-maze, although this effect is more variable, and may depend on the time interval between the two sessions on the elevated plus-maze (Holmes and Rodgers, 1998; Voikar et al., 2004). Factor analysis has been conducted on the elevated plus-maze task in mice and rats, as well as on the light↔dark exploration task, open field activity, hole board, and the mouse defensive battery (Lister, 1987b; File et al., 1993; Rodgers and Johnson, 1995; Griebel et al., 1996; Chaouloff et al., 1997; Ramos et al., 1998; Wall and Messier, 2000; Holmes et al., 2003b; Henderson et al., 2004). Explanations for the change in elevated plus-maze performance across repeated tests include different underlying factors mediating the locomotor, risk assessment, and anxiety components of the task; acquisition of fear across repeated experiences; different levels of stress on repeated trials; and a shift in the innate conflict between the tendency of mice to explore a novel environment and the aversive properties of the open arms (Crawley 1989; File 1997a; File et al., 1993, 1994; Holmes and Rodgers 1998, 1999; Holmes et al., 2003b).

These first publications on order of testing, using different experimental designs and various inbred strains, reported order-independent results on some but not all tasks. As more studies emerge, order of testing effects will be more definitively understood. The safest approach at present is to compare mutant to wildtype littermates. As long as the behavioral tests are run in the same sequence for all genotypes and sexes, using group-housed littermates, the environment and prior experience should have a similar impact on all genotypes.

SEARCHING THE SKIES FOR YOUR CONSTELLATION

How Many Tests?

The exact number of tests to conduct for complete behavioral phenotyping is unknown—or more precisely, infinite. When you invest years in generating the mutation, you want to maximize the knowledge to be gained from analyzing every aspect of the consequences of the mutation. Luckily you are not alone. Collaborators are out there to join you in the pursuit of anatomical, pathological, biochemical, physiological, and behavioral phenotypes of your mouse. First choose the optimal assays, including behavioral parameters from among the wealth of available tasks described throughout this book. Then divide the labor between tests you want to run in your own laboratory, versus assays you prefer to approach collaboratively by partnering with the experts.

No single behavioral task is universally applicable. Gerry Dawson of the Merck Neuroscience Research Centre in Terlings Park, England, aptly described his sinking feeling as conversations with molecular biologists inevitably led to the question, "What is the gold-standard test for learning and memory in mice?" (Dawson, 2001). Or even worse, "How can I tell if my mouse is schizophrenic?" (Dawson, 2001). One behavioral test will not fit all situations and scientific questions, just as one food will not satisfy all of the nutritional needs of all people (at least not since manna in the desert; see Exodus:16). Complex mouse behavioral tests that are presently considered "gold standards" include the Morris water maze and fear conditioning for learning and memory, the elevated plus-maze and light↔dark exploration to model anxiety, tail suspension and forced swim to model antidepressant actions, intravenous and two-bottle choice self-administration and conditioned place preference to measure the reinforcing properties of drugs and alcohol, resident–intruder for aggression, and prepulse inhibition as a measure of sensorimotor gating deficits in schizophrenia. If an exciting discovery is corroborated in a second or third task within the behavioral domain, without artifactual confounds, the investigators can take it to the bank.

CONCLUSIONS

Now you may be asking, "Do I really have to do all this?" Maybe not. For any specific behavioral task, there are a few highly relevant control measures, and many less relevant controls. A mouse model of amyotrophic lateral sclerosis may focus on the balance beam, footprint test, and rotarod, with only a few measures of general health as controls. Drug abuse models employing a two-bottle choice test may involve very few sensory and motor demands, since the experiment takes place in the home cage, requires minimal locomotor activity, and includes the water bottle as the control for general drinking ability. Anxiety tasks such as the Vogel thirst lick conflict test, and learning and memory tasks such as fear conditioning, similarly do not depend on extensive motor functions. Sensory control experiments can be restricted to those required by the specific task, such as hearing and pain perception for fear conditioning, or vision for a light-cued operant task. The visible platform test is a built-in control for vision and swimming abilities in the Morris water maze hidden platform task. Rearranging the sequence of administration of the control tests may also lighten the workload. Brave scientists do the critical experiment first. If the results are positive, they go back and conduct the careful control experiments afterward. Wise, brave scientists know that some of their most exciting discoveries are disproved by running the proper controls. The true gold standard in scientific research is that a discovery is widely replicated. I tell members of my laboratory that their first data merit discussion at our lab meeting, when their finding replicates it is publishable, and if the same result is found three times, then I believe it. Strategies described in this chapter and throughout this book will increase the likelihood that the behavioral phenotype of your mouse is real.

BACKGROUND LITERATURE

Anagnostopoulos AV, Sharp JJ, Mobraaten LE, Eppig JT, Davisson MT (2001). Availability and characterization of transgenic and knockout mice with behavioral manifestations: where to look and what to search for. *Behavioural Brain Research* **125**: 33–37.

Bailey KR, Rustay NR, Crawley JN (2006). Behavioral phenotyping of transgenic and knockout mice. *ILAR Journal*, Special Issue on Phenotyping of Genetically Engineered Mice, **47**: 124–131.

McIlwain KL, Merriweather MY, Yuva-Paylor LA, Paylor R (2001). The use of behavioral test batteries: Effects of training history. *Physiology and Behavior* **73**: 705–717.

Society for Neuroscience (2003). Short Course II Mouse Behavioral Phenotyping. JN Crawley, Editor, Washington DC.

Wahlsten D (2000) Planning genetic experiments: Power and sample size. In *Neurobehavioral Genetics: Methods and Applications*, B Jones, P Mormede, Eds. CRC Press, Boca Raton, FL, pp. 31–42.

Wahlsten D, Metten P, Phillips TJ, Boehm SL, Burkhart-Kasch S, Dorow J, Doerksen S, Downing C, Fogarty J, Rodd-Henricks K, Hen R, McKinnon CS, Merrill CM, Nolte C, Schalomon M, Schlumbohn JP, Sibert JR, Wenger CD, Dudek BC, Crabbe JC (2003c). Different data from different labs: Lessons from studies of gene-environment interaction. *Journal of Neurobiology* **54**: 283–311.

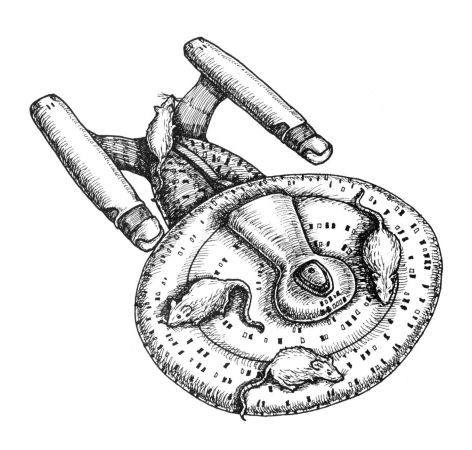

14

The Next Generation

What's next? Science advances in leaps off the springboards of new technologies. Behavioral phenotyping of transgenic and knockout mice provides a powerful tool for rapid advances in biomedical research. The foregoing chapters described the mutant mouse technology, methods for behavioral phenotyping of transgenic and knockout mice, and examples of successful identification of the behavioral consequences of targeted mutations in genes expressed in the brain. This last chapter speculates on future directions, based on currently evolving techniques.

CONDITIONAL AND INDUCIBLE TRANSGENICS AND KNOCKOUTS

Conventional mutations are "global," that is, ubiquitously distributed. Lack of tissue specificity is a fundamental problem inherent in the conventional transgenic and knockout technology. When the goal of generating the mutation is to study the role of the gene in the brain, the biochemical or physiological actions of the gene in other organs may confound the behavioral phenotype. For example, mutation of the gene for a neurotrophic factor in both the brain and the kidney could present a behavioral phenotype of reduced open field activity due to poor kidney functions, buildup of circulating toxins, and debilitating sickness.

Techniques for *conditional* mutations are increasingly available. The targeted gene is placed under the control of a promoter that is specific to a tissue that expresses that promoter gene. A frequently used method for tissue-specific conditional mutations is designed around a Cre/lox system. Cre recombinase linked to the targeted gene is controlled by a promoter that is selective for the tissue in which the gene is to be expressed (Byrne and Ruddle, 1989; Lakso et al., 1992; Rajewsky et al., 1998). Ideally the mutation is expressed only in brain neurons relevant to the behavioral hypothesis. Promoters relatively selective for brain and/or neurons include nestin (Magdelano

What's Wrong With My Mouse?   By Jacqueline N. Crawley
Copyright © 2007 John Wiley & Sons, Inc.

et al., 2002) and neuron specific enolase (Sakai et al., 2002), although these are also expressed in pancreas and some neuroendocrine tumors (Humphrey et al., 2003; Portela-Gomes et al., 2004). Most conditional mutations are present from early stages in development. In some lucky cases, an anatomically selective promoter also happens to be expressed later in development. For example, robust expression of a conditional mutation in a kainite receptor begins at 4 weeks of age in hippocampal CA3 pyramidal cells (Nakazawa et al., 2002).

Mark Mayford at the University of California at San Diego identified the promoter for calcium-calmodulin-dependent kinase IIα (CaMKIIα), which is expressed only in neurons of the mouse neocortex, hippocampus, amygdala, and basal ganglia (Mayford et al., 1996). Using the CaMKIIα promoter, deletion of the NMDAR1 glutamate receptor was specific to the CA1 pyramidal cells of the hippocampus. These conditional mutant mice exhibited impairments on the Morris water task and impaired hippocampal place cell activity during navigation of an L-shaped linear track (Tsien et al., 1996a; McHugh et al., 1996). Also generated on a CaMKII promoter, knockout mice conditionally deficient in the gene for NCAM, a neural cell adhesion molecule mediating neuronal migration and synaptic plasticity, showed delayed acquisition of some components of the Morris water maze (Bukalo et al., 2004). Using another neuroanatomically specific promoter, NMDA receptor inactivation only in CA3 neurons of the hippocampus produced deficits in spatial learning on the first exposure to the task, but not when the location of the reinforcer was familiar (Nakazawa et al., 2003). Richard Palmiter at the University of Washington in Seattle generated several conditional knockouts using the promoter for dopamine β-hydroxylase (Thomas and Palmiter, 1997; Weinshenker et al., 2001), the synthetic enzyme that converts dopamine to norepinephrine and epinephrine. Dopamine β-hydroxylase is expressed only in neurons of the locus coeruleus, hindbrain, sympathetic nervous system, and adrenal gland. Robert Steiner at the University of Washington employed the *Dbh* promoter to drive a transgene for galanin, generating galanin overexpressing mice with sevenfold higher levels of galanin mRNA in the locus coeruleus, where galanin normally coexists with norepinephrine (Hohmann et al., 1997; Holmes et al., 1999). Mice overexpressing galanin on the *Dbh* promoter displayed 4- to 10-fold elevations of galanin in the hippocampus and cortex (Wrenn et al., 2002), and deficits in the more difficult components of learning and memory tasks including the probe trial on the Morris water maze, social transmission of food preference, and trace cued fear conditioning, without detectable changes in relevant control behaviors (Steiner et al., 2001; Kinney et al., 2002; Wrenn et al., 2003 Figures 6.8, 6.13). David Wynick at the University of Bristol in the England generated galanin transgenic mice on a *c-Ret* gene to conditionally overexpress this neuropeptide in primary afferent dorsal root ganglion neurons, demonstrating elevated thresholds for pain sensitivity (Holmes et al., 2003f).

Conventional mutations are also "lifetime," that is, continuously present from the blastula stage of development. Conventional mutations are ideal for mouse models of human hereditary diseases, in which the mutation is also present continuously from the beginning. However, a temporally constant mutation is not an ideal research tool for investigating fundamental questions about gene function. Compensation by other genes over the course of early development may rescue the function of the mutated gene. Overcompensation may reverse the predicted phenotype. The expected phenotypic deficit due to the mutation is thus masked by compensatory biological mechanisms during development.

How big a problem are compensatory genes? Think about it: How does evolution work? Random mutations occur all the time. A mutation in a critical gene kills the organism. Survival depends on effective backup systems for essential life functions. The individual who has a redundant gene that compensates for the loss of the mutated gene is the individual who survives and reproduces. The backup system is passed on to future generations. When a new mutation causes a failure in the first backup system, a second backup system evolves. Multiple redundant genes may be extremely common, considering the long evolutionary history of mammalian physiology. A knockout mouse missing only one of the several redundant genes will therefore show a normal phenotype. One solution to this problem is to generate double- or triple-knockouts of the redundant genes, to remove the sources of compensation and reveal the phenotype.

The continuous presence of the conventional mutation is a major limitation of the present transgenic and knockout technology for studying the behavioral consequences of targeted gene mutation. There are three potential outcomes of a conventional targeted gene mutation (Figure 14.1):

1. If the gene is critical for survival, then the mutation is lethal. The homozygous mutants die as embryos or a few days after birth. The mice never reach an age at which behavioral testing can be performed.
2. The best-case scenario is that the gene is not critical for survival and has an important function. The behavioral phenotype is revealed by appropriate testing. Proof that the targeted gene mutation is responsible for the phenotype is obtained by a "rescue" experiment, described below.
3. The behavioral phenotype appears to be normal on all of the tests that the investigator designs.

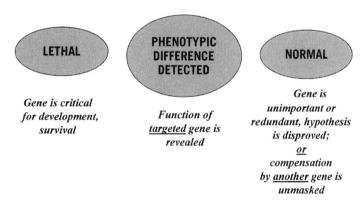

Figure 14.1 *Three potential outcomes of a gene knockout. If the gene is critical for survival, the mutation will be lethal. The embryo will not mature, the fetus will not develop, the mouse will not be born, or the neonate will not survive. If the gene is important for the function under investigation, but not critical for survival, the phenotype will be expressed and detectable. If the gene is not important for the function under investigation, the mouse will be normal. Lack of detectable abnormalities in the phenotype may reflect compensatory genes that take over the function of the mutated gene during development, for example, alternate signal transduction proteins, or redundant neurotransmitters that contribute to the regulation of feeding behaviors. While a detectable phenotypic difference is usually the desired outcome of a knockout experiment, the detected phenotype could conceivably be due to the compensatory gene. A complete absence of the predicted abnormalities can lead to interesting investigations of developmental mechanisms in multiple regulatory systems.*

The third outcome is the most difficult to interpret. Perhaps the tests are not sufficiently sensitive. Perhaps different classes of tests are needed. Perhaps the gene is truly not important for the hypothesized behavior. Or maybe compensatory events over the course of development have masked the behavioral phenotype of the mutation.

Mechanisms triggering compensatory events are a fascinating question of great interest to developmental biologists. Molecular genetics approaches such as DNA microarrays, described below, may discover the redundant genes and pathways. Assays for levels of candidate compensatory genes and gene products may reveal the site of compensation. Double- or even triple-knockouts, with mutations in the primary gene of interest and in the putative redundant gene(s), may unmask the compensatory mechanism. Another approach is pharmacological inactivation of the candidate compensatory mechanism, for example, administration of a drug that blocks a putative compensatory receptor subtype.

The ideal solution is to design a mutation that the investigator can switch on and off. The mutation is expressed only at at the time of the experiment, rather than throughout development. Many smart molecular geneticists are addressing this challenging concept (Gossen et al., 1995; Gingrich and Roder, 1998; Blau and Rossi, 1999). *Inducible* mutations include a drug-sensitive element in the DNA construct, that permits the mutation to be activated or inactivated by a drug treatment (Figure 14.2). The first inducible system was a fusion protein consisting of Cre recombinase and the estrogen-binding domain of the murine estrogen receptor, including a *lacZ* reporter gene (Zhang et al., 1996). Tamoxifen treatment induced the mutation but unfortunately also had toxic effects. Ecdysone-induced gene expression was effective in *Drosophila* and showed some promise in mice (No et al., 1996). A strategy that is presently widely used is induction or inactivation of a tetracycline regulatory system (Blau and Rossi, 1999) (Figure 14.2). Hermann Bujard and co-workers at the University of Heidelberg worked out a construct in which administration of the antibiotic drug doxycycline activates a tetracycline-controlled transactivator that induces transcription of the mutated gene in *Drosophila* (Gossen et al., 1995). In mice, a reverse tetracycline-controlled transactivator has been used that is inhibited by treatment with doxycycline (Kistner et al., 1996). Doxycycline is continuously administered in the drinking water. The mutation is turned on when the doxycycline treatment stops (Haberman et al., 1998). Several days without doxycycline treatment may be necessary for the mutation to be induced. Deactivating the mutation by reinstating doxycycline treatment may take days to weeks. Toxicity is a major problem, requiring careful studies of the minimum dose required to regulate the mutation. Other antibiotics and other inducible systems are being tested, in attempts to minimize toxicity and reduce the on–off time lag.

When the timing and toxicity issues are effectively addressed, inducible conditional mutants offer the best research tool to convey both anatomical and temporal specificity of the mutation. Mark Mayford first described behavioral phenotypes with tissue-selective promoters for conditional/inducible transgenic mice. Suppression of the transgene for an activated calcium-independent form of calcium-calmodulin-dependent kinase II (CaMKII) selectively in the forebrain was achieved with a forebrain-specific promoter combined with a tetracycline transactivator (tTA) system (Mayford et al., 1996). Hippocampal long-term potentiation was absent, and performance on the Barnes maze was impaired in transgenic mice expressing the CaMKII transgene. When the mice were treated with doxycycline, 2 mg/ml in the drinking water for 4 weeks, CaMKII was no longer expressed in the forebrain. During doxycycline treatment,

The Tet-On System

Use the Tet-on system to analyze genes you want off most of the time (such as cytotoxic genes). Cells can be maintained under normal conditions without inducer (Dox) until you are ready. Add Dox to turn expression on and analyze the results of expression.

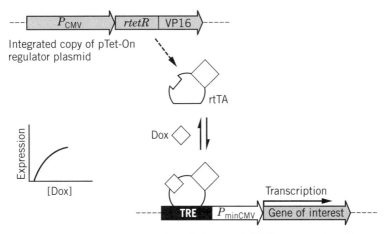

The Tet-Off System

Use the Tet-off system for analyzing genes you want expressed most of the time (essential genes, housekeeping genes, etc.). Add Tc or Dox to turn expression off and analyze the results of the absence of expression.

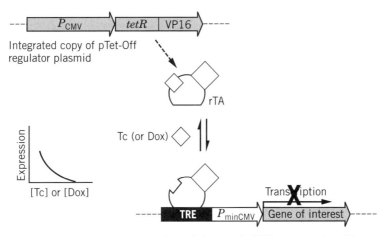

Figure 14.2 *Clontech technology for inducible mutation.* (Diagram kindly contributed by Clontech Laboratories, Inc.)

the memory deficit was reversed and the transgenics reached the criterion for choice accuracy at levels equal to wildtype controls.

Eric Nestler and colleagues at Yale University in New Haven, Connecticut succeeded in generating neuron-specific inducible transgenic mice with doxycycline-sensitive transgenes for two transcription factors, ΔFosB and CREB (Chen et al., 1998). A modified tetracycline-regulated system was used, under the control of a neuron-specific enolase promoter. Expression of the transgene appeared only in the striatum, cerebellum, and the CA1 region of the hippocampus. Now at the University of Texas Southwestern in Dallas, Nestler and co-workers have worked out many of the conditions for doxycycline administration in inducible transgenic mice that overexpress ΔFosB, providing evidence that ΔFosB increases sensitivity to drugs of abuse (Nestler et al., 2001). Marina Picciotto and co-workers at Yale generated mice with a mutation in nicotinic acetylcholine receptor subunit beta 2, on a conditional neuron specific enolase promoter and Tet-off inducibility (King et al., 2003). Ron Duman and co-workers at Yale used a conditional Tet system to confer temporally regulated expression of c-Jun in the striatum, demonstrating decreased conditioned place preference to cocaine (Peakman et al., 2003). Duman and co-workers similarly used a tetracycline transactivator and a neuron specific enolase promoter to generate a cAMP response element-binding protein (CREB) mouse that showed reduced cocaine-induced locomotion (Sakai et al., 2002). Forefront research on conditional/inducible systems is presently one of the most exciting areas of molecular genetics. Approaches described above are being refined, and new ideas are being tested (Sauer, 1998; Fishman, 1998; Morozov et al., 2003). Technical challenges remain, including the generation of the complex DNA construct, temporal delays in the effective on and off times, and side effects of the doxycycline treatment that affect behaviors. Expanded breeding requirements present a practical limitation. Crossing mice with the Tet-off mutation to mice with the conditional targeted gene mutation requires an average of 16 pups to yield 1 null mutant and 1 wildtype control. If these difficulties are resolved, and inducible conditional knockouts become feasible in many laboratories, the inducible conditional targeted mutation technology is likely to become *the* major research tool for behavioral genetics, for many areas of neuroscience, and throughout biomedical research.

VIRAL VECTOR GENE DELIVERY

Microinjection of a viral vector incorporating the mutation is an interesting new alternative to inducible, conditional transgenics to study genes mediating behaviors (Nair and Young, 2002: Lowenstein and Castro, 2002; Hsich et al., 2002; Mata et al., 2003). This technique confers a high level of both anatomical and temporal selectivity. The gene of interest is inserted into the genome of an inactivated, non-replicating adenoassociated virus (AAV), lentivirus, or similar viral vector that produces negligible toxicity to the brain and presents no health risks to the investigators (St. George, 2003; Tenenbaum et al., 2004). Quantities of the virus are grown in culture. High titers of the viral vector are microinjected into the relevant brain region. Within a few days, the gene appears in neurons adjacent to the microinjection site. An anatomical marker, usually green fluorescent protein (GFP), which is also incorporated into the viral vector, allows visualization of the location of the microinjection. In situ hybridization allows quantitation and visualization of the anatomical location of the expressed gene. In one of the earliest studies in rats, Eric Nestler and colleagues at Yale University microinjected a viral vector containing the gene for GluR1, an AMPA glutamate receptor, into the rat ventral tegmentum (Carlezon et al., 1997). These rats showed increased responses to the stimulating and rewarding properties of morphine. Expression of the newly incorporated gene may persist robustly for weeks to months. Katumi Sumikawa and co-workers at the University of California at Irvine employed an antisense DNA sequence against the NMDAR1 glutamate receptor, cloned into an adenovirus shuttle vector (Kammescheidt et al., 1997). The virus cells were microinjected into the CA1 area of the hippocampus. The hippocampal cells transcribed the antisense gene for 5 weeks. Thomas McCown at the University of North Carolina employed adeno-associated viral vector administration of the gene for galanin in the inferior colliculus, producing robust overexpression of galanin for at least six months, and conferring long-lasting resistance to seizures (Haberman et al., 2003).

Viral vector delivery of genes has been successful in studies of behavior in rats and mice, and in reversing symptoms in mouse models of human genetic diseases (Mata et al., 2003). Glial fibrillary acid transgene transfer in a herpes simplex viral vector disrupted development of astrocytes in the cerebellum, resulting in ataxic mice (Delaney et al., 1996). Lentivirus delivery of the phosphodiesterase beta gene rescued retinal photoreceptors from degeneration in retinitis pigmentosa mutant mice (Takahashi, 2004). Adenoviral vector administration of the olfactory marker protein gene rescued the olfactory deficits in null mutant mice deficient in this gene (Youngentob et al., 2004). Lentivirus delivery of the genes for the antioxidant enzymes glyoxalase 1 and glutathione reductase 1 microinjected into the cingulate cortex of 129S6/SvEvTac and C57BL/6J mice increased center time in an open field (Hovatta et al., 2005). Phenylketonuria mice lost their hyperphenylalanine and hypolocomotion phenotype within two weeks of AAV-phenylalanine hydroxylase treatment (Mochizuki et al., 2004). Adenovirus overexpression of NAC-1 protein in the rat nucleus accumbens prevented behavioral sensitization to cocaine (Mackler et al., 2000). Viral delivery of the gene for the vasopressin receptor V1a into the ventral forebrain of the socially promiscuous meadow vole produced a significant enhancement in partner preference (Lim et al., 2004). Viral vector delivery of the serotonin receptor 5-HT$_{1B}$ gene into the nucleus accumbens increased the rewarding properties of cocaine on conditioned place preference (Neumaier et al., 2002). Duman and co-workers used the complimentary approaches of AAV gene transfer and inducible transgenics to demonstrate an antidepressant-like action of mCREB (Newton et al., 2002). AAV administration of leptin reduced body weight and white fat deposits in rats, without affecting food consumption (Dhillon et al., 2001). Viral delivery of the gene for β-endorphin into the spinal cord reduced inflammatory pain, without affecting baseline nociception (Finegold et al., 1999). AAV delivery of the *parkin* gene into the substantia nigra rescued the loss of dopamine neurons and motor dysfunctions, associated with alpha-synuclein overexpression 13 weeks later, in a rat model of Parkinson's disease (Yamada et al., 2005). AAV delivery of insulin-like growth factor 1 into hindlimb muscles extended survival and delayed rotarod deficits in a mouse model of amyotrophic lateral sclerosis (Kaspar et al., 2003). Amyloid overexpressing transgenic mice immunized with AAV-Aβ 1–42 displayed improved performance on the Morris water maze probe trial (Zhang et al., 2003). AAV administration of Homer 2b, a synaptic assembly protein, rescued the deficit in ethanol place preference in Homer 2 knockout mice (Szumlinski et al., 2005). These and many other publications demonstrating beneficial outcomes of gene delivery lend credence to the potential therapeutic value of gene therapy in human diseases.

RNA SILENCING

Biology depends on checks and balances. First there is the basic machinery, next there is an on–off switch, and eventually there are a multitude of fine-tuners. Neurons, with their input dendrites and output axons, are the machinery of synaptic neurotransmission. Glutamate is the main excitatory neurotransmitter and GABA is the main inhibitory neurotransmitter in the mammalian brain. These two neurochemicals speed up and slow down neuronal firing rates, respectively. Catecholamines, neuropeptides, and many other transmitters, along with receptors, postsynaptic signal transduction mechanisms, and neurotransmitter reuptake transporters, each contribute to modulating the actions of individual neurons in the circuitry mediating behaviors. Similarly, DNA is the machinery of genes, while RNA transcribes and translates the DNA sequence into biologically functional proteins. Fine-tuning of gene expression works through regulatory elements in gene sequences that are activated or inhibited by RNA promoter and repressor sequences. Small RNA sequences bind directly to messenger RNA, to interfere with transcriptional RNAs and silence their expression (Sohail, 2004; Zamore and Haley, 2005). Double-stranded RNA (dsRNA) generates a protein-RNA complex that degrades its corresponding mRNA sequence (Zamore, 2001). Small interfering RNAs (siRNA) are shorter lengths of 21–23 nucleotides generated from the dsRNA sequence during interference (Caplen, 2004; Sohail, 2004). MicroRNAs (miRNA) are a subset of 21–23 nucleotide sequences that form hairpin loop structures (Caplen, 2004). Originally discovered in the nematode worm *Caenorhabditis elegans*, silencing of mRNA by the process of RNA interference was demonstrated in many species including mammals (Sioud, 2004). Thousands of siRNAs and miRNA have been identified. Target-specific siRNA gene silencers are commercially available from companies such as Ambion in Austin, Texas (*www.ambion.com*), and Santa Cruz Biotechnology, Inc. (*www.scbt.com*), which offer their inventory by categories including neurobiology, membrane receptors, growth factors, and homeodomain proteins. Endogenous siRNA appears to be a fundamental biological mechanism to turn genes on and off in a specific cell type during development, physiology, and behavior. Figure 14.3 illustrates the sequence by which siRNA inactivates gene expression.

Synthetic siRNAs and miRNAs have tremendous potential for therapeutic intervention (Thakker et al., 2006). These short sequences are easily manufactured, and administered by direct injection or viral vector delivery (Sioud, 2004). Interference with cancer and virus replication could be accomplished by peripheral treatment with siRNAs and miRNAs. Aberrant developmental genes could be silenced by siRNAs given prenatally. Several proof-of-principle examples have been published. Epstein-Barr virus expresses miRNAs that regulate infection of human cells (Pfeffer et al., 2004). miRNA 196 repressed the developmental homeobox gene HOXB8 in mouse embryonic cell cultures (Yekta et al., 2004). In mammal brain, silencing of developmental genes has been demonstrated in vitro (Zeringue and Constantine-Paton, 2004). RNA interference of neuroligin gene expression produced a loss of synapses in cultured rat hippocampal neurons (Chih et al., 2005). Radial migration of the developing rat neocortex was disrupted by miRNA of the doublecortin gene (Bai et al., 2003). Several successes are reported for RNA silencing of genes for neurological diseases (Forte et al., 2005). Loss of coordination and balance on the rotarod was reduced by silencing the *huntingtin* gene with an shHD in a mouse model of Huntington's disease, delaying the progression of motor loss at weeks 10 and 18 of age (Harper et al., 2005).

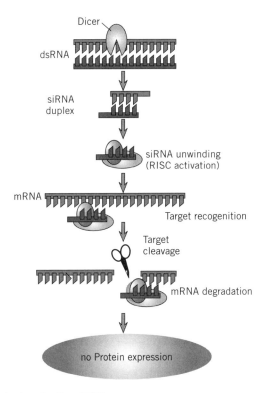

Mechanism of action of RNAi.

Double–stranded RNA is introduced into a cell and gets chopped up by the enzyme dicer to form siRNA. siRNA then binds to the RISC complex and is unwound. The anitsense RNA complexed with RISC binds to its corresponding mRNA which is then cleaved by the enzyme slicer rendering it inactive.
(http://www.bioteach.ubc.ca/MolecularBiology/AntisendeRNA/)

Figure 14.3 *Posttranscriptional gene silencing by RNA interference.* [From the National Center for Biotechnology Information, National Institutes of Health, (*http://www.ncbi.nlm.nih.gov/Class/NAWBIS/ Modules/RNA/rna13a.html*).]

Turning gene transcription off and on with small molecules offers a remarkable new approach to treating diseases. Theoretically, specificity is high. However, some weak sequence matching and cross-reactivity to nontargeted RNA sequences may limit the specificity of RNA silencing, potentially causing side effects in other neurons and organs (Jackson and Linsley, 2004). Delivery of the siRNA or miRNA to the target tissue and nowhere else presents another major challenge. As with any new technology, development of appropriate translational models and therapeutic formulations will take time.

NEW BREEDING STRATEGIES

Speed Congenics

Breeding the targeted gene mutation into a single genetic background such as C57BL/6J or a 129 inbred strain produces a *congenic line*. Congenic breeding strategies are the

TABLE 14.1   Speed Congenic Breeding of a Mutation[a]

Backcross Generation	Average % D/R Segments ± SD	% D/R Segments in "Best" Male	% Recipient Genome of "Best" Male
F_1	100 ± 0	100	50
N_2	50.00 ± 7.07	38.32	80.84
N_3	19.16 ± 4.38	11.93	94.03
N_4	5.98 ± 2.44	1.95	99.03
N_5	0.98 ± 0.98	−0	−100

Source: From Markel et al. (1997), p. 280.
[a]Microsatellite marker-assisted breeding can speed the development of a congenic line of mice bearing a mutation, from the standard 10 generations down to 5 generations. This prediction assumes that 20 male carriers of each target gene are genotyped for the mutation and that each generation is propagated from one mutation-positive male chosen for breeding.

recommended method for maintaining a line of transgenic or knockout mice (Silver, 1995; Banbury Conference, 1997; Wehner and Silva, 1997; Markel et al., 1997). Repeated backcrosses with C57BL/6J mice are described in Chapter 2. Congenic strain production time is significantly reduced as increasingly dense maps of the mouse genome become available. Marker-assisted congenic production, or *speed congenic breeding*, uses microsatellite markers to follow the inheritance of the chromosomal region or gene of interest (Markel et al., 1997; Bolivar et al., 2001a; Wong 2002). Optimal breeder mice are selected by both genotype and phenotype. The number of generations to construct the congenic strain is reduced from approximately 10 generations in traditional congenic breeding strategies to approximately 5 generations in marker-assisted congenic breeding strategies (Table 14.1). Comparison of phenotypes on two different genetic backgrounds, generated by congenic breeding of a targeted mutation into two different inbred strains, has proved successful in identifying chromosomal loci for blood pressure regulation (Jeffs et al., 2000), open field locomotion (Bolivar et al., 2001b), and obesity (Estrada-Smith et al., 2004).

The first *cloned mice* were reported in 1998 (Wakayama et al., 1998). Fertile mice were subsequently cloned by nuclear transfer, including transfer of nuclei from olfactory sensory neurons into oocytes (Eggan et al., 2004). Reported malformations of cloned mice include hypertrophic placentas and increased body weights; however, these abnormalities did not transmit to progeny in one study (Shimozawa et al., 2002). Increased body weight and delayed developmental milestones were detected, but normal activity, motor skills, learning and memory were seen in mice cloned from adult cumulus cells (Tamashiro et al., 2000), A mutation sustained in generations of cloned mice would eliminate the need for breeding strategies entirely.

TRANSGENIC AND KNOCKOUT RATS

The vast majority of rodent behavioral neuroscience research has been in rats. The larger brain size of the rat allows more accurate stereotaxic surgery, focused microinjection of drugs or viral vectors, and discrete neuroanatomical lesions. Most behavioral tasks were designed for rats. Rats excel at complex operant learning and memory tasks. The estrus cycle of the female rat follows a regular temporal sequence, ideal for detailed neuroendocrinological manipulations. For behavioral neuroscience research, it would be better to have transgenic and knockout rats, as opposed to mice.

Targeted gene mutation was historically developed in mice because immunology was the first field to embrace the technology, and the mouse is the classic animal used in immunology. The 129 mouse embryonic stem cells lines were discovered to be superlative in transmitting the mutation by colonizing large proportions of the developing embryo. There is no theoretical reason that precludes application of the same techniques to rats. Breeding requires more cage space, due to the approximately 10-fold greater size of rats as compared to mice. More effective rat embryonic stem cell lines and more extensive knowledge of the rat genome will enhance our ability to generate transgenic and knockout rats.

Successes have been reported for the generation of transgenic rats with interesting phenotypes. Transgenic rats overexpressing the gene for neuropeptide Y displayed lower levels of stress-induced anxiety-like behaviors on the elevated plus-maze (Thorsell et al., 2000). At older ages these neuropeptide Y transgenics continued to show reduced anxiety-like responses to restraint stress on the elevated plus-maze (Carvajal et al., 2004). Arginine vasopressin overexpressing transgenic rats were impaired in restoring blood pressure after hemorrhage-induced hypotension (Tachikawa et al., 2003). The transgenic hypertensive rat, *TGR(mREN2)27*, carrying an additional mouse gene for renin, a kidney hormone, shows altered blood pressure and circadian rhythms (Witte et al., 1998). Rats transgenic for the circadian gene *Per1* have been generated to test circadian entrainment cues such as meal pattern (Davidson et al., 2003). A transgenic rat incorporating an antisense sequence against angiotensinogen displayed higher anxiety-like behaviors in the elevated plus-maze, fewer light↔dark transitions, reduced social interaction, more self-grooming, and less center time in the open field (Voigt et al., 2005). The human presenilin 1 gene was recently expressed in the brain of transgenic rats (Czech et al., 1998). Presenilin transgenic rats were generally normal on the Morris water maze, showing slightly slower acquisition during training (Echevarria et al., 2004). Orthinine decarboxylase overexpressing transgenic rats displayed neuroprotection against transient focal cerebral ischemia (Lukkarinen, 1998). Superoxide dismutase transgenic rats showed a fivefold increase in copper zinc superoxide dismutase activity and neuroprotection against transient global cerebral ischemia (Chan et al., 1998). Infusion of wildtype hSOD1 into the spinal cord of superoxide dismutase transgenic rats was somewhat effective in delaying disease progression (Turner et al., 2005).

RODENT IMAGING

Both rats and mice are increasingly used in specially designed rodent magnetic resonance imaging (MRI) magnets, to investigate brain structural changes after treatments including targeted gene mutations. Functional imaging studies in rodents use evolving technologies including blood oxygen level dependent contrast pharmacological MRI (Steward et al., 2005) and manganese-enhanced MRI (Korestky and Silva, 2004; Aoki et al., 2004). Dopamine transporter knockout mice and Alzheimer's model transgenic mice overexpressing amyloid protein are examples of successful imaging of transgenic and knockout mice (Cyr et al., 2005; Falangola et al., 2005). This emerging technology is likely to provide valuable neuroanatomical correlates of behavioral phenotypes in transgenic and knockout mice and rats.

QUANTITATIVE TRAIT LOCI ANALYSIS

Before there were transgenics and knockouts, the main tools of behavioral genetics were natural mutants and selective breeding (Henderson, 1989; Silver, 1995). The ancient Egyptians recognized physical and behavioral traits in cats that led to their breeding specialized lines of cats to guard the royal temples. Gregor Mendel first worked out the mathematics of genetics by breeding selected traits such as flower color and fruit shape in the pea plant in the mid-1800s. Purebred dogs with desired behavioral traits, such as sheepherding, hunting, and sled pulling, have been maintained for centuries. Many spontaneous behavioral traits have been identified and bred in laboratory rats and mice, as described throughout this book.

When a natural trait is expressed in one strain but not in another, or in one human population but not another, matings between the two strains or populations yield pedigrees that illustrate the inheritance of the genetic mutation. An example of the Swedish family tree used to identify a gene linked to early onset Alzheimer's disease is shown in Figure 14.4. Countries that maintain careful medical records and include relatively isolated populations, such as Iceland, are ideal for tracing pedigrees and analyzing chromosomal loci linked to hereditary diseases (McInnis, 1999; Gulcher et al., 2001; Helgason et al., 2005).

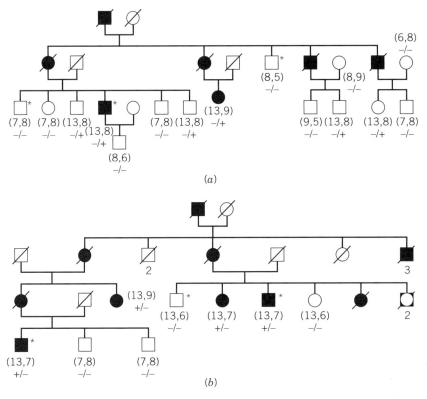

Figure 14.4 Pedigree for the Swedish mutation that revealed a polymorphism in the gene for amyloid precursor protein that is linked to Alzheimer's disease. [From Mullan et al. (1992), p. 346.]

Genetic linkage analysis derives from the observation that two genes close to each other on a chromosome are inherited together within a family (Risch, 1992). Two genes that are far apart on a chromosome are more likely to be inherited independently because of random homologous recombinations over generations. Gene one is the unidentified DNA sequence linked to the disease. Gene two is an identified microsatellite marker, with a known chromosomal location near gene one, that is visualized on a gel or DNA microchip array.

The frequency of recombination (the recombination fraction, θ) is a function of the distance between the two genes. The null hypothesis is that no linkage exists between the two genes: $\theta = \frac{1}{2}$ describes the null hypothesis, wherein the probability of recombination is as low as when the genes are on two different chromosomes. This is the same as the probability of random inheritance of the allele from each parent, $\frac{1}{2}$. When the locus for a gene is on a chromosomal segment for which many nearby genetic markers have been identified, the gene linked to the disease can be mapped to a very small region of DNA.

The big discoveries in behavioral genetics began with new mathematical tools to analyze pedigrees. The *lod score* is the logarithm of the probability L, the ratio of the probability that two traits in a pedigree are linked to the probability of no linkage (Risch, 1992). A lod score of 3, an abbreviation for $1/10^3$, represents a one in a thousand chance that an observed linkage is a chance event. When a marker locus on a chromosome shows a lod score of 3 or greater for a given pedigree of an inherited disease, the chances are very good that a gene located near that marker covaries with the disease, meaning the gene is linked to the disease.

Most behavioral traits are determined by multiple genes (Plomin et al., 1994; Churchill et al., 2004) as well as environmental factors (Goodman et al., 2004). Eric Lander and David Botstein at the Whitehead Institute, Massachusetts Institute of Technology, proposed mapping complex genetic traits with *restriction fragment length polymorphisms* (RFLPs) in 1986 (Lander and Botstein, 1986a, b). Lander and co-workers developed MAPMAKER, a computer software program for constructing genetic linkage maps (Lander et al., 1987). Calculations of linkage maps for complex traits in tomatoes, mice, and humans were pioneered by Lander and co-workers (Botstein et al., 1980; Lander and Botstein, 1989; Paterson et al., 1991; Lander and Schork, 1994; Cardon et al., 1994; Dietrich et al., 1995; Plomin, 1995). The first human neurological gene to be discovered with this mapping strategy was the gene linked to Huntington's disease (Gusella et al., 1983). Psychiatric diseases such as schizophrenia and autism show linkage to a large number of chromosomal loci, each of which may contribute to the disease and/or represent susceptibility factors (Cloninger et al., 1998; Muhle et al., 2004).

Quantitative trait loci (QTL) are chromosomal locations that influence a continuously distributed trait (Johnson et al., 1992; Neumann, 1992; Belknap et al., 1996, Wehner et al. 2001; Bolivar et al., 2001a; Phillips et al. 2002; Abiola et al., 2003; Figure 14.5). QTL analysis is designed to identify the chromosomal loci of genes that contribute to a multigenically determined trait. Advances in high-density physical mapping of the mouse genome, and identification of polymorphisms in microsatellite sequences dispersed throughout the mouse genome, have greatly increased the sensitivity of the QTL technique. Statistical programs such MapMaker/QTL and RI Manager are advancing the field of quantitative trait loci analysis of mouse behavioral genetics (Plomin et al., 1991; Manly and Elliott, 1991; Johnson et al., 1992; Neumann, 1992; Darvasi, 1998). Mapmaker software and instructions are freely available online from

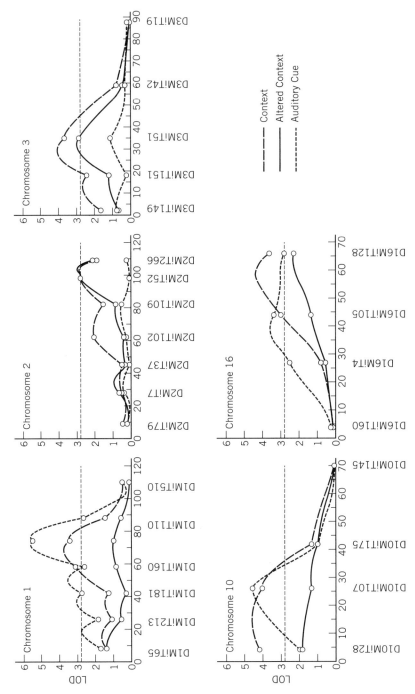

Figure 14.5 *Quantitative trait loci analysis for cued and contextual conditioning. propagated from one mutation-positive male chosen for breeding. Lod plot of contextual learning (dashed line), freezing to the altered context (solid line), and freezing to the auditory cue (dotted line). The lod score is plotted on the vertical axis. The threshold of 2.8 is indicated by the horizontal dotted line. Genetic distances along the chromosome (in centimorgans) are plotted on the horizontal axis with markers used for the assessments. Maximum likelihood positions are indicated. [From Wehner et al. (1997), p. 332.]*

the Broad Institute, affiliated with the Whitehead/Massachusetts Institute of Technology Center for Genome Research (*http://www.broad.mit.edu/tools/software.html*). After coarse mapping to a chromosomal region of 10 to 30 centimorgans (cM), additional breeding and finer mapping (1–5 cM) defines the critical region (Abiola et al., 2003). Candidate genes within the critical region may be revealed, using available databases such as WebQTL (http://webqtl.org/links.html), GenBank (*http://www.ncbi.nlm.nih.gov/Genbank*), Gene Ontology (http://www.geneontology.org), and the Complex Traits Consortium (Abiola et al., 2003; Churchill et al., 2004; *http://www.complextrait.org*). Confirmation of the identity of the candidate gene is then accomplished by various methods, including the generation and phenotyping of a knockout or knock-in mouse with a mutation in the candidate gene (Abiola et al., 2003), and DNA microarray comparisons of the genomes of the two strains (Hoffman and Tabakoff, 2005), as described below. Creative applications of new genetic technologies have further advanced the power of QTLs. An entire chromosome can be substituted from a donor strain such as A/J into the genetic background of a host strain such as C57BL/6J (Belknap, 2003; Singer et al., 2004; Flint et al., 2005).

QTL approaches have been extensively applied to the study of genes linked to behavioral traits in mice. The QTL approach begins with the identification of two inbred strains of mice that differ on the behavioral trait of interest. For example, one strain is highly aggressive while the other is submissive, or one strain shows high levels of anxiety-like behaviors on the elevated plus-maze or on light↔dark exploration while the other shows low levels, or one strain is susceptible to audiogenic seizures while the other is not, or the two strains develop sensorimotor reflexes at very different ages (Neumann and Collins, 1991; Crabbe et al., 1994b; Mathis et al., 1995; Flint et al., 1995; Le Roy et al., 1999). Ideally the two strains are extremely different on the trait, to maximize the power of the analysis.

The two strains are mated, and the F_2 offspring are tested for the trait, for example, elevated plus-maze open arm entries $\div$ total arm entries. A very large number of F_2 mice are tested, both males and females, usually 500 to 1000 individuals. DNA samples from all of the F_2 mice are then genotyped. Generally, the mice at the top and bottom 5% of the distribution will define the relevant population, so it is sometimes sufficient to genotype only F_2 mice that show behavior traits at the each of the extremes of the F_2 distribution. Reliable linkage results are obtained when the original strains were sufficiently different on the behavioral trait, sufficient number of F_2 mice are phenotyped and genotyped, and the microsatellite markers are sufficiently closely spaced.

A second QTL approach to mouse behavioral genetics is QTL analysis of *recombinant inbred strains of mice* (RIS). Rather than genotyping hundreds of F_2 offspring, some behavioral geneticists use established substrains, created from an original cross between two inbred strains of mice known to differ on a behavioral trait (Bailey, 1971; Nesbitt, 1992; Blizard, 1992; Belknap et al., 1996; Belknap, 1998; Wehner et al., 2001). The "high" and "low" strains were mated, and specific lines of the offspring were chosen and inbred. After 20 generations, each line became a genetically inbred substrain. RI sets are valuable if they contain a large number of substrains. The Jackson Laboratory Web site presently lists availability of 42 recombinant inbred strains from an original BXD cross of C57BL/6J × DBA/2, and 30 recombinant inbred strains from original AXB and BXA crosses of A/J × C57BL/6J (Nesbitt, 1992; Higgins and Paigen, 1997; Sampson et al., 1998).

The advantage of the RIS approach is the continuous availability of mice with minimal variability in the behavioral trait and virtually identical genotypes. The disadvantage is the smaller number of genotypically different individuals for linkage analysis. RIS are like human populations on a widely spaced chain of islands, each inbreeding in isolation over many generations. F_2s are like the grandchildren of hemophiliac Hapsburgs who married non-hemophiliacs, with large numbers of descendents spread throughout Europe. Each approach is useful in its own way for discovering genes linked to unusual traits.

QTL analyses of behavioral traits using both the F_2 and QTL approaches have identified many interesting candidate genes in mice. Obesity QTLs on mouse chromosomes 2 and 17 affect fat depots in adipose tissues of male mice (Taylor and Phillips, 1997). Genes associated with patterns of cerebellar foliation were worked out by QTL analysis (Wahlsten and Andison, 1991). Numbers of neurons in mouse brain are linked to a QTL site on chromosome 11 (Williams et al., 1998). Susceptibility to audiogenic seizures and drug-induced seizures shows several chromosomal linkages (Neumann and Collins, 1991; Martin et al., 1992). Locomotor responses to cocaine are linked to loci on 10 difference chromosomes (Phillips et al., 1998; Kelly et al., 2003). Open field activity and elevated plus-maze performance are linked to several chromosomal loci (Flint et al., 1995; Henderson et al., 2004; Singer et al., 2005). Light↔dark anxiety-related exploratory behavior and response to diazepam are linked to several chromosomal loci (Mathis et al., 1995). Genes associated with responses to restraint stress were revealed with QTL analysis (Tarricone et al., 1995). Ethanol preference drinking is linked to a site on chromosome 3 (Belnap et al., 1997). Behavioral responses to ethanol, sensitivity and tolerance to ethanol-induced hypothermia, and changes in corticosterone levels in response to ethanol are linked to multiple chromosomal loci (Crabbe et al., 1983, 1994b; Roberts et al., 1995; Crabbe, 2002; Hitzemann et al., 2004). Hyperlocomotion in response to cocaine, methamphetamine, and phencyclidine is associated with several loci (Alexander et al., 1996; Grisel et al., 1997; Phillips et al., 1998; Jones et al., 1999; Janowsky et al., 2001; Gil and Boyle, 2003). Morphine preference drinking is linked to loci on chromosomes 1, 6, and 10 (Berrettini et al., 1994; Ferraro et al., 2005). Saccharin intake is linked to a locus on chromosome 12 (Belknap et al., 1992). Diet selection for macronutrients identified fat preference genes on chromosomes 8, 18, and X (Smith-Richards et al., 2002), while a flavor preference locus was identified on chromosome 8 (Bolivar and Flaherty, 2004). Natural variation in retinal ganglion neurons is linked to a locus on chromosome 11 (Williams et al., 1998). Cued and contextual fear conditioning is linked to several chromosomal loci (Wehner et al., 1997; Caldarone et al., 1997; Radcliffe et al., 2000; Figure 14.5).

The ultimate goal of linkage analyses is the identification of the gene responsible for the trait. After the first QTL analysis is completed, each of the identified linkage sites is more finely mapped, using additional markers at the identified loci, to more precisely define the chromosomal site. Additional crosses and behavioral phenotyping are often conducted to enhance mapping. F_2 individuals that show the extremes of the phenotype are used to successively breed F_3, F_4, and F_5 generations for further genotyping. Various cloning and expression strategies, along with a large dose of good luck, allows the gene to be discovered, and its product identified.

QTL and targeted gene mutation are complementary techniques. Genes identified by QTL approaches provide new hypotheses about the genetic basis for a given behavioral trait. To test the hypotheses that a gene discovered by QTL is linked to a behavior, the gene is mutated and the phenotype expressed by the knockout mouse is evaluated. If the linked gene is responsible for the behavioral trait, then the knockout mouse should show the aberrant behavioral trait. This combination of "forward" genetics, from phenotype to gene, and "reverse" genetics, from gene to phenotype, provides an extremely powerful set of research tools (Rinchik et al., 1991; Takahashi et al., 1994; Tecott and Wehner, 2001; Bucan and Abel, 2002).

DNA MICROARRAYS FOR GENE EXPRESSION AND SINGLE NUCLEOTIDE POLYMORPHISMS

Each cell of an organism contains the complete genome, but only a subset of genes is expressed and active in a given cell type at a given moment. *Differential expression* defines the function of the cell, tissue, or organ. Increasingly sophisticated technologies allow us to identify the set of genes expressed in particular neurons and anatomical regions of the mouse brain. DNA microarrays are chips or glass slides printed with tens of thousands of genes. Arrays for almost the entire human genome are under development for use in research and clinical settings (Geschwind, 2003; Chaudhuri, 2005). Serial analysis of gene expression (SAGE) detects low abundance transcripts (Boheler and Stern., 2003). Protein microarrays offer the next logical step for detecting the outcome of aberrant gene expression (Leuking et al., 2005; Ho et al., 2005). Arrays for single nucleotide polymorphisms (SNPs), representing point mutations in genes that vary in the normal population, are another version of DNA microarrays that identify allelic differences between individuals that may predict disease susceptibility. Over a million human SNPs from 269 DNA samples representing 4 distinct human populations are accessible through the international HapMap project (Altshuler et al., 2005; *www.hapmap.org*). A set of 1638 SNP markers have been analyzed across 102 mouse strains, including strains used as background for targeted gene mutations and wild-derived strains (Petkov et al., 2004).

DNA microarrays are an increasingly useful tool for mouse behavioral geneticists. Microarrays compare gene expression between inbred strains of mice, knockouts versus wildtype, or vehicle versus drug treatment, in whole brain or in selected brain regions (Mineur et al., 2004; Chesler et al., 2005; Barr and Gao, 2005; Aarnio et al., 2005). Laser capture dissection and single cell amplification increase the homogeneity of cells from small brain regions (Hinkle and Eberwine, 2003; Yuferov et al., 2005). Successes in expression profiling have identified important genes for behavioral characteristics. Microarray analyses revealed genes that are overexpressed or underexpressed in mouse brain regions mediating aggression, cage restraint stress, anxiety, seizures, circadian rhythms, cognitive decline in aged mice and in a mouse model of Huntington's disease, responses to cocaine, amphetamine, ethanol consumption, conditioned taste aversion, and memory consolidation during fear conditioning (Sandburg et al., 2000; McClung and Nestler, 2003; Duffield, 2003; Feldker et al., 2003; Verbitsky et al., 2004; Funada et al., 2004; Ponomarev et al., 2004; Hitzemann et al., 2004; Levenson et al., 2004; Murata et al., 2005; Morten et al., 2005a; Hoffman and Tabakoff, 2005; Hovatta et al., 2005). Downstream changes in gene expression were identified by DNA microarrays in corticotropin releasing factor transgenic mice (Peeters et al., 2004), and mutant mouse models of Fragile X mental retardation (Brown et al., 2001).

In humans, a SNP in the neuropeptide Y gene was associated with alcohol addiction in a human Nordic population of alcoholics (Mottagui-Tabar et al., 2005). Serotonin transporter polymorphisms are associated with depression and alcoholism (Heinz et al., 2000; Serretti and Artioli, 2004; Malholtra et al., 2004; Taylor et al., 2005). Large-scale array profiling is ongoing to discover genes for schizophrenia and Alzheimer's disease (Pongrac et al., 2002; Katsel et al., 2005; Ho et al., 2005). This small sampling of the emerging literature highlights the wealth of questions that can be addressed with DNA expression profiling and SNP associations (Mirnics and Pevsner, 2004). Technical issues being worked out include replicability of findings across experiments and laboratories, and bioinformatics methods for understanding common functions in clusters of co-expressing genes (Shannon et al., 2003; Reimers, 2005). Very large arrays including the full mouse genome, multiple SNPs for a variety of genes, and human disease genes, make this a technology with widespread applications to basic research questions in mouse behavioral genetics, and for diagnostic screening to detect disease-related genes in clinical settings (Petkov et al., 2004; Aarnio et al., 2005).

CHEMICAL-INDUCED AND RADIATION-INDUCED RANDOM MUTAGENESIS

Forward genetics encompasses DNA microarray, QTL, and mutagenesis approaches, conceptually opposite to the targeted gene mutation *reverse genetics* strategy (Rinchik et al., 1991; Takahashi et al., 1994; Kasarskis et al., 1998; Hrabe de Angelis and Balling, 1998; Tecott and Wehner, 2001; Bucan and Abel, 2002). Large-scale random mutations of many loci are produced by X-ray irradiation and by mutagenic chemicals. These are "shotgun" approaches, perhaps more analogous to machine-guns spraying bullets randomly throughout the genome, as opposed to a single directed bullet in a targeted gene mutation. For chemical mutagenesis, one male mouse is treated with the mitogen. The treated male is then mated with several females. Germ-line transmission in the sperm of the treated male will pass along the random chromosomal mutation(s) to the offspring. Each offspring may receive only one mutation, or none. Large numbers of offspring of the irradiated or chemically treated mice are then screened for particular phenotypes of interest.

Oak Ridge National Laboratory in Knoxville, Tennessee, and the Medical Research Council Radiobiological Research Unit in Edinburgh, Scotland and in Harwell, England, initiated radiation-induced mutagenesis programs in the laboratory mouse after World War II, to begin to understand the effects of radiation on genes in humans (Green and Roderick, 1966; Silver, 1995). Irradiation of mice with X rays induces mutations at a rate of 13 to 50×10^{-5} per locus, usually in the form of large deletions or complex rearrangements of large segments of the chromosome (Rinchik et al., 1991; Russell and Rinchik, 1993; Silver, 1995; Takahashi et al., 1998). To produce smaller DNA segment mutations, preferably point mutations, a chemical approach was initiated. Bill and Lee Russell, Don Carpenter, and Pat Hunsicker at Oak Ridge National Laboratory tested over 50 different chemical mutagens and identified several that were optimal for induction of a large number of single-point mutations in mouse sperm cells (Rinchik et al., 1991; Russell et al., 1992). The alkylating agents N-ethyl-N-nitrosourea (ENU), chlorambucil (CHL), and melphalan (Russell et al., 1992; Rinchik et al., 1993; Brown, 1998) were effective chemical mutagens. ENU is highly mutagenic, inducing mutations at a frequency of 150×10^{-5} per locus (Popp et al., 1983; Silver, 1995).

Because ENU produces point mutations rather than large chromosomal deletions, it has become the mutagen of choice for behavioral genetics (Russell et al., 1990;

Takahashi et al., 1994; Schimenti and Bucan, 1998, Tarantino et al., 2000; Keays and Nolan, 2003). A rapid, simple behavioral task is the first screen to detect the desired behavioral phenotype in one of the mutant offspring. That mouse is bred to generate a line of mice expressing the mutation, for further study and to subsequently clone the gene. ENU mutagenesis was successfully applied to the phenotype of abnormal circadian activity rhythm in mice by Joe Takahashi and co-workers at Northwestern University. (Vitaterna et al., 1994; Takahashi et al., 1994; Figure 14.6). Mutant mice from the F_1 of an ENU-treated male with aberrant circadian patterns of running wheel activity were detected. One individual with highly disorganized circadian activity was used to create a line of mice for further breeding and positional cloning, using the circadian activity phenotype. This approach led to the discovery of *clock*, the first mammalian gene found to regulate a circadian rhythm (King et al., 1997b; Figure 14.6). A *Clock*, polymorphism in humans is associated with the morningness–eveningness preference in normal adults (Katzenberg et al., 1998; Archer et al., 2003; Mishima et al., 2005).

Large-scale chemical mutagenesis projects were initiated in the early 2000s in several countries, to discover genes underlying a broad range of phenotypes (Godinho and Nolan, 2005; Figure 14.7). The National Institutes of Health in the United States funded three Mutagenesis Centers, focused on genes underlying the development and functions of the nervous system (Williams et al., 2003; O'Brien and Frankel, 2003; Goldowitz et al., 2004; Clark et al., 2004; Table 14.2). The Tennessee Mouse Genome Consortium (*www.tnmouse.org*) includes screens for aggression, aging, auditory, drug abuse, ethanol, epilepsy, eye, learning and memory, nociception, and social behaviors. The Neuromutagenesis Facility at The Jackson Laboratory (*http://nmf.jax.org*) includes phenotyping for acoustic startle, auditory brainstem response, development, eye and vision, gait, gestation, and seizure threshold. The Neurogenomics Project at Northwestern University (*http://www.genome.northwestern.edu/neuro*) tests for phenotypes in the domains of neuroendocrine/stress, learning and memory, psychostimulant response, circadian rhythms, and vision. Discoveries, archived at Web site *www.neuromice.org*, include genes affecting neuromuscular, retinal, nociceptive, cocaine-sensitivity, visual, and circadian phenotypes. In cases where germ-line transmission was established, lines of mice are available to the research community.

The EUMORPHIA project in Europe (*www.eumorphia.org*) has a broader mandate, including the development of standardized phenotyping methods for clinical chemistry, cardiovascular, renal, sensory, skeletal muscle, cancer, bone, behavior and cognition domains. Standard operating procedures are freely available in the EMPReSS database. For example, the EMPReSS description of the mouse tail suspension task (*http://empress.har.mrc.ac.uk/EMPReSS/servlet/EMPReSS.Frameset*) includes the lab environment, recommended equipment, software settings, data analysis, and references. Training courses in phenotyping assays are designed to spread knowledge of standardized mouse testing among the next generation of mouse geneticists. Mouse models of human diseases will be shared among the research community. Representative publications from large-scale mouse mutagenesis discoveries include a neural tube defect gene (Curtin et al., 2003), ataxia gene *af4* (Isaacs et al., 2003), a limb paralysis mutation on chromosome 15 (Buchner et al., 2005), *szt1* seizure susceptibility gene Otto et al., 2004), and 61 putative mutations with phenotypic defects in electroretinograms (Pinto et al., 2004).

High-throughput screening incorporates a high risk for false negatives (Crabbe and Morris, 2004; Goldowitz et al., 2004). It is easy to miss a behavioral phenotype if only

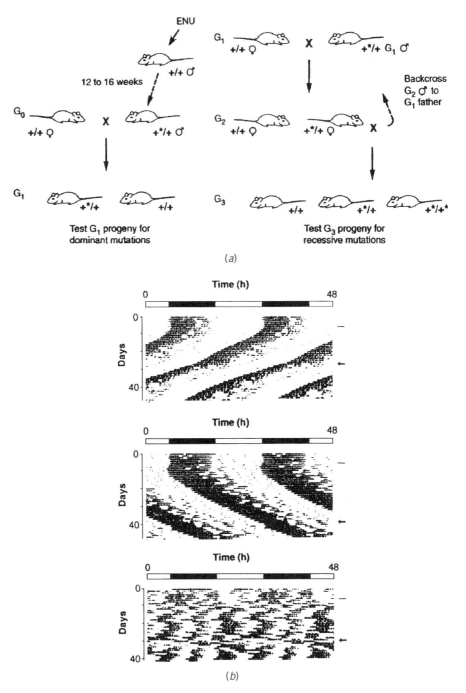

Figure 14.6 *Chemical mutagenesis induces random mutations throughout the genome. (a) Large numbers of mutations are generated by administration of the chemical mutagen to a male mouse. A specific phenotype of interest is assayed in each offspring. Individuals with phenotypes of interest are then bred, to maintain the mutation in a line of mice. (b) clock, the gene for circadian wheel running activity, was identified from an ENU strategy combined with actagram phenotyping. [From Takahashi (1994), p. 1727.]*

TABLE 14.2 ENU Phenotypes[a]

Center Name, Location, URL	Genomic Approach	Phenotypic Screens	Center Director
	Neurogenomics Project		
Northwestern University Center for Functional Genomics Evanston, IL http://www.genome.northwestern edu/neuro	Dominant G1 screen and a recessive G3 screen are utilized. Phenotypic screens focus on five primary domains: learning and memory, behavioral responses to stress, responses to psychostimulants, circadian rhythmicity, and vision.	Overt Hearing, Elevated plus-maze, Open field behavior, Fear conditioning, Cocaine response, Fundus electroretinogram, Circadian rhythms	Joseph S. Takahashi
	Neuroscience Mutagenesis Facility		
Jackson Laboratory Bar Harbor, ME http://nmf.jax.org/index.html	Three-generation backcross breeding scheme to recover dominant, semidominant, and recessive mutations. Phenotypic screens focus on identifying mutations affecting: motor function, seizure threshold, hearing, vision, and neurodevelopment.	Seizure threshold, Auditory, Vision, and Eye, Overt gait, Home cage activity	Wayne N. Frankel
	Neuromutagenesis Project		
Tennessee Mouse Genome Consortium (TMGC) Memphis, TN http://tnmouse.org/neuromutagenesis	Regional mutagenesis, covering regions on chromosomes 10, 14, 15, 19, and X. Phenotypic screens include: motor and sensory function, learning and memory, neurohistology, aging, alcohol response, abused drug response, visual function, epilepsy, and social behavior.	Aging, Hearing, Cocaine response, Ethanol response, Seizure threshold, Eye, General behavior, Neurohistology, Social behavior	Daniel Goldowitz

[a]Identified by *Neuromice.org*, an NIH-sponsored mutagenesis and phenotyping consortium consisting of The Tennessee Mouse Genome Consortium, The Jackson Laboratory, and Northwestern University that provides novel mutant lines of mice to the neuroscience community.
Source: Kindly contributed by Dr. Jeana Yates, Northwestern University School of Medicine, Chicago

Observational Assessment
A shortcut to neurological and behavioral gene function?

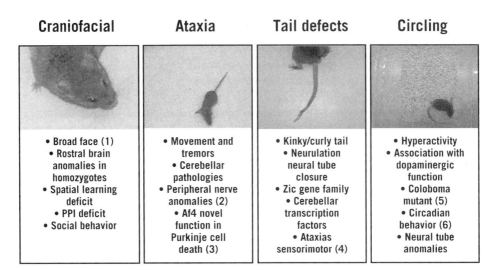

Craniofacial	Ataxia	Tail defects	Circling
• Broad face (1) • Rostral brain anomalies in homozygotes • Spatial learning deficit • PPI deficit • Social behavior	• Movement and tremors • Cerebellar pathologies • Peripheral nerve anomalies (2) • Af4 novel function in Purkinje cell death (3)	• Kinky/curly tail • Neurulation neural tube closure • Zic gene family • Cerebellar transcription factors • Ataxias sensorimotor (4)	• Hyperactivity • Association with dopaminergic function • Coloboma mutant (5) • Circadian behavior (6) • Neural tube anomalies

Figure 14.7 *ENU phenotypes identified by the European mutagenesis and phenotyping consortium, EUMORPHIA. [Kindly contributed by Patrick Nolan, Mammalian Genetics Unit, Harwell, England.]*

a few quick tasks are run in mutants on one genetic background at one age. The value of the approach depends on the choice of assays. Focusing on one simple, robust, replicable assay such as wheel running to detect circadian rhythm abnormalities, or an electroretinogram to detect visual defects, has been a successful approach (Turek and Takahashi, 1994; Pinto et al., 2004; Godinho and Nolan, 2006; Vitaterna et al., 2006). After a phenotypic abnormality has been discovered, three further avenues must be pursued to bring the discovery to fruition. The mutation must be germ line, for transmission to the offspring, which occurs in only a small fraction of ENU mutants. Follow-up testing of several subsequent generations must replicate the phenotype. More intensive behavioral analyses, such as described in Chapter 13, will reveal the true nature of the abnormalities observed in the superficial screen. Finally, fine mapping and cloning is conducted to identify the gene mutation that caused the phenotype. The large investment required for the mutagenesis strategy, coupled with the relatively small number of confirmed mutant phenotypes with germline transmission and identified genes, has raised uncertainties about future funding of large-scale mutagenesis centers in the United States. The Mammalian Genetics Unit in Harwell, England (*http://www.mut.har.mrc.ac.uk*) continues to discover large numbers of interesting behavioral phenotypes with germ-line transmission, including learning and memory and motor dysfunctions (Figure 14.7).

ES LINES FOR EVERY MOUSE GENE

Several large international organizations are devoted to the development and distribution of mutant lines of mice and mutant lines of embryonic stem cells (Austin et al., 2004; Auwerx et al., 2004; Goldowitz et al., 2004; Table 14.2). Neuromice (*www.neuromice.org*) develops and distributes new mouse models for neuroscience research. The International Gene Trap Consortium (*www.igtc.ca, www.genetrap.org*) seeks to generate a full set of embryonic stem cell lines with a null mutation in almost every gene of the mouse genome. Established as part of an International Mouse Mutagenesis Consortium, this collaborative project includes investigators from Australia, Canada, France, South Korea, Japan, Germany, United Kingdom, and United States, from universities, research foundations, and companies (Austin et al., 2004; Auwerx et al., 2004). Gene trapping generates random insertional mutations. A reporter gene or marker is activated when inserted into an endogenous gene. The gene trap vector is introduced into an embryonic stem cell line by electroporation. Once made, the cell line can be maintained in a repository. Availability is advertised through a database within the NCBI Genome Survey Sequences (*www.ncbi.nlm.nih.gov/dbGSS/index.html*), a division of GenBank, and can be searched by BLAST. Investigators wishing to generate mice with that mutation can theoretically request the embryonic cell line from the repository, and use the cells to generate chimeras and a line of knockout mice with a null mutation in their gene of interest. Novel developmental genes have been identified by retinoic acid-inducible gene traps (Chen et al., 2004). Bay Genomics in San Francisco (*http://baygenomics.ucsf.edu/*) has a large searchable database available. Lexicon Genetics and the Texas Institute for Genomic Medicine at Texas A&M University are in the process of compiling a comprehensive mouse embryonic stem cell library containing 350,000 cell lines (*http://www.lexicon-genetics.com/index.php*).

GENE THERAPY

"Genetic Medicine—When Will It Come to the Drugstore?" asked a provocative news headline in *Science* magazine in 1998 (Wittung-Stafshede, 1998). As many genes for many human genetic diseases were identified and cloned, the concept of *gene replacement therapy* moved closer to reality. More than 300 clinical trials of genetic therapies were in progress or approved in 1998, mostly against cancers, cystic fibrosis, and AIDS (Anderson, 1998; Adams and Emerson, 1998; Table 14.3). "Ready or Not? Human ES Cells Head Toward the Clinic" reflects the news in 2005 realities (Vogel, 2005). Progress in realizing the promise of this powerful therapeutic approach continues slowly (Guttmacher and Collins, 2002). Gene transfer efficiency remains low in patients, and long-term expression of the gene product in the targeted tissue remains problematic (Vile, 1996; Anderson, 1998; Gottesman, 2003; Trent and Alexander, 2004). Safety remains the primary concern. Sadly, 18-year-old Jesse Gelsinger died in a gene therapy trial at the University of Pennsylvania in 1999. In 2002, children in a gene therapy trial in France for severe combined immunodeficient disease developed leukemia. Clinical trials with gene delivery are being approached with great caution. Improved technology to ensure safety is essential.

The major technical issue is delivering the therapeutic gene to the correct location. The goal is to get the therapeutic gene specifically to the tissue that is affected by the disease. The therapeutic gene may be either an authentic full-length DNA sequence to replace the incorrect DNA of the disease gene, or a short sequence of antisense DNA, siRNA, or miRNA that binds to the incorrect sequence to block production of the incorrect disease protein (Friedmann, 1997; Wittung-Stafshede, 1998; Mata et al.,

TABLE 14.3   By 1998, over 300 Clinical Protocols Involved Testing of Gene Therapy Protocols in over 3000 Patients[a]

	Number	Percentage of total
Types of gene therapy clinical protocols[a]		
Type		
Therapy	200	(86%)
Marker	30	(13%)
Nontherapeutic[b]	2	(1%)
Total	232	(100%)
Disease targets for therapeutic gene therapy clinical protocols		
Target		
Cancer	138	(69%)
Genetic diseases	33	(16.5%)
CF	16	
Other[c]	17	
AIDS	23	(11.5%)
Other[d]	6	(3%)
Total	200	(100%)

Source: From Anderson (1998), p. 28.
Note: Cancer, cystic fibrosis (CF), and immune deficiency diseases such as AIDS were the initial target diseases. The five "other" indicated are peripheral artery diseases, rheumatoid arthritis, arterial restenosis, cubital tunnel syndrome, and coronary artery disease.
[a]Roughly 60% of all protocols use retroviral vectors, 20% use non-viral delivery systems, 10% use adenoviral vectors and the remainder use other viral vectors.
[b]A "nontherapeutic" protocol means a nontherapeutic portion of a non-gene-therapy clinical protocol.
[c]These 17 include 12 other monogenic diseases.
[d]The five "other" are peripheral artery disease, rheumatoid arthritis, arterial restenosis, cubital tunnel syndrome and coronary artery disease (2).

2003; Tinsley et al., 2004). The gene may be delivered in cells extracted from the patient or in a gene transfer vector (Figure 14.8). The gene transfer vector contains the therapeutic gene construct, which is inserted into the genome of the transfer vector by homologous recombination. Choice of vector appears to be pivotal to the success of the technique. Viruses are proving useful as vectors. Viruses are very small packages that self-replicate and are easily cultured. The right strain of virus will contain a constituent that allows entry only into a specific type of tissue. Viral vectors under consideration for delivering genes into the nervous system include lentivirus, herpes simplex, adenovirus, and adeno-associated virus (Tinsley et al., 2004; Tomanin and Scarpa, 200; Verma and Weitzman, 2004).

Gene delivery into the human nervous system presents a challenging set of technical issues (Ridet et al., 1999; Lowenstein et al., 1999; Suhr and Gage, 1999). Transport must cross the blood–brain barrier, a fine capillary network that excludes most large molecules and chemical structures of certain compositions and ionic charges. Because neurons are already fully differentiated and do not divide, retroviruses do not easily enter neurons. Gene therapy for the nervous system has focused on adenovirus delivery approaches. Adenovirus vectors are being tested for delivery of therapeutic genes into brain tumors, and for neurodegenerative diseases such as Parkinson's disease (Barkats et al., 1998). *Herpes simplex viruses* appear to be useful for delivery of

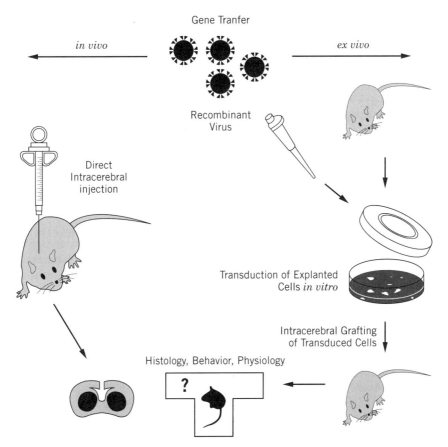

Figure 14.8 *Gene transfer strategies. Ex vivo approaches graft cells that have previously been infected with viral vectors into the target tissue. For example, tumor cells may be removed from the patient through biopsy, infected with a viral vector containing a cancer-killing gene, then the transfected tumor cells are injected back into the tumor to kill the remaining tumor cells. In vivo approaches inject the recombinant virus into the target tissue. For example, the viral vector is injected directly into the brain tumor mass.* [From Blömer et al. (1996), p. 1398.]

therapeutic genes to the nervous system (Ho and Sapolsky, 1997; Blömer et al., 1996). When injected into muscle, recombinant adenovirus is retrogradely transported into the axons of motor neurons, suggesting an application in neuromuscular diseases such as amyotrophic lateral sclerosis and vascular smooth muscle cell proliferation after arterial injury (Chang and Leiden, 1996; Barkats et al., 1998). Human adult astrocytes transfected with a therapeutic gene under negative control of a tetracycline-based regulatory system offer promise (Ridet et al., 1999). Adeno-associated virus is a ubiquitous parvovirus that was discovered as a contaminant in adenovirus stock, does not cause a known disease in humans, and requires a helper adenovirus for replication (Phillips et al., 1997). For therapeutic gene delivery, all viral coding sequences are removed to prevent replication, and replaced by the gene delivery cassette, conferring a high level of safety and stability for up to a year in vivo (Cooper, 1996; Haberman et al., 2003; Burger et al., 2005). Route of administration is a major question, for example, intrathecal versus intraparenchymal for administration into the spinal cord (Mannes et al., 1998), and intraventricular versus directly into a brain region (Brenner 1996; Carlezon et al., 1997; Haberman et al., 1998; Betz et al., 1998; Burger et al., 2005). Optimal rates and volumes for intracerebral infusion are being tested in rodent models, and methods for monitoring the effectiveness are being evaluated.

Therapeutic gene delivery in humans presents safety and efficacy challenges above the requirements for mouse research. Retroviruses (Figure 14.9.) include mouse leukemia virus and human immunodeficiency virus. Retroviruses splice a copy of their own genome into the chromosome of the cells they enter. A permanent copy of the transferred gene is thus integrated into the genome of the human cell where it will theoretically function and continue to replicate. Disadvantages of retroviruses include their relative nonspecificity for a single type of target cell and randomness of chromosomal location for insertion of the transferred gene.

Adenoviruses include the common cold virus and herpes simplex virus (Russell and Hirata, 1998). Adenoviruses infect human cells readily and replicate quickly, producing large quantities of the transferred gene in vivo (Hajjar et al., 1998; Ye et al., 1999) The adenovirus vector delivers the therapeutic gene into the nucleus but does not insert the therapeutic gene into the chromosome (Figure 14.10). Therefore the therapeutic gene does not replicate beyond the time course of survival of the adenovirus in the human cells. The therapeutic window for the inserted gene is relatively short. This temporary fix may be sufficient for some diseases, for example, to mount a large, rapid immune response to a pathogen.

Adeno-associated viral vectors, discussed above, show promise in mouse and rat models of Parkinson's disease, Huntington's disease, seizures, spinal cord injury, and Alzheimer's disease, among others (Kirik and Bjorklund, 2003; Haberman et al., 2003; Mata et al., 2003; Tinsley and Eriksson, 2004; Richichi et al., 2004; Ruitenberg et al., 2004; Maingay et al., 2005; Burger et al., 2005). A member of the parvovirus family, adeno-associated viruses enter the nucleus and can integrate into specific chromosomal sites, and have the advantage of transducing both dividing cells and nondividing neurons (Buning et al., 2004; McCarty et al., 2004). Methods to scale up production of adeno-associated viral vectors containing therapeutic genes, to manufacture sufficient quantities for clinical trials, are in progress (Farson et al., 2004).

Transgenic and knockout mouse models of human genetic diseases provide elegant systems to test gene therapies for efficacy (Shuldiner, 1996; Jucker and Ingram, 1997). Examples of good models are described throughout this book, particularly at the end of each chapter on behavioral domains (Chapters 3–11). Large compendiums of available mouse models of human diseases have been assembled (Bedell et al., 1997; Bolivar et al., 2000b; *www.jax.org*). The ideal transgenic models insert the human disease gene into the mouse genome. The ideal knockout models inactivate the gene in the mouse that is homologous to the gene mutated in the human disease. The success of these models rests on using a robust phenotype to represent a symptom of the human disease, to serve as the marker for measuring the efficacy of the gene therapy treatment. Pathology, neuroanatomy, neurochemistry, neurophysiology, and behavior are some of the tools for measuring phenotypic responses to a proposed gene therapy. In all cases, the value of the mutant model in evaluating therapeutics rests on the specificity, robustness, and replicability of the phenotype, as shown in the frontpiece illustration at the beginning of this chapter.

Successes with gene transfer have been reported in several diseases and rodent model systems. Restoration of tumor suppressor gene expression was seen after retroviral

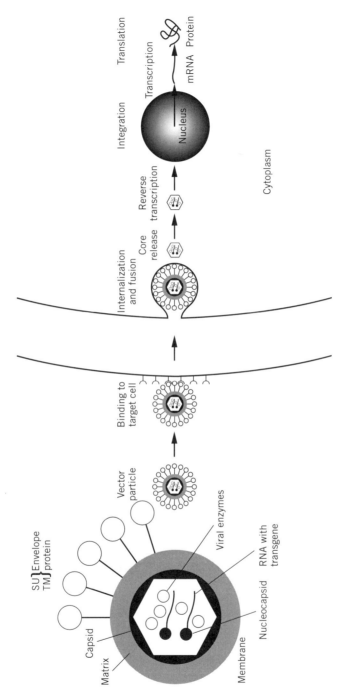

Figure 14.9 *Retrovirus vector contains the transgene that replaces the viral RNA. The virus particle binds to the target cell receptor and fuses with the target cell membrane, releasing the viral RNA into the target host cell. Reverse transcription copies the RNA into a double strand of DNA that integrates into the host chromosomal DNA. The incorporated transgene is replicated along with the rest of the host DNA in subsequent cell divisions. Retroviruses have minimal risk, as they are first rendered nonpathogenic. The disadvantages of retrovirus delivery of targeted gene therapies include their random insertion into the host genome, maintaining expression over many successive replications, and their inability to transduce nondividing cells such as neurons.* [From Anderson (1998), p. 27.]

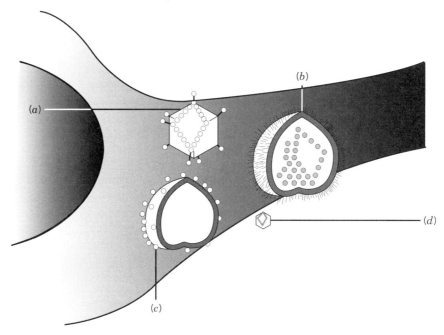

Figure 14.10 *Adenovirus vectors deliver the transgene into the nucleus of the host cell, without direct insertion into the chromosome. The inactivated adenovirus enters nerve cells and synthesizes product for the lifetime of the virus particle. The small size of these particles limits the quantity of therapeutic DNA that can be loaded. The lack of replication limits the duration of therapeutic window. (a) Adenovirus, a pathogen that commonly causes respiratory ailments, is capable of infecting neurons. In its naturally occurring form, it can damage cells and evoke a strong immune response, so care must be taken to inactivate it and minimize its immunogenicity. (b) Herpes simplex virus type 1, the agent that causes common cold sores, is especially able to carry genes into nerve cells. (Between outbreaks, the cold sore virus often remains dormant within sensory neurons.) Unfortunately, like adenoviruses, unmodified herpes viruses damage cells and cause an immune response. (c) Retrovirus incorporates its genes into the DNA of the host cells. Many retroviruses infect only cells that divide regularly and thus cannot be used to treat neurons. Others in the lentivirus family, which includes HIV, the AIDS virus, infect cells that do not divide, and so these retroviruses may prove useful in gene therapy for the nervous system. (d) Adeno-associated virus does not damage infected nerve cells or induce an immune response. It is also much more compact than other viruses under study for gene therapy. Consequently, it may be more successful at traversing the small pores that allow few substances to cross the blood–brain barrier.* [Adapted from Ho and Sapolsky (1997), p. 118.]

delivery of *p53* genes in non-small-cell lung cancer (Roth, 1998). Retrovirus transfer of the human glucocerebrosidase gene rescued the deficiency of this enzyme in fibroblasts from Gaucher's disease (Aran et al., 1996). Although retroviral treatment of severe-combined immunodeficiency disease resulted in three deaths due to leukemia, retroviral treatment of chronic granulomatous disease appears promising (Graz et al., 2005). Leptin gene therapy reduced body fat and partially corrected the diabetes phenotype of the *ob/ob* mouse (Muzzin et al., 1996; Chen et al., 1996b; Figure 14.11, and see Chapter 7). Glial cell neurotrophic factor gene delivery in myoblasts prevented motor neuron loss in a mouse model of familial amyotrophic lateral sclerosis (Mohajeri et al., 1999). Astrocytes modified with retrovirus to overexpress brain-derived neurotrophic factor conferred neuroprotection in a rat model of Parkinson's disease (Hoshimoto

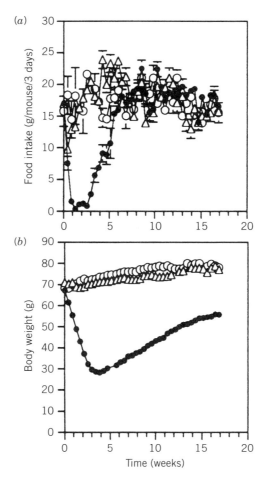

Figure 14.11 *A leptin rescue experiment. The gene for leptin, an adipose tissue protein, was transfected into an adenovirus vector and injected into ob/ob mice, a natural mutant that develops obesity due to a mutation in the leptin gene. Food consumption and body weight declined for several weeks in the ob/ob mice treated with the leptin gene (black circles), as compared to ob/ob mice treated with nontransfected adenovirus (white triangles) or with phosphate-buffered saline vehicle (white circles). Rescue experiments involve transfecting the missing gene back into the genome by conventional transgenic techniques, or by the viral vector technique, or by administering the gene product. If an aberrant behavioral phenotype is reversed by these treatments, proof is established that the observed aberrant behavioral phenotype in the mutant is due to the hypothesized natural mutation targeted gene mutation.* [From Muzzin et al. (1996), p. 14806].

et al., 1995; Bauer et al., 1999). Fibroblast growth factor-1 gene delivery in endothelial cells implanted into rat brain on postnatal day 7 protected the 10-day-old rat from hippocampal and striatal cell death induced by quinolinate, a neurotoxin (Hossain et al., 1998). Conditioned place preference to morphine was enhanced in rats given excess GluR1 glutamate receptors by adenovirus microinjection in the ventral tegmentum (Carlezon et al., 1997; Figure 14.12). Dopamine-deficient mutant mice, lacking tyrosine hydroxylase in dopaminergic neurons, were successfully treated with bilateral striatal injections of adenovirus containing the human tyrosine hydroxylase gene,

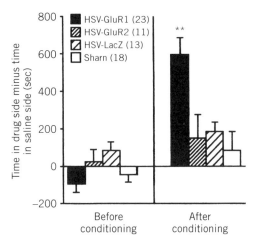

Figure 14.12   *Microinjection of the GluR1 glutamate receptor in a herpes simplex vector into the ventral tegmentum enhances sensitivity to morphine in a place preference test.* [From Carlezon et al., 1997.]

rescuing the feeding deficits and partially restoring locomotor deficits in the dopaminergic mutant mice (Szczypka et al., 1999). Adeno-associated viral vector delivery of the gene for insulin-like growth factor slowed the motor loss in a mouse model of amyotrophic lateral sclerosis (Kaspar et al., 2003; Figure 14.13). Viral vector delivery is being applied to mouse models of Huntington's and Parkinson's diseases (Kirik and Bjorklund, 2003). Lentivirus delivery of ApoE genes reduced amyloid plaque burden in the PDAPP mouse model of Alzheimer's disease (Dodart et al., 2005).

Other alternatives for gene delivery in humans are under investigation. Genes may be delivered in synthetic *lipoplexes*, small fatty liposome spheres (Yang et al., 1997). "Naked DNA," injected directly into the bloodstream, may work as an antigen for immunization against infectious diseases and some forms of cancer (Friedmann, 1997). Hematopoietic stem cells transduced with retroviral vectors containing therapeutic genes provide a new application for bone marrow transplantation (Chu et al., 1998). Each of these approaches is currently being refined, to enhance its specificity for a particular type of human cell target, to extend the half-life of the inserted gene, and to amplify the expression of gene product (Tuszynski and Gage, 1996; Teresa Girao da Cruz et al., 2005).

Embryonic stem cells deliver the full complement of normal genes (Gearhart, 2004; Keller, 2005). The potential for curing human diseases by replacing defective genes with healthy genes is enormous. For nervous system repair, pluripotent stem cells have been differentiated in culture into neurons and glia (McKay, 2004). An early example of pluripotent embryonic stem cell therapy for a central nervous system disease is illustrated in Figure 14.14. A mouse model of Sandhoff disease was developed by Rick Proia at the National Institute of Diabetes & Digestive & Kidney Diseases. Targeted mutation of the *HexB* gene generated null mutants deficient in β-hexosaminidase, the enzyme that degrades gangliosides (Norflus et al., 1998). Mice deficient in *HexB* accumulate gangliosides in central neurons, show a rapid time course of neuronal degeneration, and display progressively deteriorating motor dysfunctions. The motor symptoms are easily quantitated by latency to fall from the rotarod (Figure 4.3). The

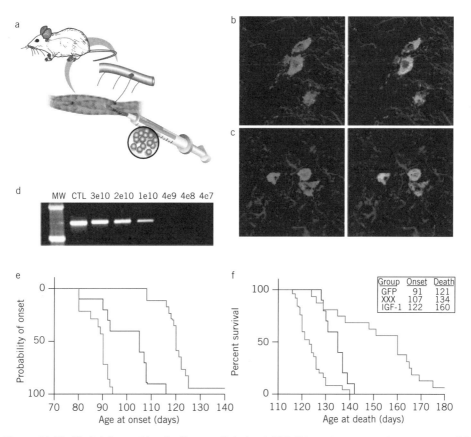

Figure 14.13 *Viral delivery of insulin-like growth factor 1 (IGF-1) to motor neurons in a mouse model of amyotrophic lateral sclerosis (ALS). (a) Adeno-associated viral vector (AAV) containing an IGF-1 construct was injected into the hindlimb quadriceps and intercostal muscles of Cu/Zn superoxide dismutase 1 (SOD-1) G93A mutant mice. Gene delivery targeted mice at 60 days of age, before the onset of symptoms. Motor symptoms included declines in grip strength and rotarod performance. Control AAV, containing a green fluorescent protein marker (GFP), was retrogradely transported into (b) lumbar and (c) thoracic motor neurons of the spinal cord, which innervate the injected muscles. (d) Polymerase chain reaction assays detected the AAV-IGF-1 in the lumbar spinal cord. (e) Onset of motor decline in the ALS mice was delayed from approximately 90 days in the AAF-GFP controls (left line), to about 110 days with AAV containing glial cell derived growth factor (middle line), and to about 120 days in AAV-IGF-1 treated mice (right line). (f) Survival of the ALS mice was extended from approximately 120 days in the AAF-GFP controls (left line), to about 130 days with AAV containing glial cell derived growth factor treatment (middle line), and up to 180 days in AAV-IGF-1 treated mice (right line). [From Kaspar et al., 2003.]*

progressive loss of motor coordination and balance in the mouse model is analogous to the progressive motor dysfunctions seen in the human disease. The rotarod task thus provides a good preclinical measure of treatment efficacy. Bone marrow transplantation of pluripotent embryonic stem cells delayed the onset of rotarod impairment (Figure 14.14), as well as extending survival time (Norflus et al., 1998). Further approaches, including delivery of the *HexB* gene through a viral vector, could quickly and easily be tested for efficacy on prevention of the rotarod deficit in the *HexB* knockout mice. Thus a simple and robust behavioral phenotype in a transgenic or knockout mouse model represents a preclinical tool for evaluating the efficacy of proposed gene therapies.

At the time of this writing, political controversy surrounds the clinical application of stem cell therapeutics (Jaenisch, 2004). Religious concerns over destroying a human embryo, and ethical concerns about the boundaries between therapeutic cloning and human cloning, have limited basic research on the feasibility of embryonic stem cell therapies. In the United States at present (2005), a small number of established stem

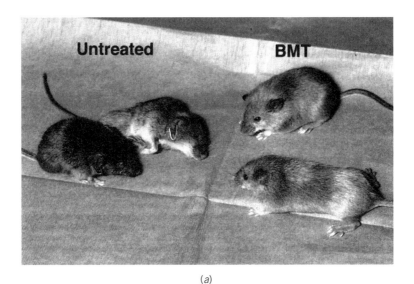

(a)

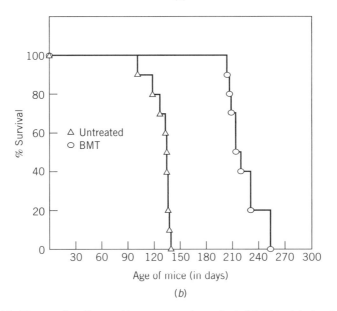

(b)

Figure 14.14 *Therapeutic efficacy of bone marrow transplants (BMT) in delaying the onset of motor deficits on the rotarod, righting reflex, and extending survival of mice deficient in β-hexosaminidase, a mouse model of Sandhoff disease.* [From Norflus et al. (1998), p. 1884].

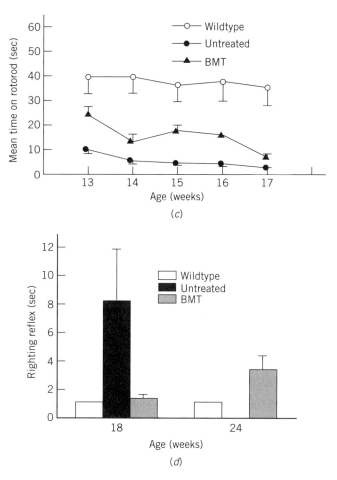

Figure 14.14 *(Continued)*

cell lines can be used in federally funded research. Californians voted for state funding of stem cell research in 2004, and other states in the United States are approving similar investments in the promise of major medical breakthroughs. Private foundations support stem cell research that targets spinal cord repair, Parkinson's disease, Alzheimer's, juvenile diabetes, and many others. Countries including Australia, Canada, France, Japan, South Korea, Switzerland, and the United Kingdom permit human stem cell research with minimal restrictions. Stem cell treatments and therapeutic cloning using nuclear transfer with stem cells lines are being evaluated in mouse and rat models of human diseases (Kerr et al., 2003; Mombaerts, 2003; McBride et al., 2004). Clinical trials ongoing in many countries will reveal the therapeutic benefits and risks of this revolutionary genetic approach to medicine.

The Human Genome Project

Initiated in 1988, the Human Genome Project estimated that the entire human genome would be sequenced by 2003 (Collins et al., 1998; Hudson, 1998; Burris et al., 1998).

The international collaborators of the Human Genome Project began by constructing the human genetic map with markers spaced at approximately 2-centiMorgan intervals, to facilitate positional cloning of every human gene (Watson, 1990; Johnson et al., 1992). New strategies and new commercial enterprises sped the process (Venter et al., 1998). The full draft sequence of the human genome was published in early 2001 by the publically funded international consortium and by a private company, Celera Genomics (Lander et al., 2001; Venter et al., 2001). One of the surprises was the downward estimate of only approximately 30,000 genes comprising the human genome (*www.genome.gov*). As few as 20,000 protein-coding genes were estimated more recently by the International Human Genome Sequencing Consortium (*Nature*, 2004).

The full draft sequence of the mouse genome was published the next year (Waterson et al., 2002), and the genomes of many other species followed. Celera Genomics Corporation contributed its entire database of genomic data on humans, rats, and mice to the free public GenBank resource on July 1, 2005 (Kaiser, 2005). Sequences and mapping information are cataloged and available at Web site *www.ncbi.nlm.nih.gov/genemap* and searchable through public Web sites with functional annotations, including GenBank (*http://www.ncbi.nlm.nih.gov/Genbank*), Gene Ontology (*http://www.ebi.ac.uk/GOA*), the European Bioinformatics Institute (*http://www.ebi.ac.uk*), and the DNA Data Bank of Japan (*http://www.ddbj.nig.ac.jp*).

Methods for detecting polymorphisms, unusual alleles derived from natural mutations, across the entire genome of each individual person may be possible in the foreseeable future. Genes for hundreds of human diseases are searchable at the Online Mendelian Inheritance in Man Web site, *http://www.ncbi.nlm.nih.gov/entrez/query. fcgi?db=OMIM*. Microarrays of thousands of genes, patterned on DNA array microchips, are commercially available (Syvanen, 1999; Collins, 1999; Lander, 1999; Lipschutz et al., 1999; Brown and Botstein, 1999; Watson and Akil, 1999). Expression arrays place complementary DNA strands directly on glass chips or microscope slides. The Affymetrix GeneChip® probe array employs synthetic oligonucleotides on the chip. A highly sensitive hybridization assay is illustrated in Figure 14.15. Theoretically, genetic screening could rapidly detect mutations and susceptibility genes for a very large number of inherited diseases (Schena et al., 1996; Service, 1998; McCabe and McCabe, 2004; Green and Pass, 2005).

Private foundations and advocacy groups are beginning to fund clinical trials of potential gene therapies, as reported in *Science* magazine (Couzin, 2005) and described above. Cystic Fibrosis Foundation Therapeutics in Bethesda, Maryland, partnered with many biotechnology companies to establish a pipeline of drug treatments, and partnered with Targeted Genetics in Seattle, Washington, to test a gene therapy. Foundation Fighting Blindness in Owings Mills, Maryland, funded research that identified 150 genes for retinal disorders such as retinitis pigmentosa, and supported research for gene therapies, cell transplants, and prosthetics to slow blindness. Its subsidiary, Neurovision Research Institute, will fund clinical trials of at least one gene therapy. Juvenile Diabetes Research Foundation has 17 ongoing clinical trials. Other leaders include the Multiple Myeloma Research Foundation in New Canaan, Connecticut, and the Amyotrophic Lateral Sclerosis Therapy Development Foundation in Calabasas Hills, California.

Pharmacogenomics applies knowledge about human genotypes to personalize drug treatments. Choosing the right cancer therapy, antidepressant, or high blood pressure

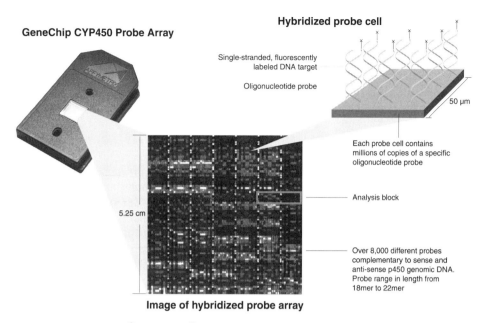

GeneChip CYP450 Probe Array

Hybridized probe cell

Single-stranded, fluorescently labeled DNA target

Oligonucleotide probe

50 µm

Each probe cell contains millions of copies of a specific oligonucleotide probe

Analysis block

5.25 cm

Over 8,000 different probes complementary to sense and anti-sense p450 genomic DNA. Probe range in length from 18mer to 22mer

Image of hybridized probe array

Figure 14.15 *Affymetrix® Gene Chip® probe array technology.* (Image courtesy of Affymetrix, Inc., Santa Clara, CA.)

drug that works best for a particular ethnic group, women versus men, children, older people, or individuals with a specific genetic polymorphism may be a life-or-death decision. Clinical drug trials are increasingly focused on subsets of patients with a genetic background that will benefit the most from the optimized drug, dose, and formulation (Service, 2005). For example, receptor subtype polymorphisms in the serotonin 2A and dopamine receptor 2 genes appear to predict the optimal choice of antipsychotic drug for a schizophrenic patient (Malholtra et al., 2004) Polymorphisms in the serotonin transporter, tryptophan hydroxylase, and the serotonin $5HT_{2A}$ gene may underlie different subtypes of depression (Enoch and Goldman, 2001; Serretti and Artioli, 2004; Taylor et al., 2005), implicating pharmacogenomic specialization in antidepressant drug prescription. "One size fits all" medicine may become an outdated concept.

Ethical Issues

Ethical issues concerning genetic testing and the public use of genetic information have been hotly debated since human genome sequences emerged (Reilly and Page, 1998; Silver, 1997; Anderson 2000; McCabe and McCabe, 2004; Taub et al., 2004; Panno, 2004). Regulatory commissions control the applications of new genetic technologies in the United States and many other countries (Berg and Singer, 1998). The National Institutes of Health Recombinant DNA Advisory Committee originally established guidelines for all federally funded research and clinical trials with recombinant DNA. In many countries, gene therapy in humans remains restricted to attempts to modify somatic cells only, disallowing the use of human embryos and precluding the passage of the therapeutic gene to subsequent generations (Gene Therapy Advisory Committee, 1999; Caplan 2003; Stock, 2005). Pharmacogenomics, in which drug treatments are customized to the patient's genotype, raises ethical issues including informed consent, genetic counseling, unequal access to treatments, discrimination by health insurance companies, and privacy concerns (Broder et al., 2002; Breckenridge et al., 2004). Applications of discoveries from the transgenic and knockout mouse literature to the treatment of human diseases remains appropriately cautious, while ethical issues are being discussed and safety issues are being resolved.

SHARE THE FUTURE

It is fitting to close this book with a plea for sharing research tools in the growing field of mouse behavioral genetics. Embryonic stem cells with defined gene mutations, inbred strains of mice, microsatellite markers, and searchable gene databases are openly available to everyone. Behavioral test equipment described in this book is commercially available or can be constructed from the methods described in the referenced primary literature. Linkage and physical maps of the mouse genome are available at *http://www.informatics.jax.org*, a Web site supported by the National Institutes of Health Mouse Genome Project and the Whitehead/Massachusetts Institute of Technology and maintained by The Jackson Laboratory. Robert Plomin and colleagues originally proposed an RI QTL cooperative data bank for recombinant inbred quantitative trait loci findings (Plomin et al., 1991), that has been realized in WebQTL, developed by Robert Williams and co-workers at the University of Tennessee (*http://www.genenetwork.org*). Databases of mutant mice are available at Web sites such as Neuromice (*www.neuromice.org*), the Mutant Mouse Resource Centers (*http://www.ncrr.nih.gov/compmed/cm_mmrrc.asp*), The European Mutant Mouse Archive (*http://www.emmanet.org/links.php*), The Jackson Laboratory's Mouse Models List, *http://jaxmice.jax.org/models/index.html*; and Taconic's Animal Models list (*http://www.taconic.com/anmodels/animlmod.htm*). Many transgenic and knockout mice are maintained in university and commercial repositories and are available to the scientific community upon request or purchase. All that is needed now is for individual investigators to contribute the transgenic and knockout lines they have generated to these repositories, to make their valuable resources available to the scientific community. When you have completed the experiments most interesting to your own laboratory, please donate breeding pairs or frozen embryos to a repository. When you discover significant phenotypes in your mutant line, enter your findings in databases such as WebQTL and the Mouse Phenome Project (*http://aretha.jax.org/pub-cgi/phenome/mpdcgi?rtn=docs/introducing*; Bogue and Grubb, 2004; Churchill et al., 2004; Crabbe et al., 2005; Moy et al., in press).

Cryopreservation of frozen mouse sperm, eggs, ovaries, and embryos has been perfected, to store lines of mutant mice. As described in the Taconic Farms newsletter, *Research Animal Review* (volume 1, number 5, 1998), there are major advantages of maintaining valuable lines in frozen storage. The most important is to keep the transgenic or knockout available for future use by the scientific community. Maintainence of breeding lines for each mutation requires more space and money than anyone can afford, as the number of transgenics and knockouts continues to escalate. It is more cost effective to store the frozen embryos than to maintain the breeding colony in the animal facility. Frozen embryos can be shipped easily, avoiding importation quarantine requirements for live animals. The resources for cryopreservation of mouse embryos are widely available, at companies including The Jackson Laboratory (Bar Harbor, Maine, *www.jax.org*), Taconic (273 Hanover Avenue, Germantown, New York 12526, 518-537-6208, *www.taconic.com*), and at the National Institutes of Health Division of Veterinary Resources (*http://dvrnet.ors.od.nih.gov/cryopreservation.asp*). Cryopreserving sperm, eggs, and ovaries containing the targeted gene mutation are other useful approaches for retaining and distributing valuable transgenic and knockout mouse lines (Sztein et al., 1998; Jackson and Abbott, 2000; Woods et al., 2004).

This book was written to share the wealth of information about mouse behavioral tests, available for use in behavioral phenotyping of transgenic and knockout mice, with our colleagues in the molecular genetics and behavioral neuroscience communities. As with any new technology, the development of rigorous methods is the first step, followed by proof of principle and evaluation of hypotheses. The transgenic and knockout technology has become increasingly sophisticated in refining methods, in testing new hypotheses, and in advancing our understanding of genes in the nervous system. It is the hope of this author that colleagues across the fields of molecular genetics, neuroscience, and behavior will continue to make their knowledge and mouse resources freely available to each other. Our collaborative efforts will then have the best chance of generating treatments and cures for the behavioral components of many catastrophic human genetic diseases.

BACKGROUND LITERATURE

Mapping Genes for Complex Traits

Aarnio V, Paananen J, Wong G (2005). Analysis of microarray studies performed in the neurosciences. *Journal of Molecular Neuroscience* **27**: 261–268.

Belknap JK, Crabbe JC, Plomin R, McClearn GE, Sampson KE, O'Toole LA, Gora-Maslak G (1992). Single-locus control of saccharin intake in BXD/Ty recombinant inbred (RI) mice: Some methodological implications for RI strain analysis. *Behavior Genetics* **22**: 81–100.

Geschwind DH (2003). DNA microarrays: Translation of the genome from laboratory to clinic. *Neurology* **2**: 275–282.

Geschwind DH, Gregg JP (2002). *Microarrays for the Neurosciences.* MIT Press, Cambridge.

Johnson TE, DeFries JC, Markel PD (1992). Mapping quantitative trait loci for behavioral traits in the mouse. *Behavior Genetics* **22**: 635–653.

Lander ES, Botstein D (1989). Mapping Mendelian factors underlying quantitative traits using RFLP linkage maps. *Genetics* **121**: 185–199.

Malhotra AK, Murphy GM, Kennedy JL (2004). Pharmacogenetics of psychotropic drug response. *American Journal of Psychiatry* **161**: 780–796.

Manly DF, Elliott RW (1991). RI Manager, a microcomputer program for analysis of data from recombinant inbred strains. *Mammalian Genome* **1**: 123–126.

Miles MF (2004). *DNA Arrays in Neurobiology.* Elsevier Academic Press, San Diego, CA.

Mineur YS, Crusio WE, Sluyter F (2004). Genetic dissection of learning and memory in mice. *Neural Plasticity* **11**: 217–240.

Nature Genetics (January 1999). Special Issue on Microarrays **21**: 1.

Neumann PE (1992). Inference in linkage analysis of multifactorial traits using recombinant inbred strains of mice. *Behavior Genetics* **22**: 665–676.

Phillips TJ, Belknap JK, Hitzemann RJ, Buck KJ, Cunningham CL, Crabbe JC (2002). Harnessing the mouse to unravel the genetics of human disease. *Genes, Brain and Behavior* **1**: 14–26.

Plomin R, McClearn GE, Gora-Maslak G, Neiderhiser JM (1991). An RI QTL cooperative data bank for recombinant inbred quantitative trait loci analysis. *Behavior Genetics* **21**: 97–98.

Science (September 2, 2005). Special Issue on Mapping RNA Form and Function, **309**: 1441–1632.

Sohail M (2004). *Gene Silencing by RNA Interference*. CRC Press, Boca Raton, FL.

Takahashi JS, Pinto LH, Vitaterna MH (1994). Forward and reverse genetic approaches to behavior in the mouse. *Science* **264**: 1724–1733.

Tecott LH, Wehner JM (2001). Mouse molecular genetic technologies. *Archives of General Psychiatry* **58**: 995–1004.

Zamore PD, Haley B (2005). Ribo-gnome: The big world of small RNAs. *Science* **309**: 1519–1524.

Gene Therapy

Anderson WF (1998). Human gene therapy. *Nature* **392:** 25–31.

Barkats M, Bilang-Bleuel A, Buc-Caron MH, Castel-Barthe MN, Corti O, Finiels F, Horellou P, Revah F, Sabate O, Mallet J (1998). Adenovirus in the brain: Recent advances of gene therapy for neurodegenerative diseases. *Progress in Neurobiology* **55**: 333–341.

Bauer M, Ueffing M, Meitinger T, Gasser T (1999). Somatic gene therapy in animal models of Parkinson's disease. *Journal of Neural Transmission* **55** (Suppl.): 131–147.

Blömer U, Naldini L, Verma IM, Trono D, Gage FH (1996). Applications of gene therapy to the CNS. *Human Molecular Genetics* **5**: 1397–1404.

Burger C, Nash K, Mandel RJ (2005). Recombinant adeno-associated viral vectors in the nervous system. *Human Gene Therapy* **16**: 781–791.

During MJ, Leone P (1995). Adeno-associated virus vectors for gene therapy of neurodegenerative disorders. *Clinical Neuroscience* **96**: 292–300.

Friedmann T (1997). Overcoming the obstacles. In *Making Gene Therapy Work: Special Report. Scientific American* **276:** 96–101.

Friedman T, Ed. (1999). *The Development of Human Gene Therapy*. Cold Spring Harbor Laboratory Press, New York.

Gene Therapy Weekly (newsletter and abstracts, published by CW Henderson Publisher, Birmingham, AL); *http://www.newsfile.com*.

Ho DY, Sapolsky RM (1997). Gene therapy for the nervous system. In *Making Gene Therapy Work: Special Report. Scientific American* **276:** 116–120.

Machida CA (2002). *Viral Vectors for Gene Therapy: Methods and Protocols*. Humana Press, Totowa, NJ.

Mata M, Glorioso JC, Fink DJ (2003). Gene transfer to the nervous system: Prospects for novel treatments directed at diseases of the aging nervous system. *Journal of Gerontology* **58A**: 1111–1118.

Nature Genetics (1999). Special Issue on The Chipping Forecast **21** (Suppl.): 1–60.

Panno J (2004). *Stem Cell Research: Medical Applications and Ethical Controversies*. Facts on File, New York.

Science (February 16, 2001). Special Issue on The Human Genome **291**.

Tuszynski MH, Gage FH (1996). Somatic gene therapy for nervous system disease. *Ciba Foundation Symposium* **196:** 85–97.

Wickstrom E, Ed. (1998). *Clinical Trials of Genetic Therapy with Antisense DNA and DNA Vectors*. Marcel Dekker, New York.

References

Aarnio V, Paananen J, Wong G (2005). Analysis of microarray studies performed in the neurosciences. *Journal of Molecular Neuroscience* **27**: 261–268.

Abbott LC, Nejad HH, Bottje WG, Hassan AS (1990). Glutathione levels in specific brain regions of genetically epileptic (tg/tg). mice. *Brain Research Bulletin* **25**: 629–631.

Abdullaev YG, Posner MI (1998). Event-related brain potential imaging of semantic encoding during processing single words. *Neuroimage* **7**: 1–13.

Abel T, Martin KC, Bartsch D, Kandel ER (1998). Memory suppressor genes: Inhibitory constraints on the storage of long-term memory. *Science* **279**: 338–341.

Abel T, Nguyen PV, Barad M, Deuel TAS, Kandel ER, Bourtchouladze R (1997). Genetic demonstration of a role for PKA in the late phase of LTP. *Cell* **88**: 615–626.

Abercrombie ED, Keefe KA, DiFrischia DS, Zigmond MJ (1989). Differential effect of stress on in vivo dopamine release in striatum, nucleus accumbens, and medial frontal cortex. *Journal of Neurochemistry* **52**: 1655–1658.

Abiola O, Angel JM, Avner P, Bachmanov AA, Belknap JK, Bennett B, Blankenhorn EP, Blizard DA, Bolivar V, Brockmann GA, Buck KJ, Bureau JF, Casley WL, Chesler EJ, Cheverud JM, Churchill GA, Cook M, Crabbe JC, Crusio WE, Darvasi A, de Haan G, Dermant P, Doerge RW, Elliot RW, Farber CR, Flaherty L, Flint J, Gershenfeld H, Gibson JP, Gu J, Gu W, Himmelbauer H, Hitzemann R, Hsu HC, Hunter K, Iraqi FF, Jansen RC, Johnson TE, Jones BC, Kempermann G, Lammert F, Lu L, Manly KF, Matthews DB, Medrano JF, Mehrabian M, Mittlemann G, Mock BA, Mogil JS, Montagutelli X, Morahan G, Mountz JD, Nagase H, Nowakowski RS, O'Hara BF, Osadchuk AV, Paigen B, Palmer AA, Peirce JL, Pomp D, Rosemann M, Rosen GD, Schalkwyk LC, Seltzer Z, Settle S, Shimomura K, Shou S, Sikela JM, Siracusa LD, Spearow JL, Teuscher C, Threadgill DW, Toth LA, Toye AA, Vadasz C, Van Zant G, Wakeland E, Williams RW, Zhang HG, Zou F; Complex Trait Consortium (2003). The nature and identification of quantitative trait loci: A community's view. *Nature Reviews Genetics* **4**: 911–916.

Accili D, Fishburn CS, Drago J, Steiner H, Lachowicz JE, Park BH, Gauda EB, Lee EJ, Cool MH, Sibley DR, Gerfen CR, Westphal H, Fuchs S (1996). A targeted mutation of the D_3 dopamine receptor gene is associated with hyperactivity in mice. *Proceedings of the National Academy of Sciences USA* **93**: 1945–1949.

Ackroff K, Sclafani A (1998). Evaluating postingestive reinforcement by nutrients on conditioned flavor preferences. *Current Protocols in Neuroscience* Wiley, New York, pp. 8.6F.1–8.6F.12.

Adams MR, Brandon EP, Chartoff EH, Idzerda R, Dorsa DM, McKnight GS (1997). Loss of haloperidol induced gene expression and catalepsy in protein kinase A-deficient mice. *Proceedings of the National Academy of Sciences USA* **94**: 12157–12161.

Adams SW, Emerson SG (1998). Gene therapy for leukemia and lymphoma. *Hematology/Oncology Clinics of North America* **12**: 631–648.

Addiction (1996). December Supplement to volume 91.

Ahima RS, Flier JS (2000). Leptin. *Annual Review of Physiology* **62**: 413–437.

Ahmed SH, Koob GF (1998). Transition from moderate to excessive drug intake: Change in hedonic set point. *Science* **282**: 298–300.

Airaksinen MS, Eilers J, Garaschuk O, Thoenen H, Konnerth A, Meyer M (1997). Ataxia and altered dendritic calcium signaling in mice carrying a targeted null mutation of the calbindin D28k gene. *Proceedings of the National Academy of Sciences USA* **94**: 1488–1493.

Aja S, Schwartz GJ, Kuhar MJ, Moran TH (2001). Intracerebroventricular CART peptide reduces rat ingestive behavior and alters licking microstructure. *American Journal of Physiology: Regulatory, Integrative and Comparative Physiology*. **280**: R1613–R1619.

Akbarian S, Bates B, Liu RJ, Skirboll SL, Pejchal T, Coppola V, Sun LD, Fan G, Kucera J, Wilson MA, Tessarollo L, Kosofsky BE, Taylor JR, Bothwell M, Nestler EJ, Aghajanian GK, Jaenisch R (2001). Neurotrophin-3 modulates noradrenergic neuron function and opiate withdrawal. *Molecular Psychiatry* **6**: 593–604.

Alberts JR, Gubernick DJ (1990). Functional organization of dyadic and triadic parent–offspring systems. In *Mammalian Parenting*, NA Krasnegor and RS Bridges, Eds. Oxford University Press, New York, pp. 416–440.

Albertson DN, Pruetz B, Schmidt CJ, Kuhn DM, Kapatos G, Bannon MJ (2004). Gene expression profile of the nucleus accumbens of human cocaine abusers: Evidence for dysregulation of myelin. *Journal of Neurochemistry* **88**: 1211–1219.

Alexander RC, Wright R, Freed W (1996). Quantitative trait loci contributing to phencyclidine-induced and amphetamine-induced locomotor behavior in inbred mice. *Neuropsychopharmacology* **15**: 484–490.

Alkire MT, Haier RJ, Fallon JH, Cahill L (1998). Hippocampal, but not amygdala, activity at encoding correlates with long-term, free recall of nonemotional information. *Proceedings of the National Academy of Sciences USA* **95**: 14506–14510.

Alleva E, Calamandrei G (1986). Odor-aversion learning and retention span in neonatal mouse pups. *Behavioral & Neural Biology* **46**: 348–357.

Almaric M, Cline EJ, Martinez JL, Bloom FE, Koob GF (1987). Rewarding properties of beta-endorphin as measured by conditioned place preference. *Psychopharmacology* **91**: 14–19.

Altafaj X, Dierssen M, Baamonde C, Martí E, Visa J, Guimerà J, Oset M, Gonzáles R, Flórez J, Fillat C, Estivill X (2001). Neurodevelopmental delay, motor abnormalities and cognitive deficits in transgenic mice overexpressing Dyrk1A (minibrain), a murine model of Down's syndrome. *Human Molecular Genetics* **10**: 1915–1923.

Altman J, Sundarshan K (1975). Postnatal development of locomotion in the laboratory rat. *Animal Behavior* **23**: 896–920.

Altshuler D, Brooks LD, Chakravarti A, Collins FS, Daly MJ, Donnelly P; International HapMap Consortium (2005). A haplotype map of the human genome. *Nature* **437**: 1299–1320.

Altschuler RA, Bobbin RP, Clopton BM, Hoffman DW (1991). *Neurobiology of Hearing: The Central Auditory System*. Raven Press, New York.

Amendola J, Verrier B, Roubertoux P, Durand J (2004). Altered sensorimotor development in a transgenic mouse model of amyotrophic lateral sclerosis. *European Journal of Neuroscience* **20**: 2822–2826.

Amir RE, Van den Veyver IB, Wan M, Tran CQ, Francke U, Zoghbi HY (1999). Rett syndrome is caused by mutations in X-linked MECP2, encoding methyl-CpG-binding protein 2. *Nature Genetics* **23**: 185–188.

Ammassari-Teule M, Fagioli S, Rossi-Arnaud C (1994). Radial maze performance and open-field behaviours in aged C57BL/6 mice: Further evidence for preserved cognitive abilities during senescence. *Physiology and Behavior* **55**: 341–345.

Anagnostaras SG, Murphy GG, Hamilton SE, Mitchell SL, Rahnama NP, Nathanson NM, Silva AJ (2003). Selective cognitive dysfunction in acetylcholine M1 muscarinic receptor mutant mice. *Nature Neuroscience* **6**: 51–58.

Anagnostopoulos AV (2002). A compendium of mouse knockouts with inner ear defects. *Trends in Genetics* **18**: 499.

Anagnostopoulos AV, Sharp JJ, Mobraaten LE, Eppig JT, Davisson MT (2001). Availability and characterization of transgenic and knockout mice with behavioral manifestations: Where to look and what to search for. *Behavioural Brain Research* **125**: 33–37.

Anand BK, Brobeck JR (1951). Hypothalamic control of food intake. *Yale Journal of Biology and Medicine* **24**: 123–140.

Anderson R, Barnes JC, Bliss TVP, Cain DP, Cambon K, Davies HA, Errington ML, Fellows LA, Gray RA, Hoh T, Stewart M, Large CH, Higgins GA (1998). Behavioural, physiological and morphological analysis of a line of apolipoprotein E knockout mouse. *Neuroscience* **85**: 93–110.

Anderson WF (1998). Human gene therapy. *Nature* **392**: 25–31.

Andreassen OA, Dedeoglu A, Klivenyi P, Beal MF, Bush AI (2000). *N*-acetyl-L-cysteine improves survival and preserves motor performance in an animal model of familial amyotrophic lateral sclerosis. *Neuroreport* **11**: 2491–2493.

Anisman H, Grimmer L, Irwin J, Remington G, Sklar LS (1979). Escape performance after inescapable shock in selectively bred lines of mice. *Journal of Comparative and Physiological Psychology* **93**: 229–241.

Antin JJ, Gibbs J, Holt J, Young RC, Smith GP (1975). Cholecystokinin elicits the complete behavioral sequence of satiety in rats. *Journal of Comparative and Physiological Psychology* **89**: 784–790.

Aoki I, Naruse S, Tanaka C (2004). Manganese-enhanced magnetic resonance imaging (MEMRI) of brain activity and applications to early detection of brain ischemia. *NMR in Biomedicine* **17**: 569–580.

Appel JB, White FJ, Holohean AM (1982). Analyzing mechanism(s) of hallucinogenic drug action with drug discrimination procedures. *Neuroscience and Biobehavioral Reviews* **6**: 529–536.

Applegate CD, Tecott LH (1998). Global increases in seizure susceptibility in mice lacking 5-HT2C receptors: A behavioral analysis. *Experimental Neurology* **154**: 522–530.

Aran JM, Light T, Gottesman MM, Pastan I (1996). Complete restoration of glucocerebrosidase deficiency in Gaucher fibroblasts using a bicistronic *MDR* retrovirus and a new selection strategy. *Human Gene Therapy* **7**: 2165–2175.

Archer SN, Robilliard DL, Skene DJ, Smits M, Williams A, Arendt J, von Schantz M (2003). A length polymorphism in the circadian clock gene *per3* is linked to delayed sleep phase syndrome and extreme diurnal preference. *Sleep* **26**: 413–415.

Arnedo MT, Salvador A, Martinez-Sanchis S, Pellicer O (2002). Similar rewarding effects of testosterone in mice rated as short and long attack latency individuals. *Addiction Biology* **7**: 373–379.

Artal P, Herreros de Tejada P, Munoz Tedo C, Green DG (1998). Retinal image quality in the rodent eye. *Visual Neuroscience* **15**: 597–605.

Aste N, Honda S, Harada N (2003). Forebrain Fos responses to reproductively related chemosensory cues in aromatase knockout mice. *Brain Research Bulletin* **60**: 191–200.

Atack JR (2003). Anxioselective compounds acting at the GABA(A) receptor benzodiazepine binding site. *Current Drug Targets CNS & Neurological Disordes* **2**: 213–232.

Auclair A, Blanc G, Glowinski J, Tassin JP (2004). Role of serotonin receptors in the d-amphetamine-induced release of dopamine: Comparison with previous data on alpha1b-adrenergic receptors. *Journal of Neurochemistry* **91**: 318–326.

Aura J, Sirvio J, Riekkinen P (1997). Methoctramine moderately improves memory but pirenzapine disrupts performance in delayed non-matching to position test. *European Journal of Pharmacology* **333**: 129–134.

Austin CP, Battey JF, Bradley A, Bucan M, Capecchi M, Collins FS, Dove WF, Duyk G, Dymecki S, Eppig JT, Grieder FB, Heintz N, Hicks G, Insel TR, Joyner A, Koller BH, Lloyd KC, Magnuson T, Moore MW, Nagy A, Pollock JD, Roses AD, Sands AT, Seed B, Skarnes WC, Snoddy J, Soriano P, Stewart DJ, Stewart F, Stillman B, Varmus H, Varticovski L, Verma IM, Vogt TF, von Melchner H, Witkowski J, Woychik RP, Wurst W, Yancopoulos GD, Young SG, Zambrowicz B (2004). The knockout mouse project. *Nature Genetics* **36**: 921–924.

Austin MC, Schultzberg M, Abbott LC, Montpied P, Evers JR, Paul SM, Crawley JN (1992). Expression of tyrosine hydroxylase in cerebellar Purkinje neurons of the mutant tottering and leaner mouse. *Molecular Brain Research* **15**: 227–240.

Ausubel FM, Brent R, Kingston RE, Moore DD, Seidman JG, Smith JA, Struhl K, Eds. (1995). *Current Protocols in Molecular Biology*, Vol. 1. Wiley, New York.

Auwerx J, Avner P, Baldock R, Ballabio A, Balling R, Barbacid M, Berns A, Bradley A, Brown S, Carmeliet P, Chambon P, Cox R, Davidson D, Davies K, Duboule D, Forejt J, Granucci F, Hastie N, de Angelis MH, Jackson I, Kioussis D, Kollias G, Lathrop M, Lendahl U, Malumbres M, von Melchner H, Muller W, Partanen J, Ricciardi-Castagnoli P, Rigby P, Rosen B, Rosenthal N, Skarnes B, Stewart AF, Thornton J, Tocchini-Valentini G, Wagner E, Wahli W, Wurst W (2004). The European dimension for the mouse genome mutagenesis program. *Nature Genetics* **36**: 925–927.

Bach ME, Hawkins RD, Osman M, Kandel ER, Mayford M (1995). Impairment of spatial but not contextual memory in CaMKII mutant mice with selective loss of hippocampal LTP in the range of the theta frequency. *Cell* **81**: 905–915.

Bahar A, Dorfman N, Dudai Y (2004). Amygdalar circuits required for either consolidation or extinction of taste aversion memory are not required for reconsolidation. *European Journal of Neuroscience* **19**: 1115–1118.

Bai F, Li X, Clay M, Lindstrom T, Skolnick P (2001). Intra- and interstrain differences in models of "behavioral despair." *Pharmacology Biochemistry and Behavior* **70**: 187–192.

Bai J, Ramos RL, Ackman JB, Thomas AM, Lee RV, LoTurco JJ (2003). RNAi reveals doublecortin is required for radial migration in rat neocortex. *Nature Neuroscience* **6**: 1277–1283.

Bailey CH, Bartsch D, Kandel ER (1996). Toward a molecular definition of long-term memory storage. *Proceedings of the National Academy of Sciences USA* **93**: 13445–13452.

Bailey DW (1971). Recombinant inbred strains: An aid to identify linkage and function of histocompatibility and other genes. *Transplantation* **11**: 325–327.

Bailey KR, Rustay NR, Crawley JN (2006). Behavioral phenotyping of transgenic and knockout mice. *ILAR Journal*, Special Issue on Phenotyping of Genetically Engineered Mice, **47**: 124–131.

Bakalyar HA, Reed RR (1990). Identification of a specialized adenylyl cyclase that may mediate odorant detection. *Science* **250**: 1403–1406.

Baker C (2004). *Behavioral Genetics.* American Association for the Advancement of Science and The Hastings Center, Washington, DC.

Baker RM, Shah MJ, Sclafani A, Bodnar RJ (2003). Dopamine D1 and D2 antagonists reduce the acquisition and expression of flavor-preferences conditioned by fructose in rats. *Pharmacology Biochemistry and Behavior* **75**: 55–65.

Bakker J, Honda S, Harada N, Balthazart J (2002a). The aromatase knock-out mouse provides new evidence that estradiol is required during development in the female for the expression of sociosexual behaviors in adulthood. *Journal of Neuroscience* **22**: 9104–9112.

Bakker J, Honda S, Harada N, Balthazart J (2002b). Sexual partner preference requires a functional aromatase (cyp19) gene in male mice. *Hormones and Behavior* **42**: 158–171.

Bale TL, Contarino A, Smith GW, Chan R, Gold LH, Sawchenko PE, Koob GF, Vale WW, Lee KF (2000). Mice deficient for corticotropin-releasing hormone receptor-2 display anxiety-like behaviour and are hypersensitive to stress. *Nature Genetics* **24**: 410–414.

Bale TL, Vale WW (2004). CRF and CRF receptors: role in stress responsivity and other behaviors. *Annual Review of Pharmacology and Toxicology* **44**: 525–557.

Bales KL, Carter CS (2003). Sex differences and developmental effects of oxytocin on aggression and social behavior in prairie voles (*Microtus ochragaster*). *Hormones and Behavior* **44**: 178–184.

Balk JH, Picetti R, Salardi A, Thirlet G, Dierich A, Depaulis A, Le Meur M, Borrelli E (1995). Parkinsonian-like locomotor impairment in mice lacking dopamine D2 receptors. *Nature* **377**: 424–428.

Balogh SA, McDowell CS, Stavnezer AJ, Denenberg VH (1999). A behavioral and neuroanatomical assessment of an inbred substrain of 129 mice with behavioral comparisons to C57BL/6J mice. *Brain Research* **836**: 38–48.

Balogh SA, Wehner JM (2003). Inbred mouse strain differences in the establishment of long-term fear memory. *Behavioural Brain Research* **140**: 97–106.

Banbury Conference (1997). Mutant mice and neuroscience: Recommendations concerning genetic background. *Neuron* **19**: 755–759.

Banko ML, Allen KM, Dolina S, Neumann PE, Seyfried TN (1997). Genomic imprinting and audiogenic seizures in mice. *Behavior Genetics* **27**: 465–475.

Bannon AW, Gunther KL, Decker MW (1995). Is epibatidine really analgesic? Dissociation of the locomotor activity, temperature, and analgesic effects of ($\pm$)-epibatidine. *Pharmacology Biochemistry and Behavior* **51**: 693–698.

Bannon AW, Seda J, Carmouche M, Francis JM, Norman MH, Karbon B, McCaleb ML (2000). Behavioral characterization of neuropeptide Y knockout mice. *Brain Research* **868**: 79–87.

Bao S, Chen L, Thompson RF (1998). Classical eyeblink conditioning in two strains of mice: Conditioned responses, sensitization, and spontaneous eyeblink. *Behavioral Neuroscience* **112**: 714–718.

Bark IC, Hahn KM, Ryabinin AE, Wilson MC (1995). Differential expression of SNAP-25 isoforms during divergent vesicle fusion events of neural development. *Proceedings of the National Academy of Sciences USA* **92**: 1510–1514.

Bark IC, Wilson MC (1994). Regulated vesicular fusion in neurons—Snapping together the details. *Proceedings of the National Academy of Science USA* **91**: 4621–4624.

Barkats M, Bilang-Bleuel A, Buc-Caron MH, Castel-Barthe MN, Corti O, Finiels F, Horellou P, Revah F, Sabate O, Mallet J (1998). Adenovirus in the brain: Recent advances of gene therapy for neurodegenerative diseases. *Progress in Neurobiology* **55**: 333–341.

Barlow C, Hirotsune S, Paylor R, Liyanage M, Eckhaus M, Collins F, Shiloh Y, Crawley JN, Ried T, Tagle D, Wynshaw-Boris A (1996). *Atm*-deficient mice: A paradigm of ataxia telangiectasia. *Cell* **86**: 159–171.

Barneoud P, Gyger M, Andres F, van der Loos H (1991). Vibrissa-related behavior in mice: Transient effect of ablation of the barrel cortex. *Behavioural Brain Research* **44**: 87–99.

Barnes CA (1979). Memory deficits associated with senescence: A neurophysiological and behavioral study in the rat. *Journal of Comparative and Physiological Psychology* **93**: 74–104.

Barnes CA (1995). Involvement of LTP in memory: Are we "searching under the street light?" *Neuron* **15**: 751–754.

Barnes CA, Markowska AL, Ingram DK, Kametani H, Spangler EL, Lemken VJ, Olton DS (1990). Acetyl-1-carnitine. **2**: Effects on learning and memory performance of aged rats in simple and complex mazes. *Neurobiology of Aging* **11**: 499–506.

Baron SP, Meltzer LT (2001). Mouse strains differ under a simple schedule of operant learning. *Behavioural Brain Research* **118**: 143–152.

Barondes SH, Jarvik ME (1964). The influence of actinomycin-D on brain RNA synthesis and on memory. *Journal of Neurochemistry* **11**: 187.

Barr AM, Lehmann-Masten V, Paulus M, Gainetdinov RR, Caron MG, Geyer MA (2004). The selective serotonin-2A receptor antagonist M100907 reverses behavioral deficits in dopamine transporter knockout mice. *Neuropsychopharmacology* **29**: 221–228.

Barr AM, Markou A, Phillips AG (2002). A "crash course" on psychostimulant withdrawal as a model of depression. *Trends in Pharmacological Sciences* **23**: 475–482.

Barr CS, Newman TK, Becker ML, Champoux M, Lesch KP, Suomi SJ, Goldman D, Higley JD (2003). Serotonin transporter gene variation is associated with alcohol sensitivity in rhesus macaques exposed to early-life stress. *Alcoholism: Clinical and Experimental Research* **27**: 812–817.

Barr G, Gao P (2005). Issues for consideration in the analysis of microarray data in behavioural studies. *Addiction Biology* **10**: 15–21.

Barrett JE, Vanover KE (2003). Assessment of learning and memory using the autoshaping of operant responding in mice. In *Current Protocols in Neuroscience*, pp. 8.5F.1–8.5F.8.

Barrett RJ, Caul WF, Huffman EM, Smith RL (1994). Drug discrimination is a continuous rather than a quantal process following training on a VI-TO schedule of reinforcement. *Psychopharmacology* **113**: 289–303.

Barski JJ, Hartmann J, Rose CR, Hoebeek F, Morl K, Noll-Hussong M, De Zeeuw CI, Konnerth A, Meyer M (2003). Calbindin in cerebellar Purkinje cells is a critical determinant of the precision of motor coordination. *Journal of Neuroscience* **23**: 3469–3477.

Barton C, Lin L, York DA, Bray GA (1995). Differential effects of enterostatin, galanin and opioids on high-fat diet consumption. *Brain Research* **702**: 55–60.

Bartoshuk LM, Beauchamp GK (1994). Chemical senses. *Annual Review of Psychology* **45**: 419–449.

Bartsch D, Casadio A, Karl KA, Serodio P, Kandel ER (1998). CREB1 encodes a nuclear activator, a repressor, and a cytoplasmic modulator that form a regulatory unit critical for long-term facilitation. *Cell* **95**: 211–223.

Basile AS, Fedorova I, Zapata A, Liu X, Shippenberg T, Duttaroy A, Yamada M, Wess J (2002). Deletion of the M5 muscarinic acetylcholine receptor attenuates morphine reinforcement and withdrawal but not morphine analgesia. *Proceedings of the National Academy of Sciences USA* **99**: 11452–11457.

Batchelder P, Lynch CB, Schneider JE (1982). The effects of age and experience on strain differences for nesting behavior in *Mus musculus. Behavior Genetics* **12**: 149–159.

Bauer M, Ueffing M, Meitinger T, Gasser T (1999). Somatic gene therapy in animal models of Parkinson's disease. *Journal of Neural Transmission* **55**: Supplement 131–147.

Baulieu EE (1989). Contragestion and other clinical applications of RU486, an antiprogesterone at the receptor. *Science* **245**: 1351–1357.

Baum MJ (2003). Activational and organizational effects of estradiol on male behavioral neuroendocrine function. *Scandanavian Journal Psychology* **44**: 213–220.

Baum MJ, Keverne EB (2002). Sex difference in attraction threshold for volatile odors from male and estrous female mouse urine. *Hormones and Behavior* 41: 213–219.

Baum MJ, Sodersten P, Vreeburg JT (1974). Mounting and receptive behavior in the ovariectomized female rat: Influence of estradiol, dihydrotestosterone, and genital anesthetization. *Hormones and Behavior* **5**: 175–190.

Baumgartner A, Graf KJ, Kurten I (1985). The dexamethasone suppression test in depression, in schizophrenia, and during experimental stress. *Biological Psychiatry* **20**: 675–679.

Baunez C, Robbins TW (1999). Effects of transient inactivation of the subthalamic nucleus by local muscimol and APV infusions on performance on the five-choice serial reaction time task in rats. *Psychopharmacology* **141**: 57–65.

Baylor DA (1996). How photons start vision. *Proceedings of the National Academy of Sciences USA* **93**: 560–565.

Beach FA, Jaynes J (1956). Studies of maternal retrieving in rats. I: Recognition of young. *Journal of Mammalogy* **37**: 177–180.

Beal MF, Ferrante RJ (2004). Experimental therapeutics in transgenic mouse models of Huntington's disease. *Nature Reviews Neuroscience* **5**: 373–384.

Beauchamp GK, Yamazaki K (2003). Chemical signaling in mice. *Biochemical Society Transactions* **31**: 147–151.

Beauchamp MH, Ormerod BK, Jhamandas K, Boegman RJ, Beninger RJ (2000). Neurosteroids and reward: allopregnanolone produces a conditioned place aversion in rats. *Pharmacology Biochemistry and Behavior* **67**: 29–35.

Beck B (2000). Neuropeptides and obesity. *Nutrition* **16**: 916–923.

Becker A, Grecksch G, Brodemann R, Kraus J, Peters B, Schroeder H, Thiemann W, Loh HH, Hollt V (2000). Morphine self-administration in mu-opioid receptor-deficient mice. *Naunyn Schmiedebergs Arch Pharmacology* **361**: 584–589.

Becker A, Grecksch G, Kraus J, Loh HH, Schroeder H, Hollt V (2002). Rewarding effects of ethanol and cocaine in mu opioid receptor-deficient mice. *Naunyn Schmiedebergs Arch Pharmacology* **365**: 296–302.

Becker JB, Breedlove SM, Crews D (1993). *Behavioral Endocrinology*. MIT Press, Cambridge.

Bedell MA, Jenkins NA, Copeland NG (1997). Mouse models of human disease. Part I: Techniques and resources for genetic analysis in mice. *Genes and Development* **11**: 1–10.

Beglopoulos V, Shen J (2006). Regulation of CRE-dependent transcription by presenilins: Prospects for therapy of Alzheimer's disease. *Trends in Pharmacological Sciences* **27**: 33–40.

Beiko J, Lander R, Hampson E, Boon F, Cain DP (2004). Contribution of sex differences in the acute stress response to sex differences in water maze performance in the rat. *Behavioural Brain Research* **151**: 239–253.

Belknap JK (1998). Effect of within-strain sample size on QTL detection and mapping using recombinant inbred mouse strains. *Behavior Genetics* **28**: 29–38.

Belknap JK (2003). Chromosome substitution strains: some quantitative considerations for genome scans and fine mapping. *Mammalian Genome* **14**: 723–732.

Belknap JK, Crabbe JC, Plomin R, McClearn GE, Sampson KE, O'Toole LA, Gora-Maslak G (1992). Single-locus control of saccharin intake in BXD/Ty recombinant inbred (RI) mice: Some methodological implications for RI strain analysis. *Behavior Genetics* **22**: 81–100.

Belknap JK, Mitchell SR, O'Toole LA, Helms ML, Crabbe JC (1996). Type I and type II error rates for quantitative trait loci (QTL) mapping studies using recombinant inbred mouse strains. *Behavior Genetics* **26**: 149–160.

Belknap JK, Richards SP, O'Toole LA, Helms ML, Phillips TJ (1997). Short-term selective breeding as a tool for QTL mapping: Ethanol preference drinking in mice. *Behavior Genetics* **27**: 55–66.

Belluscio L, Gold GH, Nemes A, Axel R (1998). Mice deficient in G_{olf} are anosmic. *Neuron* **20**: 69–81.

Belzung C, Misslin R, Vogel E (1989). Behavioural effects of the benzodiazepine receptor partial agonist RO 16–6028 in mice. *Psychopharmacology* **97**: 388–391.

Bengel D, Murphy DL, Andrews AM, Wichems CH, Feltner D, Heils A, Mossner R, Westphal H, Lesch KP (1998). Altered brain serotonin in homeostasis and locomotor sensitivity to 3,4-methylenedioxymethamphetamine ("Ecstasy") in serotonin transporter-deficient mice. *Molecular Pharmacology* **53**: 649–655.

Benjamin J, Li L, Patterson C, Greenberg BD, Murphy DL, Hamer D (1996). Population and familial association between the D4 dopamine receptor gene and measures of novelty seeking. *Nature Genetics* **12**: 81–84.

Bensadoun JC, Brooks SP, Dunnett SB (2004). Free operant and discrete trial performance of mice in the nine-hole box apparatus: Validation using amphetamine and scopolamine. *Psychopharmacology* **174**: 396–405.

Ben-Yosef T, Belyantseva IA, Saunders TL, Hughes ED, Kawamoto K, Van Itallie CM, Beyer LA, Halsey K, Gardner DJ, Wilcox ER, Rasmussen J, Anderson JM, Dolan DF, Forge A, Raphael Y, Camper SA, Friedman TB (2003). Claudin 14 knockout mice, a model for autosomal recessive deafness DFNB29, are deaf due to cochlear hair cell degeneration. *Human Molecular Genetics* **12**: 2049–2061.

Berenthal BI, Campos JJ (1984). A reexamination of fear and its determinants on the visual cliff. *Psychophysiology* **21**: 413–417.

Berg P, Singer M (1998). Regulating human cloning. *Science* **282**: 413.

Bergenson SE, Warren RK, Crabbe JC, Metten P, Erwin VG, Belknap JK (2003). Chromosomal loci influencing chronic alcohol withdrawal severity. *Mammalian Genome* **14**: 454–463.

Berger-Sweeney J (2003a). The cholinergic basal forebrain system during development and its influence on cognitive processes: Important questions and potential answers. *Neuroscience and Biobehavioral Reviews* **27**: 401–411.

Berger-Sweeney J (2003b). Using mice to model cognitive deficits in neurologic disorders: Narrowing in on Rett syndrome. *Current Neurology and Neuroscience Reports* **3**: 185–187.

Berger-Sweeney J, Hohmann CF (1997). Behavioral consequences of abnormal cortical development: Insights into developmental disabilities. *Behavioural Brain Research* **86**: 121–142.

Berger-Sweeney J, McPhie DL, Arters JA, Greenan J, Oster-Granite ML, Neve RL (1999). Impairments in learning and memory accompanied by neurodegeneration in mice transgenic for the carboxyl-terminus of the amyloid precursor protein. *Molecular Brain Research* **66**: 150–162.

Bernstein IL, Goehler LE, Bouton ME (1983). Relative potency of food and drinks as targets in aversion conditioning. *Behavioral and Neural Biology* **37**: 134–148.

Berrendero F, Kieffer BL, Maldonado R (2002). Attenuation of nicotine-induced antinociception, rewarding effects, and dependence in mu-opioid receptor knock-out mice. *Journal of Neuroscience* **22**: 10935–10940.

Berrettini W (2004). Bipolar disorder and schizophrenia: Convergent molecular data. 5 Neuromolecular Medicine **5**: 109–117.

Berrettini W, Ferraro TN, Alexander RC, Buchberg AM, Vogel WH (1994). Quantitative trait loci mapping of three loci controlling morphine preference using inbred mouse strains. *Nature Genetics* **7**: 54–58.

Berthold AA (1849). Transplantation der Hoden. *Archives of Anatomy and Physiology*: 42–46.

Betz AL, Shakui P, Davidson BL (1998). Gene transfer to rodent brain with recombinant adenoviral vectors: Effects of infusion parameters, infectious titer, and virus concentration on transduction volume. *Experimental Neurology* **150**: 136–142.

Bhalla US, Iyengar R (1999). Emergent properties of networks of biological signalling pathways. *Science* **283**: 381–387.

Bi S, Scott KA, Kopin AS, Moran TH (2004). Differential roles for cholecystokinin-A receptors in energy balance in rats and mice. *Endocrinology* **145**: 3873–3880.

Bielsky IF, Hu SB, Szegda KL, Westphal H, Young LJ (2004). Profound impairment in social recognition and reduction in anxiety-like behavior in vasopressin V1A receptor knockout mice. *Neuropsychopharmacology* **29**: 483–493.

Bignami G (1996). Economical test methods for developmental neurobehavioral toxicity. *Environmental Health Perspectives* **104** (Suppl 2): 285–298.

Black RW, Albiniak T, Davis M, Schumpert J (1973). A preference in rats for cues associated with intoxication. *Bulletin of the Psychonomic Society* **2**: 423–424.

Blakeman KH, Hao JX, Xu XJ, Jacoby AS, Shine J, Crawley JN, Iismaa T, Wiesenfeld-Hallin Z (2003). Hyperalgesia and increased neuropathic pain-like responses in mice lacking galanin receptor 1 receptors. *Neuroscience* **117**: 221–227.

Blanchard DC, Blanchard RJ (1969). Crouching as an index of fear. *Journal of Comparative and Physiological Psychology* **67**: 370–375.

Blanchard DC, Blanchard RJ (1988). Ethoexperimental approaches to the biology of emotion. *Annual Review of Psychology* **39**: 43–68.

Blanchard DC, Blanchard RJ, Griebel G (2005). Defensive responses to predator threat in the rat and mouse. *Current Protocols in Neuroscience*. Wiley, New York, pp. 8.19.1–8.20.

Blanchard DC, Griebel G, Blanchard RJ (2001). Mouse defensive behaviors: Pharmacological and behavioral assays for anxiety and panic. *Neuroscience and Biobehavioral Reviews* **25**: 205–218.

Blanchard DC, Griebel G, Blanchard RJ (2003a).The mouse defense test battery: Pharmacological and behavioral assays for anxiety and panic. *European Journal of Pharmacology* **463**: 97–116.

Blanchard RJ, Blanchard DC (1977). Aggressive behavior in the rat. *Behavioral Biology* **21**: 197–224.

Blanchard RJ, Blanchard DC, Agullana R, Weiss SM (1991). Twenty-two kHz alarm cries to presentation of a predator, by laboratory rats living in visible burrow systems. *Physiology and Behavior* **50**: 967–972

Blanchard RJ, Hebert MA, Ferrari P, Palanza P, Figueira R, Blanchard DC, Parmigiani S (1998). Defensive behaviors in wild and laboratory (Swiss) mice: The mouse defense test battery. *Physiology and Behavior* **65**: 201–209.

Blanchard RJ, Wall PM, Blanchard DC (2003b). Problems in the study of rodent aggression. *Hormones and Behavior* **44**: 161–170.

Blasberg ME, Robinson S, Henderson LP, Clark AS (1998). Inhibition of estrogen-induced sexual receptivity by androgens: Role of the androgen receptor. *Hormones and Behavior* **34**: 283–293.

Blasi G, Mattay VS, Bertolino A, Elvevag B, Callicott JH, Das S, Kolachana BS, Egan MF, Goldberg TE, Weinberger DR (2005). Effect of catechol-*O*-methyltransferase val158met genotype on attentional control. *Journal of Neuroscience* **25**: 5038–5045.

Blau HM, Rossi FMV (1999). Tet B or not tet B: Advances in tetracycline-inducible gene expression. *Proceedings of the National Academy of Sciences USA* **96**: 797–799.

Bliss TVP (1998). The saturation debate. *Science* **281**: 1975–1976.

Blizard DA (1992). Recombinant-inbred strains: General methodological considerations relevant to the study of complex characters. *Behavior Genetics* **22**: 621–633.

Blizard DA (1981). The Maudsley Reactive and Nonreactive strains: A North American perspective. *Behavior Genetics* **11**: 469–489.

Blizard DA, Bailey D (1979). Genetic correlation between open-field activity and defecation: analysis with the CXB recombinant-inbred strains. *Behavior Genetics* **9**: 349–57.

Blömer U, Naldini L, Verma IM, Trono D, Gage FH (1996). Applications of gene therapy to the CNS. *Human Molecular Genetics* **5**: 1397–1404.

Bloom FE, Reilly JF, Redwine JM, Wu CC, Young WG, Morrison JH (2005). Mouse models of human neurodegenerative disorders: Requirements for medication development. *Archives of Neurology* **62**: 185–187.

Bloom FE, Young WG (1994). *Brain Browser*. Academic Press, New York.

Blum K, Cull JG, Braverman ER, Comings DE (1996). Reward deficiency syndrome. *American Scientist* **84**: 132–145.

Blumstein LK, Crawley JN (1983). Further characterization of a simple automated exploratory model for the anxiolytic effects of benzodiazepines. *Pharmacology Biochemistry and Behavior* **18**: 37–40.

Blundell J (1979). Hunger, appetite and satiety—Constructs in search of identities. In *Proceedings of the British Nutrition Foundation*, M Turner, Ed. Applied Sciences, London, pp. 21–42.

Blundell JE (1977). Is there a role of serotonin (5–hydroxytryptamine) in feeding? *International Journal of Obesity* **1**: 15–42.

Bluthe RM, Gheusi G, Dantzer R (1993). Gonadal steroids influence the involvement of arginine vasopressin in social recognition in mice. *Psychoneuroendocrinology* **18**: 323–335.

Boder E (1975). Ataxia-telangiectasia: Some historic, clinical and pathologic observations. *Birth Defects* **11**: 255–270.

Bodnar RJ (2004). Endogenous opioids and feeding behavior: A 30-year historical perspective. *Peptides* **25**: 697–725.

Bodyak N, Slotnick B (1999). Performance of mice in an automated olfactometer: Odor detection, discrimination and odor memory. *Chemical Senses* **24**: 637–645.

Boehm SL II, Schafer GL, Phillips TJ, Browman KE, Crabbe JC (2000). Sensitivity to ethanol-induced motor incoordination in 5-HT(1B) receptor null mutant mice is task-dependent: implications for behavioral assessment of genetically altered mice. *Behavioral Neuroscience* **114**: 401–409.

Bogue MA, Grubb SC (2004). The mouse phenome project. *Genetica* **122**: 71–74.

Boheler KR, Stern MD (2003). The new role of SAGE in gene discovery. *Trends in Biotechnolology* **21**: 55–58.

Boillíe S, Cleveland DW (2004). Gene therapy for ALS delivers. *Trends in Neuroscience* **27**: 235–238.

Bolivar VJ, Caldarone BJ, Reilly AA, Flaherty L (2000a). Habituation of activity in an open field: A survey of inbred strains and F1 hybrids. *Behavior Genetics* **30**: 285–293.

Bolivar V, Cook M, Flaherty L (2000b). List of transgenic and knockout mice: behavioral profiles. *Mammalian Genome* **11**: 260–274.

Bolivar V, Cook M, Flaherty L (2001b). Mapping of quantitative trait loci with knockout/congenic strains. *Genome Research* **11**: 1549–1552.

Bolivar VJ, Flaherty L (2003). Assessing autism-like behaviors in inbred strains of mice. *Society for Neuroscience* 318.13.

Bolivar VJ, Flaherty L (2004). Genetic control of novel food preference in mice. *Mammalian Genome* **15**: 195–198.

Bolivar VJ, Manley K, Messer A (2003). Exploratory activity and fear conditioning abnormalities develop early in R6/2 Huntington's disease transgenic mice. *Behavioral Neuroscience* **117**: 1233–1242.

Bolivar VJ, Pooler O, Flaherty L (2001a). Inbred strain variation in contextual and cued fear conditioning behavior. *Mammalian Genome* **12**: 651–656.

Bond TL, Neumann PE, Mathieson WB, Brown RE (2002). Nest building in nulligravid, primigravid and primiparous C57BL/6J and DBA/2J mice (*Mus musculus*). *Physiology and Behavior* **75**: 551–555.

Booth DA (1985). Food-conditioned eating preferences and aversions with interoceptive elements: Conditioned appetites and satieties. *Annals of the New York Academy of Sciences* **443**: 22–41.

Borsini F (1995). Role of the serotonergic system in the forced swimming test. *Neuroscience and Biobehavioral Reviews* **19**: 377–395.

Bothe GWM, Bolivar VJ, Vedder MG, Geistfeld JG (2004). Genetic and behavioral differences among five inbred mouse strains commonly used in the production of transgenic and knockout mice. *Genes, Brain and Behavior* **3**: 149–157.

Botstein D, White RL, Skolnick M, Davis RW (1980). Construction of a genetic linkage map in man using restriction fragment length polymorphisms. *American Journal of Human Genetics* **32**: 314–331.

Boulay D, Depoortere R, Rostene W, Perrault G, Sanger DJ (1999). Dopamine D_3 receptor agonists produce similar decreases in body temperature and locomotor activity in D_3 knockout and wildtype mice. *Neuropharmacology* **38**: 555–565.

Boulton AA, Baker GB, Wu PW, Eds. (1992). *Animal Models of Drug Addiction.* Humana, Totowa, NJ.

Bourtchuladze R, Frenguelli B, Blendy J, Cioffi D, Schultz G, Silva AJ (1994). Deficient long-term memory in mice with a targeted mutation of the cAMP-responsive element-binding protein. *Cell* **79**: 59–68.

Boutrel B, Franc B, Hen R, Hamon M, Adrien J (1999). Key role of 5-HT_{1B} receptors in the regulation of paradoxical sleep as evidenced in 5-HT_{1B} knock-out mice. *Journal of Neuroscience* **19**: 3204–3212.

Bouvet D, Bovet-Nitti F, Oliverio A (1969). Genetic aspects of learning and memory in mice. *Science* **163**: 139–149.

Bouwknecht JA, Paylor R (2002). Behavioral and physiological mouse assays for anxiety: A survey in nine mouse strains. *Behavioural Brain Research* **136**: 489–501.

Bouwknecht JA, van der Gugten J, Hijzen TH, Maes RA, Hen R, Olivier B (2001). Male and female 5-HT(1B) receptor knockout mice have higher body weights than wildtypes. *Physiology and Behavior* **74**: 507–516.

Bowers BJ, Collins AC, Tritto T, Wehner JM (2000). Mice lacking PKC gamma exhibit decreased anxiety. *Behavior Genetics* **30**: 111–121.

Bowers BJ, McClure-Begley TD, Keller JJ, Paylor R, Collins AC, Wehner JM (2005). Deletion of the alpha7 nicotinic receptor subunit gene results in increased sensitivity to several behavioral effects produced by alcohol. *Alcoholism: Clinical and Experimental Research* **29**: 295–302.

Bowers BJ, Owen EH, Collins AC, Abeliovich A, Tonegawa S, Wehner JM (1999). Decreased ethanol sensitivity and tolerance development in γ-protein kinase C null mutant mice is dependent on genetic background. *Alcoholism: Clinical and Experimental Research* **23**: 387–397.

Boyce-Rustay JM, Risinger FO (2003). Dopamine D3 receptor knockout mice and the motivational effects of ethanol. *Pharmacology Biochemistry and Behavior* **75**: 373–379.

Bozarth MA, Wise RA (1981). Intracranial self-administration of morphine into the ventral tegmental area in rats. *Life Sciences* **28**: 551–555.

Bozon B, Kelly A, Josselyn SA, Silva AJ, Davis S, Laroche S (2003). MAPK, CREB and zif268 are all required for consolidation of recognition memory. *Philosophical Transactions of the Royal Society B: Biological Science* **358**: 805–814.

Bradberry CW, Roth RH (1989). Cocaine increases extracellular dopamine in rat nucleus accumbens and ventral tegmental area as shown by in vivo microdialysis. *Neuroscience Letters* **103**: 97–102.

Bradley A, Hasty P, Davis A, Ramirez-Solis R (1992). Modifying the mouse: Design and desire. *BioTechnology* **10**: 534–539.

Bradley DW, Joyce D, Murphy EH, Nash BM, Porsolt RD, Summerfield A, Twyman WA (1968). Amphetamine-barbiturate mixture: Effects on the behaviour of mice. *Nature* **220**: 187–188.

Bradwejn J, Koszycki D (2001). Cholecystokinin and panic disorder: Past and future clinical research strategies. *Scandanavian Journal of Clinical and Laboratory Investigation* **234**: 19–27.

Braff DL, Geyer MA (1990). Sensorimotor gating and schizophrenia: Human and animal studies. *Archives of General Psychiatry* **47**: 181–188.

Braff DL, Light GA (2004). Preattentional and attentional cognitive deficits as targets for treating schizophrenia. *Psychopharmacology* **174**: 75–85.

Braff DL, Stone C, Callaway E, Geyer MA, Glick ID, Bali L (1978). Prestimulus effects on human startle reflex in normals and schizophrenics. *Psychophysiology* **15**: 339–343.

Brain PF, Mainardi D, Parmigiani S, Eds. (1989). *House Mouse Aggression: A Model for Understanding the Evolution of Social Behaviour*. Chur, Switzerland: Harwood Academic.

Brain P, Poole A (1974). Some studies on the use of "standard opponents" in intermale aggression testing in TT albino mice. *Behaviour* **50**: 100–110.

Branchi I, Bichler Z, Berger-Sweeney J, Ricceri L (2003). Animal models of mental retardation: From gene to cognitive function. *Neuroscience and Biobehavioral Reviews* **27**: 141–153.

Branchi I, Ricceri L (2002). Transgenic and knock-out mouse pups: The growing need for behavioral analysis. *Genes, Brain and Behavior* **1**: 135–141.

Branchi I, Ricceri L (2005). Scoring learning and memory in developing rodents. *Current Protocols in Toxicology* 13.11.1–13.11.14.

Branchi I, Sanatucci D, Puopolo M, Alleva E (2004). Neonatal behaviors associated with ultrasonic vocalizations in mice (*Mus musculus*): A slow-motion analysis. *Developmental Psychobiology* **44**: 37–44.

Branchi I, Santucci D, Vitale A, Alleva E (1998). Ultrasonic vocalizations by infant laboratory mice: A preliminary spectrographic characterization under different conditions. *Developmental Psychobiology* **33**: 249–256.

Brandon EP, Logue SF, Adams MR, Qi M, Sullivan SP, Matsumoto AM, Dorsa DM, Wehner JM, McKnight GS, Idzerda RL (1998). Defective motor behavior and neural gene expression in RIIβ-protein kinase A mutant mice. *Journal of Neuroscience* **18**: 3639–3649.

Braveman NS, Bronstein P, Eds. (1985). *Experimental Assessments and Clinical Applications of Conditioned Taste Aversions*. New York Academy of Sciences, New York.

Bray GA (1987). Hypothalamic and genetic obesity: an appraisal of the autonomic hypothesis and the endocrine hypothesis. *International Journal of Obesity* 8 (Suppl 1): 33–43.

Breckenridge A, Lindpaintner K, Lipton P, McLeod H, Rothstein M, Wallace H (2004). Pharmacogenetics: Ethical problems and solutions. *Nature Reviews Genetics* **5**: 676–680.

Bredt DS, Snyder SH (1994). Nitric oxide: A physiologic messenger molecule. *Annual Review of Biochemistry* **63**: 175–195.

Bredy TW, Grant RJ, Champagne DL, Meaney MJ (2003). Maternal care influences neuronal survival in the hippocampus of the rat. *European Journal of Neuroscience* **18**: 2903–2909.

Breer H, Wanner I, Strotmann J (1996). Molecular genetics of mammalian olfaction. *Behavior Genetics* **26**: 209–219.

Breier A (1989). Experimental approaches to human stress research: Assessment of neurobiological mechanisms of stress in volunteers and psychiatric patients. *Biological Psychiatry* **26**: 438–462.

Breier A (1996). *The New Pharmacotherapy of Schizophrenia*. American Psychiatric Press, Washington, DC.

Brennan PA, Keverne EB (2004). Something in the air? New insights into mammalian pheromones. *Current Biology* **14**: R81-R89.

Brennan PA, Schellinck HM, De La Riva C, Kendrick DM, Keverne EB (1998). Changes in neurotransmitter release in the main olfactory bulb following an olfactory conditioning procedure in mice. *Neuroscience* **87**: 583–590.

Brennan TJ, Seeley WW, Kilgard M, Schreiner CE, Tecott LH (1997). Sound-induced seizures in serotonin 5-HT$_{2C}$ receptor mutant mice. *Nature Genetics* **16**: 387–390.

Brenner MK (1996). Gene transfer and therapeutic drug monitoring. *Therapeutic Drug Monitoring* **18**: 322–327.

Brink EE, Modianos DT, Pfaff DW (1980). Ablations of lumbar epaxial musculature: Effects on lordosis behavior of female rats. *Brain Behavior and Evolution* **17**: 67–88.

Brioni JD, McGaugh JL (1988). Post-training administration of GABAergic antagonists enhances retention of aversively motivated tasks. *Psychopharmacology* **96**: 505–510.

Broadhurst PL (1961). Analysis of maternal effects in the inheritance of behavior. *Animal Behavior* **9**: 129–141.

Broder S, Caplan A, Evans WE (2002). Therapeutic horizons—The human genome. *Journal of the American Pharmacetical Association (Wash)* **42**(5 Suppl 1): S22–S23.

Brodkin ES, Carlezon WA, Haile CN, Kosten TA, Heninger GR, Nestler EJ (1998). Genetic analysis of behavioral, neuroendocrine, and biochemical parameters in inbred rodents: Initial studies in Lewis and Fischer 344 rats and in A/J and C57BL/6J mice. *Brain Research* **805**: 55–68.

Brodkin ES, Hagemann A, Nemetski SM, Silver LM (2004). Social approach-avoidance behavior of inbred mouse strains towards DBA/2 mice. *Brain Research* **1002**: 151–157.

Brodkin ES, Sankoorikal GM, Kaercher KA, Boon CJ, Lee JK (2006). A mouse model system for genetic analysis of sociability: C57BL/6J versus BALB/cJ inbred mouse strains. *Biological Psychiatry* **59**: 415–423.

Brody SA, Conquet F, Geyer MA (2003). Disruption of prepulse inhibition in mice lacking mGluR1. *European Journal of Neuroscience* **18**: 3361–3366.

Brody SA, Geyer MA (2004). Interactions of the mGluR5 gene with breeding and maternal factors on startle and prepulse inhibition in mice. *Neurotoxicology Research* **6**: 79–90.

Bronson SK, Smithies o(1994). Altering mice by homologous recombination using embryonic stem cells. *Journal of Biological Chemistry* **269**: 27155–27158.

Brooks, SP, Pask T, Jones L, Dunnett SB (2004). Behavioral profiles of inbred mouse strains used as transgenic background: 1. Motor tests. *Genes, Brain and Behavior* **3**: 206–215.

Brown GL, Ebert MH, Goyer PF, Jimerson DC, Klein WJ, Bunney WE, Goodwin FK (1982). Aggression, suicide, and serotonin: Relationships to CSF amine metabolites. *American Journal of Psychiatry* **139**: 741–746.

Brown JR, Ye H, Bronson RT, Dikkes P, Greenberg ME (1996). A defect in nurturing in mice lacking the immediate early gene *fosB*. *Cell* **86**: 297–309.

Brown PO, Botstein D (1999). Exploring the new world of the genome with DNA microarrays. *Nature Genetics* **21** (January Suppl): 33–38.

Brown SD (1998). Mouse models of genetic disease: New approaches, new paradigms. *Journal of Inherited Metabolic Disease* **21**: 532–539.

Brown V, Jin P, Ceman S, Darnell JC, O'Donnell WT, Tenenbaum SA, Jin X, Feng Y, Wilkinson KD, Keene JD, Darnell RB, Warren ST (2001). Microarray identification of FMRP-associated brain mRNAs and altered mRNA translational profiles in fragile X syndrome. *Cell* **107**: 477–487.

Brudzynski SM, Chiu EM (1995). Behavioural responses of laboratory rats to playback of 22 kHz ultrasonic calls. *Physiology and Behavior* **57**: 1039–1044.

Brunelli SA (2005). Selective breeding for an infant phenotype: rat pup ultrasonic vocalization (USV). *Behavior Genetics* **35**: 53–65.

Brunelli SA, Myers MM, Asekoff SL, Hofer MA (2002). Effects of selective breeding for infant rat ultrasonic vocalization on cardiac responses to isolation. *Behavioral Neuroscience* **116**: 612–623.

Brunelli SA, Shindledecker RD, Hofer MA (1989). Early experience and maternal behavior in rats. *Developmental Psychobiology* **22**: 295–314.

Brunelli SA, Vinocur DD, Soo-Hoo D, Hofer MA (1997). Five generations of selective breeding for ultrasonic vocalization (USV): Responses in N:NIH strain rats. *Developmental Psychobiology* **31**: 255–265.

Brunner D, Nestler E, Leahy E (2002). In need of high-throughput behavioral systems. *Drug Discovery Today* **7**(18 Suppl): S107–S112.

Brusa R, Zimmermann F, Koh DS, Feldmeyer D, Gass P, Seeburg PH, Sprengel R (1995). Early-onset epilepsy with postnatal lethality associated with an editing-deficient GluR-B allele in mice. *Science* **270**: 1677–1680.

Bucan M, Abel T (2002). The mouse: Genetics meets behaviour. *Nature Reviews Genetics* **3**: 114–123.

Buccafusco JJ (2000). *Methods of Behavioral Analysis in Neuroscience*. CRC Press, Boca Raton, FL.

Buchner DA, Seburn KL, Frankel WN, Meisler MH (2004). Three ENU-induced neurological mutations in the pore loop of sodium channel Scn8a (Na(v)1.6) and a genetically linked retinal mutation, rd13. *Mammalian Genome* **15**: 344–351.

Buchner DA, Trudeau M, Meisler MH (2003). SCNM1, a putative RNA splicing factor that modifies disease severity in mice. *Science* **301**: 967–969.

Buck KJ, Rademacher BLS, Metten P, Crabbe JC (2002). Mapping murine loci for physical dependence on ethanol. *Psychopharmacology* **160**: 398–407.

Buck LB (2004). The search for odorant receptors. *Cell* **166**: S117–S119.

Buckner RL, Raichle ME, Miezin FM, Petersen SE (1996). Functional-anatomical studies of the recall of pictures and words from memory. *Journal of Neuroscience* **16**: 6219–6235.

Bukalo O, Fentrop N, Lee AY, Salmen B, Law JW, Wotjak CT, Schweizer M, Dityatev A, Schachner M (2004). Conditional ablation of the neural cell adhesion molecule reduces precision of spatial learning, long-term potentiation, and depression in the CA1 subfield of mouse hippocampus. *Journal of Neuroscience* **24**: 1565–1577.

Bulloch K, Hamburger RN, Loy R (1982). Nest-building behavior in two cerebellar mutant mice: Staggerer and Weaver. *Behavioral and Neural Biology* **36**: 94–97.

Buning H, Braun-Falco M, Hallek M (2004). Progress in the use of adeno-associated viral vectors for gene therapy. *Cells Tissues Organs* **177**: 139–150.

Bunsey M, Eichenbaum H (1995). Selective damage to the hippocampal region blocks long-term retention of a natural and nonspatial stimulus-stimulus association. *Hippocampus* **5**: 542–556.

Burkhalter JE, Balster RL (1979). Behavioral teratology evaluation of trichloromethane in mice. *Neurobehavioral Toxicology* **1**: 199–205.

Burnett AL, Lowenstein CJ, Bredt DS, Chang TSK, Synder SH (1992). Nitric oxide: A physiological mediator of penile erection. *Science* **257**: 401–403.

Burns-Cusato M, Scordalakes EM, Rissman EF (2004). Of mice and missing data: What we know (and need to learn) about male sexual behavior. *Physiology and Behavior* **83**: 217–232.

Burright EN, Orr HT, Clark HB (1997). Mouse models of human CAG repeat disorders. *Brain Pathology* **7**: 965–977.

Burris J, Cook-Deegan R, Alberts B (1998). The Human Genome Project after a decade: Policy issues. *Nature Genetics* **20**: 333–335.

Burt DR (2003). Reducing GABA receptors. *Life Sciences* **73**: 1741–1758.

Burton MJ, Cooper SJ, Popplewell DA (1981). The effect of fenfluramine on the microstructure of feeding and drinking in the rat. *British Journal of Pharmacology* **72**: 621–633.

Bushnell PJ (1998). Behavioral approaches to the assessment of attention in animals. *Psychopharmacology* **138**: 231–259.

Bussey TJ, Saksida LM, Rothblat LA (2001). Discrimination of computer-graphic stimuli by mice: a method for the behavioral characterization of transgenic and gene-knockout models. *Behavioral Neuroscience* **115**: 957–960.

Butler AA, Cone RD (2002). The melanocortin receptors: Lessons from knockout models. *Neuropeptides* **36**: 77–84.

Butler AA, Marks DI, Fan W, Kuhn CM, Bartolome M, Cone RD (2001). Melanocortin-4 receptor is required for acute homeostatic responses to increased dietary fat. *Nature Neuroscience* **4**: 605–611.

Buttini M, Yu GQ, Shockley K, Huang Y, Jones B, Masliah E, Mallory M, Yeo T, Longo FM, Mucke L (2002). Modulation of Alzheimer-like synaptic and cholinergic deficits in transgenic mice by human apolipoprotein E depends on isoform, aging, and overexpression of amyloid beta peptides but not on plaque formation. *Journal of Neuroscience* **22**: 10539–10548.

Byrne GW, Ruddle FH (1989). Multiplex gene regulation: A two-tiered approach to transgene regulation in transgenic mice. *Proceedings of the National Academy of Sciences USA* **86**: 5473–5477.

Cabib S, Puglisi-Allegra S, Ventura R (2002). The contribution of comparative studies in inbred strains of mice to the understanding of the hyperactive phenotype. *Behavioural Brain Research* 130: 103–109.

Cabib S, Pascucci T, Ventura R, Romano V, Puglisi-Allegra S (2003). The behavioral profile of severe mental retardation in a genetic mouse model of phenylketonuria. *Behavior Genetics* **33**: 301–310.

Caine SB, Negus SS, Mello NK, Patel S, Bristow L, Kulagowski J, Vallone D, Saiardi A, Borrelli E (2002). Role of dopamine D2-like receptors in cocaine self-administration: Studies with D2 receptor mutant mice and novel D2 receptor antagonists. *Journal of Neuroscience* **22**: 2977–2988.

Calamandrei G, Wilkinson LS, Keverne EB (1992). Olfactory recognition of infants in laboratory mice: Role of noradrenergic mechanisms. *Physiology and Behavior* **52**: 901–907.

Calatayud F, Belzung C (2001). Emotional reactivity in mice, a case of nongenetic heredity? *Physiology and Behavior* **74**: 355–362.

Caldarone BJ, Karthigeyan K, Harrist A, Hunsberger JG, Wittmack E, King SL, Jatlow P, Picciotto MR (2003). Sex differences in response to oral amitriptyline in three animal models of depression in C57BL/6J mice. *Psychopharmacology* **170**: 94–101.

Caldarone B, Saavedra C, Tartaglia K, Wehner JM, Dudek BC, Flaherty L (1997). Quantitative trait loci affecting contextual conditioning in mice. *Nature Genetics* **17**: 335–337.

Caldwell JD, Johns JM, Faggin BM, Senger MA, Pedersen CA (1994). Infusion of an oxytocin antagonist into the medial preoptic area prior to progesterone inhibits sexual receptivity and increases rejection in female rats. *Hormones and Behavior* **28**: 288–302.

Caldwell JD, Prange AJ, Pedersen CA (1986). Oxytocin facilitates the sexual receptivity of estrogen-treated female rats. *Neuropeptides* **7**: 175–189.

Calhoun JB (1962a). Behavioral Sink. In *Roots of Behavior*. EL Bliss, Ed. Harper and Row, New York, pp. 295–315.

Calhoun JB (1962b). Population density and social pathology. *Scientific American* **206**: 139–149.

Calne DB, Snow BJ, Lee C (1992). Criteria for diagnosing Parkinson's disease. *Annals of Neurology* **32** (Suppl): S125–S127.

Camille N, Coricelli G, Sallet J, Pradat-Diehl P, Duhamel JR, Sirigu A (2004). The involvement of the orbitofrontal cortex in the experience of regret. *Science* **304**: 1167–1170.

Campbell IL, Gold LH (1996). Transgenic modeling of neuropsychiatry disorders. *Molecular Psychiatry* **1**: 105–120.

Campfield LA, Smith FJ, Burn P (1998). Strategies and potential molecular targets for obesity treatment. *Science* **280**: 1383–1387.

Canastar A, Maxson SC (2003). Sexual aggression in mice: effects of male strain and of female estrous state. *Behavior Genetics* **33**: 521–528.

Cannon CM, Bseikri MR (2004). Is dopamine required for natural reward? *Physiology and Behavior* **81**: 741–748.

Cao W, Burkholder T, Wilkins L, Collins AC (1993). A genetic comparison of behavioral actions of ethanol and nicotine in the mirrored chamber. *Pharmacology Biochemistry and Behavior* **45**: 803–809.

Capecchi MR (1989). Altering the genome by homogous recombination. *Science* **244**: 1288–1292.

Capecchi MR (1994). Targeted gene replacement. *Scientific American* **270**: 52–59.

Caplen NJ (2004). Gene therapy progress and prospects. Downregulating gene expression: The impact of RNA interference. *Gene Therapy* **11**: 1241–1248.

Capone F, Puopolo M, Branchi I, Alleva E (2002). A new easy accessible and low-cost method for screening olfactory sensitivity in mice: Behavioural and nociceptive responses in male and female CD-1 mice upon exposure to millipede aversive odour. *Brain Research Bulletin* **58**: 193–202.

Cappell H, LeBlanc AE (1977). In *Food Aversion Learning*, NW Milgram, L Krames, TM Alloway, Eds. Plenum, New York, pp. 133–167.

Capuron L, Dantzer R (2003). Cytokines and depression: the need for a new paradigm. *Brain, Behavior, and Immunity* **17** (Suppl 1): S119–S124.

Cardon LR, Smith SD, Fulker DW, Kimberling WJ, Pennington BF, DeFries JC (1994). Quantitative trait locus for reading disability on chromosome 6. *Science* **266**: 276–279.

Carelli RM (2004). Nucleus accumbens cell firing and rapid dopamine signaling during goal-directed behaviors in rats. *Neuropharmacology* **47** (Suppl) 1: 180–189.

Carew TJ (1996). Molecular enhancement of memory formation. *Neuron* **16**: 5–8.

Carlezon WA, Boundy VA, Haile CN, Lane SB, Kalb RG, Neve RL, Nestler EJ (1997). Sensitization to morphine induced by viral-mediated gene transfer. *Science* **277**: 812–814.

Carlezon WA, Thome J, Olson VG, Lane-Ladd SB, Brodkin ES, Hiroi N, Duman RS, Neve RL, Nestler EJ (1998). Regulation of cocaine reward by CREB. *Science* **282**: 2272–2274.

Carlier M, Roubertoux P, Cohen-Salmon C (1983). Early development in mice: I. Genotype and post-natal maternal effects. *Physiology and Behavior* **30**: 837–844.

Carlson GA, Borchelt Dr, Dake A, Turner S, Danielson V, Coffin JD, Eckman C, Meiners J, Nilsen SP, Younkin SG, Hsiao KK (1997). Genetic modification of the phenotypes produced by amyloid precursor protein overexpression in transgenic mice. *Human Molecular Genetics* **6**: 1951–1959.

Carney JM, Landrum RW, Cheng MS, Seale TW (1991). Establishment of chronic intravenous drug administration in the C57BL/6J mouse. *NeuroReport* **2**: 477–480.

Carola V, D'Olimpio F, Brunamonti E, Mangia F, Renzi P. Evaluation of the elevated plus-maze and open-field tests for the assessment of anxiety-related behaviour in inbred mice (2002). *Behavioural Brain Research* **134**: 49–57.

Carter CS (1992). Oxytocin and sexual behavior. *Neuroscience and Biobehavioral Reviews* **16**: 131–144.

Carter CS (2003). Developmental consequences of oxytocin. *Physiology and Behavior* **79**: 383–397.

Carter CS, Braver TS, Barch DM, Botvinick MM, Noll D, Cohen JD (1998). Anterior cingulate cortex, error detection, and the online monitoring of performance. *Science* **280**: 747–749.

Carter CS, Kirkpatrick B, Eds. (1997). *The Integrative Neurobiology of Affiliation*. Annals of the New York Academy of Sciences **807**, New York.

Carter RJ, Lione LA, Humby T, Mangiarini L, Mahal A, Bates GP, Morton AJ, Dunnett SB (1999). Characterization of progressive motor deficits in mice transgenic for the human Huntington's disease mutation. *Journal of Neuroscience* **19**: 3248–3257.

Carter RJ, Morton J, Dunnett SB (2003). Motor coordination and balance in rodents. *Current Protocols in Neuroscience*. Wiley, New York, pp. 8.12.1–8.12.14.

Carvajal CC, Vercauteren F, Dumont Y, Michalkiewicz M, Quirion R (2004). Aged neuropeptide Y transgenic rats are resistant to acute stress but maintain spatial and non-spatial learning. *Behavioural Brain Research* **153**: 471–480.

Carver EA, Issel-Tarver L, Rine J, Olsen AS, Stubbs L (1998). Location of mouse and human genes corresponding to conserved canine olfactory receptor gene subfamilies. *Mammalian Genome* **9**: 349–354.

Cases O, Seif I, Grimsby J, Gaspar P, Chen K, Pournin S, Muller U, Aguet M, Babinet C, Shih JC, De Maeyer E (1995). Aggressive behavior and altered amounts of brain serotonin and norepinephrine in mice lacking MAOA. *Science* **268**: 1763–1766.

Castane A, Valjent E, Ledent C, Parmentier M, Maldonado R, Valverde O (2002). Lack of CB1 cannabinoid receptors modifies nicotine behavioural responses, but not nicotine abstinence. *Neuropharmacology* **43**: 857–867.

Castner SA, Xiao L, Becker JB (1993). Sex differences in striatal dopamine: In vivo microdialysis and behavioral studies. *Brain Research* **610**: 127–134.

Cayouette M, Behn D, Sendtner M, Lachapelle P, Gravel C (1998). Intraocular gene transfer of ciliary neurotrophic factor prevents death and increases responsiveness of rod photoreceptors in the *retinal degeneration slow* mouse. *Journal of Neuroscience* **18**: 9282–9283.

Cederbaum S (2002). Phenylketonuria: An update. *Current Opinions in Pediatrics* **14**: 702–706.

Chambers RA, Moore J, McEvoy JP, Levin ED (1996). Cognitive effects of neonatal hippocampal lesions in a rat model of schizophrenia. *Neuropsychopharmacology* **15**: 587–594.

Champagne FA, Francis DD, Mar A, Meaney MJ (2003). Variations in maternal care in the rat as a mediating influence for the effects of environment on development. *Physiology and Behavior* **79**: 359–371.

Champtiaux N, Changeux JP (2004). Knockout and knockin mice to investigate the role of nicotinic receptors in the central nervous system. *Progress in Brain Research* **145**: 235–251.

Chan PH, Kawase M, Murakami K, Chen SF, Li Y, Calagui B, Reola L, Carlson E, Epstein CJ (1998). Overexpression of SOD1 in transgenic rats protects vulnerable neurons against ischemic damage after global cerebral ischemia and reperfusion. *Journal of Neuroscience* **18**: 8292–8299.

Chang B, Hawes NL, Hurd RE, Davisson MT, Nusinowitz S, Heckenlively JR (2002). Retinal degeneration mutants in the mouse. *Vision Research* **42**: 517–525.

Chang B, Smith RS, Hawes NL, Anderson MG, Zabaleta A, Savinova O, Roderick TH, Heckenlively JR, Davisson MT, John SWM (1999). Interacting loci cause severe iris atrophy and glaucoma in DBA/2J mice. *Nature Genetics* **21**: 405–409.

Chang MW, Leiden JM (1996). Gene therapy for vascular proliferative disorders. *Seminars in Interventional Cardiology* **1**: 185–193.

Chao J, Nestler EJ (2004). Molecular neurobiology of drug addiction. *Annual Review of Medicine* **55**: 113–132.

Chaouloff F, Durand M, Mormede P (1997). Anxiety- and activity-related effects of diazepam and chlordiazepoxide in the rat light/dark and dark/light tests. *Behavioral Brain Research* **85**: 27–35.

Chapillon P, Lalonde R, Jones N, Caston J (1998). Early development of synchronized walking on the rotarod in rats. Effects of training and handling. *Behavioural Brain Research* **93**: 77–81.

Chapman AG, Woodburn VL, Woodruff GN, Meldrum BS (1996). Anticonvulsant effect of reduced NMDA receptor expression in audiogenic DBA/2 mice. *Epilepsy Research* **26**: 25–35.

Chapouthier G, Bondoux D, Martin B, Desforges C, Launay JM (1991). Genetic difference in sensitivity to β-carboline: Evidence for the involvement of brain benzodiazepine receptors. *Brain Research* **553**: 342–346.

Chapouthier G, Launay JM, Venault P, Breton C, Roubertoux PI, Crusio WE (1998). Genetic selection of mouse lines differing in sensitivity to a benzodiazepine receptor inverse agonist. *Brain Research* **787**: 85–90.

Charnas LR, Nussbaum RL (1994). The oculocerebrorenal syndrome of Lowe (Lowe syndrome). In *The Metabolic Basis of Inherited Disease*, CR Scriver, AL Beaudet, WS Sly, D Valle, Eds. McGraw-Hill, New York, pp. 3705–3716.

Charney DS, Manji HK (2004). Life stress, genes, and depression: multiple pathways lead to increased risk and new opportunities for intervention. *Science* STKE **225**: 5.

Chaudhuri JD (2005). Genes arrayed out for you: the amazing world of microarrays. *Medical Science Monitor.* **11**: RA52–RA62.

Chelser EJ, Lu L, Shou S, Qu Y, Gu J, Wang J, Hsu HC, Mountz JD, Baldwin NE, Langston MA, Threadgill DW, Manly KF, Williams RW (2005). Complex trait analysis of gene expression uncovers polygenic and pleiotropic networks that modulate nervous system function. *Nature Genetics* **37**: 209–210.

Chen C, Kim JJ, Thompson RF, Tonegawa S (1996a). Hippocampal lesions impair contextual fear conditioning in two strains of mice. *Behavioral Neuroscience* **110**: 1177–1180.

Chen G, Chen KS, Knox J, Inglis J, Bernard A, Martin SJ, Justice A, McConlogue L, Games D, Freedman SB, Morris RGM (2000). A learning deficit related to age and β-amyloid plaques in a mouse model of Alzheimer's disease. *Nature* **408**: 975–978.

Chen G, Koyama K, Yuan X, Lee Y, Zhou YT, O'Doherty R, Newgard CB, Unger RH (1996b). Disappearance of body fat in normal rats induced by adenovirus-mediated leptin gene therapy. *Proceedings of the National Academy of Sciences USA* **93**: 14795–14799.

Chen J, Kelz MB, Zeng G, Sakai N, Steffen C, Shockett PE, Picciotto MR, Duman RS, Nestler EJ (1998). Transgenic animals with inducible, targeted gene expression in brain. *Molecular Pharmacology* **54**: 495–503.

Chen J, Paredes W, Lowinson JH, Gardner EL (1990). Δ^9-Tetracannabinol enhances presynaptic dopamine efflux in medial prefrontal cortex. *European Journal of Pharmacology* **190**: 259–262.

Chen KS, Nishimura MC, Armanini MP, Crowley C, Spencer SD, Phillips HS (1997). Disruption of a single allele of the nerve growth factor gene results in atrophy of basal forebrain cholinergic neurons and memory deficits. *Journal of Neuroscience* **17**: 7288–7296.

Chen L, Bao S, Lockard JM, Kim JK, Thompson RF (1996). Impaired classical eyeblink conditioning in cerebellar-lesioned and Purkinje cell degeneration (pcd) mutant mice. *Journal of Neuroscience* **16**: 2829–2838.

Chen L, Bao S, Thompson RF (1999). Bilateral lesions of the interpositus nucleus completely prevent eyeblink conditioning in Purkinje cell-degeneration mutant mice. *Behavioral Neuroscience* **113**: 204–210.

Chen L, Toth M (2001). Fragile X mice develop sensory hyperreactivity to auditory stimuli. *Neuroscience* **103**: 1043–1050.

Chen P, Williams SM, Grove KL, Smith MS (2004). Melanocortin 4 receptor-mediated hyperphagia and activation of neuropeptide Y expression in the dorsomedial hypothalamus during lactation. *Journal of Neuroscience* **24**: 5091–5100.

Chen RZ, Akbarian S, Tudor M, Jaenisch R (2001). Deficiency of methyl-CpG binding protein-2 in CNS neurons results in a Rett-like phenotype in mice. *Nature Genetics* **27**: 327–331.

Chen W, Kelly MA, Opitz-Araya X, Thomas RE, Low MJ, Cone RD (1997). Exocrine gland dysfunction in MC5–R-deficient mice: Evidence for coordinated regulation of exocrine gland function by melanocortin peptides. *Cell* **91**: 789–798.

Chen Y, Hu C, Hsu CK, Zhang Q, Bi C, Asnicar M, Hsiung HM, Fox N, Slieker LJ, Yang DD, Heiman ML, Shi Y (2002). Targeted disruption of the melanin-concentrating hormone receptor-1 results in hyperphagia and resistance to diet-induced obesity. *Endocrinology* **143**: 2469–2477.

Chen YT, Liu P, Bradley A (2004). Inducible gene trapping with drug-selectable markers and Cre/loxP to identify developmentally regulated genes. *Molecular and Cellular Biology* **24**: 9930–9941.

Cheng CM, Joncas G, Reinhardt RR, Farrer R, Quarles R, Janssen J, McDonald MP, Crawley JN, Powell-Braxton L, Bondy CA (1998). Biochemical and morphometric analyses show that myelination in the insulin-like growth factor 1 null brain is proportionate to its neuronal composition. *Journal of Neuroscience* **18**: 5673–5681.

Cheng J, Dutra A, Takesono A, Garrett-Beal L, Schwartzberg PL (2004). Improved generation of C57BL/6J mouse embryonic stem cells in a defined serum-free media. *Genesis* **39**: 100–104.

Cherry JA (1993). Measurement of sexual behavior: Controls for variables. *Methods in Neuroscience* **15**: 3–15.

Chiavegatto S, Nelson RJ (2003). Interaction of nitric oxide and serotonin in aggressive behavior. *Hormones and Behavior* **44**: 233–241.

Chih B, Engelman H, Scheiffele P. Control of excitatory and inhibitory synapse formation by neuroligins. *Science* **307**: 1324–1328.

Cho YH, Giese KP, Tanila H, Silva AJ, Eichenbaum H (1998). Abnormal hippocampal spatial representations in αCaMKII^{T286A} and CREB$^{\alpha\Delta}$-mice. *Science* **279**: 867–872.

Choi DW (1997). Background genes: Out of sight, but not out of brain. *Trends in Neuroscience* **20**: 499–500.

Chou SY, Lee YC, Chen HM, Chiang MC, Lai HL, Chang HH, Wu YC, Sun CN, Chien CL, Lin YS, Wang SC, Tung YY, Chang C, Chern Y (2005). CGS21680 attenuates symptoms of Huntington's disease in a transgenic mouse model. *Journal of Neurochemistry* **93**: 310–320.

Chouinard ML, Gallagher MA, Yasuda RP, Wolfe BB, McKinney M (1995). Hippocampal muscarinic receptor function in spatial learning-impaired aged rats. *Neurobiology of Aging* **16**: 955–963.

Chronwall BM, DiMaggio DA, Massari VJ, Pickel VM, Ruggiero DA, O'Donohue TL (1985). The anatomy of neuropeptide Y-containing neurons in the rat brain. *Neuroscience* **15**: 1159–1181.

Chrousos GF (1998). Stressors, stress, and neuroendocrine integration of the adaptive response. *Annals of the New York Academy of Sciences USA* **851**: 311–335.

Chu P, Lutzko C, Stewart AK, Dube ID (1998). Retrovirus-mediated gene transfer into human hematopoietic stem cells. *Journal of Molecular Medicine* **76**: 184–192.

Churchill GA, Airey DC, Allayee H, Angel JM, Attie AD, Beatty J, Beavis WD, Belknap JK, Bennett B, Berrettini W, Bleich A, Bogue M, Broman KW, Buck KJ, Buckler E, Burmeister M, Chesler EJ, Cheverud JM, Clapcote S, Cook MN, Cox RD, Crabbe JC, Crusio WE, Darvasi A, Deschepper CF, Doerge RW, Farber CR, Forejt J, Gaile D, Garlow SJ, Geiger H, Gershenfeld H, Gordon T, Gu J, Gu W, de Haan G, Hayes NL, Heller C, Himmelbauer H, Hitzemann R, Hunter K, Hsu HC, Iraqi FA, Ivandic B, Jacob HJ, Jansen RC, Jepsen KJ, Johnson DK, Johnson TE, Kempermann G, Kendziorski C, Kotb M, Kooy RF, Llamas B, Lammert F, Lassalle JM, Lowenstein PR, Lu L, Lusis A, Manly KF, Marcucio R, Matthews D, Medrano JF, Miller DR, Mittleman G, Mock BA, Mogil JS, Montagutelli X, Morahan G, Morris DG, Mott R, Nadeau JH, Nagase H, Nowakowski RS, O'Hara BF, Osadchuk AV, Page GP, Paigen B, Paigen K, Palmer AA, Pan HJ, Peltonen-Palotie L, Peirce J, Pomp D, Pravenec M, Prows DR, Qi Z, Reeves RH, Roder J, Rosen GD, Schadt EE, Schalkwyk LC, Seltzer Z, Shimomura K, Shou S, Sillanpaa MJ, Siracusa LD, Snoeck HW, Spearow JL, Svenson K, Tarantino LM, Threadgill D, Toth LA, Valdar W, de Villena FP, Warden C, Whatley S, Williams RW, Wiltshire T, Yi N, Zhang D, Zhang M, Zou F; Complex Trait Consortium (2004). The Collaborative Cross, a community resource for the genetic analysis of complex traits. *Nature Genetics* **36**: 1133–1137.

Clapcote SJ, Roder JC (2004). Survey of embryonic stem cell line source strains in the water maze reveals superior reversal learning of 129S6/SvEvTac mice. *Behavioural Brain Research* **152**: 35–48.

Clark AT, Goldowitz D, Takahashi JS, Vitaterna MH, Siepka SM, Peters LL, Frankel WN, Carlson GA, Rossant J, Nadeau JH, Justice MJ: Implementing large-scale ENU mutagenesis screens in North America. *Genetica* **122**: 51–64.

Clark JD, Tempel BL (1998). Hyperalgesia in mice lacking the Kv1.1 potassium channel gene. *Neuroscience Letters* **251**: 121–124.

Clement Y, Chapouthier G (1998). Biological bases of anxiety. *Neuroscience and Biobehavioral Reviews* **5**: 623–633.

Choleris E, Kavaliers M, Pfaff DW (2004). Functional genomics of social recognition. *Journal of Neuroendocrinology* **16**: 383–389.

Cloninger CR, Kaufmann CA, Faraone SV, Malaspina D, Svrakic DM, Harkavy-Friedman J, Suarez BK, Matise TC, Shore D, Lee H, Hampe CL, Wynne D, Drain C, Markel PD, Zambuto CT, Schmitt K, Tsuang MT (1998). Genome-wide search for schizophrenia susceptibility loci: The NIMH genetics initiative and millennium consortium. *American Journal of Medical Genetics* **81**: 275–281.

Cocke R, Moynihan JA, Cohen N, Grota LJ, Ader R (1993). Exposure to conspecific alarm chemosignals alters immune responses in BALB/c mice. *Brain, Behavior, and Immunity* **7**: 36–46.

Cohen NJ, Eichenbaum H (1994). *Memory, Amnesia, and the Hippocampal System*, 2nd ed. MIT Press, Cambridge.

Cohen RM, Kang A, Gulick C (2001). Quantitative trait loci affecting the behavior of A/J and CBA/J intercross mice in the elevated plus-maze. *Mammalian Genome* **12**: 501–507.

Cohen-Salmon C, Carlier M, Roubertoux P, Jouhaneau J, Semal C, Paillette M (1985). Differences in patterns of pup care in mice. V. Pup ultrasonic emissions and pup care behavior. *Physiology and Behavior* **35**: 167–174.

Cole BJ, Robbins TW, Everitt BJ (1988). Lesions of the dorsal noradrenergic bundle simultaneously enhance and reduce responsivity to novelty in a food preference test. *Brain Research Reviews* **13**: 325–349.

Colgan PW (1983). *Comparative Social Recognition*. Wiley, New York.

Collins FS (1999). Microarrays and macroconsequences. *Nature Genetics* **21** (January Suppl): 2–3.

Collins FS, Patrinos A, Jordan E, Chakravarti A, Gesteland R, Walters L (1998). New Goals for the U.S. Human Genome Project: *1998–2003. Science* **282**: 682–689.

Collins RJ, Weeks JR, Cooper MM, Good PI, Russell RR (1984). Prediction of abuse liability of drugs using IV self-administration by rats. *Psychopharmacology* **82**: 6–13.

Collinson N, Kuenzi FM, Jarolimek W, Maubach KA, Cothliff R, Sur C, Smith A, Otu FM, Howell O, Atack JR, McKernan RM, Seabrook GR, Dawson GR, Whiting PJ, Rosahl TW (2002). Enhanced learning and memory and altered GABAergic synaptic transmission in mice lacking the alpha 5 subunit of the GABAA receptor. *Journal of Neuroscience* **22**: 5572–5580.

Colmers WF, Wahlstedt C (1993). *The Biology of Neuropeptide Y and Related Peptides*. Humana, Totowa, NJ, 1993.

Colvin JS, Bohne BA, Harding GW, McEwen DG, Ornitz DM (1996). Skeletal overgrowth and deafness in mice lacking fibroblast growth factor receptor 3. *Nature Genetics* **12**: 390–397.

Comery TA, Harris JB, Willems PJ, Oostra BA, Irwin SA, Weiler IJ, Greenough WT (1997). Abnormal dendritic spines in fragile X knockout mice: maturation and pruning deficits. *Proceedings of the National Academy of Sciences USA* **94**: 5401–5404.

Contarino A, Dellu F, Koob GF, Smith GW, Lee KF, Vale W, Gold LH (1999a). Reduced anxiety-like and cognitive performance in mice lacking the corticotropin-releasing factor receptor 1. Brain Research 835: 1–9.

Contarino A, Heinrichs SC, Gold LH (1999b). Understanding CRF neurobiology: Contributions from mutant mice. *Neuropeptides* **33**: 1–12.

Contarino A, Picetti R, Matthes HW, Koob GF, Kieffer BL, Gold LH (2002). Lack of reward and locomotor stimulation induced by heroin in mu-opioid receptor-deficient mice. *European Journal of Pharmacology* **446**: 103–109.

Cook MN, Bolivar VJ, McFadyen MP, Flaherty L (2002). Behavioral differences among 129 substrains: Implications for knockout and transgenic mice. *Behavioral Neuroscience* **116**: 600–611.

Cook MN, Williams RW, Flaherty L (2001). Anxiety-related behaviors in the elevated zero-maze are affected by genetic factors and retinal degeneration. *Behavioral Neuroscience* **115**: 468–476.

Cooke B, Hegstrom CD, Villeneuve LS, Breedlove SM (1998). Sexual differentiation of the vertebrate brain: principles and mechanisms. *Frontiers in Neuroendocrinology* **19**: 323–362.

Cooper JR, Bloom FE, Roth RH (1996). *The Biochemical Basis of Neuropharmacology*, 7th ed. Oxford University Press, New York.

Cooper MJ (1996). Noninfectious gene transfer and expression systems for cancer gene therapy. *Seminars in Oncology* **23**: 172–187.

Cooper SJ (1992). CCK-monoamine interactions and satiety in rodents. In *Multiple Cholecystokinin Receptors in the CNS*. CT Dourish, SJ Cooper, SD Iversen, LL Iversen, Eds. Oxford University Press, Oxford, pp. 260–279.

Cooper SJ, Dourish CT, Barber DJ (1996). Effects of the benzodiazepine receptor inverse agonist Ro 15–4513 on the ingestion of sucrose and sodium saccharin solutions: A microstructural analysis of licking behavior. *Behavioral Neuroscience* **110**: 559–566.

Cooper SJ, Dourish CT, Barber DJ (1990). Reversal of the anorectic effect of (+)-fenfluramine in the rat by the selective cholecystokinin receptor antagonist MK-329. *British Journal of Pharmacology* **1**: 65–70.

Cooper SJ, Francis J (1993). A microstructural analysis of the effects of presatiation on feeding behavior in the rat. *Physiology and Behavior* **53**: 413–416.

Coquelin A (1992). Urine-marking by female mice throughout their reproductive cycle. *Hormones and Behavior* **26**: 255–271.

Corcoran KA, Lu Y, Turner RS, Maren S (2002). Overexpression of hAPPswe impairs rewarded alternation and contextual fear conditioning in a transgenic mouse model of Alzheimer's disease. *Learning and Memory* **9**: 243–252.

Cordero-Erausquin M, Marubio LM, Klink R, Changeux JP (2000). Nicotinic receptor function: New perspectives from knockout mice. *Trends in Pharmacological Sciences* **21**: 211–217.

Cornish JL, Kalivas PW (2001). Cocaine sensitization and craving: differing roles for dopamine and glutamate in the nucleus accumbens. *Journal of Addictive Diseases* **20**: 43–54.

Corrigall WA, Coen KM (1989). Nicotine maintains robust self-administration in rats on a limited access schedule. *Psychopharmacology* **99**: 473–478.

Corrigall WA, Coen KM (1991). Opiate antagonists reduce cocaine but not nicotine self-administration. *Psychopharmacology* **104**: 171–176.

Corwin RL, Gibbs J, Smith GP (1991). Increased food intake after type A but not type B cholecystokinin receptor blockade. *Physiology and Behavior* **50**: 255–258.

Corwin RL, Robinson JK, Crawley JN (1993). Galanin antagonists block galanin-induced feeding in the hypothalamus and amygdala of the rat. *European Journal of Neuroscience* **5**: 1528–1533.

Cory-Slechta DA, Crofton KM, Foran JA, Ross JF, Sheets LP, Weiss B, Mileson B (2001). Methods to identify and characterize developmental neurotoxicity for human health risk assessment. I: Behavioral effects. *Environ Health Perspect.* **109** (Suppl 1): 79–91.

Cossu G, Ledent C, Fattore L, Imperato A, Bohme GA, Parmentier M, Fratta W (2001). Cannabinoid CB1 receptor knockout mice fail to self-administer morphine but not other drugs of abuse. *Behavioural Brain Research* **118**: 61–65.

Costa RM, Federov NB, Kogan JH, Murphy GG, Stern J, Ohno M, Kucherlapati R, Jacks T, Silva AJ (2002). Mechanisms for the learning deficits in a mouse model of neurofibromatosis type 1. *Nature* **415**: 526–530.

Costa RM, Silva AJ (2003). Mouse models of neurofibromatosis type 1: Bridging the GAP. *Trends in Molecular Medicine* **9**: 19–23.

Costall B, Domeney AM, Hughes J, Kelly ME, Naylor RJ, Woodruff GN (1991). Anxiolytic effects of CCK-B antagonists. *Neuropeptides* **19** (Suppl): 65–73.

Costantini F, Lacy E (1981). Introduction of a rabbit beta-globin gene into the mouse germ line. *Nature* **294**: 92–94.

Cotman CW, Poon WW, Rissman RA, Blurton-Jones M (2005). The role of caspase cleavage of tau in Alzheimer disease neuropathology. *Journal of Neuropathology and Experimental Neurology.* **64**: 104–112.

Counts SE, Perez SE, Ginsberg SD, De Lacalle S, Mufson EJ (2003). Galanin in Alzheimer disease. *Molecular Interventions* **3**: 137–156.

Coussons-Read ME, Crnic LS (1996). Behavioral assessment of the Ts65Dn mouse, a model for Down syndrome: Altered behavior in the elevated plus-maze and open field. *Behavior Genetics* **26**: 7–13.

Couzin J (2005). Advocating the clinical way. *Science* **308**: 940–942.

Cox PA, Sachs OW (2002). Cycad neurotoxins, consumption of flying foxes, and ALS-PDC disease in Guam. *Neurology* **58**: 956–959.

Crabbe JC (2002). Genetic contributions to addiction. *Annual Review of Psychology* **53**: 435–462.

Crabbe JC, Belknap JK, Buck KJ (1994a). Genetic animal models of alcohol and drug abuse. *Science* **264**: 1715–1723.

Crabbe JC, Belknap JK, Mitchell SR, Crawshaw LI (1994b). Quantitative trait loci mapping of genes that influence the sensitivity and tolerance to ethanol-induced hypothermia in BXD recombinant inbred mice. *Journal of Pharmacology and Experimental Therapeutics* **269**: 184–192.

Crabbe JC, Cotnam CJ, Cameron AJ, Schlumbohm JP, Rhodes JS, Metten P, Wahlsten D (2003a). Strain differences in three measures of ethanol intoxication in mice: The screen, dowel and grip strength tests. *Genes, Brain and Behavior* **2**: 201–213.

Crabbe JC, Gallaher EJ, Cross SJ, Belknap JK (1998). Genetic determinants of sensitivity to diazepam in inbred mice. *Behavioral Neuroscience* **112**: 668–677.

Crabbe JC, Harris RA (1991). *The Genetic Basis of Alcohol and Drug Actions.* Plenum, New York.

Crabbe JC, Kosobud A, Young ER, Janowsky JS (1983). Polygenic and single-gene determination of responses to ethanol in BXD/Ty recombinant inbred mouse strains. *Neurobehavioral Toxicology and Teratology* **5**: 181–187.

Crabbe JC, Metten P, Cameron AJ, Wahlsten D (2005). An analysis of the genetics of alcohol intoxication in inbred mice. *Neuroscience and Biobehavioral Reviews* **28**: 785–802.

Crabbe JC, Metten P, Yu CH, Schlumbohm JP, Cameron AJ, Wahlsten D (2003b). Genotypic differences in ethanol sensitivity in two tests of motor incoordination. *Journal of Applied Physiology* **95**: 1338–1351.

Crabbe JC, Morris RJ (2004). Festina lente: Late-night thoughts on high-throughput screening of mouse behavior. *Nature Neuroscience* **7**: 1175–1179.

Crabbe JC, Phillips TJ (2003). Pharmacogenetic studies of alcohol self-administration and withdrawal. *Psychopharmacology* **174**: 539–560.

Crabbe JC, Phillips TJ, Buck, KJ, Cunningham CL, Belknap JK (1999a). Identifying genes for alcohol and drug sensitivity: Recent progress and future directions. *Trends in Pharmacological Sciences* **22**: 174–179.

Crabbe JC, Wahlsten D, Dudek BC (1999b). Genetics of mouse behavior: Interactions with laboratory environment. *Science* **284**:1670–1672.

Crawley JN (1981). Neuropharmacologic specificity of a simple animal model for the behavioral actions of benzodiazepines. *Pharmacology Biochemistry and Behavior* **15**: 695–699.

Crawley JN (1989). Animal models of anxiety. *Current Opinion in Psychiatry* **2**: 773–776.

Crawley JN (1995). Biological actions of galanin. *Regulatory Peptides* **59**: 1–16.

Crawley JN (1996). Unusual behavioral phenotypes of inbred mouse strains. *Trends in Neurosciences* **19**: 181–182.

Crawley JN (2003). Behavioral phenotyping of rodents. *Comparative Medicine* **53**:140–146.

Crawley JN (2004). Designing mouse behavioral tasks relevant to autistic-like behaviors. *Mental Retardation and Developmental Disabilities Research Reviews* **10**: 248–258.

Crawley JN, Austin MC, Fiske SM, Martin B, Consolo S, Berthold M, Langel Ü, Fisone G, Bartfai T (1990). Activity of centrally administered galanin fragments on stimulation of feeding behavior and on galanin receptor binding in the rat hypothalamus. *Journal of Neuroscience* **10**: 3695–3700.

Crawley JN, Belknap JK, Collins A, Crabbe JC, Frankel W, Henderson N, Hitzemann RJ, Maxson SC, Miner LL, Silva AJ, Wehner JM, Wynshaw-Boris A, Paylor R (1997a). Behavioral phenotypes of inbred mouse strains: Implications and recommendations for molecular studies. *Psychopharmacology* **132**: 107–124.

Crawley JN, Contrera JF (1976). Intraventricular 6-hydroxydopamine lowers isolation-induced fighting behavior in male mice. *Pharmacology Biochemistry and Behavior* **4**: 381–384.

Crawley JN, Corwin RL (1994). Biological actions of cholecystokinin. *Peptides* **15**: 731–755.

Crawley JN, Davis LG (1982). Baseline exploratory activity predicts anxiolytic responsiveness to diazepam in five strains of mouse. *Brain Research Bulletin* **8**: 609–612.

Crawley JN, Gerfen CR, McKay R, Rogawski MW, Sibley DR, Skolnick P, Eds. (1997b). *Current Protocols in Neuroscience*. Wiley, New York.

Crawley JN, Glowa JR, Majewska MD, Paul SM (1986). Anxiolytic activity of an endogenous adrenal steroid. *Brain Research* **398**: 382–385.

Crawley JN, Goodwin FK (1980). Preliminary report of a simple animal behavior model for the anxiolytic effects of benzodiazepines. *Pharmacology Biochemistry and Behavior* **13**: 167–170.

Crawley JN, Kiss JZ (1985). Paraventricular nucleus lesions abolish the inhibition of feeding induced by systemic cholecystokinin. *Peptides* **6**: 927–935.

Crawley JN, Mufson EJ, Hohmann JG, Teklemichael D, Steiner RA, Holmberg K, Xu ZQ, Blakeman KH, Xu XJ, Wiesenfeld-Hallin Z, Bartfai T, Hokfelt T (2002). Galanin overexpressing transgenic mice. *Neuropeptides* **36**: 145–156.

Crawley JN, Paylor R (1997). A proposed test battery and constellations of specific behavior paradigms to investigate the behavioral phenotypes of transgenic and knockout mice. *Hormones and Behavior* **31**: 197–211.

Crawley JN, Schleidt WM, Contrera JF (1975). Does social environment decrease propensity to fight in male mice? *Behavioral Biology* **15**: 73–83.

Crawley JN, Schwaber JS (1984). Abolition of the behavioral effects of cholecystokinin following bilateral radiofrequency lesions of the paraventricular subdivision of the nucleus tractus solitarius. *Brain Research* **295**: 289–299.

Crawley JN, Stivers JA, Hommer DW, Skirboll LR, Paul SM (1986). Antagonists of central and peripheral behavioral actions of cholecystokinin octapeptide. *Journal of Pharmacology and Experimental Therapeutics* **236**: 320–330.

Creese I, Iversen SD (1973). Blockage of amphetamine-induced motor stimulation and stereotypy in the adult rat following neonatal treatment with 6-hydroxydopamine. *Brain Research* **55**: 369–382.

Crestani F, Keist R, Fritschy JM, Benke D, Vogt K, Prut L, Bluthmann H, Mohler H, Rudolph U (2002). Trace fear conditioning involves hippocampal alpha5 GABA(A) receptors. *Proceedings of the National Academy of Sciences USA* **99**: 8980–8985.

Crowley JJ, Jones MD, O'Leary OF, Lucki I (2004). Automated tests for measuring the effects of antidepressants in mice. *Pharmacology Biochemistry and Behavior* **78**: 269–274.

Crusio W, Ed. (1996a). The neurobehavioral genetics of aggression. *Behavior Genetics*, Special Issue **26**: 459–533.

Crusio WE (1996b). Gene-targeting studies: New methods, old problems. *Trends in Neurosciences* **19**: 186–187.

Crusio WE, Gerlai RT, Eds. (1999). *Handbook of Molecular-Genetic Techniques for Brain and Behavior Research*. Elsevier, New York.

Crusio WE, Schwegler H, Brust I (1995). Covariations between hippocampal mossy fibres and working and reference memory in spatial and non-spatial radial maze tasks in mice. *European Journal of Neuroscience* **5**: 1413–1420.

Cryan JF, Mombereau C (2004). In search of a depressed mouse: utility of models for studying depression-related behavior in genetically modified mice. *Molecular Psychiatry* **9**: 326–357.

Cryan JF, Mombereau C, Vassout A (2005). The tail suspension test as a model for assessing antidepressant activity: Review of pharmacological and genetic studies in mice. *Neuroscience and Biobehavioral Reviews* **29**: 571–625.

Cryan JF, Hoyer D Markou A (2003). Withdrawal from chronic amphetamine induces depressive-like behavioral effects in rodents. *Biological Psychiatry* **54**: 49–58.

Cryan JF, Markou A, Lucki I (2002). Assessing antidepressant activity in rodents: Recent development and future needs. *Trends in Pharmacological Sciences* **23**: 238–245.

Csernansky JG, Grace AA, Eds. (1998). New models of the pathophysiology of schizophrenia. *Schizophrenia Bulletin*, Special Issue **24**: 185–317.

Cui Z, Wang H, Tan Y, Zaia KA, Zhang S, Tsien JZ (2004). Inducible and reversible NR1 knockout reveals crucial role of the NMDA receptor in preserving remote memories in the brain. *Neuron* **41**: 781–793.

Cummings CJ, Sun Y, Opal P, Antalffy B, Mestril R, Orr HT, Dillmann WH, Zoghbi HY (2001). Over-expression of inducible HSP70 chaperone suppresses neuropathology and improves motor function in SCA1 mice. *Human Molecular Genetics* **10**: 1511–1518.

Cunningham CL, Ferree NK, Howard MA (2003). Apparatus bias and place conditioning with ethanol in mice. *Psychopharmacology* **170**: 409–422.

Cunningham CL, Howard MA, Gill SJ, Rubinstein M, Low MJ, Grandy DK (2000). Ethanol-conditioned place preference is reduced in dopamine D2 receptor-deficient mice. *Pharmacology Biochemistry and Behavior* **67**: 693–699.

Currie PJ (1993). Differential effects of NE, CLON, and 5-HT on feeding and macronutrient selection in genetically obese (*ob/ob*) and lean mice. *Brain Research Bulletin* **32**: 133–142.

Curtin JA, Quint E, Tsipouri V, Arkell RM, Cattanach B, Copp AJ, Henderson DJ, Spurr N, Stanier P, Fisher EM, Nolan PM, Steel KP, Brown SD, Gray IC, Murdoch JN (2003). Mutation of Celsr1 disrupts planar polarity of inner ear hair cells and causes severe neural tube defects in the mouse. *Current Biology* **13**: 1129–1133.

Curzon G, Gibson EL, Oluyomi A (1997). Appetite suppression by commonly used drugs depends on 5-HT receptors but not on 5-HT availability. *Trends in Pharmacological Sciences* **18**: 191–192.

Curzon P, Kim DJB, Decker MW (1994). Effect of nicotine, lobeline, and mecamylamine on sensory gating in the rat. *Pharmacology Biochemistry and Behavior* **49**: 877–882.

Cyr M, Caron MG, Johnson GA, Laakso A (2005). Magnetic resonance imaging at microscopic resolution reveals subtle morphological changes in a mouse model of dopaminergic hyperfunction. *Neuroimage* **26**: 83–90.

Czech C, Lesort M, Tremp G, Terro F, Blanchard V, Schombert B, Carpentier N, Dreisler S, Bonici B, Takashima A, Moussaoui S, Hugon J, Pradier L (1998). Characterization of human presenilin 1 transgenic rats: Increased sensitivity to apoptosis in primary neuronal cultures. *Neuroscience* **87**: 325–336.

Dacanay RJ, Riley AL (1982). The UCS preexposure effect in taste aversion learning: Tolerance and blocking are drug specific. *Animal Learning and Behavior* **10**: 91–96.

Damasio H, Grabowski T, Frank R, Galaburda AM, Damasio AR (1994). The return of Phineas Gage: Clues about the brain from the skull of a famous patient. *Science* **264**: 1102–1105.

Damassa DA, Smith ER, Tennent B, Davidson JM (1977). The relationship between circulating testosterone levels and male sexual behavior in rats. *Hormones and Behavior* **8**: 275–286.

D'Amato FR, Moles A (2001). Ultrasonic vocalizations as an index of social memory in female mice. *Behavioral Neuroscience* **115**: 834–840.

D'Amato FR, Scalera E, Sarli C, Moles A (2005). Pups call, mothers rush: Does maternal responsiveness affect the amount of ultrasonic vocalizations in mouse pups? *Behavior Genetics* **35**: 103–112.

D'Amour FE, Smith DL (1941). A method for determining loss of pain sensation. *Journal of Pharmacology and Experimental Therapeutics* **41**: 419–424.

Damsma G, Pfaus JG, Wenkstern D, Phillips AG, Fibiger HC (1992). Sexual behavior increases dopamine transmission in the nucleus accumbens and striatum of male rats: Comparison with novelty and locomotion. *Behavioral Neuroscience* **106**: 181–191.

Danguir J, Nicolaidis S (1980). Cortical activity and sleep in the rat lateral hypothalamic syndrome. *Brain Research* **185**: 305–321.

Daniels GM, Buck KJ (2002). Expression profiling identifies strain-specific changes associated with ethanol withdrawal in mice. *Genes, Brain and Behavior* **1**: 35–45.

Dantzer R (1998). Vasopressin, gonadal steroids and social recognition. *Progress in Brain Research* **119**: 409–414.

Darvasi A (1998). Experimental strategies for the genetic dissection of complex traits in animal models. *Nature Genetics* **18**: 19–24.

Darwin C (1839–1843). *Zoology of the Voyage of H.M.S. Beagle, Part 3 Birds*. Smith, Elder and Company, London.

Darwin C (1872). *The Expression of Emotions in Man and the Animals*. John Murray, London.

Das P, Howard V, Loosbrock N, Dickson D, Murphy MP, Golde TE (2003). Amyloid-β immunization effectively reduces amyloid deposition in FcRγ$^{-/-}$ knock-out mice. *Journal of Neuroscience* **23**: 8532–8538.

Dauge V, Sebret A, Beslot F, Matsui T, Roques BP (2001). Behavioral profile of CCK2 receptor-deficient mice. *Neuropsychopharmacology* **25**: 690–698.

D'Aulaire I, D'Aulaire EP (1962). *D'Aulaire's Book of Greek Myths*. Doubleday, New York.

Davidson AJ, Poole AS, Yamazaki S, Menaker M (2003). Is the food-entrainable circadian oscillator in the digestive system? *Genes Brain and Behavior* **2**: 32–39.

Davidson JM (1966). Activation of male rat's sexual behavior by intracerebral implantation of androgen. *Endocrinology* **79**: 783–794.

Davis HP, Rosenzweig MR, Bennett EL, Squire LR (1980). Inhibition of cerebral protein synthesis: Dissociation of nonspecific effects and amnesic effects. *Behavioral and Neural Biology* **28**: 99–104.

Davis JA, Naruse S, Chen H, Eckman C, Younkin S, Price DL, Borchelt D, Sisodia SS, Wong PC (1998). An Alzheimer's disease-linked PS1 variant rescues the developmental abnormalities of PS1-deficient embryos. *Neuron* **20**: 603–609.

Davis JD, Smith GP (1988). Analysis of lick rate measures the positive and negative feedback effects of carbohydrates on eating. *Appetite* **11**: 229–238.

Davis M (1986). Pharmacological and anatomical analysis of fear conditioning using the fear-potentiated startle paradigm. *Behavioral Neuroscience* **100**: 814–824.

Davis M, Falls WA, Campeau S, Kim M (1993). Fear-potentiated startle: A neural and pharmacological analysis. *Behavioral Brain Research* **58**: 175–198.

Davis M, Rainnie D, Cassell M (1994). Neurotransmission in the rat amygdala related to fear and anxiety. *Trends in Neuroscience* **17**: 208–214.

Davis M, Walker DL, Lee Y (1997). Roles of the amygdala and bed nucleus of the stria terminalis in fear and anxiety measured with the acoustic startle reflex. Possible relevance to PTSD. *Annals of the New York Academy of Sciences* **821**: 305–331.

Davis RR, Franks JR (1989). Design and construction of a noise-exposure chamber for small animals. *Journal of the Acoustic Society of America* **58**: 963–966.

Davis RR, Kozel P, Erway LC (2003). Genetic influences in individual susceptibility to noise: A review. *Noise Health* **5**: 19–28.

Davis S, Rodger J, Hicks A, Mallet J, Laroche S (1996). Brain structure and task-specific increase in expression of the gene encoding syntaxin 1B during learning in the rat: A potential molecular marker for learning-induced synaptic plasticity in neural networks. *European Journal of Neuroscience* **8**: 2068–2074.

Davisson MT, Schmidt C, Reeves RH, Irving NG, Akeson BC, Harris BS, Bronson RT (1993). Segmental trisomy as a mouse model for Down syndrome. The phenotypic mapping of Down syndrome and other aneuploid conditions. *Progress in Clinical and Biological Research*, **384**: 117–133.

Dawson GR, Bayley P, Channell S, Iversen SD (1994). A comparison of the effects of the novel muscarinic receptor agonists L-689,660 and AF102B in tests of reference and working memory. *Psychopharmacology* **113**: 361–368.

Dawson GR, Seabrook GR, Zheng H, Smith DW, Graham S, O'Dowd G, Bowery BJ, Boyce S, Trumbauer ME, Chen HY, van der Ploeg LHT, Sirinathsinghji DJS (1999). Age-related cognitive deficits, impaired long-term potentiation and reduction in synaptic marker density in mice lacking the β-amyloid precursor protein. *Neuroscience* **90**: 1–13.

Dawson GR, Tricklebank MD (1995). Use of the elevated plus-maze in the search for novel anxiolytic agents. *Trends in Pharmacological Sciences* **16**: 33–36.

Dawson TM, Snyder SH (1994). Gases as biological messengers: Nitric oxide and carbon monoxide in the brain. *Journal of Neuroscience* **14**: 5147–5159.

Day HD, Seay BM, Hale P, Hendricks D (1982). Early social deprivation and the ontogeny of unrestricted social behavior in the laboratory rat. *Developmental Psychobiology* **15**: 47–59.

De Angelis L, File SE (1979). Acute and chronic effects of three benzodiazepines in the social interaction anxiety test in mice. *Psychopharmacology* **64**: 127–129.

DeBold JF, Miczek KA (1984). Aggression persists after ovariectomy in female rats. *Hormones and Behavior* **18**: 177–190.

Decker MW, Majchrzak MJ (1992). Effects of systemic and intracerebroventricular administration of mecamylamine, a nicotinic cholinergic antagonist, on spatial memory in rats. *Psychopharmacology* **107**: 530–534.

Deeb SS (1993). Genetic determinants of visual functions. *Current Opinions in Neurobiology* **3**: 506–512.

De Felipe C, Herrero JF, O'Brien JA, Palmer JA, Doyle CA, Smith AJH, Laird JMA, Belmonte C, Cervero F, Hunt SP (1998). Altered nociception, analgesia and aggression in mice lacking the receptor for substance P. *Nature* **392**: 394–397.

DeFries JC, Hegmann JP, Halcomb RA (1974). Response to 20 generations of selection for open-field activity in mice. *Behavioral Biology* **11**: 481–495.

DeFries JC, Wilson JR, Erwin VG, Petersen DR (1989). LS X SS recombinant inbred strains of mice: Initial characterization. *Alcoholism: Clinical and Experimental Research* **13**: 196–200.

Delaney CL, Brenner M, Messing A (1996). Conditional ablation of cerebellar astrocytes in postnatal transgenic mice. *Journal of Neuroscience* **16**: 6908–6918.

Del Punta K, Leinders-Zufall T, Rodriguez I, Jukam D, Wysocki CJ, Ogawa S, Zufall F, Mombaerts P (2002). Deficient pheromone responses in mice lacking a cluster of vomeronasal receptor genes. *Nature* **419**: 70–74.

De Luca A, Giuditta A (1997). Role of a transcription factor (CREB) in memory processes. *Rivista di Biologia* **90**: 371–384.

Demas GE, Nelson RJ, Krueger BK, Yarowsky PJ (1996). Spatial memory deficits in segmental trisomic Ts65Dn mice. *Behavioral Brain Reearch* **82**: 85–92.

Denenberg VH (1999). Commentary: Is maternal stimulation the mediator of the handling effect in infancy? *Developmental Psychobiology* **34**: 1–3.

Denenberg VH, Morton JRC (1962). Effects of environmental complexity and social groupings upon modification of emotional behavior. *Journal of Comparative and Physiological Psychology* **55**: 242–246.

Deng C, Wynshaw-Boris A, Zhou F, Kuo A, Leder P (1996). Fibroblast growth factor receptor 3 is a negative regulator of bone growth. *Cell* **84**: 911–921.

Derave W, Van Den Bosch L, Lemmens G, Eijnde BO, Robberecht W, Hespel P (2003). Skeletal muscle properties in a transgenic mouse model for amyotrophic lateral sclerosis: effects of creatine treatment. *Neurobiology of Disease* **13**: 264–272.

Derbyshire SW, Jones AK, Gyulai F, Clark S, Townsend D, Firestone LL (1997). Pain processing during three levels of noxious stimulation produces differential patterns of central activity. *Pain* **73**: 431–445.

Deroche V, Caine SB, Heyser CJ, Polis I, Koob GF, Gold LH (1997). Differences in the liability to self-administer intravenous cocaine between C57BL/6 × SJL and BALB/cByJ mice. *Pharmacology Biochemistry and Behavior* **57**: 429–440.

Deroche-Gamonet V, Sillaber I, Aouizerate B, Izawa R, Jaber M, Ghozland S, Kellendonk C, Le Moal M, Spanagel R, Schutz G, Tronche F, Piazza PV (2003). The glucocorticoid receptor as a potential target to reduce cocaine abuse. *Journal of Neuroscience* **23**: 4785–4790.

D'Ercole AJ, Ye P, O'Kusky JR (2002). Mutant mouse models of insulin-like growth factor actions in the central nervous system. *Neuropeptides* **36**: 209–220, 2002.

Desimone R (1992). The physiology of memory: Recordings of things past. *Science* **258**: 245–246.

De Souza EB, Grigoriadis DE (1994). Corticotropin-releasing factor: Physiology, pharmacology and role in central nervous system and immune disorders. In *Psychopharmacology: The Fourth Generation of Progress*, FE Bloom, DJ Kupfer, Eds. Raven, New York, pp. 505–517.

Detke MJ, Rickels M, Lucki I (1995). Active behavior in the rat forced swimming test differentially produced by serotonergic and noradrenergic antidepressants. *Psychopharmacology* **121**: 66–72.

Deutch AY (1993). Prefrontal cortical dopamine systems and the elaboration of functional corticostriatal circuits. *Journal of Neural Transmission* **91**: 197–221.

Deutch AY, Cameron DS (1992). Pharmacological characterization of dopamine systems in the nucleus accumbens core and shell. *Neuroscience* **46**: 49–56.

Deutch AY, Clark WA, Roth RH (1990). Prefrontal cortical dopamine depletion enhances the responsiveness of mesolimbic dopamine neurons to stress. *Brain Research* **521**: 311–315.

Deutch AY, Tam SY, Roth RH (1985). Footshock and conditioned stress increase 3,4-dihydroxyphenylacetic acid (DOPAC). in the ventral tegmental area but not substantia nigra. *Brain Research* **333**: 143–146.

DeVries AC, Young WS, Nelson RJ (1997). Reduced aggressive behaviour in mice with targeted disruption of the oxytocin gene. *Journal of Neuroendocrinology* **9**: 363–368.

DeVries GJ (1990). Sex differences in neurotransmitter systems. *Journal of Neuroendocrinology* **2**: 1–13.

Dewachter I, Van Leuven F (2002). Secretases as targets for the treatment of Alzheimer's disease: The prospects. *The Lancet Neurology* **1**: 409–416.

De Waele JP, Papachristou DN, Gianoulakis C (1992). The alcohol-preferring C57BL/6 mice present an enhanced sensitivity of the hypothalamic β-endorphin system to ethanol than the alcohol-avoiding DBA/2 mice. *Journal of Pharmacology and Experimental Therapeutics* **261**: 788–794.

De Zeeuw CI, Strata P, Voogd J (1998). *The Cerebellum: From Structure to Control*. Elsevier Science, North Holland, Amsterdam.

Dhillon H, Kalra SP, Prima V, Zolotukhin S, Scarpace PJ, Moldawer LL, Muzyczka N, Kalra PS (2001). Central leptin gene therapy suppresses body weight gain, adiposity and serum insulin without affecting food consumption in normal rats: A long-term study. *Regulatory Peptides* **99**: 69–77.

D'Hooge R, Nagels G, Franck F, Bakker CE, Reyniers E, Storm K, Kooy RF, Oostra BA, Willems PJ, De Deyn PP (1997). Mildly impaired water maze performance in male *Fmr1* knockout mice. *Neuroscience* **76**: 367–376.

D'Hooge R, Nagels G, Westland CE, Mucke L, De Deyn PP (1996). Spatial learning deficit in mice expressing human 751-amino acid β-amyloid precursor protein. *NeuroReport* **7**: 2807–2881.

Diagnostic and Statistical Manual of Mental Disorders (1994). Fourth Edition, American Psychiatric Association, Washington, DC.

Di Chiara G (1998). A motivational learning hypothesis of the role of mesolimbic dopamine in compulsive drug use. *Journal of Psychopharmacology* **12**: 54–67.

Di Chiara G, Bassareo V, Fenu S, De Luca MA, Spina L, Cadoni C, Acquas E, Carboni E, Valentini V, Lecca D (2004). Dopamine and drug addiction: The nucleus accumbens shell connection. *Neuropharmacology* **47** (Suppl 1): 227–241.

Di Chiara G, Imperato A (1988). Drugs abused by humans preferentially increase synaptic dopamine concentrations in the mesolimbic system of freely moving rats. *Proceedings of the National Academy of Sciences USA* **85**: 5274–5278.

Di Chiara G, North RA (1992). Neurobiology of opiate abuse. *Trends in Pharmacological Sciences* **13**: 185–193.

Di Ciano P, Blaha CD, Phillips AG (1998). Conditioned changes in dopamine oxidation current in the nucleus accumbens of rats by stimuli paired with self-administration or yoked administration of *d*-amphetamine. *European Journal of Neuroscience* **10**: 1121–1127.

Dicket CA, Gordon MN, Mason JE, Wilson NJ, Diamond DM, Guzowski JF, Morgan D (2004). Amyloid suppresses induction of genes critical for memory consolidation in APP + PS1 transgenic mice. *Journal of Neurochemistry* **88**: 434–442.

Dierssen M, Fillat C, Crnic L, Arbones M, Florez J, Estivill X (2001). Murine models for Down syndrome. *Physiology and Behavior* **73**: 859–871.

Dierssen M, Fotaki V, Martínez de Lagrán M, Gratacós M, Arbonés M, Fillat C, Estivill X (2002). Neurobehavioral development of two mouse lines commonly used in transgenic studies. *Pharmacology, Biochemistry and Behavior* **73**: 19–25.

Dietrich WF, Copeland NG, Gilbert DJ, Miller JC, Jenkins NA, Lander ES (1995). Mapping the mouse genome: Current status and future prospects. *Proceedings of the National Academy of Sciences USA* **92**: 10849–10853.

Diez M, Danner S, Frey P, Sommer B, Staufenbiel M, Wiederhold KH, Hokfelt T (2003). Neuropeptide alterations in the hippocampal formation and cortex of transgenic mice over-expressing beta-amyloid precursor protein (APP) with the Swedish double mutation (APP23). *Neurobiology of Disease* **14**: 579–594.

Ding Z, Hardy CO, Thöny B (2004). State-of-the-art 2003 on PKU gene therapy. *Molecular Genetics and Metabolism* **81**: 3–8.

Dirks A, Groenink L, Schipholt MI, van der Gugten J, Hijzen TH, Geyer MA, Olivier B (2002). Reduced startle reactivity and plasticity in transgenic mice overexpressing corticotropin-releasing hormone. *Biological Psychiatry* **51**: 583–590.

Dobkin C, Rabe A, Dumas R, El Idrissi A, Haubenstock H, Brown WT (2000). *Fmr1* knockout mouse has a distinctive strain-specific learning impairment. *Neuroscience* **100**: 423–429.

Dodart JC, Bales KR, Paul SM (2003). Immunotherapy for Alzheimer's disease: Will vaccination work? *Trends in Molecular Medicine* **9**: 85–87.

Dodart JC, Marr RA, Koistinaho M, Gregersen BM, Malkani S, Verma IM, Paul SM (2005). Gene delivery of human apolipoprotein E alters brain Abeta burden in a mouse model of Alzheimer's disease. *Proceedings of the National Academy of Sciences USA* **102**: 1211–1216.

Dodart JC, Mathis C, Bales KR, Paul SM (2002). Does my mouse have Alzheimer's disease? *Genes Brain and Behavior* **1**: 142–155.

Dodart JC, Mathis C, Bales KR, Paul SM, Ungerer A (2000). Behavioral deficits in APP(V717F) transgenic mice deficient for the apolipoprotein E gene. *Neuroreport* **11**: 603–607.

Doetschman TC (1991). Gene targeting in embryonic stem cells. *Biotechnology* **16**: 89–101.

Domino EF (1998). Tobacco smoking and nicotine neuropsychopharmacology: Some future research directions. *Neuropsychopharmacology* **18**: 456–468.

Dong Y, Saal D, Thomas M, Faust R, Bonci A, Robinston T, Malenka RC (2004). Cocaine-induced potentiation of synaptic strength in dopamine neurons: Behavioral correlates in GluRA(−/−) mice. *Proceedings of the National Academy of Sciences USA* **101**: 14282–14287.

Donny EC, Caggiula AR, Mielke MM, Jacobs KS, Rose C, Sved AF (1998). Acquisition of nicotine self-administration in rats: the effects of dose, feeding schedule, and drug contingency. *Psychopharmacology* **136**: 83–90.

Donovan SL, Mamounas LA, Andrews AM, Blue ME, McCasland JS (2002). GAP-43 is critical for normal development of the serotonergic innervation in forebrain. *Journal of Neuroscience* **22**: 3543–3552.

Doty RL (1986). Odor-guided behavior in mammals. *Experientia* **15**: 257–271.

Doty RL, Bagla R, Kim N (1999). Physostigmine enhances performance on an odor mixture discrimination test. *Physiology and Behavior* **65**: 801–804.

Dourish CT (1992). Behavioural analysis of the role of CCK$_A$ and CCK$_B$ receptors in the control of feeding in rodents. In *Multiple Cholecystokinin Receptors in the CNS*, CT Dourish, SJ Cooper, SD Iversen, LL Iversen, Eds. Oxford University Press, Oxford, pp. 234–253.

Dowling JE (1998). *Creating Mind: How the Brain Works*. Norton, New York.

Downing C, Rodd-Henricks KK, Flaherty L, Dudek BC. (2003). Genetic analysis of the psychomotor stimulant effect of ethanol.*Genes, Brain and Behavior* **2**: 140–151.

Drago J, Padungchaichot P, Accili D, Fuchs S (1998). Dopamine receptors and dopamine transporter in brain function and addictive behaviors: Insights from targeted mouse mutants. *Developmental Neuroscience* **20**: 188–203.

Drazen DL, Woods SC (2003). Peripheral signals in the control of satiety and hunger. *Current Opinion in Clinical Nutrition and Metabolic Care* **6**: 621–629.

Driscoll LL, Carroll JC, Moon J, Crnic LS, Levitsky DA, Strupp BJ (2004). Impaired sustained attention and error-induced stereotypy in the aged Ts65Dn mouse: A mouse model of Down syndrome and Alzheimer's disease. *Behavioral Neuroscience* **118**: 1196–1205.

Drugan RC, Basile AS, Ha JH, Healy D, Ferland RJ (1997). Analysis of the importance of controllable versus uncontrollable stress on subsequent behavioral and physiological functioning. *Brain Research Protocols* **2**: 69–74.

Drugan RC, Maier SF, Skolnick P, Paul SM, Crawley JN (1985). An anxiogenic benzodi-azepine receptor ligand induces learned helplessness. *European Journal of Pharmacology* **113**: 453–457.

Drugan RC, Ryan SM, Minor TR, Maier SF (1984). Librium prevents the analgesia and shuttle-box escape deficit typically observed following inescapable shock. *Pharmacology Biochemistry and Behavior* **21**: 749–754.

Drugan RC, Skolnick P, Paul SM, Crawley JN (1989). A pretest reliably predicts performance in two animal models of inescapable stress. *Pharmacology Biochemistry and Behavior* **33**: 649–654.

Duan W, Guo Z, Jiang H, Ladenheim B, Xu X, Cadet JL, Mattson MP (2004). Paroxetine retards disease onset and progression in Huntingtin mutant mice. *Annals of Neurology* **55**: 590–594.

Dubnau J, Tully T (1998). Gene discovery in *Drosophila*: New insights for learning and memory. *Annual Review of Neuroscience* **21**: 407–444.

Dubuisson D, Dennis SG (1977). The formalin test: A quantitative study of the analgesic effects of morphine, meperidine, and brain stem stimulation in rats and cats. *Pain* **4**: 161–174.

Dudai Y (1989). *The Neurobiology of Memory: Concepts, Findings, Trends*. Oxford University Press, New York.

Duffield GE (2003). DNA microarray analyses of circadian timing: The genomic basis of bio-logical time. *Journal of Neuroendocrinology* **15**: 991–1002.

Dulac C, Axel R (1995). A novel family of genes encoding the putative pheromone receptors in mammals. *Cell* **83**: 195–206.

Dulawa SC, Geyer MA (2000).Effects of strain and serotonergic agents on prepulse inhibition and habituation in mice. *Neuropharmacology* **39**: 2170–2179.

Dulawa SC, Grandy DK, Low MJ, Paulus MP, Geyer MA (1999). Dopamine D4 receptor-knock-out mice exhibit reduced exploration of novel stimuli. *Journal of Neuroscience* **19**: 9550–9556.

Dulawa SC, Hen R, Scearce-Levie K, Geyer (1997). Serotonin 1B receptor modulation of startle reactivity, habituation, and prepulse inhibition in wild-type and serotonin 1B knockout mice. *Psychopharmacology* **132**: 125–134.

Dunlap JC (1996). Genetic and molecular analysis of circadian rhythms. *Annual Review of Genetics* **30**: 579–601.

Dunn AJ (1988). Stress related activation of cerebral dopaminergic systems. In *The Mesolimbic Dopamine System*. PW Kalivas, CB Nemeroff, Eds. Annals of the New York Academy of Sciences **537**. New York, pp. 188–205.

Dunn AJ, File SE (1987). Corticotropin-releasing factor has an anxiogenic action in the social interaction test. *Hormones and Behavior* **21**: 193–202.

Dunnett S (1993). The role and repair of forebrain cholinergic systems in short-term memory: Studies using the delayed matching-to-position task in rats. *Advances in Neurology* **59**: 53–65.

During MJ, Leone P (1995). Adeno-associated virus vectors for gene therapy of neurodegener-ative disorders. *Clinical Neuroscience* **3**: 292–300.

Dziewczapolski G, Menalled LB, Garcia MC, Mora MA, Gershanik OS, Rubinstein M (1998). Opposite roles of D1 and D5 dopamine receptors in locomotion revealed by selective anti-sense oligonucleotides. *NeuroReport* **9**: 1–5.

Ebstein RP, Novick O, Umansky R, Priel B, Osher Y, Blaine D, Bennett ER, Nemanov L, Katz M, Belmaker RH (1996). Dopamine D4 receptor (D4DR) Exon III polymorphisms associated with the human personality trait of novelty seeking. *Nature Genetics* **12**: 78–80.

Echeverria V, Ducatenzeiler A, Dowd E, Janne J, Grant SM, Szyf M, Wandosell F, Avila J, Grimm H, Dunnett SB, Hartmann T, Alhonen L, Cuello AC (2004). Altered mitogen-activated protein kinase signaling, tau hyperphosphorylation and mild spatial learning dysfunction in transgenic rats expressing the beta-amyloid peptide intracellularly in hippocampal and cortical neurons. *Neuroscience* **129**: 583–592.

Eddy EM, Washburn TF, Bunch DO, Goulding EH, Gladen BC, Lubahan DB, Korach KS (1996). Targeted disruption of the estrogen receptor gene in male mice causes alteration of spermatogenesis and infertility. *Endocrinology* **137**: 4796–4805.

Edwards DA, Burge KG (1973). Olfactory control of the sexual behavior of male and female mice. *Physiology and Behavior* **11**: 867–872.

Edwards GL, Ladenheim EE, Ritter RC (1986). Dorsal hindbrain participation in cholecystokinin-induced satiety. *American Journal of Physiology* **251**: R971–R977.

Egan MF, Goldberg TE, Kolachana BS, Callicott JH, Mazzanti CM, Straub RE, Goldman D, Weinberger DR (2001). Effect of COMT Val108/158 Met genotype on frontal lobe function and risk for schizophrenia. *Proceedings of the National Academy of Sciences USA* **98**: 6917–6922.

Egan MF, Weinberger DR (1997). Neurobiology of schizophrenia. *Current Opinions in Neurobiology* **7**: 701–707.

Eggan K, Baldwin K, Tackett M, Osborne J, Gogos J, Chess A, Axel R, Jaenisch R (2004). Mice cloned from olfactory sensory neurons. *Nature* **428**: 44–49.

Eichelman BS, Thoa NB (1973). The aggressive monoamines. *Biological Psychiatry* **6**: 143–164.

Eichenbaum H (2002). *The Cognitive Neuroscience of Memory: An Introduction.* Oxford University Press, New York.

Eichenbaum H, Fagan A, Mathews P, Cohen NJ (1988). Hippocampal system dysfunction and odor discrimination learning in rats: Impairment or facilitation depending on representational demands. *Behavioral Neuroscience* **102**: 3531–3542.

Eichenbaum H, Schoenbaum G, Young G, Bunsey M (1996). Functional organization of the hippocampal memory system. *Proceedings of the National Academy of Sciences USA* **93**: 13500–13507.

Eichenbaum H, Stewart C, Morris RGM (1990). Hippocampal representation in place learning. *Journal of Neuroscience* **10**: 3531–3542.

Eilam R, Peter Y, Elson A, Rotman G, Shiloh Y, Groner Y, Segal M (1998). Selective loss of dopaminergic nigro-striatal neurons in brains of *Atm*-deficient mice. *Proceedings of the National Academy of Sciences USA* **95**: 12653–12656.

Einat H, Manji HK, Belmaker RH (2003a). New approaches to modeling bipolar disorder. *Psychopharmacology Bulletin* **37**: 47–63.

Einat H, Manji HK, Gould TD, Du J, Chen G (2003b). Possible involvement of the ERK signaling cascade in bipolar disorder: behavioral leads from the study of mutant mice. *Drug News Perspect.* **16**: 453–463.

Eisenberg J (1963). The behavior of heteromyid rodents. *University of California Publications in Zoology* **69**: 1–114.

Eisenstein J, Greenberg A (2003). Ghrelin: Update 2003. *Nutrition Reviews* **61**: 101–104.

Eklund A (1996). The effects of inbreeding on aggression in wild male house mice (*Mus Domesticus*). *Behaviour* **133**: 883–901.

Elgersma Y, Sweatt JD, Giese KP (2004). Mouse genetic approaches to investigating calcium/calmodulin-dependent protein kinase II function in plasticity and cognition. *Journal of Neuroscience* **24**: 8410–8415.

El-Ghundi M, O'Dowd BF, Erclik M, George SR (2003). Attenuation of sucrose reinforcement in dopamine D1 receptor deficient mice. *European Journal of Neuroscience* **17**: 851–862.

El Mallakh RS, El-Masri MA, Huff MO, Li XP, Decker S, Levy RS (2003). Intracerebroventricular administration of ouabain as a model of mania in rats. *Bipolar Disorders* **5**: 362–365.

El Yacoubi M, Bouali S, Popa D, Naudon L, Leroux-Nicollet I, Hamon M, Costentin J, Adrien J, Vaugeois JM (2003). Behavioral, neurochemical, and electrophysiological characterization of a genetic mouse model of depression. *Proceedings of the National Academy of Sciences USA* **100**: 6227–6232.

Ellenbroek BA, Cools AR (1990). Animal models with construct validity for schizophrenia. *Behavioural Pharmacology* **1**: 469–490.

Elliot EE, Sibley DR, Katz JL. (2003). Locomotor and discriminative-stimulus effects of cocaine in dopamine D5 receptor knockout mice. *Psychopharmacology* **169**: 161–168.

Ellsworth JD, Morrow BA, Roth RH (2001). Prenatal cocaine exposure increases mesoprefrontal dopamine neuron responsivity to mild stress. *Synapse* **42**: 80–83.

Elmer GI, Pieper JO, Negus SS, Woods JH (1998). Genetic variance in nociception and its relationship to the potency of morphine-induced analgesia in thermal and chemical tests. *Pain* **75**: 129–140.

Elmer GI, Pieper JO, Rubinstein M, Low MJ, Grandy DK, Wise RA (2002). Failure of intravenous morphine to serve as an effective instrumental reinforcer in dopamine D2 receptor knock-out mice. *Journal of Neuroscience* **22** (RC224): 1–6.

Engel SR, Lyons CR, Allan AM (1998). 5-HT$_3$ receptor over-expression decreases ethanol self-administration in transgenic mice. *Psychopharmacology* **140**: 243–248.

Engelmann M, Ebner K, Wotjak CT, Landgraf R (1998). Endogenous oxytocin is involved in short-term olfactory memory in female rats. *Behavioral Brain Research* **90**: 89–94.

Engelmann M, Wotjak CT, Landgraf R (1995). Social discrimination procedure: An alternative method to investigate juvenile recognition abilities in rats. *Physiology and Behavior* **58**: 315–321.

Enoch MA, Goldman D (2001). The genetics of alcoholism and alcohol abuse. *Current Psychiatry Reports* **3**: 144–151.

Epping-Jordan MP, Watkins SS, Koob GF, Markou A (1998). Dramatic decreases in brain reward function during nicotine withdrawal. *Nature* **393**: 76–79.

Epstein AN (1971). The lateral hypothalamic syndrome: Its implications for the physiological psychology of hunger and thirst. In *Progress in Physiological Psychology*, Vol. 4, E Stellar, JM Sprague, Eds. Academic, New York.

Erickson JC, Clegg KE, Palmiter RD (1996a). Sensitivity to leptin and susceptibility to seizures of mice lacking neuropeptide Y. *Nature* **381**: 415–421.

Erickson JC, Hollopeter G, Palmiter RD (1996b). Attenuation of the obesity syndrome of *ob/ob* mice by the loss of neuropeptide Y. *Science* **274**: 1704–1707.

Erskine MS (1989). Solicitation behavior in the estrous female rat: A review. *Hormones and Behavior* **23**: 473–502.

Erven A, Skynner MJ, Okumuru K, Takebayashi S, Brown SD, Steel KP, Allen ND (2002). A novel stereocilia defect in sensory hair cells of the deaf mouse mutant Tasmanian devil. *European Journal of Neuroscience* **16**: 1433–1441.

Erway LC, Shiau YW, Davis RR, Krieg EF (1996). Genetics of age-related hearing loss in mice. III. Susceptibility of inbred and F1 hybrid strains to noise-induced hearing loss. *Hearing Research* **93**: 181–187.

Erway LC, Willott JF, Archer JR, Harrison DE (1993). Genetics of age-related hearing loss in mice. I. Inbred and F1 hybrid strains. *Hearing Research* **65**: 125–132.

Erwin VG, Heston WD, McClearn GE, Deitrich RA (1976). Effect of hypnotics on mice genetically selected for sensitivity to ethanol. *Pharmacology Biochemistry and Behavior* **4**: 679–683.

Escorihuela RM, Vallina IF, Martinez-Cue C, Baamonde C, Dierssen M, Tobena A, Florez J, Fernandez-Teruel A (1998). Impaired short- and long-term memory in Ts65Dn mice, a model for Down syndrome. *Neuroscience Letters* **247**: 171–174.

Estape N, Steckler T (2002). Cholinergic blockade impairs performance in operant DNMTP in two inbred strains of mice. *Pharmacology Biochemistry and Behavior* **72**: 319–334.

Estep DQ, Lanier DL, Dewsbury DA (1975). Copulatory behavior and nest-building behavior in the wild house mouse (*mus musculus*). *Learning and Behavior* **3**: 329–336.

Estrada-Smith D, Castellani LW, Wong H, Wen PZ, Chui A, Lusis AJ, Davis RC (2004). Dissection of multigenic obesity traits in congenic mouse strains. *Mammalian Genome* **15**: 14–22.

Ettenberg A (2004). Opponent process properties of self-administered cocaine. *Neuroscience and Biobehavioral Reviews* **27**: 721–728.

Evans BL, Smith SB (1997). Analysis of esterification of retinoids in the retinal pigmented epithelium of the $Mitf^{vit}$ (vitiligo). mutant mouse. *Molecular Vision* **3**: 11–16.

Everitt BJ (1990). Sexual motivation: A neural and behavioural analysis of the mechanisms underlying appetitive and copulatory responses of male rats. *Neuroscience and Biobehavioral Reviews* **15**: 217–232.

Everitt BJ, Robbins TW (1997). Central cholinergic systems and cognition. *Annual Review of Psychology* **48**: 649–684.

Fahrbach WE, Morrell JI, Pfaff DW (1985). Possible role for endogenous oxytocin in estrogen-facilitated maternal behaviour in rats. *Neuroendocrinology* **40**: 526–532.

Fairbanks CA, Wilcox GL (1997). Acute tolerance to spinally administered morphine compares mechanistically with chronically induced morphine tolerance. *Journal of Pharmacology and Experimental Therapeutics* **282**: 1408–1417.

Falangola MF, Lee SP, Nixon RA, Duff K, Helpern JA (2005). Histological co-localization of iron in Abeta plaques of PS/APP transgenic mice. *Neurochem Res.* **30**: 201–205.

Falls WA (2002). Fear-potentiated startle in mice. *Current Protocols in Neuroscience* 8.11B.1–8.11B.16.

Falls WA, Carlson S, Turner JG, Willott JF (1997). Fear-potentiated startle in two strains of inbred mice. *Behavioral Neuroscience* **111**: 855–861.

Fan W, Boston BA, Kesterson RA, Hruby VJ, Cone RD (1997). Role of melanocortinergic neurons in feeding and the *agouti* obesity syndrome. *Nature* **385**: 165–168.

Fanselow MS (1980). Conditional and unconditional components of post-shock freezing. *Pavlovian Journal of Biological Sciences* **15**: 177–182.

Fanselow MS (1990). Factors governing one-trial contextual conditioning. *Animal Learning and Behavior* **18**: 264–270.

Fanselow MS (1994). Neural organization of the defensive behavior system responsible for fear. *Psychonomic Bulletin Review* **1**: 429–438.

Fanselow MS, Gale GD (2003). The amygdala, fear, and memory. *Annals of the New York Academy of Sciences* **985**: 125–134.

Fanselow MS, Poulos AM (2005). The neuroscience of mammalian associative learning. *Annual Review of Psychology* **56**: 207–234.

Fanselow MS, Tighe TJ (1988). Contextual conditioning with massed versus distributed unconditional stimuli in the absence of explicit conditional stimuli. *Journal of Experimental Psychology* **14**: 187–199.

Farbman AI (1994). The cellular basis of olfaction. *Endeavour* **18**: 2–8.

Farson D, Harding TC, Tao L, Liu J, Powell S, Vimal V, Yendluri S, Koprivnikar K, Ho K, Twitty C, Husak P, Lin A, Snyder RO, Donahue BA (2004). Development and characterization of a cell line for large-scale, serum-free production of recombinant adeno-associated viral vectors. *The Journal of Gene Medicine* **6**: 1369–1381.

Fattore L, Martellotta MC, Cossu G, Fratta W. Gamma-hydroxybutyric acid: An evaluation of its rewarding properties in rats and mice. *Alcohol* **20**: 247–256.

Feldker DE, de Kloet ER, Kruk MR, Datson NA (2003). Large-scale gene expression profiling of discrete brain regions: Potential, limitations, and application in genetics of aggressive behavior. *Behavior Genetics* **33**: 537–548.

Ferguson GD, Anagnostaras SG, Silva AJ, Herschman HR (2000a). Deficits in memory and motor performance in synaptotagmin IV mutant mice. *Proceedings of the National Academy of Sciences USA* **97**: 5598–5603.

Ferguson JN, Aldag JM, Insel TR, Young LJ (2001). Oxytocin in the medial amygdala is essential for social recognition in the mouse. *Journal of Neuroscience* **21**: 8278–8285.

Ferguson JN, Young LJ, Hearn EF, Matzuk MM, Insel TR, Winslow JT (2000b). Social amnesia in mice lacking the oxytocin gene. *Nature Genetics* **25**: 284–288.

Ferguson JN, Young LJ, Insel TR (2002). The neuroendocrine basis of social recognition. *Frontiers in Neuroendocrinology* **23**: 200–224.

Ferraro TN, Golden GT, Smith GG, Martin JF, Schwebel CL, Doyle GA, Buono RJ, Berrettini WH (2005). Confirmation of a major QTL influencing oral morphine intake in C57 and DBA mice using reciprocal congenic strains. *Neuropsychopharmacology* **30**: 742–746.

Ferris CF (2000). Adolescent stress and neural plasticity in hamsters: A vasopressin-serotonin model of inappropriate aggressive behaviour. *Experimental Physiology* **85**: 85S–90S.

Fienberg AA, Hiroi N, Mermelstein PG, Song WJ, Synder GL, Nishi A, Cheramy A, O'Callaghan JP, Miller DB, Cole DG, Corbett R, Haile CN, Cooper DC, Onn SP, Grace AA, Ouimet CC, White FJ, Hyman SE, Surmeier DJ, Girault JA, Nestler EJ, Greengard P (1998). DARPP-32: Regulator of the efficacy of dopaminergic transmission. *Science* **281**: 838–842.

File SE (1980). The use of social interaction as a method for detecting anxiolytic activity of chlordiazepoxide-like drugs. *Journal of Neuroscience Methods* **2**: 219–238.

File SE (1981). Animal test of anxiety. In *Recent Advances in Neuropsycho-pharmacology*, B Angrist, Ed. Pergamon, New York, pp. 241–251.

File SE (1985). Animal models for predicting clinical efficacy of anxiolytic drugs: Social behavior. *Neuropsychobiology* **13**: 55–62.

File SE (1990). One-trial tolerance to the anxiolytic effects of chlordiazepoxide in the plus-maze. *Psychopharmacology* **100**: 281–282.

File SE (1996). Recent developments in anxiety, stress, and depression. *Pharmacology Biochemistry and Behavior* **54**: 3–12.

File SE (1997a). Animal tests of anxiety. In *Current Protocols in Neuroscience*. Wiley, New York, pp. 8.3.1–8.3.15.

File SE (1997b). Anxiolytic action of a neurokinin 1 receptor antagonist in the social interaction test. *Pharmacology Biochemistry and Behavior* **58**: 747–752.

File SE, Ed. (1998). Neurobehavioural adaptation to neurotoxicity. *Pharmacology Biochemistry and Behavior* **59**: 775–1092.

File SE, Andrews N, Wu PY, Zharkovsky A, Zangrossi H (1992). Modification of chlordiazepoxide's behavioural and neurochemical effects by handling and plus-maze experience. *European Journal of Pharmacology* **218**: 9–14.

File SE, Guardiola-Lemaitre BJ (1988). 1-Fenfluramine in tests of dominance and anxiety in the rat. *Neuropsychobiology* **20**: 205–211.

File SE, Hyde JR (1978). Can social interaction be used to measure anxiety? *British Journal of Pharmacology* **62**: 19–24.

File SE, Mahal A, Mangiarini L, Bates GP (1998). Striking changes in anxiety in Huntington's disease transgenic mice. *Brain Research* **805**: 234–240.

File SE, Wardbill AG (1975a). The reliability of the hole-board apparatus. *Psychopharmacologia* **44**: 47–51.

File SE, Wardbill AG (1975b). Validity of head-dipping as a measure of exploration in a modified hole-board. *Psychopharmacology* **44**: 53–59.

File SE, Zangrossi H, Sanders FL, Mabbutt PS (1994). Raised corticosterone in the rat after exposure to the elevated plus-maze. *Psychopharmacology* **113**: 543–546.

File SE, Zangrossi H, Viana M, Graeff FG (1993). Trial 2 in the elevated plus-maze: A different form of fear? *Psychopharmacology* **111**: 491–494.

Finegold AA, Mannes AJ, Iadarola MJ (1999). A paracrine paradigm for in vivo gene therapy in the central nervous system: Treatment of chronic pain. *Human Gene Therapy* **10**: 1251–1257.

Finitzo T, Albright K, O'Neal J (1998). The newborn with hearing loss: Detection in the nursery. *Pediatrics* **102**: 1452–1460.

Fink-Jensen A, Fedorova I, Wortwein G, Woldbye DP, Rasmussen T, Thomsen M, Bolwig TG, Knitowski KM, McKinzie DL, Yamada M, Wess J, Basile A (2003). A role for M5 muscarinic acetylcholine receptors in cocaine addiction. *Journal of Neuroscience Research* **74**: 91–96.

Finn DA, Rutledge-Gorman MT, Crabbe JC (2003). Genetic animal models of anxiety. *Neurogenetics* **4**: 109–135.

Fiorino DF, Coury A, Fibiger HC, Phillips AG (1993). Electrical stimulation of reward sites in the ventral tegmental area increases dopamine transmission in the nucleus accumbens of the rat. *Behavioural Brain Research* **55**: 131–141.

Fiorino DF, Coury A, Phillips AG (1997). Dynamic changes in nucleus accumbens dopamine efflux during the Coolidge effect in male rats. *Journal of Neuroscience* **17**: 4849–4855.

Fish EW, Faccidomo S, Gupta S, Miczek KA (2004). Anxiolytic-like effects of escitalopram, citalopram, and R-citalopram in maternally separated mouse pups. *Journal of Pharmacology and Experimental Therapeutics* **308**: 474–480.

Fishman GI (1998). Timing is everything in life: Conditional transgene expression in the cardiovascular system. *Circulation Research* **82**: 837–844.

Flaherty CF, Mitchell C (1999). Absolute and relative rewarding properties of fructose, glucose, and saccharin mixtures as reflected in anticipatory contrast. *Physiology and Behavior* **66**: 841–853.

Flanagan LM, Dohanics J, Verbalis JG, Stricker EM (1992). Gastric motility and food intake in rats after lesions of hypothalamic paraventricular nucleus. *American Journal of Physiology* **32**: R39–R44.

Flexner JB, Flexner LB, Stellar E (1963). Memory in mice as affected by intracerebral puromycin. *Science* **141**: 57–59.

Flint J, Corley R, DeFries JC, Fulker DW, Gray JA, Miller S, Collins AC (1995). A simple genetic basis for a complex psychological trait in laboratory mice. *Science* **269**: 1432–1435.

Flint J, Valdar W, Shifman S, Mott R (2005). Strategies for mapping and cloning quantitative trait genes in rodents. *Nature Reviews Genetics* **6**: 271–286.

Flood JF, Morley JE (1998). Learning and memory in the SAMP8 mouse. *Neuroscience and Biobehavioral Reviews* **22**: 1–20.

Flood JF, Rosenzweig MR, Jarvik ME (1978). Memory: Modification of anisomycin-induced amnesia by stimulants and depressants. *Science* **199**: 324–326.

Forte A, Cipollaro M, Cascino A, Galderisi U (2005). Small interfering RNAs and antisense oligonucleotides for treatment of neurological diseases. *Curr Drug Targets* **6**: 21–29.

Fox EA, Phillips RJ, Baronowski EA, Byerly MS, Jones S, Powley TL (2001). Neurotrophin-4 deficient mice have a loss of vagal intraganglionic mechanoreceptors from the small intestine and a disruption of short-term satiety. *Journal of Neuroscience* **21**: 8602–8615.

Fox MW (1965a). The visual cliff test for the study of visual depth perception in the mouse. *Animal Behavior* **13**: 232–233.

Fox WM (1965b). Reflex-ontogeny and behavioral development of the mouse. *Animal Behavior* **13**: 234–241.

Francis DD, Szegda K, Campbell G, Martin WD, Insel TR (2003). Epigenetic sources of behavioral differences in mice. *Nature Neuroscience* **6**: 445–446.

Francis DD, Young LJ, Meaney MJ, Insel TR (2002). Naturally occurring differences in maternal care are associated with the expression of oxytocin and vasopressin (V1a) receptors: Gender differences. *Journal of Neuroendocrinology* **14**: 349–353.

Frankel WN (1998). Mouse strain backgrounds: More than black and white. *Neuron* **20**: 183.

Frankland PW, Bontempi B, Talton LE, Kaczmarek L, Silva AJ (2004a). The involvement of the anterior cingulate cortex in remote contextual fear memory. *Science* **304**: 881–883.

Frankland PW, Wang Y, Rosner B, Shimizu T, Balleine BW, Dykens EM, Ornitz EM, Silva AJ (2004b). Sensorimotor gating abnormalities in young males with fragile X syndrome and *Fmr1*-knockout mice. *Molecular Psychiatry* **9**: 417–425.

Fraser HF, Harris LS (1967). Narcotic and narcotic antagonist analgesics. *Annual Review of Pharmacology* **7**: 277–300.

Fride E, Foox A, Rosenberg E, Faigenboim M, Cohen V, Barda L, Blau H, Mechoulam R (2003). Milk intake and survival in newborn cannabinoid CB1 receptor knockout mice: Evidence for a "CB3" receptor. *European Journal of Pharmacology* **461**: 27–34.

Friedman JM (2002). The function of leptin in nutrition, weight, and physiology. *Nutrition Reviews* **60**: S1–S14.

Friedman JM, Halaas JL (1998). Leptin and the regulation of body weight in mammals. *Nature* **395**: 763–770.

Friedmann T (1997). Overcoming the obstacles. In *Making Gene Therapy Work: Special Report. Scientific American* **276**: 96–101.

Friedmann T, Ed. (1999). *The Development of Human Gene Therapy*. Cold Spring Harbor Laboratory Press, New York.

Frish C, Dere E, De Souza Silva MA, Gödecke A, Schrader J, Huston JP (2000). Superior water maze performance and increase in fear-related behavior in the endothelian nitric oxide synthase-deficient mouse together with monoamine changes in cerebellum and ventral striatum. *Journal of Neuroscience* **20**: 6694–6700.

Froehlich W, Li TK (1994). Opioid involvement in alcohol drinking. *Annals of the New York Academy of Sciences* **739**: 156–167.

Fuchs PN, Roza C, Sora I, Uhl G, Raja SN (1999). Characterization of mechanical withdrawal responses and effects of μ-, δ- and κ-opioid agonists in normal and μ-opioid receptor knockout mice. *Brain Research* **821**: 480–486.

Funada M, Sato M, Makino Y, Wada K (2002). Evaluation of rewarding effect of toluene by the conditioned place preference procedure in mice. *Brain Research Protocols* **10**: 47–54.

Funada M, Zhou X, Satoh M, Wada K (2004). Profiling of methamphetamine-induced modifications of gene expression patterns in the mouse brain. *Annals of the New York Academy of Sciences* **1025**: 76–83.

Gainetdinov RR, Wetsel WC, Jones SR, Levin ED, Jaber M, Caron MG (1999). Role of serotonin in the paradoxical calming effect of psychostimulants on hyperactivity. *Science* **283**: 397–401.

Galef BG (2002). Social learning of food preferences in rodents: Rapid appetitive learning. In *Current Protocols in Neuroscience*: 8.5D.1–8.5D8.

Galef BG, Wigmore SW (1983). Transfer of information concerning distant foods: A laboratory investigation of the "information-centre" hypothesis. *Animal Behavior* **31**: 748–758.

Gallagher M (2002). *Handbook of Psychology, Biological Psychology*. Wiley, New York.

Gallagher M, Burwell R, Burchinal M (1993). Severity of spatial learning impairment in aging: Development of a learning index for performance in the Morris water maze. *Behavioral Neuroscience* **107**: 618–626.

Galsworthy MJ, Amrein I, Kuptsov PA, Poletaeva II, Zinn P, Rau A, Wyssotski A, Lipp H-P (2005). A comparison of wild-caught wood mice and bank voles in the Intellicage: Assessing exploration, daily activity patterns and place learning paradigms. *Behavioural Brain Research* **157**: 211–217.

Games D, Adams D, Alessandrini R, Barbour R, Berthelette P, Blackwell C, Carr T, Clemens J, Donaldson T, Gillespie F, Guido T, Hagoplan S, Johnson-Wood K, Khan K, Lee M, Leibowitz P, Lieberburg I, Little S, Masliah E, McConlogue L, Montoya-Zavala M, Mucke L, Paganini L, Penniman E, Power M, Schenk D, Seubert P, Snyder B, Soriano F, Tan H, Vitale J, Wadsworth S, Wolozin B, Zhao J (1995). Alzheimer-type neuropathology in transgenic mice overexpressing V717F beta-amyloid precursor protein. *Nature* **373**: 523–527.

Ganten D, Pfaff D (1986). *Neurobiology of Oxytocin. Current Topics in Neuroendocrinology*, Vol. 6. Springer, New York.

Gantois I, Bakker CE, Reyniers E, Willemsen R, D'Hooge R, De Deyn PP, Oostra BA, Kooy RF (2001). Restoring the phenotype of fragile X syndrome: Insight from the mouse model. *Current Molecular Medicine* **1**: 447–455.

Garcia J (1989). Food for Tolman: Cognition and cathexis in concert. In *Aversion, Avoidance and Anxiety*, T Archer, LG Nilsson, Eds. Lawrence Erlbaum Associates, Hillsdale, NJ, pp. 45–85.

Garcia J, Kimmeldorf DJ, Koelling RA (1955). Conditioned aversion to saccharin resulting from exposure to gamma radiation. *Science* **122**: 157–158.

Garcia MF, Gordon MN, Hutton M, Lewis J, McGowan E, Dickey CA, Morgan D, Arendash GW (2004). The retinal degeneration (rd) gene seriously impairs spatial cognitive performance in normal and Alzheimer's transgenic mice. *Neuroreport* **15**: 73–77.

Garey J, Kow LM, Huynh W, Ogawa S, Pfaff DW (2002). Temporal and spatial quantitation of nesting and mating behaviors among mice housed in semi-natural environment. *Hormones and Behavior* **42**: 294–306.

Garner JP, Weisker SM, Dufour B, Meneh JA (2004). Barbering (fur and whisker trimming) by laboratory mice as a model of human trichotillomania and obsessive-compulsive spectrum disorders. *Comparative Medicine* **54**: 216–224.

Garrett KM, Niekrasz I, Haque D, Parker KM, Seale TW (1998). Genotype differences between C57BL/6 and A inbred mice in anxiolytic and sedative actions of diazepam. *Behavior Genetics* **28**: 125–136.

Gasparini L, Ongini E, Wenk G (2004). Non-steroidal anti-inflammatory drugs (NSAIDs) in Alzheimer's disease: old and new mechanisms of action. *Journal of Neurochemistry* **91**: 521–536.

Gaveriaux-Ruff C, Kieffer BL (2002). Opioid receptor genes inactivated in mice: The highlights. *Neuropeptides* **36**: 62–71.

Gavériaux-Ruff C, Matthes HWD, Peluso J, Kieffer BL (1998). Abolition of morphine-immunosuppression in mice lacking the μ-opioid receptor gene. *Proceedings of the National Academy of Sciences USA* **95**: 6326–6330.

Gearhart J (2004). New human embryonic stem-cell lines—More is better. *New England Journal of Medicine* **350**: 1275–1276.

Gearhart JD, Singer HS, Moran TH, Tiemeyer M, Oster-Granite ML, Coyle JT (1986). Mouse chimeras composed of trisomy 16 and normal (2N) cells: Preliminary studies. *Brain Research Bulletin* **16**: 815–824.

Geinisman Y, Ganeshina O, Yoshida R, Berry RW, Disterhoft JF, Gallagher M (2004). Aging, spatial learning, and total synapse number in the rat CA1 stratum radiatum. *Neurobiology of Aging* **25**: 407–416.

Geller L, Kula T, Seifter J (1962). The effects of chlordiazepoxide and chlorpromazine on a punishment discrimination. *Psychopharmacologia* **3**: 374–385.

Gene Therapy Advisory Committee (1999). Report on the potential use of gene therapy in utero. *Human Gene Therapy* **10**: 689–692.

George FR, Goldberg SR (1989). Genetic approaches to the analysis of addiction processes. *Trends in Pharmacological Sciences* **10**: 78–83.

Gerlai R (1996). Gene-targeting studies of mammalian behavior: Is it the mutation or the background genotype? *Trends in Neurosciences* **19**: 177–181.

Gerlai R, Adams B, Fitch T, Chaney S, Baez M (2002). Performance deficits of mGluR8 knockout mice in learning tasks: the effects of null mutation and the background genotype. *Neuropharmacology* **43**: 235–249.

Gerlai R, Millen KJ, Herrup K, Fabien K, Joyner AL, Roder J (1996). Impaired motor learning performance in cerebellar En-2 mutant mice. *Behavioral Neuroscience* **110**: 126–133.

Gershenfeld HK, Neumann PE, Mathis C, Crawley JN, Li X, Paul SM (1997). Mapping quantitative trait loci for open-field behavior in mice. *Behavior Genetics* **27**: 201–210.

Gershenfeld HK, Paul SM (1997). Mapping quantitative trait loci for fear-like behaviors in mice. *Genomics* **46**: 1–8.

Gershon ES (1990). Genetics. In *Manic-Depressive Illness*. FK Goodwin, KR Jamison, Eds. Oxford University Press, New York, pp. 373–401.

Geschwind DH (2003). DNA microarrays: Translation of the genome from laboratory to clinic. *Neurology* **2**: 275–282.

Getchell TV, Margolis FL, Getchell ML (1984). Perireceptor and receptor events in vertebrate olfaction. *Progress in Neurobiology* **23**: 317–345.

Gewirtz JC, Davis M (1998). Application of Pavlovian higher-order conditioning to the analysis of the neural substrates of fear conditioning. *Neuropharmacology* **37**: 453–459.

Geyer MA, Ellenbroek B (2003). Animal behavior models of the mechanisms underlying antipsychotic atypicality. *Progress in Neuropsychopharmacology and Biological Psychiatry* **27**: 1071–1079.

Geyer MA, McIlwain KL, Paylor R (2002). Mouse genetic models for prepulse inhibition: An early review. *Molecular Psychiatry* **7**: 1039–1053.

Geyer MA, Krebs-Thomson K, Braff DL, Swerdlow NR (2001). Pharmacological studies of prepulse inhibition models of sensorimotor gating deficits in schizophrenia: A decade in review. *Psychopharmacology* **156**: 117–154.

Geyer MA, Swerdlow NR (1998). Measurement of startle response, prepulse inhibition, and habituation. In Crawley JN, Gerfen CR, McKay R, Rogawski MA, Sibley DR, Skolnick P, Eds. *Current Protocols in Neuroscience*: 8.7.1–8.7.15.

Geyer MA, Swerdlow NR, Mansbach RS, Braff DL (1990). Startle response models of sensori-motor gating and habituation deficits in schizophrenia. *Brain Research Bulletin* **25**: 485–498.

Gheusi G, Bluthé RM, Goodall G, Dantzer R (1994). Social and individual recognition in rodents: Methodological aspects and neurobiological basis. *Behavioural Processes* **33**: 59–88.

Gibbs J, Smith GP, Greenberg D (1993). Cholecystokinin: A neuroendocrine key to feeding behavior. In *Hormonally Induced Changes in Mind and Brain*, Academic Press, New York, pp. 51–69.

Gibbs J, Young RC, Smith GP (1973). Cholecystokinin decreases food intake in rats. *Journal of Comparative and Physiological Psychology* **84**: 488–495.

Giese KP (1999). The use of targeted point mutants in the study of learning and memory. In Crusio WE and Gerlai RT, Eds., *Handbook of Molecular-Genetic Techniques for Brain and Behavior Research*, Elsevier Science, Amsterdam, pp. 305–314.

Giese KP, Friedman E, Telliez JB, Federov NB, Wines M, Feig LA, Silva AJ (2001). Hippocampus-dependent learning and memory is impaired in mice lacking the Ras-guanine-nucleotide releasing factor 1 (Ras-GRF1). *Neuropharmacology* **41**: 791–800.

Gill KJ, Boyle AE (2003). Confirmation of quantitative trait loci for cocaine-induced activation in the AcB/BcA series of recombinant congenic strains. *Pharmacogenetics* **13**: 329–338.

Gilman S (1998). Imaging the brain. *New England Journal of Medicine* **338**: 889–897.

Gingrich JA, Ansorge MS, Merker R, Weisstaub N, Zhou M (2003). New lessons from knockout mice: The role of serotonin during development and its possible contribution to the origins of neuropsychiatric disorders. *CNS Spectr.* **8**: 572–577.

Gingrich JR, Roder J (1998). Inducible gene expression in the nervous system of transgenic mice. *Annual Review of Neuroscience* **21**: 377–405.

Giros B, Jaber M, Jones SR, Wightman RM, Caron MG (1996). Hyperlocomotion and indifference to cocaine and amphetamine in mice lacking the dopamine transporter. *Nature* **379**: 606–612.

Glickstein M (1969). Organization of the visual pathways. *Science.* **164**: 917–926.

Glimcher PW, Margolin DH, Giovino AA, Hoebel BG (1984). Neurotensin: A new "reward peptide." *Brain Research* **291**: 119–124.

Glowa JR (1986). Some behavioral effects of d-amphetamine, cocaine, nicotine, and caffeine on schedule-controlled responding in the mouse. *Neuropharmacology* **25**: 1127–1135.

Glowa JR, Crawley J, Suzdak PD, Paul SM (1988). Ethanol and GABA receptor complex: Studies with the partial inverse benzodiazepine receptor agonist Ro 15–4513. *Pharmacology Biochemistry and Behavior* **31**: 767–772.

Glynn D, Bortnick RA, Morton AJ (2003). Complexin II is essential for normal neurological function in mice. *Human Molecular Genetics* **12**: 2431–2448.

Goate AM, Edenberg HJ (1998). The genetics of alcoholism. *Current Opinion in Genetics and Development* **8**: 282–286.

Godinho SIH, Nolan PM (2005). The role of mutagenesis in defining genes in behaviour. *European Journal of Human Genetics*, Special Issue, **14**: 651–659.

Goeders NE, Smith JE (1987). Intracranial self-administration methodologies. *Neuroscience and Biobehavioral Reviews* **11**: 319–329.

Goldman-Rakic PS, Selemon LD (1997). Functional and anatomical aspects of prefrontal pathology in schizophrenia. *Schizophrenia Bulletin* **23**: 437–458.

Goldowitz D, Frankel WN, Takahashi JS, Holtz-Vaterna M, Bult C, Kibbe WA, Snoddy J, Li Y, Pretel S, Yates J, Swanson DJ (2004). Large-scale mutagenesis of the mouse to understand the genetic bases of nervous system structure and function. *Molecular Brain Research* **132**: 105–115.

Goldowitz D, Wahlsten D, Wimer RE, Eds. (1992). *Techniques for the Genetic Analysis of Brain and Behavior: Focus on the Mouse*. Elsevier, Amsterdam.

Gomeza J, Shannon H, Kostenis E, Felder C, Zhang L, Brodkin J, Grinberg A, Sheng H, Wess J (1999). Pronounced pharmacologic deficits in M2 muscarinic acetylcholine receptor knockout mice. *Proceedings of the National Academy of Sciences USA* **96**: 1692–1697.

Gonzalez LE, File SE, Overstreet DH (1998). Selectively bred lines of rats differ in social interaction and hippocampal 5-HT$_{1A}$ receptor function: A link between anxiety and depression? *Pharmacology Biochemistry and Behavior* **59**: 787–792.

Gonzales-Rios F, Vlaiculescu A, Ben Natan L, Protais P, Constentin J (1986). Dissociated effects of apomorphine on various nociceptive responses in mice. *Journal of Neural Transmission* **67**: 87–103.

Goodman M, New A, Siever L (2004). Trauma, genes, and the neurobiology of personality disorders. *Annals of the New York Academy of Sciences* **1032**: 104–116.

Goodson S, Halford J, Blundell J (2004). Direct, continuous behavioral analysis of drug action on feeding. *Current Protocols in Neuroscience* 8.6C.1–8.6C.11.

Goodwin FK, Jamison KR (1990). *Manic-Depressive Illness.* Oxford University Press, New York.

Goodwin SF, Del Vecchio M, Velinzon K, Hogel C, Russell SRH, Tully T, Kaiser K (1997). Defective learning in mutants of the *Drosophila* gene for a regulatory subunit of cAMP-dependent protein kinase. *Journal of Neuroscience* **17**: 8817–8827.

Gordon JW, Ruddle FH (1981). Integration and stable germ line transmission of genes injected into mouse pronuclei. *Science* **214**: 1244–1246.

Gorski JA, Balogh SA, Wehner JM, Jones KR (2003). Learning deficits in forebrain-restricted brain-derived neurotrophic factor mutant mice. *Neuroscience* **121**: 341–354.

Gossen M, Freundlieb S, Bender G, Muller G, Hillen W, Bujard H (1995). Transcriptional activation by tetracyclines in mammalian cells. *Science* **268**: 1766–1769.

Gottesman MM (2003). Cancer gene therapy: An awkward adolescence. *Cancer Gene Therapy* **10**: 501–508.

Gotz J, Chen F, Barmettler R, Nitsch RM (2001). Tau filament formation in transgenic mice expressing P301L *tau*. *Journal of Biological Chemistry* **276**: 529–534.

Gould TD, Gottesman II (2005). Psychiatric endophenotypes and the development of valid animal models. *Genes, Brain and Behavior* **5**: 113–119.

Gould TJ, Wehner JM (1999). Genetic influences on latent inhibition. *Behavioral Neuroscience* **113**: 1291–1296.

Gozes I, Bachar M, Bardea A, Davidson A, Rubinraut S, Fridkin M, Giladi E (1997). Protection against developmental retardation in apolipoprotein E-deficient mice by a fatty neuropeptide: Implications for early treatment of Alzheimer's disease. *Journal of Neurobiology* **33**: 329–342.

Grahame NJ, Cunningham CL (1995). Genetic differences in intravenous cocaine self-administration between C57BL/6J and DBA/2J mice. *Psychopharmacology* **122**: 281–291.

Grailhe R, Waeber C, Dulawa SC, Hornung JP, Zhuang X, Brunner D, Geyer MA, Hen R (1999). Increased exploratory activity and altered response to LSD in mice lacking the 5-HT$_{5A}$ receptor. Neuron **22**: 581–591.

Granholm AC, Ford KA, Hyde LA, Bimonte HA, Hunter CL, Nelson M, Albeck D, Sanders LA, Mufson EJ, Crnic LS (2002). Estrogen restores cognition and cholinergic phenotype in an animal model of Down syndrome. *Physiology and Behavior* **77**: 371–385.

Grant EC, MacIntosh JH (1963). A comparison of the social postures of some common laboratory rodents. *Behaviour* **21**: 246–259.

Grant SGN, O'Dell TJ, Karl KA, Stein PL, Soriano P, Kandel ER (1992). Impaired long-term potentiation, spatial learning, and hippocampal development in *fyn* mutant mice. *Science* **258**: 1903–1910.

Grass S, Crawley JN, Xu XJ, Wiesenfeld-Hallin Z (2003). Reduced spinal cord sensitization to C-fibre stimulation in mice overexpressing galanin. *European Journal of Neuroscience* **17**: 1829–1832.

Gravel RA, Clarke JTR, Kaback MM, Mahuran D, Sandhoff K, Suzuki K (1995). The G$_{M2}$ gangliosides. In *The Metabolic and Molecular Basis of Inherited Diseases*, CR Scriver, AL Beaudet, WS Sly, D Valle, Eds. McGraw-Hill, New York, pp. 2839–2879.

Gray JA (1982). *The Neuropsychology of Anxiety: An Inquiry into the Functions of the Septo-hippocampal System.* Oxford University Press, New York.

Graz M, Galun E, Moullier P (2005). The twelfth annual meeting of the European Society of Gene Therapy. *Molecular Therapy* **11**: 178–179.

Green DG, Herreros de Tejada P, Glover MJ (1994). Electrophysiological estimates of visual sensitivity in albino and pigmented mice. *Visual Neuroscience* **11**: 919–925.

Green EL, Roderick TH (1966). Mutant genes and linkages. In *Biology of the Laboratory Mouse*, EL Green, Ed. McGraw-Hill, New York. pp 87–150.

Green S, Walter P, Kuman V, Krust A, Bornert JM, Argos P, Chambon P (1986). Human oestrogen receptor cDNA: Sequence, expression and homology to *v-erb-A*. *Nature* **320**: 134–139.

Greenberg BD, McMahon FJ, Murphy DL (1998). Serotonin transporter candidate gene studies in affective disorders and personality: Promises and potential pitfalls. *Molecular Psychiatry* **3**: 186–189.

Grice DE, Halmi KA, Fichter MM, Strober M, Woodside DB, Treasure JT, Kaplan AS, Magistretti PJ, Goldman D, Bulik CM, Kaye WH, Berrettini WH (2002). Evidence for a susceptibility gene for anorexia nervosa on chromosome 1. *American Journal of Human Genetics* **70**: 787–792.

Griebel G, Belzung C, Perrault G, Sanger DJ (2000). Differences in anxiety-related behaviours and in sensitivity to diazepam in inbred and outbred strains of mice. *Psychopharmacology* **148**: 164–170.

Griebel G, Blanchard DC, Blanchard RJ (1996). Evidence that the behaviors in the mouse defense test battery relate to different emotional states: A factor analytic study. *Physiology and Behavior* **60**: 1255–1260.

Griebel G, Blanchard DC, Jung A, Lee JC, Masuda CK, Blanchard RJ (1995). Further evidence that the Mouse Defense Test Battery is useful for screening anxiolytic and panicolytic drugs: Effects of acute and chronic treatment with alprazolam. *Neuropharmacology* **34**: 1625–1633.

Griebel G, Perrault G, Sanger DJ (1997). CCK receptor antagonists in animal models of anxiety: Comparison between exploration tests, conflict procedures and a model based on defensive behaviours. *Behavioural Pharmacology* **8**: 549–560.

Griebel G, Perrault G, Simiand J, Cohen C, Granger P, Depoortere H, Francon D, Avenet P, Schoemaker H, Evanno Y, Sevrin M, George P, Scatton B (2003). SL651498, a GABAA receptor agonist with subtype-selective efficacy, as a potential treatment for generalized anxiety disorder and muscle spasms. *CNS Drug Reviews* **9**: 3–20.

Griebel G, Simiand J, Steinberg R, Jung M, Gully D, Roger P, Geslin M, Scatton B, Maffrand JP, Soubrie P (2002). 4-(2-Chloro-4-methoxy-5-methylphenyl)-*N*-[(1S)-2-cyclopropyl-1-(3-fluoro-4-methylphenyl)ethyl]5-methyl-N-(2-propynyl)-1, 3-thiazol-2-amine hydrochloride (SSR125543A), a potent and selective corticotrophin-releasing factor(1) receptor antagonist. II. Characterization in rodent models of stress-related disorders. *Journal of Pharmacology and Experimental Therapeutics* **301**: 333–345.

Griffin WC 3rd, Middaugh LD (2003). Acquisition of lever pressing for cocaine in C57BL/6J mice: Effects of prior Pavlovian conditioning. *Pharmacology Biochemistry and Behavior* **76**: 543–549.

Griffiths RR, Findley JD, Brady JV, Gutcher K, Robinson WW (1975). Comparison of progressive-ratio performance maintained by cocaine, methylphenidate and secobarbitol. *Psychopharmacology* **43**: 81–83.

Grillon C, Morgan CA, Southwick SM, Davis M, Charney DS (1996). Baseline startle amplitude and prepulse inhibition in Vietnam veterans with posttraumatic stress disorder. *Psychiatry Research* **64**: 169–178.

Grisel JE, Belknap JK, O'Toole LA, Helms ML, Wenger CD, Crabbe JC (1997). Quantitative trait loci affecting methamphetamine responses in BXD recombinant inbred mouse strains. *Journal of Neuroscience* **17**: 745–754.

Grootendorst J, Bour A, Vogel E, Kelche C, Sullivan PM, Dodart JC, Bales K, Mathis C (2005). Human apoE targeted replacement mouse lines: h-apoE4 and h-apoE3 Mice differ on spatial memory performance and avoidance behavior. *Behavioural Brain Research* **159**: 1–14.

Gross C, Zhuang X, Stark K, Ramboz S, Oosting R, Kirby L, Santarelli L, Beck S, Hen R (2002). Serotonin1A receptor acts during development to establish normal anxiety-like behaviour in the adult. *Nature* **416**: 396–400.

Guillot PV, Carlier M, Maxson SC, Roubertoux PL (1995). Intermale aggression tested in two procedures, using four inbred strains of mice and their reciprocal congenics: Y chromosomal implications. *Behavior Genetics* **25**: 357–360.

Guillot PV, Chapouthier G (1998). Intermale aggression, GAD activity in the olfactory bulbs and Y chromosome effect in seven inbred mouse strains. *Behavioural Brain Research* **90**: 203–206.

Gulcher JR, Kong A, Stefansson K (2001). The role of linkage studies for common diseases. *Current Opinion in Genetics and Development* **11**: 264–267.

Gura T (1997). Obesity sheds its secrets. *Science* **275**: 751–753.

Gurney ME, Fleck TJ, Himes CS, Hall ED (1998). Riluzole preserves motor function in a transgenic model of familial amyotrophic lateral sclerosis. *Neurology* **50**: 62–66.

Gusella JF, Wexler NS, Conneally PM, Naylor SL, Anderson MA, Tanzi RE, Watkins PC, Ottina K, Wallace MR, Sakaguchi AY (1983). A polymorphic DNA marker genetically linked to Huntington's disease. *Nature* **306**: 234–248.

Gustafsson JA (2003). What pharmacologists can learn from recent advances in estrogen signalling. *Trends in Pharmacological Sciences* **24**: 479–485.

Guttmacher AE, Collins FS (2002). Genomic medicine—A primer. *New England Journal of Medicine* **347**: 1512–1520.

Guy J, Hendrich B, Holmes M, Martin JE, Bird A (2001). A mouse *Mecp-2* null mutation causes neurological symptoms that mimic Rett syndrome. *Nature Genetics* **27**: 322–326.

Haberman RP, McCown RJ, Samulski RJ (1998). Inducible long-term gene expression in brain with adeno-associated virus gene transfer. *Gene Therapy* **5**: 1604–1611.

Haberman R, Samulski J, McCown TJ (2003). Attenuation of seizures and neuronal death by adeno-associated virus vector galanin expression and secretion. *Nature Medicine* **9**: 1076–1080.

Hahn ME, Hewitt JK, Schanz N, Weinreb L, Henry A (1997). Genetic and developmental influences on infant mouse ultrasonic calling. I. A diallel analysis of the calls of 3-day olds. *Behavior Genetics* **27**: 133–143.

Hahn ME, Karkowski L, Weinreb L, Henry A, Schanz N, Hahn EM (1998). Genetic and developmental influences on infant mouse ultrasonic calling. II. Developmental patterns in the calls of mice 2–12 days of age. *Behavior Genetics* **28**: 315–325.

Hahn ME, Schanz N (1996). Issues in the genetics of social behavior: Revisited. *Behavior Genetics* **26**: 463–470.

Hajjar RJ, Schmidt U, Matsui T, Guerrero JL, Lee KH, Gwathmey JK, Dec GW, Semigran MJ, Rosenzweig A (1998). Modulation of ventricular function through gene transfer in vivo. *Proceedings of the National Academy of Sciences USA* **95**: 5251–5256.

Halaas JL, Gajiwala KS, Maffei M, Cohen SL, Chait BT, Rabinowitz D, Lallone RL, Burley SK, Friedman JM (1995). Weight-reducing effects of the plasma protein encoded by the *obese* gene. *Science* **269**: 543–546.

Halford J, Blundell J (1998). Direct and continuous behavioral analyses of drug action on feeding. In *Current Protocols in Neuroscience*. Wiley, New York, pp. 8.6C.1–8.6C.7.

Hall CS (1934). Emotional behavior in the rat. *Journal of Comparative and Physiological Psychology* **18**: 385–403.

Hall FS, Goeb M, Li XF, Sora I, Uhl GR (2004a). Micro-opioid receptor knockout mice display reduced cocaine conditioned place preference but enhanced sensitization of cocaine-induced locomotion. *Molecular Brain Research* **121**: 123–130.

Hall FS, Li XF, Goeb M, Roff S, Hoggatt H, Sora I, Uhl GR (2003). Congenic C57BL/6 mu opiate receptor (MOR) knockout mice: Baseline and opiate effects. *Genes, Brain and Behavior* **2**: 114–121.

Hall FS, Li XF, Sora I, Xu F, Caron M, Lesch KP, Murphy DL, Uhl GR (2002). Cocaine mechanisms: enhanced cocaine, fluoxetine and nisoxetine place preferences following monoamine transporter deletions. *Neuroscience* **115**: 153–161.

Hall FS, Sora I, Drgonova J, Li XF, Goeb M, Uhl GR (2004b).Molecular mechanisms underlying the rewarding effects of cocaine. *Annals of the New York Academy of Sciences* **1025**: 47–56.

Hall FS, Sora I, Uhl GR (2001). Ethanol consumption and reward are decreased in mu-opiate receptor knockout mice. *Psychopharmacology* **154**: 43–49.

Hall WG, Bryan TE (1980). The ontogeny of feeding in rats. II. Independent ingestive behavior. *Journal of Comparative and Physiological Psychology* **94**: 746–756.

Hammer RE, Palmiter RD, Brinster RL (1984). Partial correction of murine hereditary growth disorder by germ-line incorporation of a new gene. *Nature* **311**: 65–67.

Hampton LL, Ladenheim EE, Akeson M, Way JM, Weber HC, Sutliff VE, Jensen RT, Wine LJ, Arnheiter H, Battey JF (1998). Loss of bombesin-induced feeding suppression in gastrin-releasing peptide receptor-deficient mice. *Proceedings of the National Academy of Sciences USA* **95**: 3188–3192.

Hampton TG, Stasko MR, Kale A, Amende I, Costa AC (2004). Gait dynamics in trisomic mice: Quantitative neurological traits of Down syndrome. *Physiology and Behavior* **82**: 381–389.

Han CJ, O'Tuathaigh CM, van Trigt L, Quinn JJ, Fanselow MS, Mongeau R, Koch C, Anderson DJ (2003). Trace but not delay fear conditioning requires attention and the anterior cingulate cortex. *Proceedings of the National Academy of Sciences USA* **100**: 13087–13092.

Handley SL, Mithani S (1984). Effects of alpha-adrenoreceptor agonists and antagonists in a maze exploration model of "fear"-motivated behaviour. *Naunyn-Schmiedeberg's Archives of Pharmacology* **327**: 1–5.

Hansen HH, Sanchez C, Meier E (1997). Neonatal administration of the selective serotonin reuptake inhibitor Lu 10–134–C increases forced swimming-induced immobility in adult rats: A putative animal model of depression? *Journal of Pharmacology and Experimental Therapeutics* **283**: 1333–1341.

Harbers K, Jahner D, Jaenisch R (1981). Microinjection of cloned retroviral genomes into mouse zygotes: Integration and expression in the animal. *Nature* **293**: 540–542.

Hargreaves K, Dubner R, Brown F, Flores C, Joris J (1988). A new and sensitive method for measuring thermal nociception in cutaneous hyperalgesia. *Pain* **32**: 77–88.

Harper SQ, Staber PD, He X, Eliason SL, Martins IH, Mao Q, Yang L, Kotin RM, Paulson HL, Davidson BL (2005). RNA interference improves motor and neuropathological abnormalities in a Huntington's disease mouse model. *Proceedings of the National Academy of Sciences USA* **102**: 5820–5825.

Harrell AV, Allan AM (2003). Improvements in hippocampal-dependent learning and decremental attention in 5-HT(3) receptor overexpressing mice. *Learning and Memory* **10**: 410–409.

Harris FM, Brecht WJ, Xu Q, Tesseur I, Kekonius L, Wyss-Coray T, Fish JD, Masliah E, Hopkins PC, Scearce-Levie K, Weisgraber KH, Mucke L, Mahley RW, Huang Y (2003). Carboxyl-terminal-truncated apolipoprotein E4 causes Alzheimer's disease-like neurodegeneration and behavioral deficits in transgenic mice. *Proceedings of the National Academy of Sciences USA* **100**: 10966–10971.

Harris RB, Mitchell TD, Yan X, Simpson JS, Redmann SM (2001). Metabolic responses to leptin in obese *db/db* mice are strain dependent. *American Journal of Physiology-Regulatory, Integrative and Comparative Physiology* **281**: R115–R132.

Harvey RJ, Depner UB, Wässle H, Ahmadi S, Heindl C, Reinold H, Smart TG, Harvey K, Schütz B, Abo-Salem OM, Zimmer A, Poisbeau P, Welzl H, Wolfer DP, Betz H, Zeilhofer HU, Müller U (2004). GlyR α 3: An essential target for spinal PGE$_2$-mediated inflammatory pain sensitization. *Science* **304**: 884–887.

Hattar S, Lucas RJ, Mrosovsky N, Thompson S, Douglas RH, Hankins MW, Lem J, Biel M, Hofmann F, Foster RG, Yau KW (2003). Melanopsin and rod-cone photoreceptive systems account for all major accessory visual functions in mice. *Nature* **424**: 76–81.

Hauser H, Gandelman R (1985). Lever pressing for pups: Evidence for hormonal influence upon maternal behavior of mice. *Hormones and Behavior* **19**: 454–468.

Hawyard MD, Pintar LE, Low MJ (2002). Selective reward deficit in mice lacking beta-endorphin and enkephalin. *Journal of Neuroscience* **22**: 8251–8258.

Hayes JM, Balkema GW (1993). Visual thresholds in mice: Comparison of retinal light damage and hypopigmentation. *Visual Neuroscience* **10**: 931–938.

Hayley S, Crocker SJ, Smith PD, Shree T, Jackson-Lewis V, Przedborski S, Mount M, Slack R, Anisman H, Park DS (2004). Regulation of dopaminergic loss by Fas in a 1-methyl-4-phenyl-1,2,3,6-tetrahydropyridine model of Parkinson's disease. *Journal of Neuroscience* **24**: 2045–2053.

Hebb D (1949). *The Organization of Behavior: A Neuropsychological Theory*. Wiley, New York.

Hebert MA, Evenson AR, Lumley LA, Meyerhoff JL (1998). Effects of acute social defeat on activity in the forced swim test: Parametric studies in DBA/2 mice using a novel measurement device. *Aggressive Behavior* **24**: 257–269.

Hedrick H, Bullock G (2004). *The Laboratory Mouse*. Academic Press, New York.

Heinrichs SC, Koob GF (1997). Application of experimental stressors in laboratory rodents. In *Current Protocols in Neuroscience*. Wiley, New York, pp. 8.4.1–8.4.14.

Heinrichs SC, Menzaghi F, Koob GF (1998). Neuropeptide Y-induced feeding and its control. *Vitamins and Hormones* **54**: 51–66.

Heinrichs SC, Merlo-Pich E, Kiczek KA, Britton KT, Koob GF (1992). Corticotropin-releasing factor antagonist reduces emotionality in socially defeated rats via direct neurotropic action. *Brain Research* **581**: 190–197.

Heinrichs SC, Min H, Tamraz S, Carmouche M, Boehme SA, Vale WW (1997). Antisexual and anxiogenic behavioral consequences of corticotropin-releasing factor overexpression are centrally mediated. *Psychoneuroendocrinology* **22**: 215–224.

Heinz A, Jones DW, Mazzanti C, Goldman D, Ragan P, Hommer D, Linnoila M, Weinberger DR (2000). A relationship between serotonin transporter genotype and in vivo protein expression and alcohol neurotoxicity. *Biological Psychiatry* **47**: 643–649.

Heinz A, Higley JD, Gorey JG, Saunders RC, Jones DW, Hommer D, Zajicek K, Suomi SJ, Lesch KP, Weinberger DR, Linnoila M (1998). In vivo association between alcohol intoxication, aggression, and serotonin transporter availability in nonhuman primates. *American Journal of Psychiatry* **155**: 1023–1028.

Heisler LK, Chu HM, Brennan TJ, Danao JA, Bajwa P, Parsons LH, Tecott LH (1998). Elevated anxiety and antidepressant-like responses in serotonin 5-HT1 A receptor mutant mice. *Proceedings of the National Academy of Sciences USA* **95**: 15049–15054.

Helgason A, Yngvadottir B, Hrafnkelsson B, Gulcher J, Stefansson K (2005). An Icelandic example of the impact of population structure on association studies. *Nature Genetics* **37**: 90–95.

Hen R (1996). Mean genes. *Neuron* **16**: 17–21.

Henderson ND (1967). Prior treatment effects on open field behavior of mice: A genetic analysis. *Animal Behavior* **15**: 364–376.

Henderson ND (1989). Interpreting studies that compare high- and low-selected lines on new characters. *Behavior Genetics* **19**: 473–502.

Henderson ND, Turri MG, DeFries JC, Flint J (2004). QTL analysis of multiple behavioral measures of anxiety in mice. *Behavior Genetics* **34**: 267–293.

Heninger GR (1995). The role of serotonin in clinical disorders. In *Psychopharmacology: The Fourth Generation of Progress*, F Bloom, D Kupfer Eds. Raven, New York. pp. 471–482.

Henry KR, Willott JF (1972). Unilateral inhibition of audiogenic seizures and Preyer reflexes. *Nature* **240**: 481–482.

Hepler DJ, Wenk GL, Cribbs BL, Olton DS, Coyle JT (1985). Memory impairments following basal forebrain lesions. *Brain Research* **346**: 8–14.

Herman JP, Guillonneau D, Dantzer R, Scatton B, Semerdjian-Rouquier L, Le Moal M (1982). Differential effects of inescapable footshocks and stimuli previously paired with inescapable footshocks on dopamine turnover in cortical and limbic areas of the rat. *Life Sciences* **30**: 2207–2214.

Hernandez L, Hoebel BG (1988). Food reward and cocaine increase extracellular dopamine in the nucleus accumbens as measured by microdialysis. *Life Sciences* **42**: 143–149.

Herson PS, Virk M, Rustay NR, Bond CT, Crabbe JC, Adelman JP, Maylie J (2003). A mouse model of episodic ataxia type-1. *Nature Neuroscience* **6**: 378–383.

Hess EJ, Jinnah HA, Kozak CA, Wilson MC (1992). Spontaneous locomotor hyperactivity in a mouse mutant with a deletion including the *Snap* gene on chromosome 2. *Journal of Neuroscience* **12**: 2865–2874.

Hess EJ, Wilson MC (1991). Tottering and leaner mutations perturb transient development expression of tyrosine hydroxylase in embryologically distinct Purkinje cells. *Neuron* **6**: 123–132.

Heyser CJ (2003). Assessment of developmental milestones in rodents. In *Current Protocols in Neuroscience*. Wiley, New York, pp. 8.18.1–8.18.15.

Heyser CJ, Fienberg AA, Greengard P, Gold LH (2000). DARPP-32 knockout mice exhibit impaired reversal learning in a discriminated operant task. *Brain Research* **867**: 122–130.

Heyser CJ, Masliah E, Samimi A, Campbell IL, Gold LH (1997). Progressive decline in avoidance learning paralleled by inflammatory neurodegeneration in transgenic mice expressing interleukin 6 in the brain. *Proceedings of the National Academy of Sciences USA* **94**: 1500–1505.

Heyser CJ, Wilson MC, Gold LH (1995). Coloboma hyperactive mutant exhibits delayed neurobehavioral developmental milestones. *Developmental Brain Research* **89**: 264–269.

Hickey MA, Chesselet MF (2003). The use of transgenic and knock-in mice to study Huntington's disease. *Cytogenetic Genome Research* **100**: 276–286.

Higgins DC, Paigen B (1997). An additional 150 SSLP markers typed for the AXB and BXA recombinant inbred mouse strains. *Mammalian Genome* **8**: 846–849.

Higgins GA (1998). From rodents to recovery: Development of animal models of schizophrenia. *CNS Drugs* **9**: 59–68.

Higgins GA, Jacobsen H (2003). Transgenic mouse models of Alzheimer's disease: Phenotype and application. *Behavioural Pharmacology* **14**: 419–438.

Higgins GA, Kew JN, Richards JG, Takeshima H, Jenck F, Adam G, Wichmann J, Kemp JA, Grottick AJ (2002). A combined pharmacological and genetic approach to investigate the role of orphanin FQ in learning and memory. *European Journal of Neuroscience* **15**: 911–922.

Higgins GA, Large CH, Rupniak HT, Barnes JC (1997). Apolipoprotein E and Alzheimer's disease: A review of recent findings. *Pharmacology Biochemistry and Behavior* **56**: 675–685.

Himms-Hagen J, Hogan S, Zaror-Behrens G (1986). Increased brown adipose tissue thermogenesis in obese (*ob/ob*) mice fed a palatable diet. *American Journal of Physiology* **250**: E274–281.

Hinkle DA, Eberwine JH (2003). Single-cell molecular biology: Implications for diagnosis and treatment of neurologic disease. *Biological Psychiatry* **54**: 413–417.

Hiroi N, Fienberg AA, Haile CN, Alburges M, Hanson GR, Greengard P, Nestler EJ (1999). Neuronal and behavioural abnormalities in striatal function in DARPP-32–mutant mice. *European Journal of Neuroscience* **11**: 1114–1118.

Hitzemann R, Bell J, Rasmussen E, McCaughran J (2001). Mapping the genes for the acoustic startle response (ASR) and prepulse inhibition of the ASR in the BXD recombinant inbred series: Effect of high-frequency hearing loss and cochlear pathology. In *Handbook of Mouse Auditory Research*, JF Willott, Ed. CRC Press, Boca Raton, FL, pp. 441–455.

Hitzemann R, Reed C, Malmanger B, Lawler M, Hitzemann B, Cunningham B, McWeeney S, Belknap J, Harrington C, Buck K, Phillips T, Crabbe J (2004). On the integration of alcohol-related quantitative trait loci and gene expression analyses. *Alcoholism, Clinical and Experimental Research* **28**: 1437–1448.

Ho DY, Sapolsky RM (1997). Gene therapy for the nervous system. In *Making Gene Therapy Work: Special Report. Scientific American* **276**: 116–120.

Ho L, Sharma N, Blackman L, Festa E, Reddy G, Pasinetti GM (2005). From proteomics to biomarker discovery in Alzheimer's disease. *Brain Research Reviews* **48**: 360–369.

Hoebel BG, Monaco AP, Hernandez L, Aulisi EF, Stanley BG, Lenard L (1983). Self-injection of amphetamine directly into the brain. *Psychopharmacology* **81**: 158–163.

Hofer MA (1996). Multiple regulators of ultrasonic vocalization in the rat. *Psychoneuroendocrinology* **21**: 203–217.

Hofer MA, Shair HN (1993). Ultrasonic vocalization, laryngeal braking, and thermogenesis in rat pups: A reappraisal. *Behavioral Neuroscience* **107**: 354–362.

Hofer MA, Shair HN (1987). Isolation distress in two-week-old rats: Influence of home cage, social companions, and prior experience with littermates. *Developmental Psychobiology* **20**: 465–476.

Hofer MA, Shair HN, Brunelli SA (2001). Ultrasonic vocalizations in rat and mouse pups. *Current Protocols in Neuroscience*. Wiley, New York, pp. 8.14.1–8.14.6.

Hoffmann O, Plesan A, Wiesenfeld-Hallin Z (1998). Genetic differences in morphine sensitivity, tolerance and withdrawal in rats. *Brain Research* **806**: 232–237.

Hoffman P, Tabakoff B (2005). Gene expression in animals with different acute responses to ethanol. *Addiction Biology* **10**: 63–69.

Hofker MH, Van Deursen J, Sklar HT (2002). *Transgenic Mouse: Methods and Protocols*. Humana Press, Totowa, NJ.

Hohmann JG, Cadd GC, Teal TH, Clifton DK, Steiner RA (1997). Transgenic mice that overexpress the galanin gene in brainstem neurons. *Society for Neuroscience* **27**: 729.15.

Hökfelt et al., Broberger C, Diez M, Xu ZQ, Shi T, Kopp J, Zhang X, Holmberg K, Landry M, Koistinaho J (1999). Galanin and NPY, two peptides with multiple putative roles in the nervous system. *Hormone and Metabolic Research* **31**: 330–334.

Holcomb LA, Gordon MN, Jantzen P, Hsiao K, Duff K, Morgan D (1999). Behavioral changes in transgenic mice expressing both amyloid precursor protein and presenilin-1 mutations: Lack of association with amyloid deposits. *Behavior Genetics*, **29**: 177–185.

Holcomb L, Gordon MN, McGowan E, Yu X, Benkovic S, Jantzen P, Wright K, Saad I, Mueller R, Morgan D, Sanders S, Zehr C, O'Campo K, Hardy J, Prada CM, Eckman C, Younkin S, Hsiao K, Duff K (1998). Accelerated Alzheimer-type phenotype in transgenic mice carrying both mutant *amyloid precursor protein* and *presenilin 1* transgenes. *Nature Medicine* **4**: 97–100.

Hole K, Olsen TJ (1993). The tail-flick and formalin tests in rodents: Changes in skin temperature as a confounding factor. *Pain* **53**: 247–254.

Holmes A (2001). Targeted gene mutation approaches to the study of anxiety-like behavior in mice. *Neuroscience and Biobehavioral Reviews* **25**: 261–273.

Holmes A, Harari AR (2003). The serotonin transporter gene-linked polymorphism and negative emotionality: Placing single gene effects in the context of genetic background and environment. *Genes, Brain and Behavior* **2**: 332–335.

Holmes A, Heilig M, Rupniak NMJ, Steckler T, Griebel G (2003a). Neuropeptide systems as novel therapeutic targets for depression and anxiety disorders. *Trends in Pharmacological Sciences* **24**: 580–588.

Holmes A, Hohmann JG, Steiner RA, Crawley JN (1999). Behavioral phenotype of transgenic mice with overexpression of the neuropeptide galanin. National Institutes of Health Research Festival Abstract K-7.

Holmes A, Hollon TR, Gleason TC, Liu Z, Dreiling J, Sibley DR, Crawley JN (2001). Behavioral characterization of dopamine D5 receptor null mutant mice. *Behavioral Neuroscience* **115**: 1129–1144.

Holmes A, Kinney JW, Wrenn CC, Li Q, Yang RJ, Ma L, Vishwanath J, Saavedra MC, Innerfield CE, Jacoby AS, Shine J, Iismaa TP, Crawley JN (2003b). Galanin GAL-R1 receptor null mutant mice display increased anxiety-like behavior specific to the elevated plus-maze. *Neuropsychopharmacology* **28**: 1031–1044.

Holmes A, Li Q, Koenig EA, Gold E, Stephenson D, Yang RJ, Dreiling J, Sullivan T, Crawley JN (2004). Phenotypic assessment of galanin overexpressing and galanin receptor R1 knockout mice in the tail suspension test for depression-related behavior. *Psychopharmacology* **178**: 276–285.

Holmes A, Li Q, Murphy DL, Gold E, Crawley JN (2003c). Abnormal anxiety-related behavior in serotonin transporter null mutant mice: The influence of genetic background. *Genes, Brain and Behavior* **2**: 365–380.

Holmes A, Murphy DL, Crawley JN (2002a). Reduced aggression in mice lacking the serotonin transporter. *Psychopharmacology* **161**: 160–167.

Holmes A, Murphy DL, Crawley JN (2003d). Abnormal behavioral phenotypes of serotonin transporter knockout mice: Parallels with human anxiety and depression. *Biological Psychiatry* **54**: 953–959.

Holmes A, Rodgers RJ (1998). Responses of Swiss-Webster mice to repeated plus-maze experience: Further evidence for a qualitative shift in emotional state? *Pharmacology Biochemistry and Behavior* **60**: 473–488.

Holmes A, Rodgers RJ (1999). Influence of spatial and temporal manipulations on the anxiolytic efficacy of chlordiazepoxide in mice previously exposed to the elevated plus-maze. *Neuroscience and Biobehavioural Reviews* **23**: 971–980.

Holmes A, Rodgers RJ (2003). Prior exposure to the elevated plus-maze sensitizes mice to the acute behavioral effects of fluoxetine and phenelzine. *European Journal of Pharmacology* **459**: 221–230.

Holmes A, Wrenn CC, Harris AP, Thayer KE, Crawley JN (2002b). Behavioral profiles of inbred strains on novel olfactory, spatial and emotional tests for reference memory in mice. *Genes, Brain and Behavior* **1**: 55–69.

Holmes A, Yang RJ, Lesch KP, Crawley JN, Murphy DL (2003e). Mice lacking the serotonin transporter exhibit 5-HT(1A) receptor-mediated abnormalities in tests for anxiety-like behavior. *Neuropsychopharmacology* **28**: 2077–2088.

Holmes A, Yang RJ, Crawley JN (2002c). Evaluation of an anxiety-related phenotype in galanin overexpressing transgenic mice. *Journal of Molecular Neuroscience* **18**: 151–165.

Holmes A, Yang RJ, Murphy DL, Crawley JN (2002d). Evaluation of antidepressant-related behavioral responses in mice lacking the serotonin transporter. *Neuropsychopharmacology* **27**: 914–923.

Holmes FE, Bacon A, Pope RJ, Vanderplank PA, Kerr NC, Sukumaran M, Pachnis V, Wynick D (2003f). Transgenic overexpression of galanin in the dorsal root ganglia modulates pain-related behavior. *Proceedings of the National Academy of Sciences USA* **100**: 6180–6185.

Holmes PV, Koprivica V, Chough E, Crawley JN (1994). Intraventricular administration of galanin does not affect behaviors associated with locus coeruleus activation in rats. *Peptides* **15**: 1303–1308.

Holtzman DM, Santucci D, Kilbridge J, Chua-Couzens J, Fontana DJ, Daniels SE, Johnson RM, Chen K, Sun Y, Carlson E, Alleva E, Epstein CJ, Mobley WC (1996). Developmental abnormalities and age-related neurodegeneration in a mouse model of Down syndrome. *Proceedings of the National Academy of Sciences USA* **93**: 1333–1338.

Homanics GE, Quinlan JJ, Firestone LL (1999). Pharmacologic and behavioral responses of inbred C57BL/6J and strain 129/SvJ mouse lines. *Pharmacology Biochemistry and Behavior* **63**: 21–26.

Hommel JD, Sears RM, Georgescu D, Simmons DL, DiLeone RJ (2003). Local gene knockdown in the brain using viral-mediated RNA interference. *Nature Medicine* **9**: 1539–1544.

Hooks MS, Duffy P, Striplin C, Kalivas PW (1994a). Behavioral and neurochemical sensitization following cocaine self-administration. *Psychopharmacology* **115**: 265–272.

Hooks MS, Jones DN, Holtzman SG, Juncos JL, Kalivas PW, Justice JB (1994b). Individual differences in behavior following amphetamine, GBR-12909, or apomorphine but not SKF-38393 or quinpirole. *Psychopharmacology* **116**: 217–225.

Hopp RMP, Ransom N, Hilsenbeck SG, Papermaster DS, Windle JJ (1998). Apoptosis in the murine *rd1* retinal degeneration is predominantly p53-independent. *Molecular Vision* **4**: 5–8.

Horinouchi Y, Akiyoshi J, Nagata A, Matsushita H, Tsutsumi T, Isogawa K, Noda T, Nagayama H (2004). Reduced anxious behavior in mice lacking the CCK2 receptor gene. *European Neuropsychopharmacology* **14**: 157–161.

Hossain MA, Fielding KE, Trescher WH, Ho T, Wilson MA, Laterra J (1998). Human FGF-1 gene delivery protects against quinolinate-induced striatal and hippocampal injury in neonatal rats. *European Journal of Neuroscience* **10**: 2490–2499.

Hossain SM, Wong BK, Simpson EM (2004). The dark phase improves genetic discrimination for some high throughput mouse behavioral phenotyping. *Genes, Brain and Behavior* **3**: 167–177.

Houpt TA, Frankmann SP (1996). TongueTwister: An integrated program for analyzing lickometer data. *Physiology and Behavior* **60**: 1277–1283.

Houpt TA, Philopena JM, Joh TH, Smith GP (1996). C-fos induction in the rat nucleus of the solitary tract correlates with the retention and forgetting of a conditioned taste aversion. *Learning and Memory* **3**: 25–30.

Hovatta I, Tennant RS, Helton R, Marr RA, Singer O, Redwine JM, Ellison JA, Schadt EE, Verma IM, Lockhart DJ, Barlow C (2005). Glyoxalase 1 and glutathione reductase 1 regulate anxiety in mice. *Nature* **438**: 662–666.

Hrabe de Angelis M, Balling R (1998). Large scale ENU screens in the mouse: Genetics meets genomics. *Mutation Research* **400**: 25–32.

Hsiao KK, Borchelt DR, Olson K, Johannsdottir R, Kitt C, Yunis W, Xu S, Eckman C, Younkin S, Price D, Iadecola C, Clark HB, Carlson G (1995). Age-related CNS disorder and early death in transgenic FVB/N mice overexpressing Alzheimer amyloid precursor protein. *Neuron* **15**: 1203–1218.

Hsiao K, Chapman P, Nilsen S, Eckman C, Harigaya Y, Younkin S, Yang F, Cole G (1996). Correlative memory deficits, Aβ elevation, and amyloid plaques in transgenic mice. *Science* **274**: 99–102.

Hsich G, Sena-Esteves M, Breakefield XO (2002). Critical issues in gene therapy for neurologic disease. *Hum Gene Ther.* **13**: 579–604.

Huang JM, Money MK, Berlin CI, Keats BJB (1995). Auditory phenotyping of heterozygous sound-responsive (+/*dn*) and deafness (*dn/dn*) mice. *Hearing Research* **88**: 61–64.

Hudson TJ (1998). The Human Genome Project: Tools for the identification of disease genes. *Clinical and Investigative Medicine* **21**: 267–276.

Huerta PT, Sun LD, Wilson MA, Tonegawa S (2000). Formation of temporal memory requires NMDA receptors within CA1 pyramidal neurons. *Neuron* **25**: 473–480.

Hughes J, Boden P, Costall B, Domeney A, Kelly E, Horwell DC, Hunter JC, Pinnock RD, Woodruff GN (1990). Development of a class of selective cholecystokinin type B receptor antagonists having potent anxiolytic activity. *Proceedings of the National Academy of Sciences USA* **87**: 6728–6732.

Humby T, Wilkinson L, Dawson G (2005). Assaying aspects of attention and impulse control in mice using the 5-choice serial reaction time task. *Current Protocols in Neuroscience*. Wiley, New York, pp. 8.5H.1–8.5H.15.

Humphrey RK, Bucay N, Beattie GM, Lopez A, Messam CA, Cirulli V, Hayek A (2003). Characterization and isolation of promoter-defined nestin-positive cells from the human fetal pancreas. *Diabetes* **52**: 2519–2525.

Hunt PS, Spear LP, Spear NE (1991). An ontogenetic comparison of ethanol-mediated taste aversion learning and ethanol-induced hypothermia in preweanling rats. *Behavioral Neuroscience* **105**: 971–983.

Hurst JL, Payne CE, Nevison CM, Marie AD, Humphries RE, Robertson DH, Cavaggioni A, Beynon RJ (2001). Individual recognition in mice mediated by major urinary proteins. *Nature* **414**: 631–634.

Huszar D, Lynch CA, Fairchild-Huntress V, Dunmore JH, Fang Q, Berkemeier LR, Gu W, Kesterson RA, Boston BA, Cone RD, Smith FJ, Campfield LA, Burn P, Lee F (1997). Targeted disruption of the melanocortin-4 receptor results in obesity in mice. *Cell* **88**: 131–141.

Ikari H, Spangler EL, Greig NH, Pei XF, Brossi A, Speer D, Patel N, Ingram DK (1995). Maze learning in aged rats is enhanced by phenserine, a novel anticholinesterase. *NeuroReport* **6**: 481–484.

Impey S, Smith DM, Obrietan K, Donahue R, Wade C, Storm DR (1998). Stimulation of cAMP response element (CRE)-mediated transcription during contextual learning. *Nature Neuroscience* **1**: 595–601.

Ingle DJ (1949). A simple means to producing obesity in the rat. *Proceedings of the Society for Experimental Biology and Medicine* **72**: 604–605.

Ingle DJ, Shein HM (1975). *Model Systems in Biological Psychiatry*. MIT Press, Cambridge.

Inoue M, Reed DR, Tordoff MG, Beauchamp GK, Bachmanov AA (2004). Allelic variation of the Tas1r3 taste receptor gene selectively affects behavioral and neural taste responses to sweeteners in the F2 hybrids between C57BL/6ByJ and 129P3/J mice. *Journal of Neuroscience* **24**: 2296–2303.

Insel TR (2001). Mouse models for autism: report from a meeting. *Mammalian Genome* **12**: 755–757.

Insel TR (2003). Is social attachment an addictive disorder? *Physiology and Behavior* **79**: 351–357.

Insel TR, Hulihan TJ (1995). A gender-specific mechanism for pair-bonding: Oxytocin and partner preference formation in monogamous voles. *Behavioral Neuroscience* **109**: 782–789.

Insel TR, Preston S, Winslow JT (1995). Mating in the monogamous male: Behavioral consequences. *Physiology and Behavior* **57**: 615–627.

International Human Genome Sequencing Consortium (2004). *Nature* **431**: 931–945.

Introini-Collison IB, Baratti CM (1992). Memory-modulatory effects of centrally acting noradrenergic drugs: Possible involvement of brain cholinergic mechanisms. *Behavioral and Neural Biology* **57**: 248–255.

Inui A (1999). Feeding and body-weight regulation by hypothalamic neuropeptides—Mediation of the actions of leptin. *Trends in Neuroscience* **22**: 62–67.

Ioffe E, Moon B, Connolly E, Friedman JM (1998). Abnormal regulation of the leptin gene in the pathogenesis of obesity. *Proceedings of the National Academy of Sciences USA* **95**: 11852–11857.

Irwin S (1968). Comprehensive observational assessment: Ia. A systematic, quantitative procedure for assessing the behavioral and physiologic state of the mouse. *Psychopharmacologia* **13**: 222–257.

Irwin S, Carlson E (1971). Single and repeated dose effects of imipramine, chlorpromazine, perphenazine, and chlordiazepoxide in the mouse. *Psychopharmacology Bulletin* **7**: 31.

Irwin SA, Idupulapati M, Gilbert ME, Harris JB, Chakravarti AB, Rogers EJ, Crisostomo RA, Larsen BP, Mehta A, Alcantara CJ, Patel B, Swain RA, Weiler IJ, Oostra BA, Greenough WT (2002). Dendritic spine and dendritic field characteristics of layer V pyramidal neurons in the visual cortex of fragile-X knockout mice. *American Journal of Medical Genetics* **111**: 140–146.

Isaacs AM, Oliver PL, Jones EL, Jeans A, Potter A, Hovik BH, Nolan PM, Vizor L, Glenister P, Simon AK, Gray IC, Spurr NK, Brown SD, Hunter AJ, Davies KE (2003). A mutation in Af4 is predicted to cause cerebellar ataxia and cataracts in the robotic mouse. *Journal of Neuroscience* **23**: 1631–1637.

Isles AR, Baum MJ, Ma D, Keverne EB, Allen ND (2001). Urinary odour preferences in mice. *Nature* **409**: 783–784.

Itoh Y, Kozakai I, Toyomizu M, Ishibashi T, Kuwano R (1998). Mapping of cholecystokinin transcription in transgenic mouse brain using *Escherichia coli* β-galactosidase reporter gene. *Development Growth and Differentiation* **40**: 395–402.

Itzhak Y, Martin JL, Ali SF (2002). Methamphetamine-induced dopaminergic neurotoxicity in mice: Long-lasting sensitization to the locomotor stimulation and desensitization to the rewarding effects of methamphetamine. *Prog Neuropsychopharmacol Biol Psychiatry* **26**: 1177–1183.

Ivell R, Russell J (1995). *Oxytocin*. Plenum, New York.

Iversen SD, Iversen LL (1981). *Behavioral Pharmacology*. Oxford University Press, New York.

Ivkovich D, Stanton ME (2001). Effects of early hippocampal lesions on trace, delay, and long-delay eyeblink conditioning in developing rats. *Neurobiology of Learning and Memory* **76**: 426–446.

Izquierdo I, Medina JH (1997). Memory formation: The sequence of biochemical events in the hippocampus and its connection to activity in other brain structures. *Neurobiology of Learning and Memory* **68**: 285–316.

Izquierdo A, Wiedholz LM, Millstein RA, Yang RJ, Bussey TJ, Saksida LM, Holmes A (2006). Genetic and dopaminergic modulation of reversal learning in a tochscreen-based operant procedure for mice. *Behavioural Brain Research* **171**: 181–188.

Jackson AL, Linsley PS (2004). Noise amidst the silence: Off-target effects of siRNAs? *Trends in Genetics* **20**: 521–524.

Jackson IJ, Abbott CM (2000). *Mouse Genetics and Transgenics: A Practical Approach*. Oxford University Press.

Jacobowitz DM, Abbott LC (1998). *Chemoarchitectonic Atlas of the Developing Mouse Brain*. CRC Press, Boca Raton, FL.

Jacobs BL, Ed. (1999). Serotonin 50th anniversary. *Neuropsychopharmacology* **21** (Special Suppl). 15.

Jacobs GH, Fenwick JC, Calderone JB, Deeb SS (1999). Human cone pigment expressed in transgenic mice yields altered vision. *Journal of Neuroscience* **19**: 3258–3265.

Jaenisch R (1988). Transgenic animals. *Science* **240**: 1468–1472.

Jaenisch R (2004). Human cloning—The science and ethics of nuclear transplantation. *New England Journal of Medicine* **351**: 2787–2791.

Jänne PA, Suchy SF, Bernard D, McDonald M, Crawley J, Grinberg A, Wynshaw-Boris A, West-phal H, Nussbaum RL (1998). Functional overlap between murine *Inpp5b* and *Ocrl1* may explain why deficiency of the murine ortholog for OCRL1 does not cause Lowe syndrome in mice. *Journal of Clinical Investigation* **101**: 2042–2053.

Janowsky A, Mah C, Johnson RA, Cunningham CL, Phillips TJ, Crabbe JC, Eshleman AJ, Belknap JK (2001). Mapping genes that regulate density of dopamine transporters and cor-related behaviors in recombinant inbred mice. *Journal of Pharmacology and Experimental Therapeutics* **298**: 634–643.

Jankowsky JL, Fadale DJ, Anderson J, Xu GM, Gonzales V, Jenkins NA, Copeland NG, Lee MK, Younkin LH, Wagner SL, Younkin SG, Borchelt DR (2004). Mutant presenilins specifically elevate the levels of the 42 residue beta-amyloid peptide in vivo : Evidence for augmentation of a 42-specific gamma secretase. *Human Molecular Genetics* **13**: 159–170.

Janus C (2003). Vaccines for Alzheimer's disease: How close are we? *CNS Drugs* **17**: 457–474.

Janus C, Pearson J, McLaurin J, Mathews PM, Jiang Y, Schmidt SD, Chishti MA, Horne P, Heslin D, French J, Mount HT, Nixon RA, Mercken M, Bergeron C, Fraser PE, St George-Hyslop P, Westaway D (2000). A beta peptide immunization reduces behavioural impairment and plaques in a model of Alzheimer's disease. *Nature* **408**: 979–982.

JAX Notes (2003). The importance of understanding substrains in the genomic age. *Jackson Laboratory JAX Notes* **491**: 1–3.

Jeffs B, Negrin CD, Graham D, Clark JS, Anderson NH, Gauguier D, Dominiczak AF (2000). Applicability of a "speed" congenic strategy to dissect blood pressure quantitative trait loci on rat chromosome 2. *Hypertension* **35**: 179–187.

Jiang M, Gold MS, Boulay G, Spicher K, Peyton M, Brabet P, Srinivasan Y, Rudolph U, Ellison G, Birnbaumer L (1998a). Multiple neurological abnormalities in mice deficient in the *G* protein G_o. *Proceedings of the National Academy of Sciences USA* **95**: 3269–3274.

Jiang YH, Armstrong D, Albrecht U, Atkins CM, Noebels JL, Eichele G, Sweatt JD, Beaudet AL (1998b). Mutation of the Angelman ubiquitin ligase in mice causes increased cytoplas-mic p53 and deficits of contextual learning and long-term potentiation. *Neuron* **21**: 799–811.

Johnson DL, Kesner RP (1994). The effects of lesions of the entorhinal cortex and the horizontal nucleus of the diagonal band of broca upon performance of a spatial location recognition task. *Behavioural Brain Research* **61**: 1–8.

Johnson KR, Gagnon LH, Webb LS, Peters LL, Hawes NL, Chang B, Zheng QY (2003). Mouse models of USH1C and DFNB18: phenotypic and molecular analyses of two new spontaneous mutations of the Ush1c gene. *Human Molecular Genetics* **12**: 3075–3086.

Johnson KR, Zheng QY, Erway LC (2000). A major gene affecting age-related hearing loss is common to at least ten inbred strains of mice. *Genomics* **70**: 171–180.

Johnson TE, DeFries JC, Markel PD (1992). Mapping quantitative trait loci for behavioral traits in the mouse. *Behavior Genetics* **22**: 635–653.

Johnson-Wood K, Lee M, Motter R, Hu K, Gordon G, Barbour R, Khan K, Gordon M, Tan H, Games D, Lieberburg I, Schenk D, Seubert P, McConlogue L (1997). Amyloid precursor protein processing and Aβ_{42} deposition in a transgenic model of Alzheimer disease. *Proceedings of the National Academy of Sciences USA* **94**: 1550–1555.

Johnston AL, File SE (1991). Sex differences in animal tests of anxiety. *Physiology and Behavior* **49**: 245–250.

Jones BC, Mormede P (1999). *Neurobehavioral Genetics: Methods and Applications.* CRC Press, Boca Raton, FL.

Jones BJ, Roberts DJ (1968). A rotarod suitable for quantitation measurements of motor inco-ordination in naïve mice. *Naunyn-Schmeidebergs Archives of Pharmacology* **259**: 211.

Jones BC, Tarantino LM, Rodriguez LA, Reed CL, McClearn GE, Plomin R, Erwin VG (1999). Quantitative-trait loci analysis of cocaine-related behaviours and neurochemistry. *Pharmacogenetics* **9**: 607–617.

Jones SR, Gainetdinov RR, Wightman RM, Caron MG (1998). Mechanisms of amphetamine action revealed in mice lacking the dopamine transporter. *Journal of Neuroscience* **18**: 1979–1986.

Joyner AL (2000). Gene *Targeting: A Practical Approach.* Oxford University Press, New York.

Jucker M, Ingram DK (1997). Murine models of brain aging and age-related neurodegenerative diseases. *Behavioural Brain Research* **85**: 1–25.

Kaiser J (2005). Celera to end subscriptions and give data to public GenBank. *Science* **308**: 775.

Kakihana R, Brown DR, McClearn GE, Tabershaw IR (1966). Brain sensitivity to alcohol in inbred mouse strains. *Science* **154**: 1574–1575.

Kalin N, Sherman J, Takahashi L (1988). Antagonism of endogenous CRH systems attenuates stress-induced freezing behavior in rats. *Brain Research* **457**: 130–135.

Kalivas PW, Duffy P (1990). Effect of acute and daily cocaine treatment on extracellular dopamine in the nucleus accumbens. *Synapse* **5**: 48–58.

Kalivas PW, Duffy P, Dilts R, Abhold R (1988). Enkephalin modulation of A10 dopamine neurons: A role in dopamine sensitization. In *The Mesolimbic Dopamine System*, PW Kalivas, CB Nemeroff, Eds. *Annals of the New York Academy of Sciences* **537**: 405–414.

Kalivas PW, Hooks MS, Sorg B (1993). The pharmacology and neural circuitry of sensitization to psychostimulants. *Behavioral Pharmacology* **4**: 315–334.

Kalivas PW, Nakamura M (1999). Neural systems for behavioral activation and reward. *Current Opinion in Neurobiology* **9**: 223–227.

Kalivas PW, Nemeroff CB, Eds. (1988). The mesolimbic dopamine system. *Annals of the New York Academy of Sciences* **537**: 1–540.

Kalivas PW, Pierce RC, Cornish J, Sorg BA (1998). A role for sensitization in craving and relapse in cocaine addiction. *Journal of Psychopharmacology* **12**: 48–53.

Kalivas PW, Szumlinski KK, Worley P (2004). Homer2 gene deletion in mice produces a phenotype similar to chronic cocaine treated rats. *Neurotoxicity Research* **6**: 385–387.

Kalra SP, Kalra PS (2004). NPY and cohorts in regulating appetite, obesity and metabolic syndrome: Beneficial effects of gene therapy. *Neuropeptides* **38**: 200–211.

Kamei J, Morita K, Miyata S, Onodera K (2003). Effects of second generation of histamine H1 antagonists, cetirizine and ebastine, on the antitussive and rewarding effects of dihydrocodeine in mice. *Psychopharmacology* **166**: 176–180.

Kammeschedit A, Kato K, Ito KI, Sumikawa K (1997). Adenovirus-mediated NMDA receptor knockouts in the rat hippocampal CA1 region. *NeuroReport* **8**: 635–639.

Kanarek RB (2004). Macronutrient selection in experimental animals. In *Current Protocols in Neuroscience*. Wiley, New York, pp. 8.6G.1–8.6G.14.

Kandel ER (2001). The molecular biology of memory storage: a dialogue between genes and synapses. *Science* **294**: 1030–1038.

Kaplan MT, Abel T (2003). Genetic approaches to the study of synaptic plasticity and memory storage. *CNS Spectrum* **8**: 597–620.

Karolyi IJ, Burrows HL, Ramesh TM, Nakajima M, Lesh JS, Seong E, Camper SA, Seasholtz AF (1999). Altered anxiety and weight gain in corticotropin-releasing hormone-binding protein-deficient mice. *Proceedings of the National Academy of Sciences USA* **96**: 11595–11600.

Kas MJ, van den Bos R, Baars AM, Lubbers M, Lesscher HM, Hillebrand JJ, Schuller AG, Pintar JE, Spruijt BM (2004). Mu-opioid receptor knockout mice show diminished food-anticipatory activity. *European Journal of Neuroscience* **20**: 1624–1632.

Kasarskis A, Manova K, Anderson KV (1998). A phenotype-based screen for embryonic lethal mutations in the mouse. *Proceedings of the National Academy of Sciences USA* **95**: 7485–7490.

Kaspar BK, Lladó J, Sherkat N, Rothstein J, Gage FH (2003). Retrograde viral delivery of IGF-1 prolongs survival in a mouse ALS model. *Science* **301**: 839—842.

Katamine S, Nishida N, Sugimoto T, Noda T, Sakaguchi S, Shigematsu K, Kataoka Y, Nakatani A, Hasegawa S, Moriuchi R, Miyamoto T (1998). Impaired motor coordination in mice lacking prion protein. *Cell and Molecular Neurobiology* **18**: 731–742.

Katsel PL, Davis KL, Haroutunian V(2005). Large-scale microarray studies of gene expression in multiple regions of the brain in schizophrenia and Alzheimer's disease. *International Review of Neurobiology* **63**: 41–82.

Katz JL, Chausmer AL, Elmer GI, Rubinstein M, Low MJ, Grandy DK (2003). Cocaine-induced locomotor activity and cocaine discrimination in dopamine D4 receptor mutant mice. *Psychopharmacology* **170**: 108–114.

Katzenberg D, Young T, Finn L, Lin L, King DP, Takahashi JS, Mignot E (1998). A *clock* polymorphism associated with human diurnal preference. *Sleep* **21**: 569–576.

Kawamata T, Akiguchi I, Yagi H, Irino M, Sugiyama H, Akiyama H, Shimada A, Takemura M, Ueno M, Kitabayashi T, Ohnishi K, Seriu N, Higuchi K, Hosokawa M, Takeda T (1997). Neuropathological studies on strains of senescence-accelerated mice (SAM) with age-related deficits in learning and memory. *Experimental Gerontology* **32**: 161–169.

Kavaliers M, Colwell DD, Choleris E, Agmo A, Muglia LJ, Ogawa S, Pfaff DW (2003). Impaired discrimination of and aversion to parasitized male odors by female oxytocin knockout mice. *Genes, Brain and Behavior* **2**: 220–230.

Keays DA, Nolan PM (2003). *N*-ethyl-*N*-nitrosourea mouse mutants in the dissection of behavioural and psychiatric disorders. *European Journal of Pharmacology* **480**: 205–217.

Keck ME, Holsboer F, Muller MB (2004). Mouse mutants for the study of corticotropin-releasing hormone receptor function: development of novel treatment strategies for mood disorders. *Annals of the New York Academy of Sciences* **1018**: 445–457.

Kehoe P, Blass EM (1986). Conditioned aversions and their memories in 5-day-old rats during suckling. *Journal of Experimental Psychology and Animal Behavior Processes* **12**: 40–47.

Keller G (2005). Embryonic stem cell differentiation: Emergence of a new era in biology and medicine. *Genes and Development* **19**: 1129–1155.

Keller JJ, Keller AB, Bowers BJ, Wehner JM (2005). Performance of α 7 nicotinic receptor null mutants is impaired in appetitive learning measured in a signaled nose poke task. *Behavioural Brain Research* **162**: 143–152.

Kelley AE (1998). Measurement of rodent stereotyped behavior. In *Current Protocols in Neuroscience*, JN Crawley, CR Gerfen, R McKay, MW Rogawski, DR Sibley, P Skolnick, Eds. Wiley, New York, pp. 8.8.1–8.8.13.

Kelley AE (2004). Memory and addiction; shared neural circuitry and molecular mechanisms. *Neuron* **44**: 161–179.

Kelliher KR, Ziesmann J, Munger SD, Reed RR, Zufall F (2003). Importance of the CNGA4 channel gene for odor discrimination and adaptation in behaving mice. *Proceedings of the National Academy of Sciences USA* **100**: 4299–4304.

Kelly JP, Wrynn AS, Leonard BE (1997). The olfactory bulbectomized rat as a model of depression: An update. *Pharmacology and Therapeutics* **74**: 299–316.

Kelly MA, Low MJ, Phillips TJ, Wakeland EK, Yanagisawa M (2003a). The mapping of quantitative trait loci underlying strain differences in locomotor activity between 129S6 and C57BL/6J mice. *Mammalian Genome.* **14**: 692–702.

Kelly MA, Rubinstein M, Phillips TJ, Lessov CN, Burkhart-Kasch S, Zhang G, Bunzow JR, Fang Y, Gerhardt GA, Grandy DK, Low MJ (1998). Locomotor activity in D2 dopamine receptor-deficient mice is determined by gene dosage, genetic background, and developmental adaptations. *Journal of Neuroscience* **18**: 3470–3479.

Kelly PH, Bondolfi L, Hunziker D, Schlecht HP, Carver K, Maguire E, Abramowski D, Wiederhold KH, Sturchler-Pierrat C, Jucker M, Bergmann R, Staufenbiel M, Sommer B (2003b). Progressive age-related impairment of cognitive behavior in APP23 transgenic mice. *Neurobiology of Aging* **24**: 365–378.

Kelsoe JR (2003). Arguments for the genetic basis of the bipolar spectrum. *Journal of Affective Disorders* **73**: 183–197.

Kendler KS (1997). The genetic epidemiology of psychiatric disorders: A current perspective. *Society for Psychiatry and Psychiatric Epidemiology* **32**: 5–11.

Kendler KS, Karkowski LM, Prescott CA (1998). Stressful life events and major depression: Risk period, long-term contextual threat, and diagnostic specifity. *Journal of Nervous and Mental Disorders* **186**: 661–669.

Kennedy JL, Farrer LA, Andreasen NC, Mayeux R, St George-Hyslop P (2003). The genetics of adult-onset neuropsychiatric disease: Complexities and conundra? *Science* **302**: 822–826.

Kennedy JL, Macciardi FM (1998). Chromosome 4 workshop. *Psychiatric Genetics* **8**: 67–71.

Kennedy JS, Bymaster FP, Schuh L, Calligaro DO, Nomikos G, Felder CC, Bernauer M, Kinon BJ, Baker RW, Hay D, Roth HJ, Dossenbach M, Kaiser C, Beasley CM, Holcombe JH, Effron MB, Breier A (2001). A current review of olanzapine's safety in the geriatric patient: From pre-clinical pharmacology to clinical data. *International Journal of Geriatric Psychiatry* **16** (Suppl 1): S33–61.

Kerr DA, Llado J, Shamblott MJ, Maragakis NJ, Irani DN, Crawford TO, Krishnan C, Dike S, Gearhart JD, Rothstein JD (2003). Human embryonic germ cell derivatives facilitate motor recovery of rats with diffuse motor neuron injury. *Journal of Neuroscience* **23**: 5131–5140.

Keverne EB, Fundele R, Narasimha M, Barton SC, Surani MA (1996). Genomic imprinting and the differential roles of parental genomes in brain development. *Developmental Brain Research* **92**: 91–100.

Kiefer SW, Hill KG, Kaczmarek HJ (1998). Taste reactivity to alcohol and basic tastes in outbred mice. *Alcoholism: Clinical and Experimental Research* **22**: 1146–1151.

Kieffer BL (1999). Opioids: First lessons from knockout mice. *Trends in Pharmacological Sciences* **20**: 19–26.

Kieffer BL, Gaveriaux-Ruff C (2002). Exploring the opioid system by gene knockout. *Progress in Neurobiology* **66**: 285–306.

Kier FJ, Molinari V (2003). "Do-it-yourself" dementia testing: Issues regarding an Alzheimer's home screening test. *Gerontologist* **43**: 295–301.

Kikusui T, Faccidomo S, Miczek KA (2005). Repeated maternal separation: Differences in cocaine-induced behavioral sensitization in adult male and female mice. *Psychopharmacology* **178**: 202–210.

Kim HJ, Jackson T, Noben-Trauth K (2003). Genetic analysis of the mouse deafness mutations varitint-waddler (Va) and jerker (Espnje). *Journal of the Association for Research in Otolaryngology* **4**: 83–90.

Kim JJ, Shih JC, Chen K, Chen L, Bao S, Maren S, Anagnostaras SG, Fanselow MS, De Maeyer E, Seif I, Thompson RF (1997). Selective enhancement of emotional, but not motor, learning in monoamine oxidase A-deficient mice. *Proceedings of the National Academy of Sciences USA* **94**: 5929–5933.

King D, Zigmond MJ, Finlay JM (1997a). Effects of dopamine depletion in the medial prefrontal cortex on the stress-induced increase in extracellular dopamine in the nucleus accumbens core and shell. *Neuroscience* **77**: 141–153.

King DL, Arendash GW (2002). Behavioral characterization of the Tg2576 transgenic model of Alzheimer's disease through 19 months. *Physiology and Behavior* **75**: 627–642.

King DP, Zhao Y, Sangoram AM, Wilsbacher LD, Tanaka M, Antoch MP, Steeves TDL, Vitaterna MH, Kornhauser JM, Lowrey PL, Turek FW, Takahashi JS (1997b). Positional cloning of the mouse circadian *clock* gene. *Cell* **89**: 641–653.

King SL, Marks MJ, Grady SR, Caldarone BJ, Koren AO, Mukhin AG, Collins AC, Picciotto MR (2003). Conditional expression in corticothalamic efferents reveals a developmental role for nicotinic acetylcholine receptors in modulation of passive avoidance behavior. *Journal of Neuroscience* **23**: 3837–3843.

King TE, Joynes RW, Payne M (1997c). The tail-flick test: II. The role of supraspinal systems and avoidance learning. *Behavioral Neuroscience* **111**: 754–767.

Kinney GG, Burno M, Campbell UC, Hernandez LM, Rodriguez D, Bristow LJ, Conn PJ (2003a). Metabotropic glutamate subtype 5 receptors modulate locomotor activity and sensorimotor gating in rodents. *Journal of Pharmacology and Experimental Therapeutics* **306**: 116–123.

Kinney JW, Starosta G, Crawley JN (2003b). Central galanin administration blocks consolidation of spatial learning. *Neurobiology of Learning and Memory* **80**: 42–54.

Kinney JW, Starosta G, Holmes A, Wrenn CC, Yang RJ, Harris AP, Long KC, Crawley JN (2002). Deficits in trace cued fear conditioning in galanin-treated rats and galanin-overexpressing transgenic mice. *Learning and Memory* **9**: 178–190.

Kirkby DL, Jones DNC, Barnes JC, Higgins GA (1996). Effects of anticholinesterase drugs tacrine and E2020, the 5-HT$_3$ antagonist ondansetron, and the H$_3$ antagonist thioperamide, in models of cognition and cholinergic function. *Behavioural Pharmacology* **7**: 513–525.

Kirik D, Bjorklund A (2003). Modeling CNS neurodegeneration by overexpression of disease-causing proteins using viral vectors. *Trends in Neuroscience* **26**: 386–392.

Kishimoto T, Radulovic J, Radulovic M, Lin CR, Schrick C, Hooshmand F, Hermanson O, Rosenfeld MG, Spiess J (2000). Deletion of crhr2 reveals an anxiolytic role for corticotropin-releasing hormone receptor-2. *Nature Genetics* 24: 415–419.

Kishimoto Y, Kawahara S, Kirino Y, Kadotani H, Nakamura Y, Ikeda M, Yoshioka T (1997). Conditioned eyeblink response is impaired in mutant mice lacking NMDA receptor subunit NR2A. *NeuroReport* **8**: 3717–3721.

Kistner A, Gossen M, Zimmermann F, Jerecic J, Ullmer C, Lübbert H, Bujard H (1996). Doxycycline-mediated quantitative and tissue-specific control of gene expression in transgenic mice. *Proceedings of the National Academy of Science* **93**: 10933–10938.

Kitanaka N, Sora I, Kinsey S, Zeng Z, Uhl GR (1998). No heroin or morphine 6β-glucuronide analgesia in μ-opioid receptor knockout mice. *European Journal of Pharmacology* **355**: R1–R3.

Kitchen I, Slowe SJ, Matthew HWD, Kieffer B (1997). Quantitative autoradiographic mapping of μ-, δ- and κ-opioid receptors in knockout mice lacking the μ-opioid receptor gene. *Brain Research* **778**: 73–88.

Kiyokawa Y, Kikusui T, Takeuchi Y, Mori Y (2005). Mapping the neural circuit activated by alarm pheromone perception by c-Fos immunohistochemistry. *Brain Research* **1043**: 145–154.

Klamer D, Palsson E, Revesz A, Engel JA, Svensson L (2004). Habituation of acoustic startle is disrupted by psychomimetic drugs: Differential dependence on dopaminergic and nitric oxide modulatory mechanisms. *Psychopharmacology* **176**: 440–450.

Kliethermes CL, Finn DA, Crabbe JC (2003). Validation of a modified mirrored chamber sensitive to anxiolytics and anxiogenics in mice. *Psychopharmacology* **169**: 190–197.

Klingensmith J, Nusse R, Perrrimon N (1994). The *Drosophila* segment polarity gene dishevelled encodes a novel protein required for response to the *wingless* signal. *Genes and Development* **8**: 118–130.

Klockgether T, Evert B (1998). Genes involved in hereditary ataxias. *Trends in Neurosciences* **21**: 413–418.

Klomberg KF, Garland T, Swallow JG, Carter PA (2002). Dominance, plasma testosterone levels, and testis size in house mice artificially selected for high activity levels. *Physiology and Behavior* **77**: 27–38.

Knutson B, Bergdorf J, Panksepp J (2002). Ultrasonic vocalizations as indices of affective states in rats. *Psychological Bulletin* **128**: 961–977.

Kobayashi DT, Chen KS (2005). Behavioral phenotypes of amyloid-based genetically modified mouse models of Alzheimer's disease. *Genes, Brain and Behavior* **4**: 173–196.

Koch M, Ehret G (1991). Parental behavior in the mouse: Effects of lesions in the entorhinal/piriform cortex. *Behavioral Brain Research* **42**: 99–105.

Koegler FH, Ritter S (1998). Galanin injection into the nucleus of the solitary tract stimulates feeding in rats with lesions of the paraventricular nucleus of the hypothalamus. *Physiology and Behavior* **63**: 521–527.

Koehl M, Battle SE, Turek FW (2003). Sleep in female mice: A strain comparison across the estrous cycle. *Sleep* **26**: 267–272.

Koepp MJ, Gunn RN, Lawrence AD, Cunningham VJ, Dagher A, Jones T, Brooks DJ, Bench CJ, Grasby PM (1998). Evidence for striatal dopamine release during a video game. *Nature* **393**: 266–268.

Kogan JH, Frankland PW, Blendy JA, Coblentz J, Marowitz Z, Schutz G, Silva AJ (1997). Spaced training induces normal long-term memory in CREB mutant mice. *Current Biology* **7**: 1–11.

Kogan JH, Frankland PW, Silva AJ (2000). Long-term memory underlying hippocampus-dependent social recognition in mice. *Hippocampus* **10**: 47–56.

König M, Zimmer AM, Steiner H, Holmes PV, Crawley JN, Brownstein MJ, Zimmer A (1996). Pain responses, anxiety and aggression in mice deficient in pre-proenkephalin. *Nature* **383**: 535–538.

Konkle AT, Baker SL, Kentner AC, Barabagallo LS, Merali Z, Bielajew C (2003). Evaluation of the effects of chronic mild stressors on hedonic and physiological responses: Sex and strain compared. *Brain Research* **992**: 227–238.

Koob GF (1992). Drugs of abuse: Anatomy, pharmacology and function of reward pathways. *Trends in Pharmacological Sciences* **13**: 177–184.

Koob GF (2003). Neuroadaptive mechanisms of addiction: studies on the extended amygdala. *European Neuropsychopharmacology* **13**: 442–452.

Koob GF, Ahmed SH, Boutrel B, Chen SA, Kenny PJ, Markou A, O'Dell LE, Parsons LH, Sanna PP (2004). Neurobiological mechanisms in the transition from drug use to drug dependence. *Neuroscience and Biobehavioral Reviews* **27**: 739–749.

Koob GF, Ehlers CL, Kupfer DJ (1989). *Animal Models of Depression*. Birkhauser, Boston.

Koob GF, Nestler EJ (1997). The neurobiology of drug addiction. *Journal of Neuropsychiatry and Clinical Neuroscience* **9**: 482–497.

Koob GF, Pettit HO, Ettenberg A, Bloom FE (1984). Effects of opiate antagonists and their quaternary derivatives on heroin self-administration in the rat. *Journal of Pharmacology and Experimental Therapeutics* **229**: 481–486.

Koob GF, Rocio M, Carrera A, Gold LH, Heyser CJ, Maldonado-Irizarry C, Markou A, Parsons LH, Roberts AJ, Schulteis G, Stinus L, Walker JR, Weissenborn R, Weiss F (1998). Substance dependence as a compulsive behavior. *Journal of Psychopharmacology* **12**: 39–48.

Koob GF, Vaccarino FJ, Amalric M, Bloom FE (1987). Positive reinforcement properties of drugs: Search for neural substrates. In *Brain Reward Systems and Abuse*, J Engel, L Oreland, Eds. Raven, New York.

Kooy RF (2003). Of mice and the fragile X syndrome. *Trends in Genetics* **18**: 148–154.

Kopin AS, Mathes WF, McBride EW, Nguyen M, Al-Haider W, Schmitz F, Bonner-Weir S, Kanarek R, Beinborn M (1999). The cholecystokinin-A receptor mediates inhibition of food intake yet is not essential for maintenance of body weight. *Journal of Clinical Investigation* **103**: 383–391.

Koretsky AP, Silva AC (2004). Manganese-enhanced magnetic resonance imaging (MEMRI). *NMR in Biomedicine* **17**: 527–531.

Kosobud AE, Crabbe JC (1990). Genetic correlations among inbred strain sensitivities to convulsions induced by 9 convulsant drugs. *Brain Research* **526**: 8–16.

Kostrzewa RM, Nowak P, Kostrzewa JP, Kostrzewa RA, Brus R (2005). Peculiarities of L-DOPA treatment of Parkinson's disease. *Amino Acids* **28**: 157–164.

Kovacs DM, Tanzi RE (1998). Monogenic determinants of familial Alzheimer's disease: Presenilin-1 mutations. *Cellular and Molecular Life Sciences* **54**: 902–909.

Kowalski TJ, Ster AM, Smith GP (1998). Ontogeny of hyperphagia in the Zucker (*fa/fa*). rat. *American Journal of Physiology* **275**: R1106–R1109.

Koy RF (2003). Of mice and the fragile X syndrome. *Trends in Genetics* **19**: 149–154.

Kozikowski AP, Johnson KM, Deschaux O, Bandyopadhyay BC, Araldi GL, Carmona G, Munzar P, Smith MP, Balster RL, Beardsley PM, Tella SR (2003). Mixed cocaine agonist/antagonist properties of (+)-methyl 4beta-(4-chlorophenyl)-1-methylpiperidine-3alpha-carboxylate, a piperidine-based analog of cocaine. *Journal of Pharmacology and Experimental Therapeutics* **305**: 143–150.

Kozorovitskiy Y, Gould E (2004). Dominance hierarchy influences adult neurogenesis in the dentate gyrus. *Journal of Neuroscience* **24**: 6755–6759.

Krakowski (2003). Violence and serotonin: Influence of impulse control, affect regulation, and social functioning. *The Journal of Neuropsychiatry and Clinical Neurosciences* **15**: 294–305.

Kramer MS, Cutler N, Feighner J, Shrivastava R, Carman J, Sramek JJ, Reines SA, Liu G, Snavely D, Wyatt-Knowles E, Hale JJ, Mills SG, MacCoss M, Swain CJ, Harrison T, Hill RG, Hefti F, Scolnick EM, Cascieri MA, Chicchi GG, Sadowski S, Williams AR, Hewson L, Smith D, Rupniak NM, et al. (1998). Distinct mechanism for antidepressant activity by blockade of central substance P receptors. *Science* **281**: 1640–1645.

Kranzler HR, Gelernter J, O'Malley S, Hernandez-Avila CA, Kaufman D (1998). Association of alcohol or other drug dependence with alleles of the mu opioid receptor gene (OPRM1). *Alcoholism Clinical and Experimental Research* **22**: 1359–1362.

Kreek MJ (1997). Opiate and cocaine addictions: Challenge for pharmacotherapies. *Pharmacology Biochemistry and Behavior* **57**: 551–569.

Kreek MJ, Nielsen DA, LaForge KS (2004). Genes associated with addiction: Alcoholism, opiate, and cocaine addiction. *Neuromolecular Medicine* **5**: 85–108.

Kreek MJ, Schlussman SD, Bart G, Steven Laforge K, Butelman ER (2004). Evolving perspectives on neurobiological research on the addictions: Celebration of the 30th anniversary of NIDA. *Neuropharmacology* **47** (Suppl 1): 324–344.

Krege JH, Hodgin JB, Couse JF, Enmark E, Warner M, Mahler JF, Sar M, Korach KS, Gustafsson JA, Smithies O (1998). Generation and reproductive phenotypes of mice lacking estrogen receptor beta. *Proceedings of the National Academy of Sciences USA* **95**: 15677–15682.

Krezel W, Ghyselinck N, Samad TA, Dupe V, Kastner P, Borrelli E, Chambon P (1998). Impaired locomotion and dopamine signaling in retinoid receptor mutant mice. *Science* **279**: 863–866.

Kreibich AS, Blendy JA (2004). cAMP response element-binding protein is required for stress but not cocaine-induced reinstatement. *Journal of Neuroscience* **24**: 6686–6692.

Kriegsfeld LJ, Dawson TM, Dawson VL, Nelson RJ, Snyder SH (1997). Aggressive behavior in male mice lacking the gene for neuronal nitric oxide synthase requires testosterone. *Brain Research* **769**: 66–70.

Kriz J, Gowing G, Julien JP (2003). Efficient three-drug cocktail for disease induced by mutant superoxide dismutase. *Annals of Neurology* **53**: 429–436.

Kruger R (2004). Genes in familial parkinsonism and their role in sporadic Parkinson's disease. *Journal of Neurology* **251** (Suppl 6): VI/2–6.

Kuczenski R, Segal D (1989). Concomitant characterization of behavioral and striatal neurotransmitter response to amphetamine using in vivo microdialysis. *Journal of Neuroscience* **9**: 2051–2065.

Kudryavtseva NN (2003). Use of the "partition" test in behavioral and pharmacological experiments. *Neuroscience and Behavioral Physiology* **33**: 461–471.

Kudwa AE, Rissman EF (2003). Double oestrogen receptor alpha and beta knockout mice reveal differences in neural oestrogen-mediated progestin receptor induction and female sexual behaviour. *Journal of Neuroendocrinology* **15**: 978–983.

Kuhar MJ, Adams S, Dominguez G, Jaworski J, Balkan B (2002). CART peptides. *Neuropeptides* **36**: 1–8.

Kuiper GGJM, Enmark E, Pelto-Huikko M, Nilsson S, Gustafsson JA (1996). Cloning of a novel estrogen receptor expressed in rat prostate and ovary. *Proceedings of the National Academy of Sciences USA* **93**: 5925–5930.

Kulkosky PJ, Gibbs J, Smith GP (1982). Behavioral effects of bombesin administration in rats. *Physiology and Behavior* **28**: 505–512.

Kuribara H, Asahi T (1997). Assessment of the anxiolytic and amnesic effects of three benzodiazepines, diazepam, alprazolam and triazolam, by conflict and non-matching to sample tests in mice. *Nihon Shinkei Seishin Yakurigaku Zasshi* **17**: 1–6.

Kurosu H, Yamamaoto M, Clark JD, Pastor JV, Nandi A, Gurnani P, McGuinness OP, Chikuda H, Yamaguchi M, Kawaguchi H, Shimomura I, Takayama Y, Herz J, Kahn CR, Rosenblatt KP, Kuro-o M (2005). Suppression of aging in mice by the hormone Klotho. *Science* **309**: 1829–1833.

Kushi A, Sasai H, Koizumi H, Takeda N, Yokoyama M, Nakamura M (1998). Obesity and mild hyperinsulinemia found in neuropeptide Y-Y1 receptor-deficient mice. *Proceedings of the National Academy of Sciences USA* **95**: 15659–15664.

Kuzmin A, Johansson B (2000). Reinforcing and neurochemical effects of cocaine: Differences among C57, DBA, and 129 mice. *Pharmacology Biochemistry and Behavior* **65**: 399–406.

Kyrkanides S, Miller JH, Brouxhon SM, Olschowka JA, Federoff HJ (2005). beta-Hexosaminidase lentiviral vectors: Transfer into the CNS via systemic administration. *Molecular Brain Research* **133**: 286–298.

Kyrkouli S, Stanley BG, Seirafi RD, Leibowitz SF (1990). Stimulation of feeding by galanin: Anatomical localization and behavioral specificity of this peptide's effects in the brain. *Peptides* **11**: 995–1001.

LaBar KS, Gatenby JC, Gore JC, LeDoux JE, Phelps EA (1998). Human amygdala activation during conditioned fear acquisition and extinction: a mixed-trial fMRI study. *Neuron* **20**: 937–945.

Ladenheim EE, Hampton LL, Whitney AC, White WO, Battey JF, Moran TH (2002). Disruptions in feeding and body weight control in gastrin-releasing peptide receptor deficient mice. *Journal of Endocrinology* **174**: 273–281

Laflamme N, Nappi RE, Drolet G, Labrie C, Rivest S (1998). Expression and neuropeptidergic characterization of estrogen receptors (ERalpha and ERbeta) throughout the rat brain: Anatomical evidence of distinct roles of each subtype. *Journal of Neurobiology* **5**: 357–378.

Lagerspetz KMJ, Lagerspetz KYH (1974). Genetic determination of aggressive behaviour. In *The Genetics of Behaviour*, JHF Van Abeelen, Ed. Elsevier, North Holland, pp. 321–346.

Lakso M, Sauer B, Mosinger B Jr, Lee EJ, Manning RW, Yu SH, Mulder KL, Westphal HA (1992). Targeted oncogene activation by site-specific recombination in transgenic mice. *Proceedings of the National Academy of Sciences USA* **89**: 6232–6236.

Lalonde R, Bensoula AN, Filali M (1995). Rotarod sensorimotor learning in cerebellar mutant mice. *Neuroscience Research* **22**: 423–426.

Lalonde R, Filali M, Bensoula AN, Lestienne F (1996). Sensorimotor learning in three cerebellar mutant mice. *Neurobiology of Learning and Memory* **65**: 113–120.

Lalonde R, Hayzoun K, Selimi F, Mariani J, Strazielle C (2003). Motor coordination in mice with hotfoot, Lurcher, and double mutations of the *Grid2* gene encoding the delta-2 excitatory amino acid receptor. *Physiology and Behavior* **80**: 333–339.

Lalonde R, Strazielle C (2003). Neurobehavioral characteristics of mice with modified intermediate filament genes. *Rev Neurosci.* **14**: 369–385.

Lam KS, Aman MG, Arnold LE (2006). Neurochemical correlates of autistic disorder: A review of the literature. *Research in Developmental Disabilities* **27**: 254–289.

Lambert KG, Gerlai R (2003). A tribute to Paul McLean: The neurobiological relevance of social behavior. *Physiology and Behavior* Special Issue **70**: 341–547.

Lamberty Y (1998). The mirror chamber test for testing anxiolytics: is there a mirror-induced stimulation? *Physiology and Behavior* **64**: 703–705.

Lamberty Y, Gower AJ (1991). Simplifying environmental cues in a Morris-type water maze improves place learning in old NMRI mice. *Behav Neural Biol.* **56**: 89–100.

Lamprecht R, Hazvi S, Dudai Y (1997). cAMP response element-binding protein in the amygdala is required for long- but not short-term conditioned taste aversion memory. *Journal of Neuroscience* **17**: 8443–8450.

Lander ES (1999). Array of hope. *Nature Genetics* **21**(Suppl): 3–5.

Lander ES, Botstein D (1986a). Mapping complex genetic traits in humans: New methods using a complete RFLP linkage map. *Cold Spring Harbor Symposium on Quantitative Biology* **51**: 49–62.

Lander ES, Botstein D (1986b). Strategies for studying heterogeneous genetic traits in humans by using a linkage map of restriction fragment length polymorphisms. *Proceedings of the National Academy of Sciences USA* **83**: 7353–7357.

Lander ES, Botstein D (1989). Mapping Mendelian factors underlying quantitative traits using RFLP linkage maps. *Genetics* **121**: 185–199.

Lander ES, Green P, Abrahamson J, Barlow A, Daly MJ, Lincoln SE, Newburg L (1987). MAP-MAKER: An interactive computer package for constructing primary genetic linkage maps of experimental and natural populations. *Genomics* **1**: 174–181.

Lander ES, et al. (2001). Initial sequencing and analysis of the human genome. *Nature* **409**: 860–921.

Lander ES, Schork NJ (1994). Genetic dissection of complex traits. *Science* **265**: 2037–2048.

Lang AP, de Angelis L (2003). Experimental anxiety and antiepileptics: The effects of valproate and vigabatrin in the mirrored chamber test. *Methods and Findings in Experimental and Clinical Pharmacology* **25**: 265–271.

Langhans N, Rindi G, Chiu M, Rehfeld JF, Ardman B, Beinborn M, Kopin AS (1997). Abnormal gastric histology and decreased acid production in cholecystokinin-B/gastrin receptor-deficient mice. *Gastroenterology* **112**: 280–286.

Larson J, Hoffman JS, Guidotti A, Costa E (2003). Olfactory discrimination learning deficits in heterozygous reeler mice. *Brain Research* **971**: 40–46.

Larson J, Lynch G (1986). Induction of synaptic potentiation in hippocampus by patterned stimulation involves two events. *Science* **232**: 985–988.

Lashley KS (1929). *Brain Mechanisms and Intelligence*. Chicago University Press, Chicago.

Lashley KS (1950). In search of the engram. In *Physiological Mechanisms in Animal Behavior*, Cambridge University Press, Cambridge, p. 478.

Lathe R (1996). Mice, gene targeting and behaviour: more than just genetic background. *Trends in Neurosciences* **19**: 183–186.

Lathe R (2004). The individuality of mice. *Genes, Brain and Behavior* **3**: 317–327.

Lauder Jm, Waage-Bauder H, Seidel H, Crawley JN, Perez A, Moy SS (2004). Differential effects of the Fragile-X mutation on gene expression and behavioral phenotype in mice on FVB and C57BL/6 backgrounds. *Society for Neuroscience* 116.2.

LaVail MM, Matthes MT, Yasumura D, Steinberg RH (1997). Variability in rate of cone degeneration in the retinal degeneration (*rd/rd*). mouse. *Experimental Eye Research* **65**: 45–50.

LaVail MM, Yasumura D, Matthes MT, Lau-Villacorta C, Unoki K, Wung CH, Steinberg RH (1998). Protection of mouse photoreceptors by survival factors in retinal degenerations. *Investigative Opthamology and Visual Science* **39**: 592–602.

Lavery KS, King TH (2003). Antisense and RNAi: Powerful tools in drug target discovery and validation. *Current Opinions in Drug Discovery and Development* **6**: 561–569.

Laviola G, Adriani W, Morley-Fletcher S, Terranova ML (2002). Peculiar response ot adolescent mice to acute and chronic stress and to amphetamine: Evidence of sex differences. *Behavioural Brain Research* **130**: 117–125.

Ledent C, Valverde O, Cossu G, Petitet F, Aubert JF, Beslot F, Böhme GA, Imperato A, Pedrazzini T, Roques BP, Vassart G, Fratta W, Parmentier M (1999). Unresponsiveness to cannabinoids and reduced addictive effects of opiates in CB_1 receptor knockout mice. *Science* **283**: 401–404.

Ledent C, Vaugeois JM, Schiffmann SN, Pedrazzini T, El Yacoubi M, Vanderhaeghen JJ, Costentin J, Heath JK, Vassart G, Parmentier M (1997). Aggressiveness, hypoalgesia and high blood pressure in mice lacking the adenosine A_{2a} receptor. *Nature* **388**: 674–678.

Lederhendler II (2003). Introduction: Behavioral neuroscience and childhood mental illness. *Annals of the New York Academy of Sciences* **1008**:1–10.

Ledermann B (2000). Embryonic stem cells and gene targeting. *Experimental Physiology* **85**: 603–613.

Ledermann B, Burki K (1991). Establishment of a germline competent C57BL/6 embryonic stem cell line. *Experimental Cell Research* **197**: 254–258.

LeDoux JE (1992). Brain mechanisms of emotion and emotional learning. *Current Opinions in Neurobiology* **2**: 191–197.

LeDoux JE (1995). Emotion: Clues from the brain. *Annual Review of Psychology* **46**: 209–235.

LeDoux JE (1996). *The Emotional Brain.* Simon and Schuster, New York.

LeDoux J (2002). *Synaptic Self: How Our Brains Become Who We Are.* Viking, New York.

Lee GH, Proenca R, Montez LM, Carroll KM, Darvishzadeh JG, Lee JI, Friedman JM (1996a). Abnormal splicing of the leptin receptor in *diabetic* mice. *Nature* **379**: 632–635.

Lee KW, Lee SH, Kim H, Song JS, Yang SD, Paik SG, Han PL (2004). Progressive cognitive impairment and anxiety induction in the absence of plaque deposition in C57BL/6 inbred mice expressing transgenic amyloid precursor protein. *Journal of Neuroscience Research* **76**: 572–580.

Lee MK, Borchelt DR, Wong PC, Sisodia SS, Price DL (1996b). Transgenic models of neurodegenerative diseases. *Current Opinion in Neurobiology* **6**: 651–660.

Lee RS, Koob GF, Henriksen SJ (1998). Electrophysiological responses of nucleus accumbens neurons to novelty stimuli and exploratory behavior in the awake, unrestrained rat. *Brain Research* **799**: 317–322.

Lee VM, Kenyon TK, Trojanowski JQ (2005). Transgenic animal models of tauopathies. *Biochim Biophys Acta* **1739**: 251–259.

Lefkowitz RJ, Inglese J, Koch WJ, Pitcher J, Attramadal H, Caron MG (1992). G-protein-coupled receptors: Regulatory role of receptor kinases and arrestin proteins. *Cold Spring Harbor Symposia on Quantitative Biology* **57**: 127–133.

Leibowitz SF (1995). Brain peptides and obesity: Pharmacologic treatment. *Obesity Research* **3**(Suppl 4): 573S–589S.

Leibowitz SF, Akabayashi A, Wang J (1998). Obesity on a high-fat diet: Role of hypothalamic galanin in neurons of the anterior paraventricular nucleus projecting to the median eminence. *Journal of Neuroscience* **18**: 2709–2719.

Leibowitz SF, Wortley KE (2004). Hypothalamic control of energy balance: Different peptides, different functions. *Peptides* **25**: 473–504.

Leigh H, Hofer MA (1973). Behavioral and physiologic effects of littermate removal on the remaining single pup and mother during the pre-weanling period in rats. *Psychosomatic Medicine* **35**: 497–508.

Leigh H, Hofer MA (1975). Long-term effects of preweanling isolation from littermates in rats. *Behavioral Biology* **15**: 173–181.

Leighty RE, Nilsson LNG, Potter H, Costa DA, Low MA, Bales KR, Paul SM, Arendash GW (2004). Use of multimetric statistical analysis to characterize and discriminate between the performance of four Alzheimer's transgenic mouse lines differing in Aβ deposition. *Behavioural Brain Research* **153**: 107–121.

Le Marec N, Lalonde R (1997). Sensorimotor learning and retention during equilibrium tests in Purkinje cell degeneration mutant mice. *Brain Research* **768**: 310–316.

Leone P, Di Chiara G (1987). Blockade of D-1 receptors by SCH 23390 antagonizes morphine- and amphetamine-induced place preference conditioning. *European Journal of Pharmacology* **135**: 251–254.

Lepicard EM, Joubert C, Hagneau I, Perez-Diaz F, Chapouthier G (2000). Differences in anxiety-related behavior and response to diazepam in BALB/cByJ and C57BL/6J strains of mice. *Pharmacology Biochemistry and Behavior* **67**: 739–748.

Le Roy I, Perez-Diaz F, Cherfouh A, Roubertoux PL (1999). Preweanling sensorial and motor development in laboratory mice: Quantitative trait loci mapping. *Developmental Psychobiology* **34**: 139–158.

Levenson JM, Choi S, Lee SY, Cao YA, Ahn HJ, Worley KC, Pizzi M, Liou HC, Sweatt JD (2004). A bioinformatics analysis of memory consolidation reveals involvement of the transcription factor c-rel. *Journal of Neuroscience* **24**: 3933–3943.

Levin BE, Govek EK, Dunn-Meynell AA (1998). Reduced glucose-induced neuronal activation in the hypothalamus of diet-induced obese rats. *Brain Research* **808**: 317–319.

Levine AS, Billington CJ (2004). Opioids as agents of reward-related feeding: A consideration of the evidence. *Physiology and Behavior* **82**: 57–61.

Levine MS, Cepeda C, Hickey MA, Fleming SM, Chesselet MF (2004). Genetic mouse models of Huntington's and Parkinson's diseases: Illuminating but imperfect. *Trends in Neuroscience* **27**: 691–697.

Lewis MH (2004). Environmental complexity and central nervous system development and function. *Mental Retardation and Developmental Disabilities Research Reviews* **10**: 91–95.

Lewis MJ, Johnson DF, Waldwman D, Leibowitz SF, Hoebel BG (2004). Galanin microinjection in the third ventricle increases voluntary ethanol intake. *Alcoholism: Clinical and Experimental Research* **28**: 1822–1828.

Li DL, Simmons RMA, Iyengar S (1998). 5HT1A receptor antagonists enhance the functional activity of fluoxetine in a mouse model of feeding. *Brain Research* **781**: 119–126.

Li HS, Borg E (1991). Age-related loss of auditory sensitivity in two mouse genotypes. *Acta Otolarygology* **111**: 827–834.

Li LL, Keverne EB, Aparicio SA, Ishino F, Barton SC, Surani MA (1999). Regulation of maternal behavior and offspring growth by paternally expressed *Peg3*. *Science* **284**: 330–333.

Li X, Schwartz PE, Rissman EF (1997). Distribution of estrogen receptor α-like immunoreactivity in rat forebrain. *Neuroendocrinology* **66**: 63–67.

Libby RT, Kitamoto J, Holme RH, Williams DS, Steel KP (2003). Cdh23 mutations in the mouse are associated with retinal dysfunctions but not retinal degeneration. *Experimental Eye Research* **77**: 731–739.

Liebenauer LL, Slotnick BM (1996). Social organization and aggression in a group of olfactory bulbectomized male mice. *Physiology and Behavior* **60**: 403–409.

Lieberman JA (2004). Dopamine partial agonists: a new class of antipsychotic. *CNS Drugs* **18**: 251–267.

Liebman JM, Cooper SJ, Eds. (1989). *The Neuropharmacological Basis of Reward*. Oxford University Press, New York.

Lijam J, Paylor R, McDonald MP, Crawley JN, Deng CX, Herrup K, Stevens KE, Maccaferri G, McBain CJ, Sussman DJ, Wynshaw-Boris A (1997). Social interaction and sensorimotor gating abnormalities in mice lacking *Dvl1 Cell* **90**: 895–905.

Lim MM, Bielsky IF, Young LJ (2005). Neuropeptides and the social brain: Potential rodent models of autism. *International Journal of Developmental Neuroscience* **23**: 235–243.

Lim MM, Wang Z, Olazabal DE, Ren X, Terwilliger EF, Young LJ (2004). Enhanced partner preference in a promiscuous species by manipulating the expression of a single gene. *Nature* **429**: 754–757.

Lin MM, Wang Z, Olazabal DE, Ren X, Terwilliger EF, Young LJ (2004). Enhanced partner preference in a promiscuous species by manipulating the expression of a single gene. *Nature* **429**: 754–757.

Lin S, Boey D, Herzog H (2004). NPY and Y receptors: Lessons from transgenic and knockout models. *Neuropeptides* **38**: 189–200.

Lin W, Arellano J, Slotnick B, Restrepo D (2004). Odors detected by mice deficient in cyclic nucleotide-gated channel subunit A2 stimulate the main olfactory system. *Journal of Neuroscience* **24**: 3703–3710.

Lin YJ, Seroude L, Benzer S (1998). Extended life-span and stress resistance in the *Drosophila* mutant methuselah. *Science* **282**: 943–946.

Lincoln DW, Paisley AC (1982). Neuroendocrine control of milk ejection. *Journal of Reproductive Fertility* **65**: 571–586.

Lincoln DW, Wakerley JB (1975). Factors governing the periodic activation of supraoptic neurosecretory cells during suckling in the rat. *Journal of Physiology* **250**: 443–461.

Lindner MD, Plone MA, Schallert T, Emerich DF (1997). Blind rats are not profoundly impaired in the reference memory Morris water maze and cannot be clearly discriminated from rats with cognitive deficits in the cued platform task. *Cognitive Brain Research* **5**: 329–333.

Lipkind D, Sakov A, Kafkafi N, Elmer GI, Benjamini Y, Golani I (2004). New replicable anxiety-related measures of wall vs center behavior of mice in the open field. *Journal of Applied Physiology* **97**: 347–359.

Lipp HP, Wolfer DP (1998). Genetically modified mice and cognition. *Current Opinions in Neurobiology* **8**: 272–280.

Lipschutz RJ, Fodor SPA, Gingeras TR, Lockhart DJ (1999). High density synthetic oligonucleotide arrays. *Nature Genetics* **21** (January Suppl): 20–25.

Lipska BK (2004). Using animal models to test a neurodevelopmental hypothesis of schizophrenia. *Journal of Psychiatry and Neuroscience*. **29**: 282–286.

Lipska BK, Jaskiw GE, Weinberger DR (1993). Postpubertal emergency of hyperresponsiveness to stress and to amphetamine after neonatal excitotoxic hippocampal damage: A potential animal model of schizophrenia. *Neuropsychopharmacology* **9**: 67–75.

Lipska BK, Swerdlow NR, Geyer MA, Jaskiw GE, Braff DL, Weinberger DR (1995). Neonatal excitotoxic hippocampal damage in rats causes post-pubertal changes in prepulse inhibition of startle and its disruption by apomorphine. *Psychopharmacology* **122**: 35–43.

Lipska BK, Weinberger DR (1995). Genetic variation in vulnerability to the behavioral effects of neonatal hippocampal damage in the rat. *Proceedings of the National Academy of Sciences USA* **92**: 8906–8910.

Lipsky RH, Goldman D (2003). Genomics and variation of ionotropic glutamate receptors. *Annals of the New York Academy of Sciences* **1003**: 22–35.

Lister RG (1987a). Interactions of Ro 15–4513 with diazepam, sodium pentobarbital and ethanol in a holeboard test. *Pharmacology Biochemistry and Behavior* **28**: 75–79.

Lister RG (1987b). The use of a plus-maze to measure anxiety in the mouse. *Psychopharmacology* **92**: 180–185.

Lister RG (1992). Benzodiazepine tolerance and dependence. In *Animal Models of Drug Addiction*, AA Boulton, GB Baker, PW Wu, Eds. Humana Press, Totowa, NJ.

Liu D, Diorio J, Day JC, Francis DD, Meaney MJ (2000). Maternal care, hippocampal synaptogenesis and cognitive development in rats. *Nature Neuroscience* **3**: 799–806.

Liu H, Mantyh PW, Basbaum AI (1997a). NMDA-receptor regulation of substance P release from primary afferent nociceptors. *Nature* **386**: 721–724.

Liu RC, Miller KD, Merzenich MM, Schreiner CE (2003). Acoustic variability and distinguishability among mouse ultrasonic vocalizations. *Journal of the Acoustical Society of America* **114**: 3412–3422.

Liu X, Gershenfeld HK (2003). An exploratory factor analysis of the tail suspension test in 12 inbred strains of mice and an F2 intercross. *Brain Research Bulletin* **60**: 223–231.

Liu Y, Hoffmann A, Grinberg A, Westphal H, McDonald MP, Miller KM, Crawley JN, Sandhoff K, Suzuki K, Proia RL (1997b). Mouse model of G_{M2} activator deficiency manifests cerebellar pathology and motor impairment. *Proceedings of the National Academy of Sciences USA* **94**: 8138–8143.

Livy DJ, Wahlsten D (1997). Retarded formation of the hippocampal commissure in embryos from mouse strains lacking a corpus callosum. *Hippocampus* **7**: 2–14.

Lloyd MA, Appel JB (1976). Signal detection theory and the psychophysics of pain: An introduction and review. *Psychosomatic Medicine* **38**: 79–94.

Logue SF, Owen EH, Rasmussen DL, Wehner JM (1997b). Assessment of locomotor activity, acoustic and tactile startle, and prepulse inhibition of startle in inbred mouse strains and F_1 hybrids: Implications of genetic background for single gene and quantitative trait loci analyses. *Neuroscience* **80**: 1075–1086.

Logue SF, Paylor R, Wehner JM (1997a). Hippocampal lesions cause learning deficits in inbred mice in the Morris water maze and conditioned-fear task. *Behavioral Neuroscience* **111**: 104–113.

Logue SF, Swartz RJ, Wehner JM (1998). Genetic correlation between performance on an appetitive-signaled nosepoke task and voluntary ethanol consumption. *Alcoholism: Clinical and Experimental Research* **22**: 1912–1920.

Loh HH, Liu HC, Vacalli A, Yang W, Chen YF, Wei LN (1998). μ Opioid receptor knockout in mice: Effects on ligand-induced analgesia and morphine lethality. *Molecular Brain Research* **54**: 321–326.

Long JM, LaPorte P, Paylor R, Wynshaw-Boris A (2004). Expanded characterization of the social interaction abnormalities in mice lacking Dvl1. *Genes, Brain and Behavior* **3**: 51–62.

Long SY (1972). Hair-nibbling and whisker-trimming as indicators of social hierarchy in mice. *Animal Behavior* **20**: 10–12.

Lonstein JS, De Vries GJ (2000). Sex differences in the parental behavior of rodents. *Neuroscience and Biobehavioral Reviews* **24**: 669–686.

Lonstein JS, Fleming AS (2001). Parental behaviors in rats and mice. In *Current Protocols in Neuroscience*. Wiley, New York, pp. 8.15.1.–8.15.26.

Lord C, Risi S, Lambrecht L, Cook EH Jr, Leventhal BL, DiLavore PC, Pickles A, Rutter M (2000). The autism diagnostic observation schedule-generic: A standard measure of social and communication deficits associated with the spectrum of autism. *Journal of Autism and Developmental Disorders* **30**: 205–223.

Lore R, Flannelly K (1977). Rat societies. *Scientific American* **236**: 106–116.

Lorenz K (1966). *On Aggression*. Methuen, London.

Lorenz KZ (1974). Analogy as a source of knowledge. *Science* **185**: 229–234.

Low K, Crestani F, Keist R, Benke D, Brunig I, Benson JA, Fritschy JM, Rulicke T, Bluethmann H, Mohler H, Rudolph U (2000). Molecular and neuronal substrate for the selective attenuation of anxiety. *Science* **290**: 131–134.

Lowe G, Nakamura T, Gold GH (1989). Adenylate cyclase mediates olfactory transduction for a wide variety of odorants. *Proceedings of the National Academy of Sciences USA* **86**: 5641–5645.

Lowenstein PR, Castro MG (2002). Progress and challenges in viral vector-mediated gene transfer to the brain. *Current Opinion in Molecular Therapeutics* **4**: 359–371.

Lowenstein PR, Thomas CE, Castro MG (1999). Politically correct gene therapy? A "clean environment" improves gene delivery to the brain! *Gene Therapy* **6**: 463–464.

Lu YM, Jia Z, Janus C, Henderson JT, Gerlai R, Wojtowicz YM, Roder JC (1997). Mice lacking metabotropic glutamate receptor 5 show impaired learning and reduced CA1 long-term potentiation (LTP). but normal CA3 LTP. *Journal of Neuroscience* **17**: 5196–5205.

Lubahn DB, Moyer JS, Golding TS, Couse JF, Korach KS, Smithies O (1993). Alteration of reproductive function but not prenatal sexual development after insertional disruption of the mouse estrogen receptor gene. *Proceedings of the National Academy of Science USA* **90**: 11162–11166.

Lubow RE, Gewirtz JC (1995). Latent inhibition in humans: Data, theory, and implications for schizophrenia. *Psychological Bulletin* **117**: 87–103.

Lucas JJ, Yamamoto A, ScearceLevie K, Saudou F, Hen F (1998). Absence of fenfluramine-induced anorexia and reduced *c-fos* induction in the hypothalamus and central amygdaloid complex of serotonin 1B receptor knock-out mice. *Journal of Neuroscience* **18**: 5537–5544.

Lucki I, Dalvi A, Mayorga AJ (2001). Sensitivity to the effects of pharmacologically selective antidepressants in different strains of mice. *Psychopharmacology* **155**: 315–322.

Lueking A, Cahill DJ, Mullner S (2005). Protein biochips: A new and versatile platform technology for molecular medicine. *Drug Discovery Today* **10**: 789–794.

Lukkarinen JA, Kauppinen RA, Grohn OH, Oja JM, Sinervirta R, Jarvinen A, Alhonen LI, Janne J (1998). Neuroprotective role of ornithine decarboxylase activation in transient focal cerebral ischemia: A study using ornithine decarboxylase-overexpressing transgenic rats. *European Journal of Neuroscience* **10**: 2046–2055.

Luo AH, Cannon EH, Wekesa KS, Lyman RF, Vandenbergh JG, Anholt RR (2002). Impaired olfactory behavior in mice deficient in the alpha subunit of G(o). *Brain Research* **941**: 62–71.

Luo M, Fee MS, Katz LC (2003). Encoding pheromonal signals in the accessory olfactory bulb of behaving mice. *Science* **299**: 1196–1201.

Lush IE (1995). The genetics of tasting in mice. VII. Glycine revisited, and the chromosomal location of Sac and Soa. *Genetics Research* **66**: 167–174.

Lydiard RB, Brawman-Mintzer O (1998). Anxious depression. *Journal of Clinical Psychiatry* **59**(Suppl 18): 10–17.

Lydon JP, DeMayo FJ, Funk CR, Mani SK, Hughes AR, Montgomery CA, Shyamala G, Conneely OM, O'Malley BW (1995). Mice lacking progesterone receptor exhibit pleiotropic reproductive abnormalities. *Genes and Development* **9**: 2266–2278.

Lynch CB (1980). Response to divergent selection for nesting behavior in *Mus musculus*. *Genetics* **96**: 757–765.

Lynch G (1986). *Synapses, Circuits, and the Beginnings of Memory*. MIT Press, Cambridge.

Lynch G (1998). Memory and the brain: Unexpected chemistries and a new pharmacology. *Neurobiology of Learning and Memory* **70**: 82–100.

Mackler SA, Korutla L, Cha XY, Koebbe MJ, Fournier KM, Bowers MS, Kalivas PW (2000). NAC-1 is a brain POZ/BTB protein that can prevent cocaine-induced sensitization in the rat. *Journal of Neuroscience* **20**: 6210–6217.

Macknin JB, Higuchi M, Lee VM, Trojanowski JQ, Doty RL (2004). Olfactory dysfunction occurs in transgenic mice overexpressing human tau protein. *Brain Research* **1000**: 174–178.

Maffei M, Halaas J, Ravussin E, Pratley RE, Lee GH, Zhang Y, Fei H, Kim S, Lallone R, Ranganathan S, Kern PA, Friedman JM (1995). Leptin levels in human and rodent: Measurement of plasma leptin and *ob* RNA in obese and weight-reduced subjects. *Nature Medicine* **1**: 1155–1161.

Magara F, Müller U, Li ZW, Lipp H-P, Weissman C, Stagljar M, Wolfer DP (1999). Genetic background changes the pattern of forebrain commissure defects in transgenic mice underexpressing the β-amyloid-precursor protein. *Proceedings of the National Academy of Sciences USA* **96**: 4656–4661.

Magdaleno S, Keshvara L, Curran T (2002). Rescue of ataxia and preplate splitting by ectopic expression of Reelin in reeler mice. *Neuron* **33**: 573–586.

Maggio JC, Whitney G (1985). Ultrasonic vocalizing by adult female mice (*Mus musculus*). *Journal of Comparative Psychology* **99**: 420–436.

Maier SF (1984). Learned helplessness and animal models of depression. *Progress in Neuropsychopharmacology and Biological Psychiatry* **8**: 435–446.

Maier SF, Watkins LR (2005). Stressor controllability and learned helplessness: the roles of the dorsal raphe nucleus, serotonin, and corticotropin-releasing factor. *Neuroscience and Biobehavioral Reviews* **29**: 829–841.

Maines MD, Polevoda B, Coban T, Johnson K, Stoliar S, Huang TJ, Panahian N, Cory-Slecta DA, McCoubrey WK (1998). Neuronal overexpression of heme oxygenase-1 correlates with an attenuated exploratory behavior and causes an increase in neuronal NADPH-diaphorase staining. *Journal of Neurochemistry* **70**: 2057–2069.

Maingay M, Romero-Ramos M, Kirik D (2005). Viral vector mediated overexpression of human alpha-synuclein in the nigrostriatal dopaminergic neurons: A new model for Parkinson's disease. *CNS Spectrums* **10**: 235–244.

Maldonado C, Rodriguez-Arias M, Aguilar MA, Minarro J (2004). GHB ameliorates naloxone-induced conditioned place aversion and physical aspects of morphine withdrawal in mice. *Psychopharmacology* **177**: 130–140.

Maldonado R (2002). Study of cannabinoid dependence in animals. *Pharmacology and Therapeutics* **95**: 153–164.

Maldonado R, Blendy JA, Tzavara E, Gass P, Roques BP, Hanoune J, Schütz G (1996). Reduction of morphine abstinence in mice with a mutation in the gene encoding CREB. *Science* **273**: 657–659.

Maldonado R, Salardi A, Valverde O, Samad TA, Roques BP, Borrelli E (1997). Absence of opiate rewarding effects in mice lacking dopamine D2 receptors. *Nature* **388**: 586–589.

Malholtra AK, Murphy GM, Kennedy JL (2004). Pharmacogenetics of psychotropic drug response. *American Journal of Psychiatry* **161**: 780–796.

Malleret G, Hen R, Guillou JL, Segu L, Buhot MC (1999). 5-HT1B receptor knock-out mice exhibit increased exploratory activity and enhanced spatial memory performance in the Morris water maze. *Journal of Neuroscience* **19**: 6157–6168.

Malmberg AB and Bannon AW (1999). Models of nociception: Hot-plate, tail flick, and formalin tests in rodents. In *Current Protocols in Neuroscience*. Wiley, New York, pp. 8.9.1–8.9.16.

Manly DF, Elliott RW (1991). RI Manager, a microcomputer program for analysis of data from recombinant inbred strains. *Mammalian Genome* **1**: 123–126.

Mannes AJ, Caudle RM, O'Connell BC, Iadarola MJ (1998). Adenoviral gene transfer to spinal cord neurons: Intrathecal vs. intraparenchymal administration. *Brain Research* **793**: 1–6.

Manning CJ, Wakeland EK, Potts WK (1992). Communal nesting patterns in mice implicate MHC genes in kin recognition. *Nature* **360**: 581–583.

Mansuy IM, Bujard H (2000). Tetracycline-regulated gene expression in the brain. *Current Opinions in Neurobiology* **10**: 593–596.

Mansuy IM, Mayford M, Jacob B, Kandel ER, Back ME (1998). Restricted and regulated overexpression reveals calcineurin as a key component in the transition from short-term to long-term memory. *Cell* **92**: 39–49.

Mansuy IM, Mayford M, Kandel ER (1999). Regulated temporal and spatial expression of mutants of CaMKII and calcineurin with the tetracycline-controlled transactivator (tTA) and reverse rTA (rtTA) systems. In *Handbook of Molecular-Genetic Techniques*, WE Crusio and RT Gerlai, Eds. Elsevier Science, Amsterdam, pp. 291–304.

Marcus MM, Nomikos GG, Svensson TH (1996). Differential actions of typical and atypical antipsychotic drugs on dopamine release in the core and shell of the nucleus accumbens. *European Neuropsychopharmacology* **6**: 29–38.

Margolis F, Getchell TV, Eds. (1988). *Molecular Neurobiology of the Olfactory System*. Plenum, New York.

Marinelli M, Barrot M, Simon H, Oberlander C, Dekeyne A, Le Moal M, Piazza PV (1998). Pharmacological stimuli decreasing nucleus accumbens dopamine can act as positive reinforcers but have a low addictive potential. *European Journal of Neuroscience* **10**: 3269–3275.

Markel P, Shu P, Ebeling C, Carlson GA, Nagle DL, Smutko JS, Moore KJ (1997). Theoretical and empirical issues for marker-assisted breeding of congenic mouse strains. *Nature Genetics* **17**: 280–284.

Markou A, Koob GF (1991). Post cocaine anhedomia. An animal model of cocaine withdrawal. *Neuropsychopharmacology* **4**: 17–26.

Markowska AL, Spangler EL, Ingram DK (1998). Behavioral assessment of the senescence-accelerated mouse (SAM P8 and R1). *Physiology and Behavior* **64**: 15–26.

Markowska AL, Stone WS, Ingram DK, Reynolds J, Gold PE, Conti LH, Pontecorvo MJ, Wenk GL, Olton DS (1989). Individual differences in aging: Behavioral and neurobiological correlates. *Neurobiology of Aging* **10**: 31–43.

Marler P, Hamilton WJ (1968). *Mechanisms of Animal Behavior*. Wiley, New York.

Marsh DJ, Hollopeter G, Huszar D, Laufer R, Yagaloff KA, Fisher SL, Burn P, Palmiter RD (1999). Response of melanocortin-4 receptor-deficient mice to anorectic and orexigenic peptides. *Nature Genetics* **21**: 119–122.

Marsh DJ, Hollopeter G, Kafer KE, Palmiter RD (1998). Role of the Y5 neuropeptide Y receptor in feeding and obesity. *Nature Medicine* **4**: 718–721.

Martellotta MC, Cossu G, Fattore L, Gessa GL, Fratta W (1998). Self-administration of the cannabinoid receptor agonist WIN 55,212-2 in drug-naïve mice. *Neuroscience* **85**: 327–330.

Martellota MC, Kuzmin A, Zvartau E, Cossu G, Gessa GL, Fratta W (1995). Isradipine inhibits nicotine intravenous self-administration in drug naïve mice. *Pharmacology Biochemistry and Behavior* **52**: 271–274.

Martin B, Chapouthier G, Motta R (1992). Analysis of B10.D2 recombinant congenic mouse strains shows that audiogenic and β-CCM-induced seizures depend on different genetic mechanisms. *Epilepsia* **33**: 11–13.

Martin B, Desforges C, Chapouthier G (1991). Comparison between patterns of convulsions induced by two β-carbolines in 10 inbred strains of mice. *Neuroscience Letters* **133**: 73–76.

Martin M, Ledent C, Parmentier M, Maldonado R, Valverde O (2000). Cocaine, but not morphine, induces conditioned place preference and sensitization to locomotor responses in CB1 knockout mice. *European Journal of Neuroscience* **12**: 4038–4046.

Martin P (1998). Animal models sensitive to anti-anxiety agents. *Acta Psychiatrica Scandinavica* **98**(Suppl 393): 74–80.

Martin WR (1967). Opioid antagonists. *Pharmacological Reviews* **19**: 463–521.

Martinez JL, Jensen RA, Messing RB, Rigter H, McGaugh JL (1981). *Endogenous Peptides and Learning and Memory Processes.* Academic Press, New York.

Martinez JL, Kesner RP, Eds. (1998). *Neurobiology of Learning and Memory.* Academic Press, New York.

Marubio LM, Gardier AM, Durier S, David D, Klink R, Arroyo-Jimenez MM, McIntosh JM, Rossi F, Champtiaux N, Zoli M, Changeux JP (2003). Effects of nicotine in the dopaminergic system of mice lacking the alpha4 subunit of neuronal nicotinic acetylcholine receptors. *European Journal of Neuroscience* **17**: 1329–1337.

Mas-Nieto M, Pommier B, Tzavara ET, Caneparo A, Da Nascimento S, Le Fur G, Roques BP, Noble F (2001). Reduction of opioid dependence by the CB(1) antagonist SR141716A in mice: Evaluation of the interest in pharmacotherapy of opioid addiction. *British Journal of Pharmacology* **132**: 1809–1816.

Masliah E, Rockenstein E, Veinbergs I, Mallory M, Hashimoto M, Takeda A, Sagara Y, Sisk A, Mucke L (2000). Dopaminergic loss and inclusion body formation in alpha-synuclein mice: implications for neurodegenerative disorders. *Science* **287**: 1265–1269.

Mason C, Sotelo C, Eds. (1997). The cerebellum: A model for construction of a cortex. *Perspectives on Developmental Neurobiology*, Special Issue **5**: 1–95.

Mastropaolo J, Nadi NS, Ostrowski NL, Crawley JN (1988). Galanin antagonizes acetylcholine on a memory task in basal forebrain-lesioned rats. *Proceedings of the National Academy of Sciences USA* **85**: 9841–9845.

Masugi M, Yokoi M, Shigemoto R, Muguruma K, Watanabe Y, Sansig G, van der Putten H, Nakanashi S (1999). Metabotropic glutamate receptor subtype 7 ablation causes deficit in fear response and conditioned taste aversion. *Journal of Neuroscience* **19**: 955–963.

Mata M, Glorioso JC, Fink DJ (2003). Gene transfer to the nervous system: Prospects for novel treatments directed at diseases of the aging nervous system. *Journal of Gerontology* **58A**: 1111–1118.

Mathis C, Neumann PE, Gershenfeld H, Paul SM, Crawley JN (1995). Genetic analysis of anxiety-related behaviors and responses to benzodiazepine-related drugs in AXB and BXA recombinant inbred mouse strains. *Behavior Genetics* **25**: 557–568.

Mathis C, Paul SM, Crawley JN (1994). Characterization of benzodiazepine-sensitive behaviors in the A/J and C57BL/6J inbred strains of mice. *Behavior Genetics* **24**: 171–180.

Matsunami H, Amrein H (2004). Taste perception: How to make a gourmet mouse. *Current Biology* **14**: R118-R120.

Matthes HWD, Maldonado R, Simonin F, Valverde O, Slowe S, Kitchen I, Befort K, Dierich A, Le Meur M, Dollé P, Tzavara E, Hanoune J, Roques BP, Kieffer BL (1996). Loss of morphine-induced analgesia, reward effect and withdrawal symptoms in mice lacking the μ-opioid-receptor gene. *Nature* **383**: 819–823.

Matthes HWD, Smadja C, Valverde O, Vonesch JL, Foutz AS, Boudinot E, Denavit-Saubié M, Severini C, Negri L, Roques BP, Maldonado R, Kieffer BL (1998). Activity of the δ-opioid receptor is partially reduced, whereas activity of the κ-receptor is maintained in mice lacking the μ- receptor. *Journal of Neuroscience* **18**: 7285–7295.

Matthews K, Robbins TW (2003). Early experience as a determinant of adult behavioural responses to reward: the effects of repeated maternal separation in the rat. *Neuroscience and Biobehavioral Reviews* **27**: 45–55.

Matthews TJ, Grigore M, Tang L, Doat M, Kow LM, Pfaff DW (1997). Sexual reinforcement in the female rat. *Journal Experimental Analysis of Behavior* **68**: 399–410.

Maxson SC (1996). Searching for candidate genes with effects on an agonistic behavior, offense, in mice. *Behavior Genetics* **26**: 471–476.

Mayeux-Portas V, File SE, Steward CL, Morris RJ (2000). Mice lacking the cell adhesion molecule Thy-1 fail to use socially transmitter cues to direct their choice of food. *Current Biology* **10**: 68–75.

Mayford M, Bach ME, Huang YY, Wang L, Hawkins RD, Kandel ER (1996). Control of memory formation through regulated expression of a CaMKII transgene. *Science* **274**: 1678–1683.

Mayorga AJ, Dalvi A, Page ME, Zimov-Levinson S, Hen R, Lucki I (2001). Antidepressant-like behavioral effects in 5-hydroxytryptamine$_{1A}$ and 5-hydroxytryptamine$_{1B}$ receptor mutant mice. *Journal of Pharmacology and Experimental Therapeutics* **298**: 1101–1107.

McAlanon GM, Dawson GR, Wilkinson LO, Robbins TW, Everitt BJ (1995). The effects of AMPA-induced lesions of the medial septum and vertical limb of the diagonal band of Broca on spatial delayed non-matching to sample and spatial learning in the water maze. *European Journal of Neuroscience* **7**: 1034–1049.

McBride JL, Behrstock SP, Chen EY, Jakel RJ, Siegel I, Svendsen CN, Kordower JH (2004). Human neural stem cell transplants improve motor function in a rat model of Huntington's disease. *Journal of Comparative Neurology* **475**: 211–219.

McCarty DM, Young SM Jr, Samulski RJ (2004). Integration of adeno-associated virus (AAV) and recombinant AAV vectors. *Annual Review of Genetics* **38**: 819–845.

McCaughran JA, Bell J, Hitzemann RJ (1999). On the relationships of high-frequency hearing loss and cochlear pathology to the acoustic startle response (ASR) and prepulse inhibition of the ASR in the BXD recombinant inbred series. *Behavior Genetics* **29**: 21–30.

McCaughran JA, Bell J, Hitzemann RJ (2000). Fear-potentiated startle response in mice: Genetic analysis of the C57BL/6J and DBA/2J intercross. *Pharmacology Biochemistry and Behavior* **65**: 301–312.

McClung CA, Nestler EJ (2003). Regulation of gene expression and cocaine reward by CREB and DeltaFosB. *Nature Neuroscience* **6**: 1208–1215.

McCormick DA, Lavond DG, Thompson RF (1982). Concomitant classical conditioning of the rabbit nictitating membrane and eyelid response: Correlations and implications. *Physiology and Behavior* **28**: 769–775.

McDonald JD, Andriolo M, Cali F, Mirisola M, Puglisi-Allegra S, Romano V, Sarkissian CN, Smith CB (2002). The phenylketonuria mouse model: A meeting review. *Mol Genet Metab* **76**: 256–261.

McDonald MP, Crawley JN (1996). Galanin receptor antagonist M40 blocks galanin-induced choice accuracy deficits on a delayed-nonmatching-to-position task. *Behavioral Neuroscience* **110**: 1025–1032.

McDonald MP, Miller KM, Li C, Deng C, Crawley JN. (2001). Motor deficits in fibroblast growth factor receptor-3 null mutant mice. *Behavioural Pharmacology* **12**: 477–486.

McDonald MP, Wong R, Goldstein G, Weintraub B, Cheng SY, Crawley JN (1998). Hyperactivity and learning deficits in transgenic mice bearing a human mutant thyroid hormone β 1 receptor gene. *Learning and Memory* **5**: 289–301.

McDonnell DP (1995). Unraveling the human progesterone receptor signal transduction pathway: Insights into antiprogestin actions. *Trends in Endocrinology and Metabolism* **6**: 133–138.

McDonnell DP, Dana SL, Hoener PA, Lieberman BA, Imhof MO, Stein RB (1995). Cellular mechanisms which distinguish between hormone- and antihormone-activated estrogen receptor. *Annals of the New York Academy of Sciences* **761**: 121–137.

McDowd JM, Filion DL, Harris MJ, Braff DL (1993). Sensory gating and inhibitory function in late-life schizophrenia. *Schizophrenia Bulletin* **19**: 733–746.

McFadyen MP, Kusek G, Bolivar VJ, Flaherty L (2003). Differences among eight inbred strains of mice in motor ability and motor learning on a rotarod. *Genes, Brain and Behavior* **2**: 214–219.

McGaugh JL (1966). Time-dependent processes in memory storage. *Science* **153**: 1351–1358.

McGaugh JL, Cahill L (1997). Interaction of neuromodulatory systems in modulating memory storage. *Behavioral Brain Research* **83**: 31–38.

McGaughy J, Sarter M (1995). Behavioral vigilance in rats: task validation and effects of age, amphetamine, and benzodiazepine receptor ligands. *Psychopharmacology* **117**: 340–357.

McGeer PL, McGeer EG (2001). Inflammation, autotoxicity and Alzheimer disease. *Neurobioliology of Aging.* **22**: 799–809.

McGuirk J, Muscat R, Willner P (1992). Effects of chronically administered fluoxetine and fenfluramine on food intake, body weight and the behavioural satiety sequence. *Psychopharmacology* **106**: 401–407.

McHugh TJ, Blum KI, Tsien JZ, Tonegawa S, Wilson MA (1996). Impaired hippocampal representation of space in CA1-specific NMDAR1 knockout mice. *Cell* **87**: 1339–1349.

McIlwain KL, Meriweather MY, Yuva-Paylor LA, Paylor R (2001). The use of behavioral test batteries: effects of training history. *Physiology and Behavior* **73**: 705–717.

McInerney-Leo A, Hadley DW, Gwinn-Hardy K, Hardy J (2005). Genetic testing in Parkinson's disease. *Movement Disorders* **20**: 1–10.

McInnis MG (1999). The assent of a nation: Genetics and Iceland. *Clinical Genetics* **55**: 234–239.

McKay (2004). Stem cell biology and neurodegenerative disease. *Philosophical Transactions of the Royal Society B: Biological Sciences* **359**: 851–856.

McKernan RM, Rosahl TW, Reynolds DS, Sur C, Wafford KA, Atack JR, Farrar S, Myers J, Cook G, Ferris P, Garrett L, Bristow L, Marshall G, Macaulay A, Brown N, Howell O, Moore KW, Carling RW, Street LJ, Castro JL, Ragan CI, Dawson GR, Whiting PJ (2000). Sedative but not anxiolytic properties of benzodiazepine are mediated by the GABA(A) receptor alpha1 subtype. *Nature Neuroscience* **3**: 587–592.

McKinney WT (1989). Basis of development of animal models in psychiatry: An overview. In *Animal Models of Depression*, GF Koob, CL Ehlers, DJ Kupfer, Eds. Birkhauser, Boston, pp. 3–17.

McLeod AL, Ritchie J, Cuello AC, Julien JP, Ribeiro-da-Silva A, Henry JL (1999). Transgenic mice over-expressing substance P exhibit allodynia and hyperalgesia which are reversed by substance P and *N*-methyl-D-aspartate receptor antagonists. *Neuroscience* **89**: 891–899.

McNamara RK, Namgung U, Routtenberg A (1996). Distinctions between hippocampus of mouse and rat: Protein F1/GAP-43 gene expression, promoter activity, and spatial memory. *Molecular Brain Research* **40**: 177–187.

McNamara RK, Skelton RW (1993). The neuropharmacological and neurochemical basis of place learning in the Morris water maze. *Brain Research Reviews* **18**: 33–49.

McNamara RK, Stumpo DJ, Morel LM, Lewis MH, Wakeland EK, Blackshear PJ, Lenox RH (1998). Effect of reduced myristoylated alanin-rich C kinase substrate expression on hippocampal mossy fiber development and spatial learning in mutant mice: Transgenic rescue and interactions with gene background. *Proceedings of the National Academy of Sciences USA* **95**: 14517–14522.

McNish KA, Gewirtz JC, Davis M (1997). Evidence of contextual fear after lesions of the hippocampus: A disruption of freezing but not fear-potentiated startle. *Journal of Neuroscience* **17**: 9353–9360.

Meaney MJ (2001). Maternal care, gene expression, and the transmission of individual differences in stress reactivity across generations. *Annual Review of Neuroscience* **24**: 1161–1192.

Mechoulam R, Parker L (2003). Cannabis and alcohol—A close friendship. *Trends in Pharmacological Sciences* **24**: 266–268.

Meincke U, Light GA, Geyer MA, Braff DL, Gouzoulis-Mayfrank E. Sensitization and habituation of the acoustic startle reflex in patients with schizophrenia. *Psychiatry Research* **126**: 51–61.

Meiri N, Masos T, Rosenblum K, Miskin R, Dudai Y (1994). Overexpression of urokinase-type plasminogen activator in transgenic mice is correlated with impaired learning. *Proceedings of the National Academy of Sciences USA* **91**: 3196–3200.

Meisel RL, Sachs BD (1994). The physiology of male sexual behavior. In *The Physiology of Reproduction*, 2nd ed., E Knobil and JD Neill, Eds. Raven, New York, pp. 3–105.

Meliska CJ, Bartke A, McGlacken G, Jensen RA (1995). Ethanol, nicotine, amphetamine, and aspartamate consumption and preferences in C57BL/6 and DBA/2 mice. *Pharmacology Biochemistry and Behavior* **50**: 619–626.

Mello NK, Negus SS (2000). Interactions between kappa opioid agonists and cocaine: Preclinical studies. *Annals of the New York Academy of Sciences* **909**: 104–132.

Meltzer CC, Smith G, DeKosky ST, Pollock BG, Mathis CA, Moore RY, Kupfer DJ, Reynolds CF (1998). Serotonin in aging, late-life depression, and Alzheimer's disease: The emerging role of functional imaging. *Neuropsychopharmacology* **18**: 407–430.

Meltzer HY, Li Z, Kaneda Y, Ichikawa J (2003). Serotonin receptors: Their key role in drugs to treat schizophrenia. *Progress in Neuropsychopharmacology and Biological Psychiatry* **27**: 1159–1172.

Melzer P, Smith CB (1998). Plasticity of cerebral metabolic whisker maps in adult mice after whisker follicle removal. I. Modifications of barrel cortex coincide with reorganization of follicular innervation. *Neuroscience* **83**: 27–41.

Menalled LB, Sison JD, Wu Y, Olivieri M, Li XJ, Li H, Zeitlin S, Chesselet MF (2002). Early motor dysfunction and striosomal distribution of huntingtin microaggregates in Huntington's disease knock-in mice. *Journal of Neuroscience* **22**: 8266–8276.

Menard J, Treit D (1999). Effects of centrally administered anxiolytic compounds in animal models of anxiety. *Neuroscience and Biobehavioral Reviews* **23**: 591–613.

Menzaghi F, Rassnick S, Heinrichs S, Baldwin H, Pich EM, Weiss F, Koob GF (1994). The role of corticotropin-releasing factor in the anxiogenic effects of ethanol withdrawal. *Annals of the New York Academy of Sciences* **739**: 176–184.

Messeri P, Eleftheriou BE, Oliverio A (1975). Dominance behavior: A phylogenetic analysis in the mouse. *Physiology and Behavior* **14**: 53–58.

Metz GA, Schwab ME (2004). Behavioral characterization in a comprehensive mouse test battery reveals motor and sensory impairments in growth-associated protein-43 null mutant mice. *Neuroscience* **129**: 563–574.

Meyer JS (1998). Behavioral assessment in developmental neurotoxicology: Approaches involving unconditioned behavior and pharmacological challenges in rodents. In *Handbook of Developmental Neurotoxicology*, W Slikker, LW Chang, Eds. Academic Press, New York.

Meziane H, Dodart JC, Mathis C, Little S, Clemens J, Paul SM, Ungerer A (1998). Memory-enhancing effects of secreted forms of the β-amyloid precursor protein in normal and amnestic mice. *Proceedings of the National Academy of Sciences USA* **95**: 12683–12688.

Michalik A, Martin JJ, Van Broeckhoven C (2004). Spinocerebellar ataxia type 7 associated with pigmentary retinal dystrophy. *European Journal of Human Genetics* **12**: 2–15.

Miczek KA (1983). Ethological analysis of drug action on aggression, defense and defeat. In *Behavioral Models and the Analysis of Drug Action*. MY Spiegelstein, A Levy, Eds. Elsevier, Amsterdam, pp. 225–239.

Miczek KA, Maxson SC, Fish EW, Faccidomo S (2001). Aggressive behavioral phenotypes in mice. *Behavioral Brain Research* **125**: 167–181.

Mieda M, Willie JT, Hara J, Sinton CM, Sakurai T, Yanagisawa M (2004). Orexin peptides prevent cataplexy and improve wakefulness in an orexin neuron-ablated model of narcolepsy in mice. *Proceedings of the National Academy of Sciences USA* **101**: 4649–4654.

Migaud M, Charlesworth P, Dempster M, Webster LC, Watabe AM, Makhinson M, He Y, Ramsay MF, Morris RGM, Morrison JH, O'Dell TJ, Grant SGN (1998). Enhanced long-term potentiation and impaired learning in mice with mutant postsynaptic density-95 protein. *Nature* **396**: 433–440.

Mihalick SM, Langlois JC, Krienke JD, Dube WV (2000). An olfactory discrimination procedure for mice. *Journal of the Experimental Analysis of Behavior* **73**: 305–318.

Militzer K, Wecker E (1986). Behaviour-associated alopecia areata in mice. *Laboratory Animals* **20**: 9–13.

Milner B, Squire LR, Kandel ER (1998). Cognitive neuroscience and the study of memory. *Neuron* **20**: 445–468.

Miner LL (1997). Cocaine reward and locomotor activity in C57BL/6J and 129/SvJ inbred mice and their F1 cross. *Pharmacology Biochemistry and Behavior* **58**: 25–30.

Miner LL, Drago J, Chamberlain PM, Donovan D, Uhl G (1995). Retained cocaine conditioned place preference in D1 receptor deficient mice. *Neuroreport* **6**: 2314–2316.

Mineur YS, Crusio WE, Sluyter F (2004). Genetic dissection of learning and memory in mice. *Neural Plasticity* **11**: 217–240.

Mineur YS, Prasol DJ, Belzung C, Crusio WE (2003). Agonistic behavior and unpredictable chronic mild stress in mice. *Behavior Genetics* **33**: 513–519.

Mineur YS, Sluyter F, de Wit S, Oostra BA, Crusio WE (2002). Behavioral and neuroanatomical characterization of the Fmr1 knockout mouse. *Hippocampus* **12**: 39–46.

Minoshima S, Giordani B, Berent S, Frey KA, Foster NL, Kuhl DE (1997). Metabolic reduction in the posterior cingulate cortex in very early Alzheimer's disease. *Annals of Neurology* **42**: 85–94.

Mirnics K, Pevsner J (2004). Progress in the use of microarray technology to study the neurobiology of disease. *Nature Neuroscience* **7**: 434–439.

Miserando MJD, Sananes CB, Melia KR, Davis M (1990). Blocking of acquisition but not expression of conditioned fear-potentiated startle by NMDA antagonists in the amygdala. *Nature* **345**: 716–718.

Mishima K, Tozawa T, Satoh K, Saitoh H, Mishima Y. The 3111T/C polymorphism of hClock is associated with evening preference and delayed sleep timing in a Japanese population sample. *American Journal of Medical and Genetics Part B: Neuropsychiatric Genetics* **133**: 101–104.

Misslin R (2003). The defense system of fear: Behavior and neurocircuitry. *Neurophysiol Clin.* **33**: 55–66.

Misslin R, Belzung C, Vogel E (1989). Behavioural validation of a light/dark choice procedure for testing antianxiety agents. *Behavioural Processes* **18**: 119–132.

Mitchner NA, Garlick C, Ben-Jonathan N (1998). Cellular distribution and gene regulation of estrogen receptor alpha and beta in the rat pituitary gland. *Endocrinology* **139**: 3976–3983.

Miyakawa T, Leiter LM, Gerber DJ, Gainetdinov RR, Sotnikova TD, Zeng H, Caron MG, Tonegawa S (2003). Conditional calcineurin knockout mice exhibit multiple abnormal behaviors related to schizophrenia. *Proceedings of the National Academy of Sciences USA* **100**: 8987–8992.

Miyakawa T, Yagi T, Kagiyama A, Niki H (1996a). Radial maze performance, open field and elevated plus-maze behaviors in Fyn-kinase deficient mice: Further evidence for increased fearfulness. *Molecular Brain Research* **37**: 145–150.

Miyakawa T, Yagi T, Kitazawa H, Yasuda M, Kawai N, Tsuboi K, Niki H (1997). Fyn-kinase as a determinant of ethanol sensitivity: Relation to NMDA-receptor function. *Science* **278**: 698–701.

Miyasaka K, Ohta M, Kanai S, Yoshida Y, Sato N, Nagata A, Matsui T, Noda T, Jimi A, Takiguchi S, Takata Y, Kawanami T, Funakoshi A (2004). Enhanced gastric emptying of a liquid gastric load in mice lacking cholecystokinin-B receptor: A study of CCK-A,B, and AB receptor gene knockout mice. *Journal of Gastroenterology* **39**: 319–323.

Miyakawa T, Yagi T, Tateishi K, Niki H (1996b). Susceptibility to drug-induced seizures of Fyn tyrosine kinase deficient mice. *NeuroReport* **7**: 2723–2726.

Miyakawa T, Yagi T, Watanabe S, Niki H (1994). Increased fearfulness of Fyn tyrosine kinase deficient mice. *Molecular Brain Research* **27**: 179–182.

Miyakawa T, Yared E, Pak JH, Huang FL, Huang KP, Crawley JN (2001). Neurogranin null mutant mice display performance deficits on spatial learning tasks with anxiety related components. *Hippocampus* **11**: 763–765.

Miyasaka K, Kanai S, Ohta M, Kawanami T, Kono A, Funakoshi A (1994). Lack of satiety effect of cholecystokinin (CCK) in a new rat model not expressing the CCK-A receptor gene. *Neuroscience Letters* **180**: 143–146.

Miyakawa T, Yagi T, Taniguchi M, Matsuura H, Tateishi K, Niki H (1995). Enhanced susceptibility of audiogenic seizures in Fyn-kinase deficient mice. *Molecular Brain Research* **28**: 349–352.

Mochizuki S, Mizukami H, Ogura T, Kure S, Ichinohe A, Kojima K, Matsubara Y, Kobayahi E, Okada T, Hoshika A, Ozawa K, Kume A (2004). Long-term correction of hyperphenylalaninemia by AAV-mediated gene transfer leads to behavioral recovery in phenylketonuria mice. *Gene Therapy* **11**: 1081–1086.

Moghaddam B, Bunney BS (1990). Acute effects of typical and atypical antipsychotic drugs on the release of dopamine from prefrontal cortex, nucleus accumbens, and striatum of the rat: An in vivo microdialysis study. *Journal of Neurochemistry* **54**: 1755–1760.

Mogil JS, Belknap JK (1997). Sex and genotype determine the selective activation of neurochemically-distinct mechanisms of swim stress-induced analgesia. *Pharmacology Biochemistry and Behavior* **56**: 61–66.

Mogil JS, McCarson KE (2000). Identifying pain genes: Bottom-up and tom-down approaches. *The Journal of Pain* **1**: 66–80.

Mogil JS, Sternberg WF, Marek P, Sadowski B, Belknap JK, Liebeskind JC (1996). The genetics of pain and pain inhibition. *Proceedings of the National Academy of Sciences USA* **93**: 3048–3055.

Mohajeri MH, Figlewicz DA, Bohn MC (1999). Intramuscular grafts of myoblasts genetically modified to secrete glial cell line-derived neurotrophic factor prevent motoneuron loss and disease progression in a mouse model of familial amyotrophic lateral sclerosis. *Human Gene Therapy* **10**: 1853–1866.

Mohajeri MH, Madani R, Saini K, Lipp H-P, Nitsch RM, Wolfer DP (2004). The impact of genetic background on neurodegeneration and behavior in seizured mice. *Genes, Brain and Behavior* **3**: 228–239.

Mohn AR, Yao WD, Caron MG (2004). Genetic and genomic approaches to reward and addiction. *Neuropharmacology* **47**(Suppl 1): 101–110.

Moles A, Kieffer BL, D'Amato FR (2004). Deficit in attachment behavior in mice lacking the mu-opioid receptor gene. *Science* **304**: 1983–1986.

Molewijk HE, van der Poel AM, Mos J, van der Heyden JAM, Oliver B (1995). Conditioned ultrasonic distress vocalizations in adult male rats as a behavioural paradigm for screening anti-panic drugs. *Psychopharmacology* **117**: 32–40.

Molinari S, Battini R, Ferrari S, Pozzi L, Killcross AS, Robbins TW, Jouvenceau A, Billard JM, Dutar P, Lamour Y, Baker WA, Cox H, Emson PC (1996). Deficits in memory and hippocampal long-term potentiation in mice with reduced calbindin D28K expression. *Proceedings of the National Academy of Sciences USA* **93**: 8028–8033.

Mombaerts P (2003). Therapeutic cloning in the mouse. *Proceedings of the National Academy of Sciences USA* **100** (Suppl 1): 11924–11925.

Mombaerts P (2004). Genes and ligands for odorant, vomeronasal and taste receptors. *Nature Reviews in Neuroscience* **5**: 263–278.

Mombereau C, Kaupmann K, Froestl W, Sansig G, van der Putten H, Cryan JF (2004). Genetic and pharmacological evidence of a role for GABA(B) receptors in the modulation of anxiety- and antidepressant-like behavior. *Neuropsychopharmacology* **29**: 1050–1062.

Monaghan EP, Glickman SE (1993). Hormones and aggressive behavior. In *Behavioral Endocrinology*, JB Becker, SM Breedlove, D Crews, Eds. MIT Press, Cambridge, pp. 261–285.

Monahan EJ, Maxson SC (1998). Y chromosome, urinary chemosignals, and an agonistic behavior (offense) of mice. *Physiology and Behavior* **64**: 123–132.

Moncada S, Ed. (1991, 1993, 1997). *The Biology of Nitric Oxide*, Vols. 1–6. Portland Press, London.

Mondragon R, Mayagoitia L, Lopez-Lujan A, Díaz JL (1987). Social structure features in three inbred strains of mice, C57BL/6J, Balb/cj, and NIH: A comparative study. *Behavioral and Neural Biology* **47**: 384–391.

Monks DA, Lonstein JS, Breedlove SM (2003). Got milk? Oxytocin triggers hippocampal plasticity. *Nature Neuroscience*: **6**: 327–328.

Montkowski A, Poettig M, Mederer A, Holsboer F (1997). Behavioural performance in three substrains of mouse strain 129. *Brain Research* **762**: 12–18.

Moody TW, Merali Z (2004). Bombesin-like peptides and associated receptors within the brain: Distribution and behavioral implications. *Peptides* **25**: 511–520.

Moran TH (2004). Gut peptides in the control of food intake: 30 years of ideas. *Physiology and Behavior* **82**: 175–180.

Moran TH, Ameglio PJ, Schwartz GJ, McHugh PR (1992). Blockade of type A, not type B, CCK receptors attenuates satiety actions of exogenous and endogenous CCK. *American Journal of Physiology* **31**: R46–R50.

Moran TH, Baldessarini AR, Salorio CF, Lowery T, Schwartz GJ (1997). Vagal afferent and efferent contributions to the inhibition of food intake by cholecystokinin. *American Journal of Physiology* **272**: R1245–R1251.

Moran TH, Capone GT, Knipp S, Davisson MT, Reeve RH, Gearhart JD (2002). The effects of piracetam on cognitive performance in a mouse model of Down's syndrome. *Physiology and Behavior* **77**: 403–409.

Moran PM, Higgins LS, Cordell B, Moser PC (1995). Age-related learning deficits in transgenic mice expressing the 751-amino acid isoform of human β-amyloid precursor protein. *Proceedings of the National Academy of Sciences USA* **92**: 5341–5345.

Moran TH, Katz LF, Plata-Salaman CR, Schwartz GJ (1998). Disordered food intake and obesity in rats lacking cholecystokinin A receptors. *American Journal of Physiology* **274**: R618–R625.

Moran TH, McHugh PR (1988). Gastric and nongastric mechanisms for satiety action of cholecystokinin. *American Journal of Physiology* **254**: R628–R632.

Moreau JL (1997). Validation of an animal model of anhedonia, a major symptom of depression. *Encephale* **23**: 280–289.

Morency MA, Beninger RJ (1986). Dopaminergic substrates of cocaine-induced place conditioning. *Brain Research* **399**: 33–41.

Moretti P, Bouwknecht JA, Teague R, Paylor R, Zoghbi HY (2005). Abnormalities of social interactions and home-cage behavior in a mouse model of Rett syndrome. *Human Molecular Genetics* **14**: 205–220.

Morgan D, Diamond DM, Gottschall PE, Ugen KE, Dickey C, Hardy J, Duff K, Jantzen P, DiCarlo G, Wilcock D, Connor K, Hatcher J, Hope C, Gordon M, Arendash GW (2000). A beta peptide vaccination prevents memory loss in an animal model of Alzheimer's disease. *Nature* **408**: 982–985.

Morgan D, Holcomb L, Saad I, Gordon M, Maines M (1998). Impaired spatial navigation learning in transgenic mice over-expressing heme oxygenase. *Brain Research* **808**: 110–112.

Mori T, Nomura M, Nagase H, Narita M, Suzuki T (2002). Effects of a newly synthesized kappa-opioid receptor agonist, TRK-820, on the discriminative stimulus and rewarding effects of cocaine in rats. *Psychopharmacology* **161**: 17–22.

Morice E, Denis C, Giros B, Nosten-Bertrand M (2004). Phenotypic expression of the targeted null-mutation in the dopamine transporter gene varies as a function of the genetic background. *European Journal of Neuroscience* **20**: 120–126.

Morley JE, Levine AS (1982). Corticotropin-releasing factor, grooming and ingestive behavior. *Life Sciences* **31**: 1459–1464.

Morozov A, Kellendonk C, Simpson E, Tronche F (2003). Using conditional mutagenesis to study the brain. *Biological Psychiatry* **54**: 1125–1133.

Morris JS, Ohman A, Dolan RJ (1998). Conscious and unconscious emotional learning in the human amygdala. *Nature* **393**: 467–470.

Morris RGM (1981). Spatial localization does not depend on the presence of local cues. *Learning and Motivation* **12**: 239–260.

Morris R (1984). Developments of water-maze procedure for studying spatial learning in the rat. *Journal of Neuroscience* Methods **11**: 47–60.

Morris RGM (2001). Episodic-like memory in animals: Psychological criteria, neural mechanisms and the value of episodic-like tasks to investigate animal models of neurodegenerative diseases. *Philosophical Transactions of the Royal Society B: Biological Sciences* **356**: 1453–1465.

Morris RGM, Garrud P, Rawlins JNP, O'Keefe J (1982). Place navigation impaired in rats with hippocampal lesions. *Nature* **297**: 681–683.

Morse HC (1978). Introduction. In *Origins of Inbred Mice*, HC Morse, Ed. Academic Press, New York, pp. 1–31.

Morton AJ, Hunt MJ, Hodges AK, Lewis PD, Redfern AJ, Dunnett SB, Jones L (2005a). A combination drug therapy improves cognition and reverses gene expression changes in a mouse model of Huntington's disease. *European Journal of Neuroscience* **21**: 855–870.

Morton AJ, Wood NI, Hastings MH, Hurelbrink C, Barker RA, Maywood ES (2005b). Disintegration of the sleep-wake cycle and circadian timing in Huntington's disease. *Journal of Neuroscience* **25**: 157–163.

Moser VC, Cheek BM, MacPhail RC (1995). A multidisciplinary approach to toxicological screening: III. Neurobehavioral toxicology. *Journal of Toxicology and Environmental Health* **45**: 173–210.

Moser VC, Tilson HA, MacPhail RC, Becking GC, Cuomo V, Frantik E, Kulig BM, Winneke G (1997). The IPCS collaborative study on neurobehavioral screening methods: II. Protocol design and testing procedures. *Neurotoxicology* **18**: 929–938.

Moskowitz AS, Terman GW, Liebeskind JC (1985). Stress-induced analgesia in the mouse: Strain comparisons. *Pain* **23**: 67–72.

Mottagui-Tabar S, Prince JA, Wahlestedt C, Zhu G, Goldman D, Heilig M (2005). A novel single nucleotide polymorphism of the neuropeptide Y (NPY) gene associated with alcohol dependence. *Alcoholism, Clinical and Experimental Research* **29**: 702–707.

Moy SS, Nadler JJ, Magnuson TR, Crawley JN (2006). Mouse models of autism spectrum disorders: The challenge for behavioral genetics. *American Journal of Medical Genetics* **142**: 40–51.

Moy SS, Nadler JJ, Perez A, Barbaro RP, Johns JM, Magnuson TR, Piven J, Crawley JN (2004). Sociability and preference for social novelty in five inbred strains: An approach to assess autistic-like behaviors in mice. *Genes, Brain and Behavior* **3**: 287–302.

Moy SS, Nadler JJ, Young NB, Perez A, Holloway LP, Barbaro RP, Barboro JR, Wilson LM, Threadgil DW, Lauder JM, Magnuson TR, Crawley JN (2007) Mouse behavioral tasks relevant to autism: Phenotypes of ten inbred stains. *Behavioural Brain Research*, in press.

Moye TB and Rudy JW (1987). Ontogenesis of trace conditioning in young rats: Dissociation of associative and memory processes. *Developmental Psychobiology* **20**: 405–414.

Muenke M, Schell U (1995). Fibroblast-growth-factor receptor mutations in human skeletal disorders. *Trends in Genetics* **11**: 308–313.

Muglia LJ, Jacobson L, Weninger SC, Karalis KP, Jeong K, Majzoub JA (2001). The physiology of corticotropin-releasing hormone deficiency in mice. *Peptides* **22**: 725–731.

Muhle R, Trentacoste SV, Rapin I (2004). The genetics of autism. *Pediatrics* **113**: 472–486.

Muir JI, Everitt BJ, Robbins TW (1996). The cerebral cortex of the rat and visual attentional function: Dissociable effects of mediofrontal, cingulate, anterior dorsolateral, and parietal cortex lesions on a five-choice serial reaction time task. *Cerebral Cortex* **6**: 470–481.

Mullan M, Crawford F, Axelman K, Houlden H, Lilius L, Winblad B, Lannfelt L (1992). A pathogenic mutation for probable Alzheimer's disease in the APP gene at the *N*-terminus of β-amyloid. *Nature Genetics* **1**: 345–347.

Mura A, Feldon J, Mintz M (1998). Reevaluation of the striatal role in the expression of turning behavior in the rat model of Parkinson's disease. *Brain Research* **808**: 48–55.

Murata S, Yoshiara T, Lim CR, Sugino M, Kogure M, Ohnuki T, Komurasaki T, Matsubara K (2005). Psychophysiological stress-regulated gene expression in mice. *FEBS Letters* **579**: 2137–2142.

Murchison CF, Zhang XY, Zhang WP, Ouyang M, Lee A, Thomas SA (2004). A distinct role for norepinephrine in memory retrieval. *Cell* **117**: 131–143.

Murphy DL, Uhl GR, Holmes A, Ren-Patterson R, Hall FS, Sora I, Detera-Wadleigh S, Lesch KP (2003). Experimental gene interaction studies with SERT mutant mice as models for human polygenic and epistatic traits and disorders. *Genes Brain and Behavior* **2**: 350–364.

Murphy DL, Wichems C, Li Q, Heils A (1999). Molecular manipulations as tools for enhancing our understanding of 5-HT neurotransmission. *Trends in Pharmacological Sciences* **20**: 246–252.

Murphy KP, Carter RJ, Lione LA, Mangiarini L, Mahal A, Bates GP, Dunnett SB, Morton AJ (2000). Abnormal synaptic plasticity and impaired spatial cognition in mice transgenic for exon 1 of the human Huntington's disease mutation. *Journal of Neuroscience* **20**: 5115–5123.

Murtra P, Sheasby AM, Hunt SP, De Felipe C (2000). Rewarding effects of opiates are absent in mice lacking the receptor for substance P. *Nature* **405**: 180–183.

Muzzin P, Eisensmith RC, Copeland KC, Woo SLC (1996). Correction of obesity and diabetes in genetically obese mice by leptin gene therapy. *Proceedings of the National Academy of Sciences USA* **93**: 14804–14808.

Myers WA (1970). Some observations on "reeler," a neuromuscular mutation in mice. *Behavior Genetics* **1**: 225–234.

Nadler JJ, Moy SS, Dold G Trang D, Simmons N, Perez A, Young NB, Barbaro RP, Piven J, Magnuson TR, Crawley JN (2004). Automated apparatus for rapid quantitation of social approach behaviors in mice. *Genes, Brain and Behavior* **3**: 303–314.

Nagahara AH, Otto T, Gallagher M (1995). Entorhinal-perirhinal lesions impair performance of rats on two versions of place learning in the Morris water maze. *Behavioral Neuroscience* **109**: 3–9.

Nagle DL, McGrail SH, Vitale J, Woolf EA, Dussault BJ, DiRocco L, Holmgren L, Montagno J, Bork P, Huszar D, Fairchild-Huntress V, Ge P, Keilty J, Eberling C, Baldini L, Gilchrist J, Burn P, Carlson GA, Moore KJ (1999). The mahogany protein is a receptor involved in suppression of obesity. *Nature* **398**: 148–152.

Nagy A, Gertsenstein M, Vintersten K, Behringer R (2002). *Manipulating the Mouse Embryo: A Laboratory Manual.* Cold Spring Harbor Laboratory, Cold Spring Harbor, NY.

Nair HP, Young LJ (2002). Application of adeno-associated viral vectors in behavioral research. *Methods* **28**: 195–202.

Nakahara D, Fuchikami K, Ozaki N, Iwasaki T, Nagatsu T (1992). Differential effect of self-stimulation on dopamine release and metabolism in the rat medial frontal cortex, nucleus accumbens and striatum studies by in vivo microdialysis. *Brain Research* **574**: 164–170.

Nakamura K, Kurasawa M (2001). Anxiolytic effects of aniracetam in three different mouse models of anxiety and the underlying mechanism. *European Journal of Pharmacology* **420**: 33–43.

Nakanishi S (1992). Molecular diversity of glutamate receptors and implications for brain function. *Science* **258**: 597–603.

Nakazawa K, Quirk MC, Chitwood RA, Watanabe M, Yeckel MF, Sun LD, Kato A, Carr CA, Johnston D, Wilson MA, Tonegawa S (2002). Requirement for hippocampal CA3 NMDA receptors in associative memory recall. *Science* **297**: 211–218.

Nakazawa K, Sun LD, Quirk MC, Rondi-Reig L, Wilson MA, Tonegawa S (2003). Hippocampal CA3 NMDA receptors are crucial for memory acquisition of one-time experience. *Neuron* **38**: 305–315.

Nalbantoglu J, Tirado-Santiago G, Lahsainl A, Poirier J, Goncalves O, Verge G, Momoll F, Welner SA, Massicotte G, Julien JP, Shapiro ML (1997). Impaired learning and LTP in mice expressing the carboxy terminus of the Alzheimer amyloid precursor protein. *Nature* **387**: 500–505.

Nally RE, Kinsella A, Tighe O, Croke DT, Fienberg AA, Greengard P, Waddington JL (2004). Ethologically based resolution of D2-like dopamine receptor agonist-versus antagonist-induced behavioral topography in dopamine- and adenosine 3',5'-monophosphate-regulated phosphoprotein of 32 kDa "knockout" mutants congenic on the C57BL/6 genetic background. *Journal of Pharmacology and Experimental Therapeutics* **310**: 1281–1287.

Narita M, Mizuo K, Mizoguchi H, Sakata M, Narita M, Tseng LF, Suzuki T (2003). Molecular evidence for the functional role of dopamine D3 receptor in the morphine-induced rewarding effect and hyperlocomotion. *Journal of Neuroscience* **23**: 1006–1012.

Nasir J, Floresco SB, O'Kusky JR, Dlewen VM, Richman JM, Zeisler J, Borowski A, Marth JD, Phillips AG, Hayden MR (1995). Targeted disruption of the Huntington's disease gene results in embryonic lethality and behavioral and morphological changes in heterozygotes. *Cell* **81**: 811–823.

Nathan BP, Yost J, Litherland MT, Struble RG, Switzer PV (2004). Olfactory function in apoE knockout mice. *Behavioural Brain Research* **150**: 1–7.

Nathan C (1992). Nitric oxide as a secretory product of mammalian cells. *FASEB Journal* **6**: 3051–3064.

Nature Genetics (1999). *Special Issue: The Chipping Forecast* **21** (Suppl): 1–60.

Navarro JF, Rivera A, Maldonado E, Cavas M, de la Calle A (2004). Anxiogenic-like activity of 3,4-methylenedioxy-methamphetamine ("Ecstasy") in the social interaction test is accompanied by an increase of c-fos expression in mice amygdala. *Progress in Neuropsychopharmacology and Biological Psychiatry* **28**: 249–254.

Nebert DW, Duffy JJ (1997). How knockout mouse lines will be used to study the role of drug-metabolizing enzymes and their receptors during reproduction and development, and in environmental toxicity, cancer, and oxidative stress. *Biochemical Pharmacology* **53**: 249–254.

Nelson RJ (2005). *An Introduction to Behavioral Endocrinology.* Sinauer, Sunderland, MA.

Nelson RJ, Chiavegatto S (2001). Molecular basis of aggression. *Trends in Neuroscience* **24**: 713–719.

Nelson RJ, Demas GE, Huang PL, Fishman MC, Dawson VL, Dawson TM, Snyder SH (1995). Behavioural abnormalities in male mice lacking neuronal nitric oxide synthase. *Nature* **378**: 383–386.

Nelson RJ, Young KA (1998). Behavior in mice with targeted disruption of single genes. *Neuroscience and Biobehavioral Reviews* **22**: 453–462.

Nemeroff CB (2004). Neurobiological consequences of childhood trauma. *Journal of Clinical Psychiatry* **65** (Suppl 1):18–28.

Nesbitt MN (1992). The value of recombinant inbred strains in the genetic analysis of behavior. In *Techniques for the Genetic Analysis of Brain and Behavior: Focus on the Mouse*, D Goldowitz, D Wahlsten, RE Wimer, Eds. Elsevier, Amsterdam, pp. 141–146.

Nestler EJ (2004). Historical review: Molecular and cellular mechanisms of opiate and cocaine addiction. *Trends in Pharmacological Sciences* **25**: 210–208.

Nestler EJ, Barrot M, Self DW (2001). DeltaFosB: a sustained molecular switch for addiction. *Proceedings of the National Academy of Sciences USA* **98**: 11042–11046.

Nestler EJ, Gould E, Manji H, Buncan M, Duman RS, Greshenfeld HK, Hen R, Koester S, Lederhendler I, Meaney M, Robbins T, Winsky L, Zalcman S (2002). Preclinical models: status of basic research in depression. *Biological Psychiatry* **S52**: 503–528.

Neumaier JF, Vincow ES, Arvanitogiannis A, Wise RA, Carlezon WA (2002). Elevated expression of 5-HT1B receptors in nucleus accumbens efferents sensitizes animals to cocaine. *Journal of Neuroscience* **22**: 10856–10863.

Neumann PE (1992). Inference in linkage analysis of multifactorial traits using recombinant inbred strains of mice. *Behavior Genetics* **22**: 665–676.

Neumann PE, Collins RL (1991). Genetic dissection of susceptibility to audiogenic seizures in inbred mice. *Proceedings of the National Academy of Sciences USA* **88**: 5408–5412.

Nevison CM, Armstrong S, Beynon RJ, Humphries RE, Hurst JL (2003). The ownership signature in mouse scent marks is involatile. *Proceedings of the Royal Society of London B: Biological Sciences* **270**: 1957–1963.

Newton SS, Thome J, Wallace TL, Shirayama Y, Schlesinger L, Sakai N, Chen J, Neve R, Nestler EJ, Duman RS (2002). Inhibition of cAMP response element-binding protein or dynorphin in the nucleus accumbens produces an antidepressant-like effect. *Journal of Neuroscience* **22**: 10883–10890.

Nicolaidis S, Danguir J, Mather P (1979). A new approach to sleep and feeding behaviors in the laboratory rat. *Physiology and Behavior* **23**: 717–722.

Nicot A, Otto T, Brabet P, DiCicco-Bloom EM (2004). Altered social behavior in pituitary adenylate cyclase-activating polypeptide (PACAP) type 1 receptor deficient mice. *Journal of Neuroscience* **24**: 8786–8795.

Nielsen DM, Carey GJ, Gold LH (2004). Antidepressant-like activity of corticotropin-releasing factor type-1 receptor antagonists in mice. *European Journal of Pharmacology* **499**: 135–146.

Nielsen DM, Derber WJ, McClellan DA, Crnic LS (2002). Alterations in the auditory startle response in Fmr1 targeted mutant mouse models of fragile X syndrome. *Brain Research* **927**: 8–17.

Niki H, Miyakawa T, Yagi T (1996). Experimental analysis of behavior of Fyn-tyrosine kinase-deficient mice. In *Brain Processes and Memory*, K Ishikawa, JJ McGaugh, H Sakara, Eds. Elsevier Science, Amsterdam, pp. 279–290.

Nilsson OG, Gage FH (1993). Anticholinergic sensitivity in the aging rat septohippocampal system as assessed in a spatial memory task. *Neurobiology of Aging* **14**: 487–497.

Nishimura F, Yoshikawa M, Kanda S, Nonaka M, Yokota H, Shiroi A, Nakase H, Hirabayashi H, Ouji Y, Birumachi J, Ishizaka S, Sakaki T (2003). Potential use of embryonic stem cells for the treatment of mouse parkinsonian models: improved behavior by transplantation of in vitro differentiated dopaminergic neurons from embryonic stem cells. *Stem Cells* **21**: 171–180.

Nishimori K, Young LJ, Guo Q, Wang Z, Insel TR, Matzuk MM (1996). Oxytocin is required for nursing but is not essential for parturition or reproductive behavior. *Proceedings of the National Academy of Sciences USA* **93**: 11699–11704.

Nishio M, Kuroki Y, Watanabe Y (2003). Role of hippocampal alpha(2A)-adrenergic receptor in methamphetamine-induced hyperlocomotion in the mouse. *Neuroscience Letters* **341**: 156–160.

No D, Yao TP, Evans RM (1996). Ecdysone-inducible gene expression in mammalian cells and transgenic mice. *Proceedings of the National Academy of Sciences USA* **93**: 3346–3351.

Noben-Trauth K, Zheng QY, Johnson KR (2003). Association of cadherin 23 with polygenic inheritance and genetic modification of sensorineural hearing loss. *Nature Genetics* **35**: 21–23.

Noble EP (1998). The D2 dopamine receptor gene: A review of association studies in alcoholism and phenotypes. *Alcohol* **16**: 33–45.

Noble F, Roques BP (2002). Phenotypes of mice with invalidation of cholecystokinin (CCK$_1$ or CCK$_2$) receptors. *Neuropeptides* **36**: 157–170.

Noebels JL (1984). Isolating single genes of the inherited epilepsies. *Annals of Neurology* **16** (Suppl): S18–21.

Noebels JL, Qiao X, Bronson RT, Spencer C, Davisson MT (1990). Stargazer: A new neurological mutant on chromosome 15 in the mouse with prolonged cortical seizures. *Epilepsy Research* **7**: 129–135.

Nomura M, Durbak L, Chan J, Smithies O, Gustafsson JA, Korach KS, Pfaff DW, Ogawa S (2002). Genotype/age interactions on aggressive behavior in gonadally intact estrogen receptor beta knockout (betaERKO) male mice. *Hormones and Behavior* **41**: 288–296.

Nonogaki K, Strack AM, Dallman MF, Tecott LH (1998). Leptin-dependent hyperphagia and type 2 diabetes in mice with a mutated serotonin 5-HT2C receptor gene. *Nature Medicine* **4**: 1152–1156.

Norflus F, Tifft CJ, McDonald MP, Goldstein G, Crawley JN, Hoffmann A, Sandhoff K, Suzuki K, Proia RL (1998). Bone marrow transplantation prolongs life span and ameliorates neurologic manifestations in Sandhoff disease mice. *Journal of Clinical Investigation* **101**: 1881–1888.

Nosten-Bertrand M, Errington ML, Murphy KPSJ, Tokugawa Y, Barboni E, Kozlova E, Michaelovich D, Morris RGM, Silver J, Stewart CL, Bliss TVP, Morris RJ (1996). Normal spatial learning despite regional inhibition of LTP in mice lacking *Thy-1*. *Nature* **379**: 826–829.

O'Brien TP, Frankel WN (2004). Moving forward with chemical mutagenesis in the mouse. *Journal of Physiology* **554**(Pt 1): 13–21.

O'Callaghan JP, Holtzman SG (1985). Quantification of the analgesic activity of narcotic antagonists by a modified hot-plate procedure. *Journal of Pharmacology and Experimental Therapeutics* **192**: 497–505.

Ogawa S, Lubahn DB, Korach KS, Pfaff DW (1997). Behavioral effects of estrogen receptor gene disruption in male mice. *Proceedings of the National Academy of Sciences USA* **94**: 1476–1481.

Ohta R, Shigemura N, Sasamoto K, Koyano K, Ninomiya Y (2003). Conditioned taste aversion learning in leptin-receptor-deficient *db/db* mice. *Neurobiology of Learning and Memory* **80**: 105–112.

Olds J (1956). A preliminary mapping of electrical reinforcing effects in the rat brain. *Journal of Comparative and Physiological Psychology* **49**: 281–285.

Olds J (1962). Hypothalamic substrates of reward. *Physiological Reviews* **42**: 554–604.

Olivier B, Molewijk E, van Oorschot R, van der Heyden J, Ronken E, Mos J (1998). Rat pup ultrasonic vocalization: Effects of benzodiazepine receptor ligands. *European Journal of Pharmacology* **358**: 117–128.

Olmstead CE (1987). Neurological and neurobehavioral development of the mutant "twitcher" mouse. *Behavioural Brain Research* **25**: 143–153.

Olton DS, Samuelson RJ (1976). Remembrance of places passed: Spatial memory in rats. *Experimental Psychology (Animal Behavior)*. **2**: 97–116.

Olszewski PK, Levine AS (2004). Minireview: Characterization of influence of central nociceptin/orphanin FQ on consummatory behavior. *Endocrinology* **145**: 2627–2632.

Orsini C, Buchini F, Piazza PV, Puglisi-Allegra S, Cabib S (2004). Susceptibility to amphetamine-induced place preference is predicted by locomotor response to novelty and amphetamine in the mouse. *Psychopharmacology* **172**: 264–270.

Otto C, Kovalchuk Y, Wolfer DP, Gass P, Martin M, Zuschratter W, Grone HJ, Kellendonk C, Tronche F, Maldonado R, Lipp HP, Konnerth A, Schutz G (2001). Impairment of mossy fiber long-term potentiation and associative learning in pituitary adenylate cyclase activating polypeptide type I receptor-deficient mice. *Journal of Neuroscience* **21**: 5520–5527.

Otto JF, Yang Y, Frankel WN, Wilcox KS, White HS (2004). Mice carrying the szt1 mutation exhibit increased seizure susceptibility and altered sensitivity to compounds acting at the m-channel. *Epilepsia* **45**: 1009–1016.

Otto T, Eichenbaum H (1992). Complementary roles of the orbital prefrontal cortex and the perirhinal-entorhinal cortices in an odor-guided delayed-nonmatching-to-sample task. *Behavioral Neuroscience* **106**: 762–775.

Otto T, Giardino ND (2001). Pavlovian conditioning of emotional responses to olfactory and contextual stimuli: a potential model for the development and expression of chemical intolerance. *Annals of the New York Academy of Sciences* **933**: 291–309.

Overstreet DH, Rezvani AH, Janowsky DS (1992). Maudsley reactive and nonreactive rats differ only in some tasks reflecting emotionality. *Physiology and Behavior* **52**: 149–152.

Oyama F, Sawamura N, Kobayashi K, Morishima-Kawashima M, Kuramochi T, Ito M, Tomita T, Maruyama K, Saido TC, Iwatsubo T, Capell A, Walter J, Grunberg J, Ueyama Y, Haas C, Ihara Y (1998). Mutant presenilin 2 transgenic mouse: Effect on an age-dependent increase of amyloid beta-protein 42 in the brain. *Journal of Neurochemistry* **71**: 313–322.

Palmer AA, Miller MN, McKinnon CS, Phillips TJ (2002). Sensitivity to the locomotor stimulant effects of ethanol and allopregnanolone is influenced by common genes. *Behavioral Neuroscience* **116**: 126–137.

Palmiter RD, Brinster RL, Hammer RE, Trumbauer ME, Rosenfeld MG, Birnberg NC, Evans RM (1982). Dramatic growth of mice that develop from eggs microinjected with metallothionein-growth hormone fusion genes. *Nature* **300**: 611–615.

Palmiter RD, Norstedt G, Gelinas RE, Hammer RE, Brinster RL (1983). Metallothionein-human GH fusion genes stimulate growth of mice. *Science* **222**: 810–814.

Panksepp J (1981). The ontogeny of play in rats. *Developmental Psychobiology* **14**: 327–332.

Panksepp J (1998). *Affective Neuroscience: The Foundations of Human and Animal Emotions.* Oxford University Press, New York.

Papaioannou VE, Behringer RR (2005). *Mouse Phenotypes: A Handbook of Mutation Analysis.* Cold Spring Harbor Laboratory Press, Cold Spring Harbor, New York.

Pape HC, Stork O (2003). Genes and mechanisms in the amygdala involved in the formation of fear memory. *Annals of the New York Academy of Sciences*: **985**: 92–105.

Paradee W, Melikian HE, Rasmussen DL, Kenneson A, Conn PJ, Warren ST (1999). Fragile X mouse: strain effects of knockout phenotype and evidence suggesting deficient amygdala function. *Neuroscience* **94**: 185–192.

Park K, Huber J, Kolta MG, Stino FK, Samaan SS, Soliman KF (2000). Behavioral responses to repeated cocaine exposure in mice selectively bred for differential sensitivity to pentobarbital. *Annals of the New York Academy of Sciences* **914**: 323–335.

Parks CL, Robinson PS, Sibille E, Shenk T, Toth M (1998). Increased anxiety of mice lacking the serotonin1A receptor. *Proceedings of the National Academy of Sciences USA* **95**: 10734–10739.

Pascoe WS, Kemler R, Wood SA (1992). Genes and functions: Trapping and targeting in embryonic stem cells. *Biochimica et Biophysica Acta* **1114**: 209–221.

Patel N, Hitzemann B, Hitzemann R (1998). Genetics, haloperidol, and the fos response in the basal ganglia: A comparison of the C57BL/6J and DBA/2J inbred mouse strains. *Neuropsychopharmacology* **18**: 480–491.

Paterson AH, Damon S, Hewitt JD, Zamir D, Rabinowitch HD, Lincoln SE, Lander ES, Tanksley SD (1991). Mendelian factors underlying quantitative traits in tomato: Comparison across species, generations, and environments. *Genetics* **127**: 181–197.

Paterson NE, Markou A (2004). Prolonged nicotine dependence associated with extended access to nicotine self-administration in rats. *Psychopharmacology* **173:** 64–72.

Pavlov IP (1927). *Conditioned Reflexes.* Oxford University Press, Oxford.

Paxinos G, Watson C (1982). *The Rat Brain in Stereotaxic Coordinates.* Academic Press, New York.

Paylor R, Crawley JN (1997). Inbred strain differences in prepulse inhibition of the mouse startle response. *Psychopharmacology* **132**: 169–180.

Paylor R, Hirotsume S, Gambello M, Yuva-Paylor L, Crawley JN, Wynshaw-Boris A (1999). Impaired learning and motor behavior in heterozygous *Pafah1b1 (Lis1)* mutant mice. *Learning and Memory,* **6**: 521–537.

Paylor R, Johnson RS, Papaioannou V, Spiegelman BM, Wehner JM (1994a). Behavioral assessment of *c-fos* mutant mice. *Brain Research* **651**: 275–282.

Paylor R, Nguyen M, Crawley JN, Patrick J, Beaudet A, Orr-Urtreger A (1998). α 7 Nicotinic receptor subunits are not necessary for hippocampal-dependent learning or sensorimotor gating: A behavioral characterization of *Acra7*-deficient mice. *Learning and Memory* **5**: 302–316.

Paylor R, Spencer CM, Yuva-Paylor LA, Pieke-Dahl S (2006). The use of behavioral test batteries, II. Effect of test interval. *Physiology and Behavior,* **87**: 95–102.

Paylor R, Tracy R, Wehner J, Rudy JW (1994b). DBA/2 and C57BL/6 mice differ in contextual fear but not auditory fear conditioning. *Behavioral Neuroscience* **108**: 810–817.

Paylor R, Zhao Y, Libbey M, Westphal H, Crawley JN (2001). Learning impairments and motor dysfunctions in adult Lhx5-deficient mice displaying hippocampal disorganization. *Physiology and Behavior* **73**: 781–792.

Peachey NS, Ball SL (2003). Electrophysiological analysis of visual function in mutant mice. *Doc Ophthalmol* **107**: 13–36.

Peakman MC, Colby C, Perrotti LI, Tekumalla P, Carle T, Ulery P, Chao J, Duman C, Steffen C, Monteggia L, Allen MR, Stock JL, Duman RS, McNeish JD, Barrot M, Self DW, Nestler EJ, Schaeffer E (2003). Inducible, brain region-specific expression of a dominant negative mutant of c-Jun in transgenic mice decreases sensitivity to cocaine. *Brain Research* **970**: 73–86.

Pedrazzi P, Cattaneo L, Valeriani L, Boschi S, Cocchi D, Zoli M (1998). Hypothalamic neuropeptide Y and galanin in overweight rats fed a cafeteria diet. *Peptides* **19**: 157–165.

Pedrazzini T, Seydoux J, Kunstner P, Aubert JF, Grouzmann E, Beermann F, Brunner HR (1998). Cardiovascular response, feeding behavior and locomotor activity in mice lacking the NPYY1 receptor. *Nature Medicine* **4**: 722–726.

Peeters PJ, Fierens FL, van den Wyngaert I, Goehlmann HW, Swagemakers SM, Kass SU, Langlois X, Pullan S, Stenzel-Poore MP, Steckler T (2004). Gene expression profiles highlight adaptive brain mechanisms in corticotropin releasing factor overexpressing mice. *Molecular Brain Research* **129**: 135–150.

Peier AM, McIlwain KL, Kenneson A, Warren ST, Paylor R, Nelson DL (Over)correction of FMR1 deficiency with YAC transgenics: Behavioral and physical features. *Human Molecular Genetics* **9**: 1145–1159.

Pelletier G (2000). Localization of androgen and estrogen receptors in rat and primate tissues. *Histol Histopathol* **15**: 1261–1270.

Pelleymounter MA, Cullen MJ, Baker MB, Hecht R, Winters D, Boone T, Collins F (1995). Effects of the *obese* gene product on body weight regulation in *ob/ob* mice. *Science* **269**: 540–549.

Pellis SM, Pellis VC, Whishaw IQ (1992). The role of the cortex in play fighting by rats: Developmental and evolutionary implications. *Brain Behavior and Evolution* **39**: 270–284.

Pellow S, Chopin P, File SE, Briley M (1985). Validation of open:closed arm entries in an elevated plus-maze as a measure of anxiety in the rat. *Journal of Neuroscience Methods* **14**: 149–167.

Perrigo G, Bronson FH (1985). Behavioral and physiological responses of female house mice to foraging variation. *Physiology and Behavior* **34**: 437–440.

Pertwee RG (2001). Cannabinoid receptors and pain. *Progress in Neurobiology* **63**: 569–611.

Peterson C, Maier SF, Seligman MEP (1995). *Learned Helplessness: A Theory for the Age of Personal Control*. Oxford University Press, New York.

Peterson DA, Dickinson HA, Leppert JT, Lee KF, Gage FH (1999). Central neuronal loss and behavioral impairment in mice lacking neurotrophin receptor p75. *Journal of Comparative Neurology* **404**: 1–20.

Petit C (2001). Usher syndrome: From genetics to pathogenesis. *Annual Review of Genomics Human Genetics* **2**: 271–297.

Petkov PM, Ding Y, Cassell MA, Zhang W, Wagner G, Sargent EE, Asquith S, Crew V, Johnson KA, Robinson P, Scott VE, Wiles MV (2004). An efficient SNP system for mouse genomic scanning and elucidating strain relationships. *Genome Research* **14**: 1806–1811.

Petrulis A, Alvarez P, Eichenbaum H (2005). Neural correlates of social odor recognition and the representation of individual distinctive social odors within entorhinal cortex and ventral subiculum. *Neuroscience* **130**: 259–274.

Pettit HO, Ettenberg A, Bloom FE, Koob GF (1984). Destruction of dopamine in the nucleus accumbens selectively attenuates cocaine but not heroin self-administration in rats. *Psychopharmacology* **84**: 167–173.

Pfaff D, Frohlich J, Morgan M (2002). Hormonal and genetic influences on arousal—Sexual and otherwise. *Trends in Neurosciences* **25**: 45–50.

Pfaff DW, Schwartz-Giblin S (1988). Cellular mechanisms of female reproductive behaviors. In *The Physiology of Reproduction*, E Knobil, J Neill, Eds. Raven, New York, pp. 1487–1568.

Pfaus JG, Mendelson SD, Phillips AG (1990). A correlational and factor analysis of anticipatory and consummatory measures of sexual behavior in the male rat. *Psychoneuroendocrinology* **15**: 329–350.

Pfaus JG, Phillips AG (1991). Role of dopamine in anticipatory and consummatory aspects of sexual behavior in the male rat. *Behavioral Neuroscience* **105**: 725–741.

Pfeffer S, Zavolan M, Grasser FA, Chien M, Russo JJ, Ju J, John B, Enright AJ, Marks D, Sander C, Tuschl T (2004). Identification of virus-encoded microRNAs. *Science* **304**: 734–736.

Phelps SM, Lydon JP, O'Malley BW, Crews D (1998). Regulation of male sexual behavior by progesterone receptor, sexual experience and androgen. *Hormones and Behavior* **34**: 294–302.

Phillips AG, Fibiger HC (1989). Neuroanatomical bases of intracranial self-stimulation: Untangling the Gordian knot. In *The Neuropharmacological Basis of Reward*, JM Liebman, SJ Cooper, Eds. Oxford University Press, New York, pp. 66–103.

Phillips AG, LePiane FG, Fibiger HC (1983). Dopaminergic mediation of reward produced by direct injection of enkephalin into the ventraltegmental area of the rat. *Life Sciences* **33**: 2505–2511.

Phillips TJ, Belknap JK, Hitzemann RJ, Buck KJ, Cunningham CL, Crabbe JC (2002). Harnessing the mouse to unravel the genetics of human disease. *Genes, Brain and Behavior* **1**: 14–26.

Phillips TJ, Crabbe JC (1991). Behavioral studies of genetic differences in alcohol action. In *The Genetic Basis of Alcohol and Drug Action*, JC Crabbe, RA Harris, Eds. Plenum, New York, pp. 25–104.

Phillips TJ, Huson MG, McKinnon CS (1998). Localization of genes mediating acute and sensitized locomotor responses to cocaine in BXD/Ty recombinant inbred mice. *Journal of Neuroscience* **18**: 3023–3034.

Piazza PV, Le Moal M (1998). The role of stress in drug self-administration. *Trends in Pharmacological Sciences* **222**: 67–74.

Picciotto MR, Caldarone BJ, King SL, Zachariou V (2000). Nicotinic receptors in the brain: Links between molecular biology and behavior. *Neuropsychopharmacology* **22**: 451–465.

Picciotto MR, Zoli M, Rimondini R, Léna C, Marubio LM, Pich EM, Fuxe K, Changeux JP (1998). Acetylcholine receptors containing the β 2 subunit are involved in the reinforcing properties of nicotine. *Nature* **391**: 173–177.

Pich EM, Epping-Jordan MP (1998). Transgenic mice in drug dependence research. *Annals of Medicine* **30**: 390–396.

Pich EM, Heinrichs SC, Rivier C, Miczek KA, Fisher DA, Koob GF (1993). Blockade of pituitary-adrenal axis activation induced by peripheral immunoneutralization of corticotropin-releasing factor does not affect the behavioral response to social defeat stress in rats. *Psychoneuroendocrinology* **18**: 495–507.

Pick CG, Cheng J, Paul D, Pasternak GW (1991). Genetic influences in opioid analgesic sensitivity in mice. *Brain Research* **566**: 295–298.

Pick CG, Peter Y, Terkel J, Gavish M, Weizman R (1996). Effect of the neuroactive steroid alpha-THDOC on staircase test behavior in mice. *Psychopharmacology* **128**: 61–66.

Pickar D (1998). The dopamine hypothesis of schizophrenia—Love it or leave it. *Molecular Psychiatry* **3**: 101–102.

Pickar D, Breier A, Hsiao JK, Doran AR, Wolkowitz OM, Pato CN, Konicki PE, Potter WZ (1990). Cerebrospinal fluid and plasma monoamine metabolites and their relation to psychosis: Implications for regional brain dysfunction in schizophrenia. *Archives of General Psychiatry* **47**: 541–648.

Pickens R, Thompson T (1968). Cocaine reinforced behavior in rats: Effects of reinforcement magnitude and fixed ratio size. *Journal of Pharmacology and Experimental Therapeutics* **161**: 122–129.

Pieke-Dahl S, Ohlemiller KK, McGee J, Walsh EJ, Kimberling WJ (1997). Hearing loss in the RBF/DnJ mouse, a proposed animal model of Usher syndrome type IIa. *Hearing Research* **112**: 1–12.

Pierce C, Kalivas P (1997). Locomotor behavior. In *Current Protocols in Neuroscience*. Wiley, New York, pp. 8.1.1–8.1.8.

Pinto LH, Enroth-Cugell C (2000). Tests of the mouse visual system. *Mammalian Genome* **11**: 531–536.

Pinto LH, Vitaterna MH, Siepka SM, Shimomura K, Lumayag S, Baker M, Fenner D, Mullins RF, Sheffield VC, Stone EM, Heffron E, Takahashi JS (2004). Results from screening over 9000 mutation-bearing mice for defects in the electroretinogram and appearance of the fundus. *Vision Research* **44**: 3335–3345.

Pinto S, Roseberry AG, Liu H, Diano S, Shanabrough M, Cai X, Friedman JM, Horvath TL (2004). Rapid rewiring of arcuate nucleus feeding circuits by leptin. *Science* **304**: 110–115.

Pitcher GM, Ritchie J, Henry JL (1999). Paw withdrawal threshold in the von Frey hair test is influenced by the surface on which the rat stands. *Journal of Neuroscience Methods* **87**: 185–193.

Pittenger C, Huang YY, Paletzki RF, Bourtchouladze R, Scanlin H, Vronskaya S, Kandel ER (2002). Reversible inhibition of CREB/ATF transcription factors in region CA1 of the dorsal hippocampus disrupts hippocampus-dependent spatial memory. *Neuron* **34**: 447–462.

Piven J (2001). The broad autism phenotype: A complementary strategy for molecular genetics studies of autism. *American Journal of Medical Genetics Part B: Neuropsychiatric Genetics* **105**: 34–35.

Plomin R (1995). Molecular genetics and psychology. *Current Directions in Psychological Science* **4**: 114–117.

Plomin R, McClearn GE, Gora-Maslak G, Neiderhiser JM (1991a). An RI QTL cooperative data bank for recombinant inbred quantitative trait loci analysis. *Behavior Genetics* **21**: 97–98.

Plomin R, McClearn GE, Gora-Maslak G, Neiderhiser JM (1991b). Use of recombinant inbred strains to detect quantitative trait loci associated with behavior. *Behavior Genetics* **21**: 99–116.

Plomin R, Owen MJ, McGuffin P (1994). The genetic basis of complex human behavior. *Science* **264**: 1733–1739.

Pogue-Geile M, Ferrell R, Deka R, Debski T, Manuck S (1998). Human novelty-seeking personality traits and dopamine D4 receptor polymorphisms: A twin and genetic association study. *American Journal of Medical Genetics* **81**: 44–48.

Polleux F, Lauder JM (2004). Toward a developmental neurobiology of autism. *Mental Retardation and Developmental Disabilities Research Reviews* **10**: 303–317.

Pongrac J, Middleton FA, Lewis DA, Levitt P, Mirnics K (2002). Gene expression profiling with DNA microarrays: Advancing our understanding of psychiatric disorders. *Neurochemical Research* **27**: 1049–1063.

Ponomarev I, Schafer GL, Blednov YA, Williams RW, Iyer VR, Harris RA. Convergent analysis of cDNA and short oligomer microarrays, mouse null mutants and bioinformatics resources to study complex traits. *Genes, Brain and Behavior* **3**: 360–368.

Pontieri FE, Tanda G, Di Chiara G (1995). Intravenous cocaine, morphine, and amphetamine preferentially increase extracellular dopamine in the "shell" as compared to the "core" of the rat nucleus accumbens. *Proceedings of the National Academy of Sciences USA* **92**: 12304–12308.

Popp RA, Bailiff EG, Show LC, Johnson FM, Lewis SE (1983). Analysis of a mouse alpha-globin gene mutation induced by ethylnitrosourea. *Genetics* **105**: 157–167.

Porciatti V, Pizzorusso T, Maffei L (1999). The visual physiology of the wild type mouse determined with pattern VEPs. *Vision Research* **39**: 3071–3081.

Porrino LJ, Esposito RU, Seeger TF, Crane AM, Pert A, Sokoloff L (1984). Metabolic mapping of the brain during rewarding self-stimulation. *Science* **224**: 306–309.

Porsolt RD, Anton G, Blaver N, Jalfre M (1978). Behavioral despair in rats: A new model sensitive to antidepressant treatments. *European Journal of Pharmacology* **47**: 379–391.

Porsolt RD, Brossard G, Roux S (1998). Models of affective illness: Behavioral despair test in rodents. *Current Protocols in Neuroscience*. Wiley, New York, pp. 5.8.1–8.5.

Porsolt RD, Le Pichon M, Jalfre M (1977). Depression: A new animal model sensitive to antidepressant treatments. *Nature* **266**: 730–732.

Portela-Gomes GM, Hacker GW, Weitgasser R (2004). Neuroendocrine cell markers for pancreatic islets and tumors. *Applied Immunohistochemistry and Molecular Morphology* **12**: 183–192.

Posner MI, Raichle ME, Eds. (1998). The neuroimaging of human brain function. *Proceedings of the National Academy of Sciences USA*, Special Issue **95**: 763–929.

Post RM, Weiss RB, Pert A (1988). Cocaine-induced behavioral sensitization and kindling: Implications for the emergence of psychopathology and seizures. In *The Mesocorticolimbic Dopamine System*, PW Kalivas, CB Nemeroff, Eds. *Annals of the New York Academy of Sciences* **537**: 292–308.

Potter H (1995). Transfection by electroporation. In *Current Protocols in Molecular Biology*, Vol. 1, FM Ausubel, R Brent, RE Kingston, DD Moore, JG Seidman, JA Smith, K Struhl, Eds. Wiley, New York, pp. 9.3.1–9.3.6.

Price DL, Sisodia SS (1998). Mutant genes in familial Alzheimer's disease and transgenic models. *Annual Review of Neuroscience* **21**: 479–505.

Price DL, Sisodia SS, Borchelt DR (1998). Genetic neurodegenerative diseases: The human illness and transgenic models. *Science* **282**: 1079–1083.

Price EO (1991). Practical considerations in the measurement of sexual behavior. *Methods in Neuroscience* **15**: 16–31.

Primeaux SD, Holmes PV (2000). Olfactory bulbectomy increases met-enkephalin- and neuropeptide-Y-like immunoreactivity in rat limbic structures. *Pharmacology Biochemistry and Behavior* **67**: 331–337.

Proust M (1956). *Remembrance of Things Past: Swann's Way*, translated by CK Scott Moncrieff. Modern Library, New York, p. 65.

Provencio I, Wong S, Lederman AB, Argamaso SM, Foster RG (1994). Visual and circadian responses to light in aged retinally degenerate mice. *Vision Research* **34**: 1799–1806.

Pruhs RJ, Quock RM (1989). Interaction between nitrous oxide and diazepam in the mouse staircase test. *Anesthesia and Analgesia* **68**: 501–505.

Puttaparthi K, Gitomer WL, Krishnan U, Son M, Rajendran B, Elliott JL (2002). Disease progression in a transgenic model of familial amyotrophic lateral sclerosis is dependent on both neuronal and non-neuronal zinc binding proteins. *Journal of Neuroscience* **22**: 8790–8796.

Quartermain D, McEwen BS (1970). Temporal characteristics of amnesia induced by protein synthesis inhibitor: Determination by shock level. *Nature* **228**: 677–678.

Qin M, Kang J, Smith CB (2002). Increased rates of cerebral glucose metabolism in a mouse model of fragile X mental retardation. *Proceedings of the National Academy of Sciences USA* **99**: 15758–15763.

Quirk GJ, Armony JL, LeDoux JE (1997). Fear conditioning enhances different temporal components of time-evoked spike trains in auditory cortex and lateral amygdala. *Neuron* **19**: 613–624.

Quock RM, Emmanouil DE, Vaughn LK, Pruhs RJ (1992). Benzodiazepine receptor mediation of behavioral effects of nitrous oxide in mice. *Psychopharmacology* **107**: 310–314.

Quock RM, Mueller JL, Vaughn LK, Belknap JK (1996). Nitrous oxide antinociception in BXD recombinant inbred mouse strains and identification of quantitative trait loci. *Brain Research* **725**: 23–29.

Rada PV, Mark GP, Hoebel BG (1998). Dopamine release in the nucleus accumbens by hypothalamic stimulation-escape behavior. *Brain Research* **782**: 228–234.

Radcliffe RA, Erwin VG, Wehner JM (1998). Acute functional tolerance to ethanol and fear conditioning are genetically correlated in mice. *Alcoholism: Clinical and Experimental Research* **22**: 1673–1679.

Radcliffe RA, Lowe MV, Wehner JM (2000). Confirmation of contextual fear conditioning QTLs by short-term selection. *Behavior Genetics* **30**: 183–191.

Rader N, Bausano M, Richards JE (1980). On the nature of the visual-cliff avoidance response in human infants. *Child Development* **51**: 61–68.

Rainbow TC, Hoffman PL, Flexner LB (1980). Studies of memory: A reevaluation in mice of the effects of inhibitors on the rate of synthesis of cerebral proteins as related to amnesia. *Pharmacology Biochemistry and Behavior* **12**: 79–84.

Rainbow T, Parsons B, McEwen BS (1982). Sex differences in rat brain oestrogen and progestin receptors. *Nature* **300**: 648–649.

Rajewsky K, Gu H, Kuhn R, Betz UA, Muller W, Roes J, Schwenk F (1998). Conditional gene targeting. *Journal of Clinical Investigation* **98**: 600–603.

Ralph RJ, Paulus MP, Fumagalli F, Caron MG, Geyer MA (2001). Prepulse inhibition deficits and perseverative motor patterns in dopamine transporter knock-out mice: Differential effects of D1 and D2 receptor antagonists. *Journal of Neuroscience* **27**: 305–313.

Ralph RJ, Varty GB, Kelly MA, Wang YM, Caron MG, Rubinstein M, Grandy DK, Low MJ, Geyer MA (1999). The dopamine D2, but not D3 or D4, receptor subtype is essential for the disruption of prepulse inhibition produced by amphetamine in mice. *Journal of Neuroscience* **19**: 4627–4633.

Ralph-Williams RJ, Lehmann-Masten V, Otero-Corchon V, Low MJ, Geyer MA (2002). Differential effects of direct and indirect dopamine agonists on prepulse inhibition: A study in D1 and D2 receptor knock-out mice. *Journal of Neuroscience* **22**: 9604–9611.

Ramboz S, Oosting R, Amara DA, Kung HF, Blier P, Mendelsohn M, Mann JJ, Brunner D, Hen R (1998). Serotonin receptor 1A knockout: An animal model of anxiety-related disorder. *Proceedings of the National Academy of Sciences USA* **95**: 14476–14481.

Ramos A, Mellerin Y, Mormede P, Chaouloff F (1998). A genetic and multifactorial analysis of anxiety-related behaviours in Lewis and SHR intercrosses. *Behavioral Brain Research* **96**: 195–205.

Rao VSN, Santos FA, Paula WG, Silva RM, Campos AR (1999). Effects of acute and repeated dose administration of caffeine and pentoxifylline on diazepam-induced mouse behavior in the hole-board test. *Psychopharmacology* **144**: 61–66.

Raphael Y, Kobayashi KN, Dootz GA, Beyer LA, Dolan EF, Burmeister M (2001). Severe vestibular and auditory impairment in three alleles of Ames waltzer (av) mice. *Hearing Research* **151**: 237–249.

Rasmussen K (1995). The role of the locus coeruleus and N-methyl-D-aspartic acid (NMDA). and AMPA receptors in opiate withdrawal. *Neuropsychopharmacology* **13**: 295–300.

Raud S, Runkorg K, Veraksits A, Reimets A, Nelovkov A, Abramov U, Matsui T, Bourin M, Volke V, Koks S, Vasar E (2003). Targeted mutation of CCK2 receptor gene modifies the behavioural effects of diazepam in female mice. *Psychopharmacology* **168**: 417–425.

Reddy PH, Williams M, Charles V, Garrett L, Pike-Buchanan L, Whetsell WO, Miller G, Tagle DA (1998). Behavioural abnormalities and selective neuronal loss in HD transgenic mice expressing mutated full-length HD cDNA. *Nature Genetics* **20**: 198–202.

Redrobe JP, Bourin M (1998). Dose-dependent influence of buspirone on the activities of selective serotonin reuptake inhibitors in the mouse forced swimming test. *Psychopharmacology* **138**: 198–206.

Redrobe JP, Dumont Y, Fournier A, Quirion R (2002). The neuropeptide Y (NPY) Y1 receptor subtype mediates NPY-induced antidepressant-like activity in the mouse forced swimming test. *Neuropsychopharmacology* **26**: 615–624.

Redrobe JP, Dumont Y, Herzog H, Quirion R (2003). Neuropeptide Y (NPY) Y2 receptors mediate behaviour in two animal models of anxiety: Evidence from Y2 receptor knockout mice. *Behavioural Brain Research* **141**: 251–255.

Reed DR, Li S, Li X, Huang L, Tordoff MG, Starling-Roney R, Taniguchi K, West DB, Ohmen JD, Beauchamp GK, Bachmanov AA (2004). Polymorphisms in the taste receptor gene (Tas1r3) region are associated with saccharin preference in 30 mouse strains. *Journal of Neuroscience* **24**: 938–946.

Reeves RH, Baxter LI, Richtsmeier JT (2001). Too much of a good thing: Mechanisms of gene action in Down syndrome. *Trends in Genetics* **17**: 83–88.

Reeves RH, Irving NG, Moran TH, Wohn A, Kitt C, Sisodia SS, Schmidt C, Bronson RT, Davisson MT (1995). A mouse model for Down syndrome exhibits learning and behaviour deficits. *Nature Genetics* **11**: 177–184.

Reich T, Edenberg HJ, Goate A, Williams JT, Rice JP, Van Eerdewegh P, Foroud T, Hesselbrock V, Schuckit MA, Bucholz K, Porjesz B, Li TK, Conneally PM, Nurnberger JI, Tishcfield JA, Crowe RR, Cloninger CR, Wu W, Shears S, Carr K, Crose C, Willig C, Begleiter H (1998). Genome-wide search for genes affecting the risk for alcohol dependence. *American Journal of Medical Genetics* **81**: 207–215.

Reidelberger RD, Varga G, Solomon TE (1991). Effects of selective cholecystokinin antagonists L364,718 and L365,260 on food intake in rats. *Peptides* **12**: 1215–1221.

Reilly PR, Page DC (1998). We're off to see the genome. *Nature Genetics* **20**: 15–17.

Reimers M (2005). Statistical analysis of microarray data. *Addiction Biology* **10**: 23–35.

Ren-Patterson RF, Cochran LW, Holmes A, Sherrill S, Huang SJ, Tolliver T, Lesch KP, Lu B, Murphy DL (2005). Loss of brain-derived neurotrophic factor gene allele exacerbates brain monoamine deficiencies and increases stress abnormalities of serotonin transporter knockout mice. *Journal of Neuroscience Research* **79**: 756–771.

Ribeiro Do Couto B, Aguilar MA, Manzanedo C, Rodriguez-Arias M, Minarro J (2003). Reinstatement of morphine-induced conditioned place preference in mice by priming injections. *Neural Plasticity* **10**: 279–290.

Ricceri L, Alleva E, Chiarotti F, Calamandrei G (1996). Nerve growth factor affects passive avoidance learning and retention in developing mice. *Brain Research Bulletin* **39**: 219–226.

Ricceri L, Berger-Sweeney J (1998). Postnatal choline supplementation in preweanling mice: Sexually dimorphic behavioral and neurochemical effects. *Behavioral Neuroscience* **112**: 1387–1392.

Rice OV, Gordon N, Gifford AN (2002). Conditioned place preference to morphine in cannabinoid CB1 receptor knockout mice. *Brain Research* **945**: 135–138.

Richardson JD, Aanonsen L, Hargreaves KM (1998). Antihyperalgesic effects of spinal cannabinoids. *European Journal of Pharmacology* **345**: 145–153.

Richichi C, Lin EJ, Stefanin D, Colella D, Ravizza T, Grignaschi G, Veglianese P, Sperk G, During MJ, Vezzani A (2004). Anticonvulsant and antiepileptogenic effects mediated by adeno-associated virus vector neuropeptide Y expression in the rat hippocampus. *Journal of Neuroscience* **24**: 3051–3059.

Richter C (1922). A behavioristic study of the activity of the rat. *Comparative Psychology Monograph* **1**: 1–55.

Richter CP (1967). Sleep and activity: Their relation to the 24-hour clock. *Research Publications of the Association for Research on Nervous and Mental Disorders* **45**: 8–29.

Richter-Levin G, Canevari L, Bliss TV (1995). Long-term potentiation and glutamate release in the dentate gyrus: Links to spatial learning. *Behavioral Brain Research* **66**: 37–40.

Ridet JL, Corti O, Pencalet P, Hanoun N, Hamon M, Philippon J, Mallet J (1999). Toward autologous *ex vivo* gene therapy for the central nervous system with human adult astrocytes. *Human Gene Therapy* **10**: 217–280.

Riley AL (1998). Conditioned flavor aversions: Assessment of drug-induced suppression of intake. In *Current Protocols in Neuroscience*, JN Crawley, CR Gerfen, R McKay, MW Rogawski, DR Sibley, P Skolnick, Eds. Wiley, New York, pp. 8.6E.1–8.6E.10.

Riley AL, Freeman FB (2004). Conditioned taste aversion: A database. *Pharmacology, Biochemistry and Behavior* **77**: 655–656.

Rinchik EM, Flaherty L, Russell LB (1993). High-frequency induction of chromosomal rearrangements in mouse germ cells by the chemotherapeutic agent chlorambucil. *Bioessays* **15**: 831–836.

Rinchik EM, Johnson DK, Margolis FL, Jackson IJ, Russell LLB, Carpenter DA (1991). Reverse genetics in the mouse and its application to the study of deafness. *Annals of the New York Academy of Sciences* **630**: 80–92.

Ripley TL, Gadd CA, De Felipe C, Hunt SP, Stephens DN (2002). Lack of self-administration and behavioural sensitisation to morphine, but not cocaine, in mice lacking NK1 receptors. *Neuropharmacology* **43**: 1258–1268.

Ripley TL, Rocha BA, Oglesby MW, Stephens DW (1999). Increased sensitivity to cocaine, and over-responding during cocaine self-administration in tPA knockout mice. *Brain Research* **826**: 117–127.

Ripoll N, David DJ, Dailley E, Hascoet M, Bourin M (2003). Antidepressant-like effects in various mice strains in the tail suspension test. *Behavioural Brain Research* **143**: 193–200.

Risch N (1992). Genetic linkage: interpreting lod scores. *Science* **255**: 803–804.

Risinger FO, Cunningham CL (2000). DBA/2J develop stronger lithium chloride-induced conditioned taste and place aversions than C57BL/6J mice. *Pharmacology Biochemistry and Behavior* **67**: 17–24.

Risinger FO, Freeman PA, Greengard P, Fienberg AA (2001). Motivational effects of ethanol in DARPP-32 knock-out mice. *Journal of Neuroscience* **21**: 340–348.

Rissman EF, Early AH, Taylor JA, Korach KS, Lubahn DB (1997a). Estrogen receptors are essential for female sexual receptivity. *Endocrinology* **138**: 507–510.

Rissman EF, Wersinger SR, Fugger HN, Foster TC (1999). Sex with knockout models: behavioral studies of estrogen receptor alpha. *Brain Research* **835**: 80–90.

Rissman EF, Wersinger SR, Taylor JA, Lubahn DB (1997b). Estrogen receptor function as revealed by knockout studies: Neuroendocrine and behavioral aspects. *Hormones and Behavior* **31**: 232–243.

Ritter RC (1984). The role of the rat in research on control of ingestion. *Journal of Animal Science* **59**: 1373–1380.

Robbins TW (1996). Refining the taxonomy of memory. *Science* **273**: 1353–1354.

Robbins TW, Everitt BJ (1996). Neurobehavioural mechanisms of reward and motivation. *Current Opinion in Neurobiology* **6**: 228–236.

Robbins TW, Everitt BJ, Marston HM, Wilkinson J, Jones GH, Page KJ (1989). Comparative effects of ibotenic acid- and quisqualic acid-induced lesions of the substantia innominata on attentional function in the rat: Further implications for the role of the cholinergic neurons of the nucleus basalis in cognitive processes. *Behavioural Brain Research* **35**: 221–240.

Roberts AJ, Finn DA, Phillips TJ, Belknap JK, Keith LD (1995). Genetic analysis of the corticosterone response to ethanol in BXD recombinant inbred mice. *Behavioral Neuroscience* **109**: 1199–1208.

Roberts AJ, McDonald JS, Heyser CJ, Kieffer BL, Matthes HW, Koob GF, Gold LH (2000). mu-Opioid receptor knockout mice do not self-administer alcohol. *Journal of Pharmacology and Experimental Therapeutics* **293**: 1002–1008.

Robertson B, Xu XJ, Hao JX, Wiesenfeld-Hallin Z, Mhlanga J, Grant G, Kristensson K (1997a). Interferon-γ receptors in nociceptive pathways: Role in neuropathic pain-related behaviour. *NeuroReport* **8**: 1311–1316.

Robertson NG, Skvorak AB, Yin Y, Weremowicz S, Johnson KR, Kovatch KA, Battey JF, Bieber FR, Morton CC (1997b). Mapping and characterization of a novel cochlear gene in human and in mouse: A positional candidate gene for a deafness disorder, DFNA9. *Genomics* **46**: 345–354.

Robinson IC (1986). Oxytocin and the milk-ejection reflex. In *Neurobiology of Oxytocin*, D Ganten, D Pfaff, Eds. *Current Topics in Neuroendocrinology*, Vol. 6, Springer, Heidelberg, pp. 153–172.

Robinson JK, Crawley JN (1993). Intraventricular galanin impairs delayed nonmatching-to-sample performance in rats. *Behavioral Neuroscience* **107**: 458–467.

Robinson TE, Becker JB (1986). Enduring changes in brain and behavior produced by chronic amphetamine administration: A review and evaluation of animal models of amphetamine psychosis. *Brain Research Reviews* **11**: 157–198.

Robinson TE, Camp DM (1990). Does amphetamine preferentially increase the extracellular concentration of dopamine in the mesolimbic system of freely moving rats? *Neuropsychopharmacology* **3**: 163–173.

Rocha BA, Fumagalli F, Gainetdinov RR, Jones SR, Ator R, Giros B, Miller GW, Caron MC (1998a). Cocaine self-administration in dopamine-transporter knockout mice. *Nature Neuroscience* **1**: 132–137.

Rocha BA, Goulding EH, O'Dell LE, Mead AN, Coufal NG, Parsons LH, Tecott LH (2002). Enhanced locomotor, reinforcing, and neurochemical effects of cocaine in serotonin 5-hydroxytryptamine 2C receptor mutant mice. *Journal of Neuroscience* **22**: 10039–10045.

Rocha BA, Scearce-Levie K, Lucas JJ, Hiroi N, Castanon N, Crabbe JC, Nestler EJ, Hen R (1998b). Increased vulnerability to cocaine in mice lacking the serotonin-1B receptor. *Nature* **393**: 175–178.

Rodd-Henricks ZA, McKinzie DL, Li TK, Murphy JM, McBride WJ (2002). Cocaine is self-administered into the shell but not the core of the nucleus accumbens of Wistar rats. *Journal of Pharmacology and Experimental Therapeutics* **303**: 1216–1226.

Rodgers RJ (1997). Animal models of 'anxiety': Where next? *Behavioural Pharmacology* **8**: 477–496.

Rodgers RJ, Davies B, Shore R. Absence of anxiolytic response to chlordiazepoxide in two common background strains exposed to the elevated plus-maze: Importance and implications of behavioural baseline (2002). *Genes, Brain and Behavior* **1**: 242–251.

Rodgers RJ, Halford JC, Nunes de Souza RL, Canto de Souza AL, Piper DC, Arch JR, Blundell JE (2000). Dose-response effects of orexin-A on food intake and the behavioural satiety sequence in rats. *Regulatory Peptides* **96**: 71–84.

Rodgers RJ, Ishii Y, Halford JCG, Blundell JE (2002). Orexins and appetite regulation. *Neuropeptides* **36**: 303–325.

Rodgers RJ, Johnson NJ (1995). Factor analysis of spatiotemporal and ethological measures in the murine elevated plus-maze test of anxiety. *Pharmacology Biochemistry and Behavior* **52**: 297–303.

Rodriguez de Fonseca F, Rubio P, Menzaghi F, Merlo-Pich E, Rivier J, Koob GF, Navarro M (1996). Corticotropin-releasing factor (CRF) antagonist [D-Phe12,Nle21,38,C alpha MeLeu37]CRF attenuates the acute actions of the highly potent cannabinoid receptor agonist HU-210 on defensive-withdrawal behavior in rats. *Journal of Pharmacology and Experimental Therapeutics* **276**: 56–64.

Rogers DC, Fisher EM, Brown SD, Peters J, Hunter AJ, Martin JE (1997). Behavioral and functional analysis of mouse phenotype: SHIRPA, a proposed protocol for comprehensive phenotype assessment. *Mammalian Genome* **8**: 711–713.

Rohrer DC, Nilaver G, Nipper V, Machida CA (1996). Genetically modified PC12 brain grafts: Survivability and inducible nerve growth factor expression. *Cell Transplant.* **5**: 57–68.

Roman FS, Marchetti E, Bouquerel A, Soumireu-Mourat B (2002). The olfactory tubing maze: A new apparatus for studying learning and memory processes in mice. *Journal of Neuroscience Methods* **117**: 173–181.

Rondi-Reig L, Dubreuil YL, Martinou JC, Delhaye-Bouchaud N, Caston J, Mariani J (1997). Fear decrease in transgenic mice over-expressing bcl-2 in neurons. *NeuroReport* **8**: 2429–2432.

Rondi-Reig L, Lemaigre-Dubreuil Y, Montecot C, Muller D, Martinou JC, Caston J, Mariani J (2001a). Transgenic mice with neuronal overexpression of bcl-2 gene present navigation disabilities in a water task. *Neuroscience* **104**: 207–215.

Rondi-Reig L, Libbey M, Eichenbaum H, Tonegawa S (2001b). CA-specific *N*-methyl-D-aspartate receptor knockout mice are deficient in solving a nonspatial transverse patterning task. *Proceedings of the National Academy of Sciences USA* **98**: 3543–3548.

Rosen JB, Schulkin J (1998). From normal fear to pathological anxiety. *Psychological Review* **105**: 325–350.

Rosenberg RN, Morris A, Zale W, Prusiner S, DiMauro S, Barchi RL, Nestler EJ (2003). *Molecular and Genetic Basis of Neurologic and Psychiatric Disease*. Butterworth Heinemann, Stoneham, MA.

Rosenblatt JS, Siegal HI, Mayer AD (1979). Progress in the study of maternal behavior in the rat: Hormonal, nonhormonal, sensory, and developmental aspects. *Advances in the Study of Behavior* **10**: 226–302.

Rosi S, Ramirez-Amaya V, Vazdarjanova A, Worley PF, Barnes CA, Wenk GL (2005). Neuroinflammation alters the hippocampal pattern of behaviorally induced Arc expression. *Journal of Neuroscience* 25: 723–731.

Roth JA (1998). Restoration of tumour suppressor gene expression for cancer. *Forum (Genova)* **8**: 368–376.

Roth RH, Tam SY, Ida Y, Yang JX, Deutch AY (1988). Stress and the mesocorticolimbic dopamine systems. In *The Mesocorticolimbic Dopamine System*, PW Kalivas, CB Nemeroff, Eds. *Annals of the New York Academy of Sciences* **537**: 138–147.

Rothman RB, Baumann MH (2003). Monoamine transporters and psychostimulant drugs. *European Journal of Pharmacology* **479**: 23–40.

Rothstein JD (2003). Of mice and men: Reconciling preclinical ALS mouse studies and human clinical trials. *Annals of Neurology* **53**: 423–426.

Roubertoux PL, Carlier M, Degrelle G, Haas-Dupertuis MC, Phillips J, Moutier R (1994). Co-segregation of intermale aggression with the pseudoautosomal region of the Y chromosome in mice. *Genetics* **136**: 225–230.

Roubertoux P, Semal C, Ragueneau S (1985). Early development in mice: II. Sensory motor behavior and genetic analysis. *Physiology and Behavior* **35**: 659–666.

Rouge-Pont F, Deroche V, Le Moal M, Piazza PV (1998). Individual differences in stress-induced dopamine release in the nucleus accumbens are influenced by corticosterone. *European Journal of Neuroscience* **10**: 3903–3907.

Roullet P, Sara S (1998). Consolidation of memory after its reactivation: Involvement of beta noradrenergic receptors in the late phase. *Neural Plasticity* **6**: 63–68.

Rowland LP, Shneider NA (2001). Amyotrophic lateral sclerosis. *New England Journal of Medicine* **344**: 1688–1700.

Rowland NE, Kalra SP (1997). Potential role of neuropeptide ligands in the treatment of overeating. *CNS Drugs* **6**: 419–426.

Rowlett JK, Tornatzky W, Cook JM, Ma C, Miczek KA (2001). Zolpidem, triazolam, and diazepam decrease distress vocalizations in mouse pups: Differential antagonism by flumazenil and beta-carboline-3-carboxylate-*t*-butyl ester (beta-CCt). *Journal of Pharmacology and Experimental Therapeutics* **297**: 247–253.

Royle SJ, Collins FC, Rupniak HT, Barnes VC, Anderson R (1999). Behavioural analysis and susceptibility to CNS injury of four inbred strains of mice. *Brain Research* **816**: 337–349.

Rubinstein M, Mogil JS, Japon M, Chan EC, Allen RG, Low MJ (1996). Absence of opioid-stress-induced analgesia in mice lacking beta-endorphin by site-directed mutagenesis. *Proceedings of the National Academy of Sciences USA* **30**: 3995–4000.

Rubinstein M, Phillips TJ, Bunzow JR, Falzone TL, Dziewczapolski G, Zhang G, Fang Y, Larson JL, McDougall JA, Chester JA, Saez C, Pugsley TA, Gershanik O, Low MJ, Grandy DK (1997). Mice lacking dopamine D4 receptors are supersensitive to ethanol, cocaine, and methamphetamine. Cell **90**: 991–1001.

Rudolph U, Mohler H (2004). Analysis of GABAA receptor function and dissection of the pharmacology of benzodiazepines and general anesthetics through mouse genetics. *Annual Review of Pharmacology and Toxicology* **44**: 475–498.

Rudy JW, Cheatle MD (1977). Odor-aversion learning in neonatal rats. *Science* **198**: 845–846.

Ruitenberg MJ, Blits B, Dijkhuizen PA, te Beek ET, Bakker A, van Heerikhuize JJ, Pool CW, Hermens WT, Boer GJ, Verhaagen J (2004). Adeno-associated viral vector-mediated gene transfer of brain-derived neurotrophic factor reverses atrophy of rubrospinal neurons following both acute and chronic spinal cord injury. *Neurobiology of Disease* **15**: 394–406.

Rushing PA, Houpt T, Henderson RP, Gibbs J (1997). High lick rate is maintained throughout spontaneous liquid meals in freely feeding rats. *Physiology and Behavior* **62**: 1185–1188.

Russell DW, Hirata RK (1998). Human gene targeting by viral vectors. *Nature Genetics* **18**: 325–330.

Russell JA, Leng G (1998). Sex, parturition and motherhood without oxytocin? *Journal of Endocrinology* **157**: 343–359.

Russell LB, Hunsicker PR, Cacheiro NL, Rinchik EM (1992). Genetic, cytogenetic, and molecular analyses of mutations induced by melphalan demonstrate high frequencies of heritable deletions and other rearrangements from exposure of postspermatogonial stages of the mouse. *Proceedings of the National Academy of Sciences USA* **89**: 6182–6186.

Russell LB, Rinchik EM (1993). Structural differences between specific-locus mutations induced by different exposure regimes in mouse spermatogonial stem cells. *Mutation Research* **288**: 187–195.

Russell LB, Russell WL, Rinchik EM, Hunsicker PR (1990). Factors affecting the nature of induced mutations. In *Biology of Mammalian Germ Cell Mutagenesis*, JW Allen, JW Bridges, BA Lyon, MF Moses, LB Russell, Eds. Cold Spring Harbor Press, Cold Spring Harbor, NY, pp. 271–289.

Rustay NR, Wahlsten D, Crabbe JC (2003). Assessment of genetic susceptibility to ethanol intoxication in mice. *Proceedings of the National Academy of Sciences USA* **100**: 2917–2922.

Rustay NR, Wrenn CC, Kinney JW, Holmes A, Bailey KR, Sullivan TL, Harris AP, Long KC, Saavedra MC, Starosta G, Innerfield CE, Yang RJ, Dreiling JL, Crawley JN (2005). Galanin impairs performance on learning and memory tasks: findings from galanin transgenic and GAL-R1 knockout mice. *Neuropeptides* **39**: 239–243.

Sackler AM, Weltman AS (1985). Effects of methylphenidate on whirler mice: An animal model for hyperkinesis. *Life Sciences* **37**: 425–431.

Sage RD (1981). Wild mice. In *The Mouse in Biomedical Research*, HL Foster, JD Small, JG Fox, Eds. Academic Press, New York, pp. 40–90.

Sago H, Carlson EJ, Smith DJ, Kilbridge J, Rubin EM, Mobley WC, Epstein CJ, Huang TT (1998). Ts1Cje, a partial trisomy 16 mouse model for Down syndrome, exhibits learning and behavioral abnormalities. *Proceedings of the National Academy of Sciences USA* **95**: 6256–6261.

Sago H, Carlson EJ, Smith DJ, Rubin EM, Crnic LS, Huang TT, Epstein CJ (2000). Genetic dissection of region associated with behavioral abnormalities in mouse models for Down syndrome. *Pediatric Research* **48**: 606–613.

Sainsbury A, Schwarzer C, Couzens M, Fetissov S, Furtinger S, Jenkins A, Cox HM, Sperk G, Hokfelt T, Herzog H (2002). Important role of hypothalamic Y2 receptors in body weight regulation revealed in conditional knockout mice. *Proceedings of the National Academy of Sciences USA* **99**: 8938–8943.

Sakaguchi S, Katamine S, Nishida N, Moriuchi R, Shigematsu K, Sugimoto T, Nakatani A, Kataoka Y, Houtani T, Shirabe S, Okada H, Hasegawa S, Miyamoto T, Noda T (1996). Loss of cerebellar Purkinje cells in aged mice homozygous for a disrupted PrP gene. *Nature* **380**: 528–531.

Sakai N, Thome J, Newton SS, Chen J, Kelz MB, Steffen C, Nestler EJ, Duman RS (2002). Inducible and brain region-specific CREB transgenic mice. *Molecular Pharmacology* **61**: 1453–1464.

Sakurai T (2005). Reverse pharmacology of orexin: From an orphan GPCR to integrative physiology. *Regulatory Peptides* **126**: 3–10.

Sakurai T, Amemiya A, Ishii M, Matsuzaki I, Chemelli RM, Tanaka H, Williams SC, Richardson JA, Koslowski GP, Wilson S, Arch JRS, Buckingham RE, Haynes AC, Carr SA, Annan RS, McNulty DE, Liu WS, Terrett JA, Eishourbagy NA, Bergsma DJ, Yanagisawa M (1998). Orexins and orexin receptors: a family of hypothalamic neuropeptides and G protein-coupled receptors that regulate feeding behavior. *Cell* **92**: 573–585.

Salamone JD (1994). The involvement of nucleus accumbens dopamine in appetitive and aversive motivation. *Behavioral Brain Research* **61**: 117–133.

Salamone JD, Correa M (2002). Motivational views of reinforcement: Implications for understanding the behavioral functions of nucleus accumbens dopamine. *Behavioural Brain Research* **137**: 3–25.

Salamone JD, Cousins MS, McCullough LD, Carriero DL, Berkowitz RJ (1994). Nucleus accumbens dopamine release increases during instrumental lever pressing for food but not free food consumption. *Pharmacology Biochemistry and Behavior* **49**: 25–31.

Salas R, Pieri F, Fung B, Dani JA, De Biasi M (2003). Altered anxiety-related responses in mutant mice lacking the beta4 subunit of the nicotinic receptor. *Journal of Neuroscience* **23**: 6255–6263.

Salinger WL, Ladrow P, Wheeler C (2003). Behavioral phenotype of the Reeler mutant mouse: Effects of *Reln* gene dosage and social isolation. *Behavioral Neuroscience* **117**: 1257–1275.

Sallinen J, Haapalinna A, Viitamaa T, Kobilka BK, Scheinin M (1998). Adrenergic α_{2C}-receptors mediate the acoustic startle reflex, prepulse inhibition, and aggression in mice. *Journal of Neuroscience* **18**: 3035–3042.

Sampson SB, Higgins DC, Elliott RW, Taylor BA, Lueders KK, Koza RA, Paigen B (1998). An edited linkage map for the AXB and BXA recombinant inbred mouse strains. *Mammalian Genome* **9**: 688–694.

Sams-Dodd F, Lipska BK, Weinberger DR (1997). Neonatal lesions of the rat ventral hippocampus result in hyperlocomotion and deficits in social behaviour in adulthood. *Psychopharmacology* **132**: 303–310.

Sanchis-Segura C, Cline BH, Marsicano G, Lutz B, Spanagel R (2004). Reduced sensitivity to reward in CB1 knockout mice. *Psychopharmacology* **176**: 223–232.

Sandberg R, Yasuda R, Pankratz DG, Carter TA, Del Rio JA, Wodicka L, Mayford M, Lockhart DJ, Barlow C (2000). Regional and strain-specific gene expression mapping in the adult mouse brain. *Proceedings of the National Academy of Sciences USA* **97**: 11038–11043.

Sanders MJ, Wiltgen BJ, Fanselow Ms (2003). The place of the hippocampus in fear conditioning. *European Journal of Pharmacology* **28**: 217–223.

Sandhoff K, Harzer K, Wassle W, Jatzkewitz H (1971). Enzyme alterations and lipid storage in three variants of Tay-Sachs disease. *Journal of Neurochemistry* **18**: 2469–2489.

Sango K, McDonald MP, Crawley JN, Mack ML, Tifft CJ, Skop E, Starr CM, Hoffmann A, Sandhoff K, Suzuki K, Proia RL (1996). Mice lacking both subunits of lysosomal beta-hexosaminidase display gangliosidosis and mucopolysaccharidosis. *Nature Genetics* **14**: 348–352.

Sango K, Yamanaka S, Hoffmann A, Okuda Y, Grinberg A, Westphal H, McDonald MP, Crawley JN, Sandhoff K, Suzuki K, Proia RL (1995). Mouse models of Tay-Sachs and Sandhoff diseases differ in neurological phenotype and ganglioside metabolism. *Nature Genetics* **11**: 170–176.

Sarna JR, Dyck RH, Whishaw IQ. (2000). The Dalila effect: C57BL6 mice barber whiskers by plucking. *Behavioural Brain Research* **108**: 39–45.

Sarter M (1987). Measurement of cognitive abilities in senescent animals. *International Journal of Neuroscience* **32**: 765–774.

Sarter M, Hagan J, Dudchenko P (1992). Behavioral screening for cognitive enhancers: From indiscriminate to valid testing. *Psychopharmacology* **107**: 461–473.

Saudou F, Amara DA, Dierich A, LeMeur M, Ramboz S, Segu L, Buhot MC, Hen R (1994). Enhanced aggressive behavior in mice lacking 5-HT$_{1B}$ receptor. *Science* **265**: 1875–1878.

Sauer B (1998). Inducible gene targeting in mice using the Cre/lox system. *Methods* **14**: 381–392.

Sauerberg P, Jeppesen L, Olesen PH, Rasmussen T, Swedberg MD, Sheardown MJ, Fink-Jensen A, Thomsen C, Thogersen H, Rimvall K, Ward JS, Calligaro DO, DeLapp NW, Bymaster FP, Shannon HE (1998). Muscarinic agonists with antipsychotic-like activity: Structure-activity relationships of 1,2,5-thiadiazole analogues with functional dopamine antagonist activity. *Journal of Medicinal Chemistry* **41**: 4378–4384.

Saura CA, Choi SY, Beglopoulos V, Malkani S, Zhang D, Shankaranarayana R, Chattarji S, Kelleher RJ, Kandel ER, Duff K, Kirkwood A, Shen J (2004). Loss of presenilin function causes impairments of memory and synaptic plasticity followed by age-dependent neurodegeneration. *Neuron* **42**: 23–36.

Savonenko AV, Xu GM, Price DL, Borchelt DR, Markowska AL (2003). Normal cognitive behavior in two distinct congenic lines of transgenic mice hyperexpressing mutant APP SWE. *Neurobiology of Disease* **12**: 194–211.

Sawchenko PE, Swanson LW, Grzanna R, Howe PRC, Polak JM, Bloom SR (1985). Co-localization of neuropeptide Y immunoreactivity in brainstem catecholaminegic neurons that project to the paraventricular nucleus of the hypothalamus. *Journal of Comparative Neurology* **241**: 138–153.

Sayah DM, Khan AH, Gasperoni TL, Smith DJ (2000). A genetic screen for novel behavioral mutations in mice. *Molecular Psychiatry* **5**: 369–377.

Schafe GE, Stein PL, Park CR, Bernstein IL (1996). Taste aversion learning in *fyn* mutant mice. *Behavioral Neuroscience* **110**: 845–848.

Schatzberg AF, Samson JA, Rothschild AJ, Bond TC, Reiger DA (1998). McLean Hospital depression research facility: Early-onset phobic disorders and adult-onset major depression. *British Journal of Psychiatry* **34**(Suppl): 29–34.

Schauwecker PE, Steward O (1997). Genetic determination of susceptibility to excitotoxic cell death: Implications for gene targeting approaches. *Proceedings of the National Academy of Sciences USA* **94**: 4103–4108.

Schechter MD, Glennon RA (1985). Cathinone, cocaine and methamphetamine: Similarity of behavioral effects. *Pharmacology Biochemistry and Behavior* **22**: 913–916.

Schellinck HM, Forestell CA, LoLordo VM (2001). A simple and reliable test of olfactory learning and memory in mice. *Chemical Senses* **26**: 663–672.

Schena M, Shalon D, Heller R, Chai A, Brown PO, Davis RW (1996). Parallel human genome analysis: Microarray-based expression monitoring of 1000 genes. *Proceedings of the National Academy of Sciences USA* **93**: 10614–10619.

Schenk D, Games D, Seubert P (2001). Potential treatment opportunities for Alzheimer's disease through inhibition of secretases and Aβ immunization. *Journal of Molecular Neuroscience* **17**: 259–268.

Schenk D, Hagen M, Seubert P (2004). Current progress in beta-amyloid immunotherapy. *Current Opinion in Immunology* **16**: 599–606.

Schicknick H, Hoffmann HJ, Schneider R, Crusio WE (1993). Genetic analysis of isolation-induced aggression in the mouse. III. Classical cross-breeding analysis of differences between two closely related inbred strains. *Behavioral and Neural Biology* **59**: 242–248.

Schiffman SS, Erickson RP (1971). A psychophysical model for gustatory quality. *Physiology and Behavior* **7**: 617–633.

Schimenti J, Bucan M (1998). Functional genomics in the mouse: Phenotype-based mutagenesis screens. *Genome Research* **8**: 698–710.

Schmahmann JD (1997). *The Cerebellum and Cognition*. Academic Press, New York.

Schmidt MJ, Sawyer BD, Perry KW, Fuller RW, Foreman MM, Ghetti B (1982). Dopamine deficiency in the weaver mutant mouse. *Journal of Neurochemistry* **2**: 376–380.

Schneider JE, Hamilton JM, Wade GN (1987). Genetic association between nest building and brown adipose tissue thermogenesis in female house mice. *Journal of Comparative Physiology B* **157**: 39–44.

Schneider R, Hoffmann HJ, Schicknick H, Moutier R (1992). Genetic analysis of isolation-induced aggression. I. Comparison between closely related inbred mouse strains. *Behavioral and Neural Biology* **57**: 198–204.

Schneider-Stock R, Epplen JT (1995). Congenic AB mice: A novel means for studying the (molecular) genetics of aggression. *Behavior Genetics* **25**: 475–482.

Schober A (2004). Classic toxin-induced animal models of Parkinson's disease: 6-OHDA and MPTP. *Cell Tissue Research* **318**: 215–224.

Schoenbaum G, Nugent S, Saddoris MP, Gallagher M (2002). Teaching old rats new tricks: Age-related impairments in olfactory reversal learning. *Neurobiology of Aging* **23**: 555–564.

Schoots AF, Crusio WE, Van Abeelen JH (1978). Zinc-induced peripheral anosmia and exploratory behavior in two inbred mouse strains. *Physiology and Behavior* **21**: 779–784.

Schrott LM, Crnic LS (1996). The role of performance factors in the active avoidance-conditioning deficit in autoimmune mice. *Behavioral Neuroscience* **110**: 486–491.

Schuldiner AR (1996). Molecular medicine: Transgenic animals. *New England Journal of Medicine* **334**: 653–655.

Schulkin J (1991). *Sodium Hunger: The Search for a Salty Taste*. Cambridge University Press, Cambridge.

Sclafani A (1995). How food preferences are learned—Laboratory animal models. *Proceedings of the Nutrition Society* **54**: 419–427.

Sclafani A, Glendinning JI (2003). Flavor preference conditioned in C57BL/6J mice by intragastric carbohydrate self-infusion. *Physiology and Behavior* **79**: 783–788.

Scordalakes EM, Imwalle DB, Rissman EF (2002). Oestrogen's masculine side: Mediation of mating in male mice. *Reproduction* **124**: 331–338.

Scott JP (1985). Investigative behavior: Toward a science of sociality. In *Leaders in the Study of Animal Behavior*, DA Dewsbury, Ed. Associated University Presses, Cranbury, NJ, pp. 389–429.

Scoville WB, Milner B (1957). Loss of recent memory after bilateral hippocampal lesions. *Journal of Neurological and Neurosurgical Psychiatry* **20**: 11–12.

Seabrook GR, Rosahl TW (1999). Transgenic animals relevant to Alzheimer's disease. *Neuropharmacology* **38**: 1–17.

Seale TW, Johnson P, Roderick TH, Carney JM, Rennert OM (1985). A single gene difference determines relative susceptibility to caffeine-induced lethality in SWR and CBA inbred mice. *Pharmacology Biochemistry and Behavior* **23**: 275–278.

Seale TW, Niekrasz I, Garrett KM (1996). Anxiolytics by ethanol, diazepam and buspirone in a novel murine behavioral assay. *Neuroreport* **7**: 1803–1808.

Segal DS, Kuczenski D (1997). An escalating dose "binge" model of amphetamine psychosis: Behavioral and neurochemical characteristics. *Journal of Neuroscience* **17**: 2551–2566.

Segal-Lieberman G, Bradley RL, Kokkotou E, Carlson M, Trombly DJ, Wang X, Bates S, Myers MG Jr, Flier JS, Maratos-Flier E (2003). Melanin-concentrating hormone is a critical mediator of the leptin-deficient phenotype. *Proceedings of the National Academy of Sciences USA* **100**: 10085–10090.

Sekuler R (1974). Spatial vision. *Annual Review of Psychology* **25**: 195–232.

Self DW, Choi KH, Simmons D, Walker JR, Smagula CS (2004). Extinction training regulates neuroadaptive responses to withdrawal from chronic cocaine self-administration. *Learning and Memory* **11**: 648–657.

Seligman MEP, Maier SF (1967). Failure to escape traumatic shock. *Journal of Experimental Psychology* **74**: 1–9.

Selkoe DJ (1998). The cell biology of beta-amyloid precursor protein and presenilin in Alzheimer's disease. *Trends in Cell Biology* **8**: 447–453.

Selko DJ (2001). Alzheimer's disease: genes, proteins, and therapy. *Physiological Reviews* **81**: 741–766.

Sen S, Villafuerte S, Nesse R, Stoltenberg SF, Hopcian J, Gleiberman L, Weder A, Burmeister M (2004). Serotonin transporter and GABAAA alpha 6 receptor variants are associated with neuroticism. *Biological Psychiatry* **55**: 244–249.

Seong E, Saunders TL, Stewart CL, Burmeister M (2004). To knockout in 129 or in C57BL/6: That is the question. *Trends in Genetics* **20**: 59–62.

Serretti A, Artioli P (2004). From molecular biology to pharmacogenetics: A review of the literature on antidepressant treatment and suggestions of possible candidate genes. *Psychopharmacology* **174**: 490–503.

Service, RF (1998). Microchip arrays put DNA on the spot. *Science* **282**: 396–399.

Service RF (2005). Going from genome to pill. *Science* **308**: 1858–1860.

Seyfried TN (1982). Developmental genetics of audiogenic seizure susceptibility in mice. In *Genetic Basis of the Epilepsies*. VE Anderson, WA Hauser, JK Penry, CF Sing, Eds. Raven, New York, pp. 198–210.

Shahbazian M, Young J, Yuva-Paylor L, Spencer C, Antalffy B, Noebels J, Armstrong D, Paylor R, Zoghbi H (2002). Mice with truncated MeCP2 recapitulate many Rett syndrome features and display hyperacetylation of histone H3. *Neuron* **35**: 243–254.

Shair HN, Brunelli SA, Masmela JR, Boone E, Hofer MA (2003). Social, thermal, and temporal influences on isolation-induced and maternally potentiated ultrasonic vocalizations of rat pups. *Developmental Psychobiology* **42**: 206–222.

Shair HN, Jasper A (2003). Decreased venous return is neither sufficient nor necessary to ultrasonic vocalizations of infant rat pups. *Behavioral Neuroscience* **117**: 840–853.

Shaldubina A, Einat H, Szechtman H, Shimon H, Belmaker RH (2002). Preliminary evaluation of oral anticonvulsant treatment in the quinpirole model of bipolar disorder. *Journal of Neural Transmission* **109**: 433–440.

Shanks N, Anisman H (1989). Strain-specific effects of antidepressants on escape deficits induced by inescapable shock. *Psychopharmacology* **99**: 122–128.

Shanks N, Anisman H (1993). Escape deficits induced by uncontrollable foot shock in recombinant inbred strains of mice. *Pharmacology Biochemistry and Behavior* **46**: 511–517.

Shannon W, Culverhouse R, Duncan J (2003). Analyzing microarray data using cluster analysis. *Pharmacogenomics* **4**: 41–52.

Shaw CA, Wilson JM (2003). Analysis of neurological disease in four dimensions: insight from ALS-PDC epidemiology and animal models. *Neuroscience and Biobehavioral Reviews* **27**: 493–505.

Shepherd JK, Grewal SS, Fletcher A, Bill DJ, Dourish CT (1994). Behavioural and pharmacological characterisation of the elevated "zero-maze" as an animal model of anxiety. *Psychopharmacology* **116**: 56–64.

Sherman JE, Roberts T, Roskam SE, Holman EW (1980). Temporal properties of the rewarding and aversive effects of amphetamine in rats. *Pharmacology Biochemistry and Behavior* **13**: 597–599.

Shimada M, Tritos NA, Lowell BB, Flier JS, Maratos-Flier E (1998). Mice lacking melanin-concentrating hormone are hypophagic and lean. *Nature* **396**: 670–674.

Shimozawa N, Ono Y, Kimoto S, Hioki K, Araki Y, Shinkai Y, Kono T, Ito M (2002). Abnormalities in cloned mice are not transmitted to the progeny. *Genesis* **34**: 203–207.

Shor-Posner G, Brennan G, Ian C, Jasaitis R, Madhu K, Leibowitz SF (1994). Meal patterns of macronutrient intake in rats with particular dietary preferences. *American Journal of Physiology* **266**: R1395–R1402.

Shyamala G, Yang X, Silberstein G, Barcellos-Hoff MH, Dale E (1998). Transgenic mice carrying an imbalance in the native ratio of A to B forms of progesterone receptor exhibit developmental abnormalities in mammary glands. *Proceedings of the National Academy of Sciences USA* **95**: 696–701.

Sills TL, Onalaja AO, Crawley JN (1998). Mesolimbic dopaminergic mechanisms underlying individual differences in sugar consumption and amphetamine hyperlocomotion in Wistar rats. *European Journal of Neuroscience* **10**: 1895–1902.

Sills TL, Vaccarino FJ (1991). Facilitation and inhibition of feeding by a single dose of amphetamine: Relationship to baseline intake and accumbens cholecystokinin. *Psychopharmacology* **105**: 329–334.

Silos-Santiago I, Fagan AM, Garber M, Fritzsch B, Barbacid M (1997). Severe sensory deficits but normal CNS development in newborn mice lacking TrkB and TrkC tyrosine protein kinase receptors. *European Journal of Neuroscience* **9**: 2045–2056.

Silva AJ, Frankland PW, Marowitz Z, Friedman E, Laszlo GS, Cioffi D, Jacks T, Bourtchuladze R (1997). A mouse model for the learning and memory deficits associated with neurofibromatosis type I. *Nature Genetics* **15**: 281–284.

Silva AJ, Kogan JH, Frankland PW, Kida S (1998). CREB and memory. *Annual Review of Neuroscience* **21**: 127–148.

Silva AJ, Paylor R, Wehner JM, Tonegawa S (1992a). Impaired spatial learning in α-calcium-calmodulin kinase II mutant mice. *Science* **257**: 206–211.

Silva AJ, Smith AM, Giese KP (1997). Gene targeting and the biology of learning and memory. *Annual Review of Genetics* **31**: 527–546.

Silva AJ, Stevens CF, Tonegawa S, Wang Y (1992b). Deficient hippocampal long-term potentiation in α-calcium-calmodulin kinase II mutant mice. *Science* **257**: 201–206.

Silver LM (1995). *Mouse Genetics. Concepts and Applications* Oxford University Press, New York.

Silver LM (1997). *Remaking Eden: Cloning and Beyond in a Brave New World.* Avon, New York.

Simiand J, Keane PE, Morre M (1984). The staircase test in mice: A simple and efficient procedure for primary screening of anxiolytic agents. *Psychopharmacology* **84**: 48–53.

Simon SA, Roper RD, Eds. (1993). *Mechanisms of Taste Transduction.* CRC Press, Boca Raton, FL.

Simpson EM, Linder CC, Sargent EE, Davisson MT, Mobraaten LE, Sharp JJ (1997). Genetic variation among 129 substrains and its importance for targeted mutagenesis in mice. *Nature Genetics* **16**: 19–27.

Singer JB, Hill AE, Burrage LC, Olszens KR, Song J, Justice M, O'Brien WE, Conti DV, Witte JS, Lander ES, Nadeau JH (2004). Genetic dissection of complex traits with chromosome substitution strains in mice. *Science* **304**: 445–447.

Singer JB, Hill AE, Nadeau JH, Lander ES (2005). Mapping quantitative trait loci for anxiety in chromosome substitution strains of mice. *Genetics* **169**: 855–862.

Sioud M (2004). Therapeutic siRNAs. *Trends in Pharmacological Sciences* **25**: 22–28.

Sisk CL, Meek LR (1997). Sexual and reproductive behaviors, In *Current Protocols in Neuroscience*, JN Crawley, CR Gerfen, R McKay, MA Rogawski, DR Sibley, P Skolnick, Eds. Wiley, New York, pp. 8.2.1–8.2.15.

Siviy SM, Atrens DM, Menendez JA (1990). Idazoxan increases rough-and-tumble play, activity and exploration in juvenile rats. *Psychopharmacology* **100**: 119–123.

Siviy SM, Panksepp J (1987). Sensory modulation of juvenile play in rats. *Developmental Psychobiology* **20**: 39–55.

Skinner BF (1938). *The Behavior of Organisms.* Appleton-Century-Crofts, New York.

Sloane SA, Shea SL, Proctor MM, Dewsbury DA (1978). Visual cliff performance in 10 species of muroid rodents. *Animal Learning and Behavior* **6**: 244–248.

Slotnick B, Hanford L, Hodos W (2000). Can rats acquire an olfactory learning set? *Journal of Experimental Psychology: Animal Behavior Processes* **26**: 399–415.

Slotnick BM, Nigrosh BJ (1975). Maternal behavior of mice with cingulate cortical, amygdala, or septal lesions. *Journal of Comparative and Physiological Psychology* **88**: 118–127.

Slotnick B, Restrepo D (2005). Olfactometry with mice. In *Current Protocols in Neuroscience*, Wiley, New York, pp. 8.20.1–8.20.4.

Sluyter F, Arseneault L, Moffitt TE, Veenema AH, de Boer S. Koolhaas JM (2003). Toward an animal model for antisocial behavior: Parallels between mice and humans. *Behavior Genetics* **33**: 563–574.

Sluyter F, Bult A, Lynch CB, van Oortmerssen GA, Koolhaas JM (1995). A comparison between house mouse lines selected for attack latency or nest-building: Evidence for a genetic basis of alternative behavioral strategies. *Behavior Genetics* **25**: 247–252.

Sluyter F, van Oortmerssen GA, de Ruiter AJ, Koolhaas JM (1996). Aggression in wild house mice: Current state of affairs. *Behavior Genetics* **26**: 489–496.

Small WS (1901). Experimental study of the mental processes of the rat. American *Journal of Psychology* **12**: 206–239.

Smith BK, Berthoud HR, York DA, Bray GA (1997). Differential effects of baseline macronutrient preferences on macronutrient selection after galanin, NPY, and an overnight fast. *Peptides* **18**: 207–211.

Smith DR, Striplin CD, Geller AM, Mailman RB, Drago J, Lawler CP, Gallagher M (1998a). Behavioural assessment of mice lacking D_{1A} dopamine receptors. *Neuroscience* **86**: 135–146.

Smith ER, Damassa DA, Davidson JM (1977). Hormone administration: Peripheral and intracranial implants. *Methods in Psychobiology* **3**: 259–279.

Smith GP (1982). The physiology of the meal. In Silverstone T, Ed. *Drugs and Appetite*. Academic, New York, pp. 1–21.

Smith GP (1995). Dopamine and food reward. *Progress in Psychobiology and Physiological Psychology* **16**: 83–144.

Smith GP (1996). The direct and indirect controls of meal size. *Neuroscience and Biobehavioral Reviews* **20**: 41–46.

Smith GP (1998a). *Satiation: From Gut to Brain*. Oxford University Press, New York.

Smith GP (1998b). Sham feeding in rats with chronic, reversible gastric fistulas. In *Current Protocols in Neuroscience*, Wiley, New York, pp. 8.6D.1–8.6D-6.

Smith GP (2000). The controls of eating: brain meanings of food stimuli. *Progress in Brain Research* **122**: 173–186.

Smith GP, Jerome C, Norgren R (1985). Afferent axons in abdominal vagus mediate satiety effect of cholecystokinin in rats. *American Journal of Physiology* **249**: R638–R641.

Smith GW, Aubry JM, Dellu F, Contarino A, Bilezikjian LM, Gold LH, Chen R, Marchuk Y, Hauser C, Bentley CA, Sawchenko PE, Koob GF, Vale W, Lee KF (1998b). Corticotropin releasing factor receptor 1–deficient mice display decreased anxiety, impaired stress response, and aberrant neuroendocrine development. *Neuron* **20**: 1093–1102.

Smith Richards BK, Belton BN, Poole AC, Mancuso JJ, Churchill GA, Li R, Volaufova J, Zuberi A, York B (2002). QTL analysis of self-selected macronutrient diet intake: Fat, carbohydrate, and total kilocalories. *Physiol Genomics* **11**: 205–217.

Smithies O (1993). Animal models of human genetic diseases. *Trends in Genetics* **9**: 112–116.

Smithies O, Kim HS (1994). Targeted gene duplication and disruption for analyzing quantitative genetic traits in mice. *Proceedings of the National Academy of Sciences USA* **91**: 3612–3615.

Society for Neuroscience (2003). *Short Course II Mouse Behavioral Phenotyping*. JN Crawley, Ed. Washington DC, 59.

Sohail M (2004). *Gene Silencing by RNA Interference*. CRC Press, Boca Raton, FL.

Sokolowski JD, Conlan AN, Salamone JD (1998). A microdialysis study of nucleus accumbens core and shell dopamine during operant responding in the rat. *Neuroscience* **86**: 1001–1009.

Sora I, Elmer G, Funada M, Pieper J, Li XF, Hall FS, Uhl GR (2001a). Mu opiate receptor gene dose effects on different morphine actions: Evidence for differential in vivo mu receptor reserve. *Neuropsychopharmacology* **25**: 41–54.

Sora I, Hall FS, Andrews AM, Itokawa M, Li XF, Wei HB, Wichems C, Lesch KP, Murphy DL, Uhl GR (2001b). Molecular mechanisms of cocaine reward: Combined dopamine and serotonin transporter knockouts eliminate cocaine place preference. *Proceedings of the National Academy of Sciences USA* **98**: 5300–5305.

Sora I, Takahashi N, Funada M, Ujike H, Revay RS, Donovan DM, Miner L, Uhl GR (1997). Opiate receptor knockout mice define μ receptor roles in endogenous nociceptive responses and morphine-induced analgesia. *Proceedings of the National Academy of Sciences USA* **94**: 1544–1549.

Sora I, Wichems C, Takahashi N, Li XF, Zeng Z, Revay R, Lesch KP, Murphy DL, Uhl GR (1998). Cocaine reward models: Conditioned place preference can be established in dopamine- and in serotonin-transporter knockout mice. *Proceedings of the National Academy of Sciences USA* **95**: 7699–7704.

South EH, Ritter RC (1988). Capsaicin application to central or peripheral vagal fibers attenuates CCK satiety. *Peptides* **9**: 601–612.

Southwick CH, Clark LH (1968). Interstrain differences in aggressive behavior and exploratory activity in inbred mice. *Communications in Behavioral Biology* **1**: 49–59.

Spear LP (1990). Neurobehavioral assessment during the early postnatal period. *Neurotoxicology and Teratology* **12**: 489–495.

Spear LP, File SE (1996). Methodological considerations in neurobehavioral teratology. *Pharmacology Biochemistry and Behavior* **55**: 455–457.

Spencer CM, Alekseyenko O, Serysheva E, Yuva-Paylor LA, Paylor R (2005). Altered anxiety-related and social behaviors in the *Fmr1* knockout mouse model of fragile X syndrome. *Genes, Brain and Behavior* **4**: 420–430.

Spencer SJ, Ebner K, Day TA (2004). Differential involvement of rat medial prefrontal cortex dopamine receptors in modulation of hypothalamic-pituitary-adrenal axis responses to different stressors. *European Journal of Neuroscience* **20**: 1008–1016.

Spielewoy C, Biala G, Roubert C, Hamon M, Betancur C, Giros B (2001). Hypolocomotor effects of acute and daily d-amphetamine in mice lacking the dopamine transporter. *Psychopharmacology* **159**: 2–9.

Spielewoy C, Gonon F, Roubert C, Fauchey V, Jaber M, Caron MG, Roques BP, Hamon M, Betancur C, Maldonado R, Giros B (2000). Increased rewarding properties of morphine in dopamine-transporter knockout mice. *European Journal of Neuroscience* **12**: 1827–1837.

Spooren WP, Vassout A, Neijt HC, Kuhn R, Gasparini F, Roux S, Porsolt RD, Gentsch C (2000). Anxiolytic-like effects of the prototypical metabotropic glutamate receptor 5 antagonist 2-methyl-6-(phenylethynyl)pyridine in rodents. *Journal of Pharmacology and Experimental Therapeutics* **295**: 1267–1275.

Spyraki C, Fibiger HC, Phillips AG (1983). Attenuation of heroin reward in rats by disruption of the mesolimbic dopamine system. *Psychopharmacology* **87**: 225–232.

Squire LR, Bloom FE, McConnell SK, Roberts JL, Spitzer NC, Zigmond MJ (2002). *Fundamental Neuroscience*, 2nd ed. Academic Press, New York.

Squire LR, Ojemann JG, Miezin FM, Petersen SE, Videen TO, Raichle ME (1992). Activation of the hippocampus in normal humans: A functional anatomical study of memory. *Proceedings of the National Academy of Sciences USA* **89**: 1837–1841.

Squire LR, Zola SM (1996). Structure and function of declarative and nondeclarative memory systems. *Proceedings of the National Academy of Sciences USA* **93**: 13515–13522.

St George JA (2003). Gene therapy progress and prospects: adenoviral vectors. *Gene Therapy* **10**: 1135–1141.

Stanley BG, Leibowitz SF (1985). Neuropeptide Y injected into the paraventricular hypothalamus: A powerful stimulant of feeding behavior. *Proceedings of the National Academy of Sciences USA* **82**: 3940–3943.

Stanton ME, Spear LP (1990). Workshop on the qualitative and quantitative comparability of human and animal developmental neurotoxicity, Work Group I report: Comparability of measures of developmental neurotoxicity in humans and laboratory animals. *Neurotoxicology and Teratology* **12**: 261–267.

Staubli U, Ivy G, Lynch G (1985). Denervation of hippocampus causes rapid forgetting of olfactory memory in rats. *Proceedings of the National Academy of Sciences USA* **81**: 5885–5887.

Staubli U, Thibault O, DiLorenzo M, Lynch G (1989). Antagonism of NMDA receptors impairs acquisition but not retention of olfactory memory. *Behavioral Neuroscience* **103**: 54–60.

Steel KP (1995). Inherited hearing defects in mice. *Annual Review of Genetics* **29**: 675–701.

Steele PM, Medina JF, Nores WL, Mauk MD (1998). Using genetic mutations to study the neural basis of behavior. *Cell* **95**: 879–882.

Steiner RA, Hohmann JG, Holmes A, Wrenn CC, Cadd G, Jureus A, Clifton DK, Luo M, Gutshall M, Ma SY, Mufson EJ, Crawley JN (2001). Galanin transgenic mice display cognitive and neurochemical deficits characteristic of Alzheimer's disease. *Proceedings of the National Academy of Sciences USA* **98**: 4184–4189.

Steinert PA, Infurna RN, Spear NE (1980). Long-term retention of a conditioned taste aversion in preweanling and adult rats. *Animal Learning and Behavior* **8**: 375–381.

Steinmayr M, André E, Conquet F, Rondi-Reig L, Delhaye-Bouchaud N, Auclair N, Daniel H, Crepel F, Mariani J, Sotelo C, Becker-André M (1998). *Staggerer* phenotype in retinoid-related orphan receptor α-deficient mice. *Proceedings of the National Academy of Sciences USA* **95**: 3960–3965.

Steinmetz JE (1998). The localization of a simple type of learning and memory: The cerebellum and classical eyeblink conditioning. *Current Directions in Psychological Science* **7**: 72–77.

Stellar E (1954). The physiology of motivation. *Psychological Review* **61**: 5–22.

Stellar E, Epstein AN (1991). Neuroendocrine factors in salt appetite. *Journal of Physiology and Pharmacology* **42**: 345–355.

Stellar JR, Stellar E (1985). *The Neurobiology of Motivation and Reward*. Springer, New York.

Stenzel-Poore MP, Heinrichs SC, Rivest S, Koob GF, Vale WW (1994). Overproduction of corticotropin-releasing factor in transgenic mice: A genetic model of anxiogenic behavior. *Journal of Neuroscience* **14**: 2579–2584.

Stern JM, Mackinnon DA (1978). Sensory regulation of maternal behavior in rats: Effects of pup age. *Developmental Psychobiology* **11**: 579–586.

Sterneck E, Paylor R, Jackson-Lewis V, Libbey M, Przedborski S, Tessarollo L, Crawley JN, Johnson PF (1998). Selectively enhanced contextual fear conditioning in mice lacking the CCAAT/enhancer binding protein delta. *Proceedings of the National Academy of Sciences USA* **95**: 10908–10913.

Stéru L, Chermat R, Thierry B, Simon P (1985). The tail suspension test: A new method for screening antidepressants in mice. *Psychopharmacology* **85**: 367–370.

Steru L, Thierry B, Chermat R, Millet B, Simon P, Porsolt RD (19987). Comparing benzodiazepines using the staircase test in mice. *Psychopharmacology* **92**: 106–109.

Stevens KE, Kem WR, Mahmir VM, Freedman R (1998). Selective α_7, nicotinic agonists normalize inhibition of auditory response in DBA mice. *Psychopharmacology* **136**: 320–327.

Steward CA, Marsden CA, Prior MJ, Morris PG, Shah Y (2005). Methodological considerations in rat brain BOLD contrast pharmacological MRI. *Psychopharmacology* **180**: 687–704.

St George-Hyslop PH, Petit A (2005). Molecular biology and genetics of Alzheimer's disease. *C R Biol.* **328**: 119–130.

Stock G (2005). Germinal choice technology and the human future. *Reproductive Biomedicine Online* (Suppl 1): 27–35.

Stork O, Welzl H, Cremer H, Schachner H (1997). Increased intermale aggression and neuroendocrine response in mice deficient for the neural cell adhesion molecule (NCAM). *European Journal of Neuroscience* **9**: 1117–1125.

Stork O, Welzl H, Wolfer D, Schuster T, Mantei N, Stork S, Hoyer D, Lipp H, Obata K, Schachner M (2000). Recovery of emotional behaviour in neural cell adhesion molecule (NCAM) null mutant mice through transgenic expression of NCAM180. *European Journal of Neuroscience* **12**: 3291–3306.

Stowers L, Holy TE, Meister M, Dulac C, Koentges G (2002). Loss of sex discrimination and male–male aggression in mice deficient for TRP2. *Science* **295**: 1493–1500.

Street VA, McKee-Johnson JW, Fonseca RC, Tempel BL, Noben-Trauth K (1998). Mutations in a plasma membrane Ca^{2+}-ATPase gene cause deafness in deafwaddler mice. *Nature Genetics* **19**: 390–394.

Strekalova T, Spanagel R, Bartsch D, Henn FA, Gass P (2004). Stress-induced anhedonia in mice is associated with deficits in forced swimming and exploration. *Neuropsychopharmacology* **29**: 2007–2017.

Stricker EM, Verbalis JG (1990). Control of appetite and satiety: Insights from biologic and behavioral studies. *Nutrition Review* **48**: 49–56.

Strohmayer AJ, Greenberg D (1994). Devazepide alters meal patterns in lean, but not obese male Zucker rats. *Physiology and Behavior* **56**: 1037–1039.

Strohmayer AJ, Smith GP (1987). The meal pattern of genetically obese (*ob/ob*) mice. *Appetite* **8**: 111–123.

Strozik E, Festing MF (1981). Whisker trimming in mice. *Laboratory Animals* **15**: 309–312.

Suhr ST, Gage FH (1999). Gene therapy in the central nervous system. *Archives of Neurology* **56**: 287–292.

Suzuki M, Mizuno A, Kodaira K, Imai M (2003). Impaired pressure sensitization in mice lacking TRPV4. *Journal of Biological Chemistry* **278**: 22664–22668.

Swann AC, Secunda SK, Stokes PE, Croughan J, Davis JM, Koslow SH, Maas JW (1990). Stress, depression, and mania: Relationship between perceived role of stressful events and clinical and biochemical characteristics. *Acta Psychiatrica Scandanavica* **81**: 389–397.

Sweeney JE, Hohmann CF, Moran TH, Coyle JT (1988). A long-acting cholinesterase inhibitor reverses spatial memory deficits in mice. *Pharmacology Biochemistry and Behavior* **31**: 141–147.

Sweeney WA, Luedtke J, McDonald MP, Overmier JB (1997). Intrahippocampal injections of exogenous beta-amyloid induce postdelay errors in an eight-arm radial maze. *Neurobiology of Learning and Memory* **68**: 97–101.

Swerdlow NR, Braff DL, Taaid N, Geyer MA (1994). Assessing the validity of an animal model of deficient sensorimotor gating in schizophrenic patients. *Archives of General Psychiatry* **51**: 139–154.

Swerdlow NR, Koob GF (1984). Restrained rats learn amphetamine-conditioned locomotion, but not place preference. *Psychopharmacology* **84**: 163–166.

Swerdlow NR, Vaccarino FJ, Amalric M, Koob GF (1986). The neural substrates for the motor-activating properties of psychostimulants: A review of recent findings. *Pharmacology Biochemistry and Behavior* **25**: 233–248.

Swiergel AH, Takahashi LK, Rubin WW, Kalin NH (1992). Antagonism of corticotropin-releasing factor receptors in the locus coeruleus attenuates shock-induced freezing in rats. *Brain Research* **587**: 263–268.

Swithers SE (2003). Do metabolic signals stimulate intake in rat pups? *Physiology and Behavior* **79**: 71–78.

Syvanen AC (1999). From gels to chips: "Minisequencing" primer extension for analysis of point mutations and single nucleotide polymorphisms. *Human Mutation* **13**: 1–10.

Szczypka MS, Mandel RJ, Donahue BA, Synder RO, Leff SE, Palmiter RD (1999). Viral gene delivery selectively restores feeding and prevents lethality of dopamine-deficient mice. *Neuron* **22**: 167–178.

Sztein J, Sweet H, Farley J, Mobraaten L (1998). Cryopreservation and orthotopic transplantation of mouse ovaries: New approach in gamete banking. *Biology of Reproduction* **58**: 1071–1074.

Szumlinski KK, Lominac KD, Oleson EB, Walker JK, Mason A, Dehoff MH, Klugman M, Cagle S, Welt K, During M, Worley PF, Middaugh LD, Kalivas PW (2005). Homer2 is necessary for EtOH-induced neuroplasticity. *Journal of Neuroscience* **25**: 7054–7061.

Tabakoff B, Bhave SV, Hoffman PL (2003). Selective breeding, quantitative trait loci analysis, and gene arrays identify candidate genes for complex drug-related behaviors. *Journal of Neuroscience* **23**: 4491–4498.

Tachikawa K, Yokoi H, Nagasaki H, Arima H, Murase T, Sugimura Y, Miura Y, Hirabayashi M, Oiso Y (2003). Altered cardiovascular regulation in arginine vasopressin-overexpressing transgenic rat. *American Journal of Physiology and Endocrinology Metabolism* **285**: E1161–E1166.

Takahashi JS (1995). Molecular neurobiology and genetics of circadian rhythms in mammals. *Annual Review of Neuroscience* **18**: 531–553.

Takahashi JS, Pinto LH, Vitaterna MH (1994). Forward and reverse genetic approaches to behavior in the mouse. *Science* **264**: 1724–1733.

Takahashi M (2004). Delivery of genes to the eye using lentiviral vectors. *Methods in Molecular Biology* **246**: 439–449.

Takahashi N, Miner LL, Sora I, Ujike H, Revay RS, Kostic V, Jackson-Lewis V, Przedborski S, Uhl GR (1997). VMAT2 knockout mice: Heterozygotes display reduced amphetamine-conditioned reward, enhanced amphetamine locomotion, and enhanced MPTP toxicity. *Proceedings of the National Academy of Sciences USA* **94**: 9938–9943.

Takahashi S, Kubota Y, Sato H (1998). Mutant frequencies in *lacZ* transgenic mice following the internal irradiation from 89Sr or the external gamma-ray irradiation. *Journal of Radiation Research* **39**: 53–60.

Takeda M, Sawano S, Imaizumi M, Fushiki T (2001). Preference for cornoil in olfactory-blocked mice in the conditioned place preference test and the two-bottle choice test. *Life Sciences* **69**: 847–854.

Takeda Y, Akasaka K, Lee S, Kobayashi S, Kawano H, Murayama S, Takahashi N, Hashimoto K, Kano M, Asano M, Sudo K, Iwakura Y, Watanabe K (2003). Impaired motor coordination in mice lacking neural recognition molecule NB-3 of the contactin/F3 subgroup. *Journal of Neurobiology* **56**: 252–265.

Talbot CJ, Radcliffe RA, Fullerton J, Hitzemann R, Wehner JM, Flint J (2003). Fine scale mapping of a genetic locus for conditioned fear. *Mammalian Genome* **14**: 223–230.

Tallman JF, Paul SM, Skolnick P, Gallager DW (1980). Receptors for the age of anxiety: Pharmacology of the benzodiazepines. *Science* **207**: 274–281.

Tanda G, Goldberg SR (2003). Cannabinoids: reward, dependence, and underlying neurochemical mechanisms—A review of recent preclinical data. *Psychopharmacology* **169**: 115–134.

Tanda G, Pontiere FE, De Chiara G (1997). Cannabinoid and heroin activation of mesolimbic dopamine transmission by a common μ_1 opioid receptor mechanism. *Science* **276**: 2048–2050.

Tang X, Orchard SM, Sanford LD (2002). Home cage activity and behavioral performance in inbred and hybrid mice. *Behavioural Brain Research* **136**: 555–569.

Tang Y-P, Shimizu E, Dube GR, Rampon C, Kerchner GA, Zhu M, Liu G, Tsien JZ (1999). Genetic enhancement of learning and memory in mice. *Nature* **401**: 63–69.

Tang Y, Shimizu E, Tsien JZ (2001). Do "smart" mice feel more pain, or are they just better learners? *Nature Neuroscience* **4**: 453–454.

Tanzi RE, Bertram L (2005). Twenty years of the Alzheimer's disease amyloid hypothesis: A genetic perspective. *Cell* **120**: 545–555.

Tarantino LM, Gould TJ, Druhan JP, Bucan M (200). Behavior and mutagenesis screens: The importance of baseline analysis of inbred strains. *Mammalian Genome* **11**: 555–564.

Tarricone BJ, Hingtgen JN, Belknap JK, Mitchell SR, Nurnberger JI (1995). Quantitative trait loci associated with the behavioral response of BXD recombinant inbred mice to restraint stress: A preliminary consideration. *Behavior Genetics* **25**: 489–495.

Tatebayashi Y, Miyasaka T, Chui DH, Akagi T, Mishima K, Iwasaki K, Fujiwara M, Tanemura K, Murayama M, Ishiguro K, Planel E, Sato S, Hashikawa T, Takashima A (2002). Tau filament formation and associative memory deficit in aged mice expressing mutant (R406W) human tau. *Proceedings of the National Academy of Sciences USA* **99**: 13896–13901.

Taylor BA, Phillips SJ (1997). Obesity QTLs on mouse chromosomes 2 and 17. *Genomics* **43**: 249–257.

Taylor WD, Steffens DC, Payne ME, MacFall JR, Marchuk DA, Svenson IK, Krishnan KR (2005). Influence of serotonin transporter promoter region polymorphisms on hippocampal volumes in late-life depression. *Archives of General Psychiatry* **62**: 537–544.

Tecott LH (2003). The genes and brains of mice and men. *American Journal of Psychiatry* **160**: 646–656.

Tecott LH, Abdallah L (2003). Mouse genetic approaches to feeding regulation: Serotonin 5-HT2C receptor mutant mice. *CNS Spectrum* **8**:584–588.

Tecott LH, Logue SF, Wehner JM, Kauer JA (1998). Perturbed dentate gyrus function in serotonin 5-HT2C receptor mutant mice. *Proceedings of the National Academy of Sciences USA* **95**: 15026–15031.

Tecott LH, Sun LM, Akana SF, Strack AM, Lowenstein DH, Dallman MF, Julius D (1995). Eating disorder and epilepsy in mice lacking 5-HT2C serotonin receptors. *Nature* **374**: 542–546.

Tecott LH, Wehner JM (2001). Mouse molecular genetic technologies. *Archives of General Psychiatry* **58**: 995–1004.

Teitelbaum P, Epstein AN (1962). The lateral hypothalamic syndrome: Recovery of feeding and drinking after lateral hypothalamic lesions. *Psychological Review* **69**: 74–90.

Tempel DL, Leibowitz SF (1990). Diurnal variations in the feeding responses to norepinephrine, neuropeptide Y and galanin in the PVN. *Brain Research Bulletin* **25**: 821–825.

Temple JL, Scordalakes EM, Bodo C, Gustafsson JA, Rissman EF (2003). Lack of functional estrogen receptor beta gene disrupts pubertal male sexual behavior. *Hormones and Behavior* **44**: 427–434.

Tenenbaum L, Chtarto A, Lehtonen E, Velu T, Brotchi J, Levivier M (2004). Recombinant AVV-mediated gene delivery to the central nervous system. *Journal of Gene Medicine* **6**: S212–S222.

Teresa Girao da Cruz M, Cardoso AL, de Almeida LP, Simoes S, Pedroso de Lima MC (2005). Tf-lipoplex-mediated NGF gene transfer to the CNS: neuronal protection and recovery in an excitotoxic model of brain injury. *Gene Therapy* **12**: 1242–1252.

Terranova ML, Laviola G (2005). Scoring of social interactions and play in mice during adolescence. *Current Protocols in Toxicology* 13.10.1–13.10.11.

Terranova ML, Laviola G, de Acetis L, Alleva E (1998). A description of the ontogeny of mouse agonistic behavior. *Journal of Comparative Psychology* **112**: 3–12.

Terranova JP, Perio A, Worms P, Le Fur G, Soubrie P (1994). Social olfactory recognition in rodents: deterioration with age, cerebral ischemia and septal lesions. *Behavioral Pharmacology* **5**: 90–98.

Terranova JP, Storme JJ, Lafon N, Perio A, Rinaldi-Carmona M, Le Fur G, Soubrie P (1996). Improvement of memory in rodents by the selective CB1 cannabinoid receptor antagonist, SR 141716. *Psychopharmacology* **126**: 165–172.

Tessari M, Pilla M, Andreoli M, Hutcheson DM, Heidbreder CA (2004). Antagonism at metabotropic glutamate 5 receptors inhibits nicotine- and cocaine-taking behaviours and prevents nicotine-triggered relapse to nicotine-seeking. *European Journal of Pharmacology* **499**: 121–133.

Tezval H, Jahn O, Todorovic C, Sasse A, Eckart K, Spiess J (2004). Cortagine, a specific agonist of corticotropin-releasing factor receptor subtype 1, is anxiogenic and antidepressive in the mouse model. *Proceedings of the National Academy of Sciences USA* **101**: 9468–9473.

Thakker DR, Hoyer D, Cryan JF (2006). Interfering with the brain: Use of RNA interference for understanding the pathophysiology of psychiatric and neurological disorders. *Pharmacology and Therapeutics* **109**: 413–438.

Thibault L, Booth DA (1998). Macronutrient-specific dietary selection in rodents and its neural bases. *Neuroscience and Biobehavioral Reviews* **23**: 457–528.

Thiele TE, Koh MT, Pedrazzini T (2002). Voluntary alcohol consumption is controlled via the neuropeptide Y Y1 receptor. *Journal of Neuroscience* **22**: RC208; 1–6.

Thiele TE, Marsh DJ, Ste-Marie L, Bernstein IL, Palmiter RD (1998). Ethanol consumption and resistance are inversely related to neuropeptide Y levels. *Nature* **396**: 366–369.

Thiele TE, Naveilhan P, Ernfors P (2004a). Assessment of ethanol consumption and water drinking by NPY Y(2) receptor knockout mice. *Peptides* **25**: 975–983.

Thiele TE, Sparta DR, Hayes DM, Fee JR (2004b). A role for neuropeptide Y in neurobiological responses to ethanol and drugs of abuse. *Neuropeptides* **38**: 235–243.

Thierry AM, Tassin JP, Blanc G, Glowinski J (1976). Selective activation of mesocortical DA system by stress. *Nature* **263**: 242–244.

Thomas GM, Huginar RL (2004). MAPK cascade signalling and synaptic plasticity. *Nature Reviews Neuroscience* **5**: 173–183.

Thomas SA, Palmiter RD (1997). Impaired maternal behavior in mice lacking norepinephrine and epinephrine. *Cell* **91**: 583–592.

Thor DH, Holloway WR (1984). Social play in juvenile rats: A decade of methodological and experimental research. *Neuroscience and Biobehavioral Reviews* **9**: 455–464.

Thorsell A, Heilig M (2002). Diverse functions of neuropeptide Y revealed using genetically modified animals. *Neuropeptides* **36**: 182–193.

Thorsell A, Michalkiewicz M, Dumont Y, Quirion R, Caberlotto L, Rimondini P, Mathé AA, Heilig M (2000). Behavioral insensitivity to stress, absent fear suppression of behavior and impaired spatial learning in transgenic rats with hippocampal neuropeptide Y overexpression. *Proceedings of the National Academy of Sciences USA* **97**: 12582–12587.

Timpl P, Spanagel R, Sillaber I, Kresse A, Reul JMHM, Stalla GK, Blanquet V, Steckler T, Holsboer F, Wurst W (1998). Impaired stress response and reduced anxiety in mice lacking a functional corticotropin-releasing hormone receptor 1. *Nature Genetics* **19**: 162–166.

Tinbergen N (1974). Ethology and stress diseases. *Science* **185**: 20–27.

Tinsley R, Eriksson P (2003). Use of gene therapy in central nervous system repair. *Acta Neurol Scan* **109**: 1–8.

Tirelli E, Laviola G, Adriani W (2003). Ontogenesis of behavioral sensitization and conditioned place preference induced by psychostimulants in laboratory rodents. *Neuroscience and Biobehavioral Reviews* **27**: 163–178.

Todtenkopf MS, Carreiras T, Melloni RH, Stellar JR (2002). The dorsomedial shell of the nucleus accumbens facilitates cocaine-induced locomotor activity during the induction of behavioral sensitization. *Behavioural Brain Research* **131**: 9–16.

Tomanin R, Scarpa M (2004). Why do we need new gene therapy viral vectors? Characteristics, limitations and future perspectives of viral vector transduction. *Current Gene Therapy.* **4**: 357–372.

Tornatzky W, Miczek KA (1993). Long-term impairment of autonomic circadian rhythms after brief intermittent social stress. *Physiology and Behavior* **53**: 983–993.

Toubas PL, Abla KA, Cao W, Logan LG, Seale TW (1990). Latency to enter a mirrored chamber: A novel behavioral assay for anxiolytic agents. *Pharmacology Biochemistry and Behavior* **35**: 121–126.

Toyota H, Dugovic C, Koehl M, Laposky AD, Weber C, Ngo K, Wu Y, Lee DH, Yanai K, Sakurai E, Watanabe T, Liu C, Barbier AJ, Turek FW, Fung-Leung WP, Lovenberg TW (2002). Behavioral characterization of mice lacking histamine H(3) receptors. *Molecular Pharmacology* **62**: 389–397.

Tran AH, Tamura R, Uwano T, Kobayashi T, Katsuki M, Matsumoto G, Ono T (2002). Altered accumbens neural response to prediction of reward associated with place in dopamine D2 receptor knockout mice. *Proceedings of the National Academy of Sciences USA* **99**: 8986–8991.

Trent RJ, Alexander IE (2004). Gene therapy: Applications and progress towards the clinic. *Internal Medicine Journal* **34**: 621–625.

Trinchese F, Liu S, Battaglia F, Walter S, Mathews PM, Arancio O (2004). Progressive age-related development of Alzheimer-like pathology in APP/PS1 mice. *Annals of Neurology* **55**: 801–814.

Trinh JV, Nehrenberg DL, Jacobsen JP, Caron MG, Wetsel WC (2003). Differential psychostimulant-induced activitation of neural circuits in dopamine transporter knockout and wild type mice. *Neuroscience* **118**: 297–310.

Trinh K, Storm DR (2003). Vomeronasal organ detects odorants in absence of signaling through main olfactory epithelium. *Nature Neuroscience* **6**: 438–440.

Trojanowski JQ, Lee VM (2003). Parkinson's disease and related alpha-synucleinopathies are brain amyloidoses. *Annals of the New York Academy of Sciences* **991**: 107–110.

Trullas R, Jackson B, Skolnick P (1989). Genetic differences in tail suspension test for evaluating antidepressant activity. *Psychopharmacology* **99**: 287–288.

Trullas R, Skolnick P (1993). Differences in fear motivated behaviors among inbred mouse strains. *Psychopharmacology* **111**: 323–331.

Tschenett A, Singewald N, Carli M, Balducci C, Salchner P, Vezzani A, Herzog H, Sperk G (2003). Reduced anxiety and improved stress coping ability in mice lacking NPY-Y2 receptors. *European Journal of Neuroscience* **18**: 143–148.

Tseng W, Guan R, Disterhoft JF, Weiss C (2004). Trace eyeblink conditioning is hippocampally dependent in mice. *Hippocampus* **14**: 58–65.

Tsien JZ, Chen DF, Gerber D, Tom C, Mercer EH, Anderson DJ, Mayford M, Kandel ER, Tonegawa S (1996*a*). Subregion- and cell type-restricted gene knockout in mouse brain. *Cell* **87**(7): 1317–1326.

Tsien JZ, Huerta PT, Tonegawa S (1996*b*). The essential role of hippocampal CA1 NMDA receptor-dependent synaptic plasticity in spatial memory. *Cell* **87**: 1327–1338.

Turner BJ, Atkin JD, Farg MA, Zang da W, Rembach A, Lopes EC, Patch JD, Hill AF, Cheema SS (2005). Impaired extracellular secretion of mutant superoxide dismutase 1 associates with neurotoxicity in familial amyotrophic lateral sclerosis. *Journal of Neuroscience* **25**: 108–117.

Turner CA, Presti MF, Newman HA, Bugenhagen P, Crnic L, Lewis MH (2001). Spontaneous stereotype in an animal model of Down syndrome: Ts65Dn mice. *Behavior Genetics* **31**: 393–400.

Turner RS, Grafton ST, Votaw JR, Delong MR, Hoffman JM (1998). Motor subcircuits mediating the control of movement velocity: A PET study. *Journal of Neurophysiology* **80**: 2162–2176.

Turri MG, Datta SR, DeFries J, Henderson ND, Flint J (2001). QTL analysis identifies multiple behavioral dimensions in ethological tests of anxiety in laboratory mice. *Current Biology* **11**: 725–734.

Tuszynski MH, Gage FH (1996). Somatic gene therapy for nervous system disease. *Growth Factors as Drugs for Neurological and Sensory Disorders, Ciba Foundation Symposium* **196**: 85–97.

Tzschentke TM (1998). Measuring reward with the conditioned place preference paradigm: A comprehensive review of drug effects, recent progress and new issues. *Progress in Neurobiology* **56**: 613–672.

Ueno N, Dube MG, Inui A, Kalra PS, Kalra SP (2004). Leptin modulates orexigenic effects of ghrelin, attenuates adiponectin and insulin levels, and selectively the dark-phase feeding as revealed by central leptin gene therapy. *Endocrinology* **145**: 4176–4184.

Uhl GR, Lin Z (2003). The top 20 dopamine transporter mutants: Structure-function relationships and cocaine actions. *European Journal of Pharmacology* **479**: 71–82.

Ukai M, Maeda H, Nanya Y, Kameyama T, Matsuno K (1998). Beneficial effects of acute and repeated administrations of σ receptor agonists on behavioral despair in mice exposed to tail suspension. *Pharmacology Biochemistry and Behavior* **61**: 247–252.

Ungerer A, Mathis C, Melan C (1998). Are glutamate receptors specifically implicated in some forms of memory processes? *Experimental Brain Research* **123**: 45–51, 1998.

Ungerleider LG (1995). Functional brain imaging studies of cortical mechanisms for memory. *Science* **270**: 769–775.

Ungerstedt U (1971). Postsynaptic supersensitivity after 6-hydroxydopamine induced degeneration of the nigro-striatal dopamine system. *Acta Physiologica Scandinavica* **367** (Suppl): 89–93.

Vaccarino FJ, Bloom FE, Koob GF (1985). Blockade of nucleus accumbens opiate receptors attenuates intravenous heroin reward in the rat. *Psychopharmacology* **85**: 37–42.

Vale W, Speiss J, Rivier C, Rivier J (1981). Characterization of a 41-residue ovine hypothalamic peptide that stimulates secretion of corticotropin and β-endorphin. *Science* **213**: 1394–1397.

Valverde O, Fournié MC, Roques BP, Maldonado R (1996). The CCK$_B$ antagonist PD-134,308 facilitates rewarding effects of endogenous enkephalins but does not induce place preference in rats. *Psychopharmacology* **123**: 119–126.

Valverde O, Ledent C, Beslot F, Parmentier M, Roques BP (2000). Reduction of stress-induced analgesia but not of exogenous opioid effects in mice lacking CB1 receptors. *European Journal of Neuroscience* **12**: 533–539.

Valverde O, Mantamadiotis T, Torrecilla M, Ugedo L, Pineda J, Bleckmann S, Gass P, Kretz O, Mitchell JM, Schutz G, Maldonado R (2004). Modulations of anxiety-like behavior and morphine dependence in CREB-deficient mice. *Neuropsychopharmacology* **29**: 1122–1133.

Valzelli L (1967). Drugs and aggressiveness. *Advances in Pharmacology* **5**: 79–108.

Valzelli L, Bernasconi S, Gomba P (1974). Effect of isolation on some behavioral characteristics in three strains of mice. *Biological Psychiatry* **9**: 329–334.

van Abeelen JHF (1963). Mouse mutants studied by means of ethological methods. I. Ethogram. *Genetica* **34**: 79–94.

Van Daal JHHM, De Kok YJM, Jenks BG, Wendelaar Bonga SE, VanAbeelen JHF (1987). A genotype-dependent hippocampal dynorphinergic mechanism controls mouse exploration. *Pharmacology Biochemistry and Behavior* **28**: 465–468.

Van Dam D, D'Hooge R, Hauben E, Reyniers E, Gantois I, Bakker CE, Oostra BA, Kooy RF, De Deyn PP (2000). Spatial learning, contextual fear conditioning and conditioned emotional response in Frm1 knockout mice. *Behavioural Brain Research* **117**: 127–136.

Vandenbergh JG (1973). Effects of central and peripheral anosmia on reproduction of female mice. *Physiology and Behavior* **10**: 257–261.

Van der Heyden JAM, Molewijk E, Olivier B (1987). Strain differences in response to drugs in the tail suspension test for antidepressant activity. *Psychopharmacology* **92**: 127–130.

van der Kooy D, O'Shaughnessy M, Mucha RF, Kalant H (1983). Motivational properties of ethanol in naïve rats as studied by place conditioning. *Pharmacology Biochemistry and Behavior* **19**: 441–445.

Van de Weerd HA, Van Loo PLP, Van Zutphen LFM, Koolhaas JM, Baumans V (1997). Nesting material as environmental enrichment has no adverse effects on behavior and physiology of laboratory mice. *Physiology and Behavior* **62**: 1019–1028.

van Dooren T, Dewachter I, Borghgraef P, van Leuven F (2005). Transgenic mouse models for APP processing and Alzheimer's disease: Early and late defects. *Subcell Biochem.* **38**: 45–63.

Van Essen DC (1979). Visual areas of the mammalian cerebral cortex. *Annual Review of Neuroscience* **2**: 227–263.

van Gaalen MM, Steckler T (2000). Behavioural analysis of four mouse strains in an anxiety test battery. *Behavioural Brain Research* **115**: 95–106.

Van Horn JD, Gold JM, Esposito G, Ostrem JL, Mattay V, Weinberger DR, Berman KF (1998). Changing patterns of brain activation during maze learning. *Brain Research* **793**: 29–38.

Van Loo PL, Van de Weerd HA, Van Zutphen LF, Baumans V (2004). Preference for social contact versus environmental enrichment in male laboratory mice. *Laboratory Animals* **38**: 178–188.

Van Oortmerssen GA (1971). Biological significance, genetics and evolutionary origin of variability in behaviour within and between inbred strains of mice (*Mus musculus*): A behaviour genetic study. *Behaviour* **38**: 1–92.

Van Oortmerssen GA, Bakker TCM (1981). Artificial selection for short and long attack latencies in wild *Mus musculus domesticus*. *Behavior Genetics* **11**: 115–126.

Van Oortmerssen GA, Busser J (1989). Studies in wild house mice: Disruptive selection on aggression as a possible force in evolution. In *House Mouse Aggression: A Model for Understanding the Evolution of Social Behaviour*, PF Brain, D Mainardi, S Parmigiani, Eds. Harwood Academic, Chur, Switzerland, pp. 87–117.

Vanover KE, Barrett JE (1998). An automated learning and memory model in mice: Pharmacological and behavioral evaluation of an autoshaped response. *Behavioural Pharmacology* **9**: 273–283.

Van Raamsdonk JM, Pearson J, Slow EJ, Hossain SM, Leavitt BR, Hayden MR (2005). Cognitive dysfunction precedes neuropathology and motor abnormalities in the YAC129 mouse model of Huntington's disease. *Journal of Neuroscience* **25**: 4169–4180.

Vanyukov MM, Moss HB, Gioio AE, Hughes HB, Kaplan BB, Tarter RE (1998). An association between a microsatellite polymorphism at the DRD5 gene and the liability to substance abuse: Pilot study. *Behavior Genetics* **28**: 75–82.

Varty GB, Hyde LA, Hodgson RA, Lu SX, McCool MF, Kazdoba TM, Del Vecchio RA, Guthrie DH, Pond AJ, Grzelak ME, Xu X, Korfmacher WA, Tulshian D, Parker EM, Higgins GA (2005). Characterization of the nociceptin receptor (ORL-1) agonist, Ro64-6198, in tests of anxiety across multiple species. *Psychopharmacology* **182**: 132–143.

Vaugeois JM, Odiévre C, Loisel L, Costentin J (1996). A genetic mouse model of helplessness sensitive to imipramine. *European Journal of Pharmacology* **316**: R1–R2.

Veasey SC, Valladares O, Fenik P, Kapfhamer D, Sanford L, Benington J, Bucan M (2000). An automated system for recording and analysis of sleep in mice. *Sleep* **23**: 1025–1040.

Veng LM, Bjugstad KB, Freed CR, Marrack P, Clarkson ED, Bell KP, Hutt C, Zawada WM (2002). Xenografts of MHC-deficient mouse embryonic mesencephalon improve behavioral recovery in hemiparkinsonian rats. *Cell Transplant.* **11**: 5–16.

Venter JC et al. (2001). The sequence of the human genome. *Science* **291**: 1304–1351.

Venter JC, Adams MD, Sutton GG, Kerlavage AR, Smith HO, Hunkapiller M (1998). Shotgun sequencing of the human genome. *Science* **280**: 1540–1542.

Verbitsky M, Yonan AL, Malleret G, Kandel ER, Gilliam TC, Pavlidis P (2004). Altered hippocampal transcript profile accompanies an age-related spatial memory deficit in mice. *Learning and Memory* **11**: 253–260.

Verkerk AJ, Pieretti M, Sutcliffe JS, Fu YH, Kuhl DP, Pizzuti A, Reiner O, Richards S, Victoria MF, Zhang FP, Eissen BE, van Oomen GJB, Blonden LAJ, Riggins GJ, Chastain JL, Kunst CB, Galjaard H, Caskey T, Nelson DL, Oostra BA, Warren ST (1991). Identification of a gene (*FMR-1*) containing a CGG repeat coincident with a breakpoint cluster region exhibiting length variation in fragile X syndrome. *Cell* **65**: 905–914.

Verma IM, Weitzman MD (2004). Gene therapy: Twenty-first century medicine. *Annual Review of Biochemistry* **74**: 711–738.

Vile RC (1996). Gene therapy for cancer—In the dock, blown off course or full speed ahead? *Cancer and Metastasis Reviews* **15**: 283–286.

Vitaterna MH, King DP, Chang AM, Kornhauser JM, Lowrey PL, McDonald JD, Dove WF, Pinto LH, Turek FW, Takahashi JS (1994). Mutagenesis and mapping of a mouse gene, *Clock*, essential for circadian behavior. *Science* **264**: 719–725.

Vitaterna MH, Pinto LH, Takahashi JS (2006). Large-scale mutagenesis and phenotypic screens for the nervous system and behavior in mice. *Trends in Neuroscience* **29**: 233–240.

Vivian JA, Miczek KA (1993). Diazepam and gepirone selectively attenuate either 20–32 or 32–64 kHz ultrasonic vocalizations during aggressive encounters. *Psychopharmacology* **112**: 66–73.

Vizi ES, Kiss JP (1998). Neurochemistry and pharmacology of the major hippocampal transmitter systems: Synaptic and nonsynaptic interactions. *Hippocampus* **8**: 566–607.

Vogel G (2005). Ready of not? Human ES cells head toward the clinic. *Science* **308**: 1534–1535.

Vogel JR, Beer B, Clody DE (1971). A simple and reliable conflict procedure for testing anti-anxiety agents. *Psychopharmacologia* **21**: 1–7.

Voigt JP, Hortnagl H, Rex A, van Hove L, Bader M, Fink H (2005). Brain angiotensin and anxiety-related behavior: The transgenic rat TGR(ASrAOGEN)680. *Brain Research* **1046**: 145–156.

Voikar V, Vasar E, Rauvala H (2004). Behavioral alterations induced by repeated testing in C57BL/6J and 129S2/Sv mice: implications for phenotyping screens. *Genes, Brain and Behavior* **3**: 27–38.

Volkmar FR, Pauls D (2003). Autism. *Lancet* **362**: 1133–1141.

Von Frisch K (1967). *The Dance Language and Orientation of Bees*. Belknap, Cambridge, MA.

Von Koch CS, Zheng H, Chen H, Trumbauer M, Thinakaran G, Van Der Ploeg LHT, Price DL, Sisodia SS (1997). Generation of APLP2 KO mice and early postnatal lethality in APLP2/APP double KO mice. *Neurobiology of Aging* **18**: 661–669.

Wagner EF, Stewart TA, Mintz B (1981a). The human beta-globin gene and a functional thymidine kinase gene in developing mice. *Proceedings of the National Academy of Sciences USA* **78**: 5016–5020.

Wagner KU, Young WS, Liu X, Ginns EI, Li M, Furth PA, Hennighausen L (1997). Oxytocin and milk removal are required for post-partum mammary-gland development. *Genes and Function* **1**: 233–244.

Wagner TE, Hoppe PC, Jollick JD, Scholl DR, Hodinka RL, Gault JB (1981b). Microinjection of a rabbit beta-globin gene into zygotes and its subsequent expression in adult mice and their offspring. *Proceedings of the National Academy of Sciences USA* **78**: 6376–6380.

Wahlsten D (1972). Genetic experiments with animal learning: A critical review. *Behavioral Biology* **7**: 143–182.

Wahlsten D (1982). Deficiency of corpus callosum varies with strain and supplier of the mice. *Brain Research* **239**: 329–347.

Wahlsten D (1999). Experimental design and statistical inference. In *Handbook of Molecular Genetic Techniques for Brain and Behavioral Research*, WE, Crusio RT Gerlai, Eds. Elsevier, Amsterdam, pp. 40–57.

Wahlsten D (2000). Planning genetic experiments: Power and sample size. In *Neurobehavioral Genetics: Methods and Applications*, B Jones, P Mormede, Eds. CRC Press, Boca Raton, FL, pp. 31–42.

Wahlsten D, Andison M (1991). Patterns of cerebellar foliation in recombinant inbred mice. *Brain Research* **557**: 184–189.

Washlsten D, Bachmanov A, Finn DA, Crabbe JC (2006). Stability of inbred mouse strain differences in behavior and brain size between laboratories and across decades. *Proceedings of the National Academy of Sciences USA* **103**: 16364–16369.

Wahlsten D, Crabbe JC (2006). Behavioral testing. In *The Mouse in Biomedical Research*, in press.

Wahlsten D, Crabbe JC, Dudek BC (2001). Behavioural testing of standard inbred and 5HT1B knockout mice: Implications of absent corpus callosum. *Behavioural Brain Research* **125**: 23–32.

Wahlsten D, Metten P, Crabbe J (2003a) A rating scale for wildness and ease of handling laboratory mice: Results for 21 inbred strains tested in two laboratories. *Genes, Brain and Behavior* **2**: 71–79.

Wahlsten D, Metten P, Crabbe JC (2003b). Survey of 21 inbred mouse strains in two laboratories reveals that BTBR T/+ *tf/tf* has severely reduced hippocampal commissure and absent corpus callosum. *Brain Research* **971**: 47–54.

Wahlsten D, Metten P, Phillips TJ, Boehm SL, Burkhart-Kasch S, Dorow J, Doerksen S, Downing C, Fogarty J, Rodd-Henricks K, Hen R, McKinnon CS, Merrill CM, Nolte C, Schalomon M, Schlumbohn JP, Sibert JR, Wenger CD, Dudek BC, Crabbe JC (2003*c*). Different data from different labs: Lessons from studies of gene-environment interaction. *Journal of Neurobiology* **54**: 283–311.

Wahlsten D, Rustay NR, Metten P, Crabbe JC (2003*d*). In search of a better mouse test. *Trends in Neuroscience* **26**: 132–136.

Wainwright P, Francey C (1987). Effect of pre-and postweaning nutrition on dietary induced obesity in B6D2F2 mice. *Physiology and Behavior* **41**: 87–91.

Wakayama T, Perry AC, Zuccotti M, Johnson KR, Yanagimachi R (1998). Full-term development of mice from enucleated oocytes injected with cumulus cell nuclei. *Nature* **394**: 369–374.

Wall PM, Messier C (2000). Ethological confirmatory factor analysis of anxiety-like behaviour in the murine elevated plus-maze. *Behavioural Brain Research* **114**: 199–212.

Walsh FS, Doherty P (1991). NCAM gene structure and function. *Seminars in Neuroscience* **3**: 271–284.

Wang H, Ng K, Hayes D, Gao X, Forster G, Blaha C, Yeomans J (2004). Decreased amphetamine-induced locomotion and improved latent inhibition in mice mutant foro the m5 muscarinic receptor gene found in the human 15q schizophrenia region. *Neuropsychopharmacology* **29**: 2126–2139.

Wang JH, Short J, Ledent C, Lawrence AJ, van den Buuse M (2003). Reduced startle habituation and prepulse inhibition in mice lacking the adenosine A2A receptor. *Behavioural Brain Research* **143**: 201–207.

Wang MW, Crombie D1, Hayes JS, Heap RB (1995). Aberrant maternal behaviour in mice treated with a progesterone receptor antagonist during pregnancy. *Journal of Endocrinology* **145**: 371–377.

Wang YM, Gainetdinov RR, Fumagalli F, Xu F, Jones SR, Bock CB, Miller GW, Wightman RM, Caron MG (1997). Knockout of the vesicular monoamine transporter 2 gene results in neonatal death and supersensitivity to cocaine and amphetamine. *Neuron* **19**: 1285–1296.

Warren ST, Nelson DL (1993). Trinucleotide repeat expansions in neurological disease. *Current Opinions in Neurobiology* **3**: 752–759.

Wassink TH, Brzustowicz LM, Bartlett CW, Szatmari P (2004). The search for autism disease genes. *Mental Retardation and Developmental Disabilities Research Reviews* **10**: 272–283.

Waterson RH, et al. (2002). Initial sequencing and comparative analysis of the mouse genome. *Nature* **420**: 520–562.

Watkins LR, Maier SF (2005). Immune regulation of central nervous system functions: From sickness responses to pathological pain. *Journal of Internal Medicine* **257**: 139–155.

Watson JD (1990). The human genome project: Past, present, and future. *Science* **248**: 44–49.

Watson SJ, Akil H (1999). Gene chips and arrays revealed: A primer on their power and uses. *Biological Psychiatry* **45**: 533–543.

Weber P, Metzger D, Chambon P (2001). Temporally controlled targeted somatic mutagenesis in the mouse brain. *European Journal of Neuroscience* **14**: 1777–1783.

Weed JL, Boone JL (1992). A MacIntosh computer system for collecting and analyzing rodent sexual behavior. *Physiology and Behavior* **52**: 183–185.

Weerts EM, Miller LG, Hood KE, Miczek KA (1992). Increased $GABA_A$-dependent chloride uptake in mice selectively bred for low aggressive behavior. *Psychopharmacology* **108**: 196–204.

Wehner JM, Keller JJ, Keller AB, Picciotto MR, Paylor R, Booker TK, Beaudet A, Heinemann SF, Balogh SA (2004). Role of neuronal nicotinic receptors in the effects of nicotine and ethanol on contextual fear conditioning. *Neuroscience* **129**: 11–24.

Wehner JM, Radcliffe RA (2004). Cued and contextual fear conditioning in mice. In *Current Protocols in Neuroscience*. Wiley, New York, pp. 8.5C.1–8.5C.14.

Wehner JM, Radcliffe RA, Bowers BJ (2001). Quantitative genetics and mouse behavior. *Annual Review of Neuroscience* **24**: 845–867.

Wehner JM, Radcliffe RA, Rosmann ST, Christensen SC, Rasmussen DL, Fulker DW, Wiles M (1997). Quantitative trait locus analysis of contextual fear conditioning in mice. *Nature Genetics* **17**: 331–334.

Wehner JM, Silva A (1996). Importance of strain differences in evaluations of learning and memory processes in null mutants. *Mental Retardation and Developmental Disabilities Research Reviews* **2**: 243–248.

Wehner JM, Sleight S, Upchurch M (1990). Hippocampal protein kinase C activity is reduced in poor spatial learners. *Brain Research* **523**: 181–187.

Weinberger DR, Berman KF (1996). Prefrontal function in schizophrenia: Confounds and controversies. *Philosophical Transactions of the Royal Society of London. Series B: Biological Sciences* **351**: 1495–1503.

Weinberger DR, Egan MF, Bertolino A, Callicott JH, Mattay VS, Lipska BK, Berman KF, Goldberg TE (2001). Prefrontal neurons and the genetics of schizophrenia. *Biological Psychiatry* **50**: 825–844.

Weiner I (2003). The "two-headed" latent inhibition model of schizophrenia: Modeling positive and negative symptoms and their treatment. *Psychopharmacology* **169**: 257–297.

Weiner I, Shadach E, Tarrasch R, Kidron R, Feldon J (1996). The latent inhibition model of schizophrenia: Further validation using the atypical neuroleptic, clozapine. *Biological Psychiatry* **40**: 834–843.

Weinshenker D, Szot P, Miller NS, Rust NC, Hohmann JG, Pyati U, White SS, Palmiter RD (2001). Genetic comparison of seizure control by norepinephrine and neuropeptide Y. *Journal of Neuroscience* **21**: 7764–7769.

Weiss C, Venkatasubramanian PN, Aguado AS, Power JM, Tom BC, Li L, Chen KS, Disterhoft JF, Wyrwicz AM (2002). Impaired eyeblink conditioning and decreased hippocampal volume in PDAPP V717F mice. *Neurobiology of Disease*. **11**: 425–433.

Weiss F, Markou A, Lorang MT, Koob GF (1992). Basal extracellular dopamine levels in the nucleus accumbens are decreased during cocaine withdrawal after unlimited-access self-administration. *Brain Research* **593**: 314–318.

Weiss JM (1971). Effects of coping behavior in different warning signal conditions on stress pathology in rats. *Journal of Comparative and Physiological Psychology* **77**: 1–13.

Weiss JM, Goodman PA, Losito BG, Corrigan S, Charry JM, Bailey WH (1981). Behavioral depression produced by an uncontrollable stressor: Relationship to norepinephrine, dopamine, and serotonin levels in various regions of rat brain. *Brain Research Reviews* **3**: 167–205.

Weizman R, Weizman R, Paz L, Backer MM, Amiri Z, Modai I, Pick CG (1999). Mouse strains differ in their sensitivity to alprazolam effect in the staircase test. *Brain Research* **839**: 58–65.

Welch BL, Welch AS (1969). Aggression and the biogenic amine neurohumors. In *Aggressive Behaviour*, S Garattini, EB Sigg, Eds. Excerpia Medica, Amsterdam, pp. 188–202.

Welker E, Armstrong-James M, Bronchti G, Ourednik W, Gheorghita-Baechler F, Dubois R, Guernsey DL, Van der Loos H, Neumann PE (1996). Altered sensory processing in the somatosensory cortex of the mouse mutant barrelless. *Science* **271**: 1864–1867.

Wellman PJ, McMahon LR (1998). Basic measures of food intake. In *Current Protocols in Neuroscience*. Wiley, New York, pp. 8.6B.1–8.6B8.

Welner SA, Dunnett SB, Salamone JD, MacLean B, Iversen SD (1988). Transplantation of embryonic ventral forebrain grafts to the neocortex of rats with bilateral lesions of nucleus basalis magnocellularis ameliorates a lesion-induced deficit in spatial memory. *Brain Research* **463**: 192–197.

Wenger GR, Schmidt C, Davisson MT (2004). Operant conditioning in the Ts65Dn mouse: Learning. *Behavior Genetics* **34**: 105–119.

Wenk GL (1997). Learning and Memory. In *Current Protocols in Neuroscience*, JN Crawley, CR Gerfen, R McKay, MW Rogawski, DR Sibley, P Skolnick, Eds. Wiley, New York, pp. 8.5.1–8.5B.11.

Wersinger SR, Ginns EI, O'Carroll AM, Lolair SJ, Young WS (2002). Vasopressin V1b receptor knockout reduces aggressive behavior in male mice. *Molecular Psychiatry* **7**: 975–984.

Wersinger SR, R Kelliher K, Zufall F, Lolait SJ, O'Carroll AM, Young WS 3rd. (2004). Social motivation is reduced in vasopressin 1b receptor null mice despite normal performance in an olfactory discrimination task. *Hormones and Behavior* **46**: 638–645.

Wersinger SR, Rissman EF (2000). Oestrogen receptor alpha is essential for female-directed chemo-investigatory behaviour but is not required for the pheromone-induced luteinizing hormone surge in male mice. *Journal of Neuroendocrinology* **12**: 103–110.

Wersinger SR, Sannen K, Villalba C, Lubahn DB, Rissman EF, De Vries GJ (1997). Masculine sexual behavior in disrupted in male and female mice lacking a functional estrogen receptor α gene. *Hormones and Behavior* **32**: 176–183.

Westerink BHC, Teisman A, de Vries JB (1994). Increase in dopamine release from the nucleus accumbens in response to feeding: A model to study interactions between drugs and naturally activated dopaminergic neurons in the rat brain. *Naunyn-Schmiedeberg's Archives of Pharmacology* **349**: 230–235.

Wexler NS, Rose EA, Housman DE (1991). Molecular approaches to hereditary diseases of the nervous system: Huntington's disease as a paradigm. *Annual Reviews Neuroscience* **14**: 503–529.

Wheeler AH, Porreca F, Cowan A (1990). The rat paw formalin test: Comparison of noxious agents. *Pain* **40**: 229–238.

Whishaw IQ, Kolb B (2004). *The Behavior of the Laboratory Rat: A Handbook with Tests.* Oxford University Press, Oxford.

Whishaw IQ, Oddie SD, McNamara RK, Harris TL, Perry BS (1990). Psychophysical methods for study of sensory-motor behavior using a food-carrying (hoarding) task in rodents. *Journal of Neuroscience Methods* **32**: 123–133.

Whishaw IQ, Tomie JA (1996). Of mice and mazes: Similarities between mice and rats on dry land but not water mazes. *Physiology and Behavior* **60**: 1191–1197.

Whiting PJ (2003). The GABAA receptor gene family: new opportunities for drug development. *Current Opinion in Drug Discovery and Development* **6**: 648–657.

Whitney G, Harder G (1986). Single-locus control of sucrose octaacetate tasting among mice. *Behavior Genetics* **16**: 559–574.

Wickstrom E, Ed. (1998). *Clinical Trials of Genetic Therapy with Antisense DNA and DNA Vectors.* Marcel Dekker, New York.

Wiedermayer CP, Myers MM, Mayford M, Barr GA (2000). Olfactory based spatial learning in neonatal mice and its dependence on CaMKII. *Neuroreport* **11**: 1051–1055.

Wiles M, Keller G (1991). Multiple hematopoietic lineages develop from embryonic stem (ES) cell lines in culture. *Development* **111**: 259–267.

Williams JR, Insel TR, Harbaugh CR, Carter CS (1994). Oxytocin administered centrally facilitates formation of a partner preference in female prairie voles (*Microtus ochrogaster*). *Journal of Neuroendocrinology* **6**: 247–250.

Williams RW, Flaherty L, Threadgill DW (2003). The math of making mutant mice. *Genes, Brain and Behavior* **2**: 191–200.

Williams RW, Strom RC, Goldowitz D (1998). Natural variation in neuron number in mice is linked to a major quantitative trait locus on chr 11. *Journal of Neuroscience* **18**: 138–146.

Willie JT, Chemelli RM, Sinton CM, Tokita S, Williams SC, Kisanuki YY, Marcus JN, Lee C, Elmquist JK, Kohlmeier KA, Leonard CS, Richardson JA, Hammer RE, Yanagisawa M (2003). Distinct narcolepsy syndromes in Orexin receptor-2 and Orexin null mice: Molecular genetic dissection of non-REM and REM sleep regulatory processes. *Neuron* **38**: 715–730.

Willner P (1984). The validity of animal models of depression. Psychopharmacology **83**: 1–16.

Willott JF (2001). *Handbook of Mouse Auditory Research: From Behavior to Molecular Biology.* CRC Press, Boca Raton, FL.

Willott JF, Bross LS (1996). Morphological changes in the anteroventral cochlear nucleus that accompany sensorineural hearing loss in DBA/2J and C57BL/6J mice. *Developmental Brain Research* **91**: 218–226.

Willott JF, Tanner L, O'Steen J, Johnson KR, Bogue MA, Gagnon L (2003). Acoustic startle and prepulse inhibition in 40 inbred strains of mice. *Behavioral Neuroscience* **117**: 716–727.

Willott JF, Turner JG, Carlson S, Ding D, Bross LS, Falls WA (1998). The BALB/c mouse as an animal model for progressive sensorineural hearing loss. *Hearing Research* **115**: 162–174.

Wilson EO (1975). *Sociobiology: The New Synthesis.* Harvard University Press, Cambridge, MA.

Winder DG, Mansuy IM, Osman M, Moallem TM, Kandel ER (1998). Genetic and pharmacological evidence for a novel, intermediate phase of long-term potentiation suppressed by calcineurin. *Cell* **92**: 25–37.

Winshaw-Boris A, Garrett L, Chen A, Barlow C (1999). Embryonic stem cells and gene targeting. In WE Crusio, RT Gerlai, Eds. *Handbook of Molecular-Genetic Techniques for Brain and Behavior Research*, Elsevier Science, Amsterdam, pp. 259–271.

Winslow JT (2003). Mouse social recognition and preference. In *Current Protocols in Neuroscience.* Wiley, New York, pp. **8**.16.1–8.16.16.

Winslow JT, Hearn EF, Ferguson J, Young LJ, Matzuk MM, Insel TR (2000). Infant vocalization, adult aggression, and fear behavior of an oxytocin null mutant mouse. *Hormones and Behavior* **37**: 145–155.

Winslow JT, Insel TR (1990). Serotonergic and catecholaminergic reuptake inhibitors have opposite effects on the ultrasonic isolation calls of rat pups. *Neuropsychopharmacology* **3**: 51–59.

Winslow JT, Insel TR (2002). The social deficits of the oxytocin knockout mouse. *Neuropeptides* **36**: 221–229.

Winslow JT, Insel TR (2004). Neuroendocrine basis of social recognition. *Current Opinions in Neurobiology.* **14**: 248–253.

Winslow JT, Miczek KA (1983). Habituation of aggression in mice: Pharmacological evidence of catecholaminergic and serotonergic mediation. *Psychopharmacology* **81**: 286–291.

Winter C, Eckler JR, Rabin RA (2004). Serotonergic/gluamatergic interactions: The effects of mGlu1/3 receptor ligands in rats trained with LSD and PCP as discriminative stimuli. *Psychopharmacology* **172**: 233–240.

Wise RA (1996a). Addictive drugs and brain stimulation reward. *Annual Review of Neuroscience* **19**: 319–340.

Wise RA (1996b). Neurobiology of addiction. *Current Opinions in Neurobiology* **6**: 243–251.

Wise RA (1989). The brain and reward. In *The Neuropharmacological Basis of Reward*, JM Liebman, SJ Cooper, Eds. Oxford University Press, New York, pp. 377–424.

Wise RA, Bozarth MA (1987). A psychomotor stimulant theory of addiction. *Psychological Review* **94**: 469–492.

Witt DM, Winslow JT, Insel TR (1992). Enhanced social interactions in rats following chronic, centrally infused oxytocin. *Pharmacology Biochemistry and Behavior 1992* **43**: 855–861.

Witte K, Schnecko A, Buijs RM, van der Vliet J, Scalbert E, Delagrange P, Guardiola-Lemaitre B, Lemmer B (1998). Effects of SCN lesions on circadian blood pressure rhythm in normotensive and transgenic hypertensive rats. *Chronobiology International* **15**: 135–145.

Wittung-Stafshede, P (1998). Genetic medicine—When will it come to the drugstore? *Science* **281**: 657–658.

Wong GT (2002). Speed congenics: Applications for transgenic and knock-out mouse strains. *Neuropeptides* **36**: 230–236.

Wong GT, Gannon KS, Margolskee RF (1996). Transduction of bitter and sweet taste by gustducin. *Nature* **381**: 796–800.

Wong ST, Trinh K, Hacker B, Chan GC, Lowe G, Gaggar A, Xia Z, Gold GH, Storm DR (2000). Disruption of the type III adenyl cyclase gene leads to peripheral and behavioral anosmia in transgenic mice. *Neuron* **27**: 487–497.

Wood ER, Agster KM, Eichenbaum H (2004). One-trial odor reward association: A form of event memory not dependent on hippocampal function. *Behavioral Neuroscience* **118**: 526–539.

Wood GK, Tomasiewicz H, Rutishauser U, Magnuson T, Quirion R, Rochford J, Srivastava LK (1998). NCAM-180 knockout mice display increased lateral ventricle size and reduced prepulse inhibition of startle. *NeuroReport* **9**: 461–466.

Woods EJ, Benson JD, Agca Y, Critser JK (2004). Fundamental cryobiology of reproductive cells and tissues. *Cryobiology* **48**: 146–156.

Woods SC (2004). Gastrointestinal satiety signals. I. An overview of gastrointestinal signals that influence food intake. *American Journal of Physiology - Gastrointestinal and Liver Physiology* **286**: G7–G13.

Woods SC, Seeley RJ, Porte D, Schwartz MW (1998). Signals that regulate food intake and energy homeostasis. *Science* **280**: 1378–1383.

Woody CD (1970). Conditioned eye blink: Gross potential activity at coronal-precruciate cortex of the cat. *Journal of Neurophysiology* **33**: 838–850.

Woolf CJ, Mannion RJ, Neumann S (1998). Null mutations lacking substance: Elucidating pain mechanisms by genetic pharmacology. *Neuron* **20**: 1063–1066.

Wortley KE, Anderson KD, Garcia K, Murray JD, Malinova L, Liu R, Moncrieffe M, Thabet K, Cox HJ, Yancopoulos GD, Wiegand SJ, Sleeman MW (2004). Genetic deletion of ghrelin does not decrease food intake but influences metabolic fuel preference. *Proceedings of the National Academy of Sciences USA* **101**: 8227–8232.

Wrenn CC (2004b). Social transmission of food preference in mice. In *Current Protocols in Neuroscience*. Wiley, New York, pp. 8.5G.1–8.5G.6.

Wrenn CC, Harris AP, Saavedra MC, Crawley JN (2003). Social transmission of food preference in mice: methodology and application to galanin-overexpressing transgenic mice. *Behavioral Neuroscience* **117**: 21–31.

Wrenn CC, Kinney JW, Marriott LK, Holmes A, Harris AP, Saavedra MC, Starosta G, Innerfield CE, Jacoby AS, Shine J, Iismaa TP, Wenk GL, Crawley JN (2004a). Learning and memory performance in mice lacking the GAL-R1 subtype of galanin receptor. *European Journal of Neuroscience* **19**: 1384–1396.

Wrenn CC, Marriott LK, Kinney JW, Holmes A, Wenk GL, Crawley JN (2002). Galanin peptide levels in hippocampus and cortex of galanin-overexpressing transgenic mice evaluated for cognitive performance. *Neuropeptides* **36**: 413–426.

Wrenn CC, Turchi JN, Schlosser S, Dreiling JL, Stephenson DA, Crawley JN (2006). Performance of galanin transgenic mice in the 5-choice serial reaction time attentional task. *Pharmacology Biochemistry and Behavior* **83**: 428–440.

Wu ZL, Thomas SA, Villacres EC, Xia Z, Simmons ML, Chavkin C, Palmiter RD, Storm DR (1995). Altered behavior and long-term potentiation in type I adenylyl cyclase mutant mice. *Proceedings of the National Academy of Sciences USA* **92**: 220–224.

Wurbel H (2002). Behavioral phenotyping enhanced–Beyond (environmental) standardization. *Genes, Brain and Behavior* **1**: 3–8.

Xu XJ, Hökfelt T, Bartfai T, Wiesenfeld-Hallin Z (2000). Galanin and spinal nociceptive mechanisms: Recent advances and therapeutic implications. *Neuropeptides* **34**: 137–147.

Xu J, Qiu Y, DeMayo FJ, Tsai SY, O'Malley BW (1998). Partial hormone resistance in mice with disruption of the steroid receptor co-activator-1 (SRC-1). gene. *Science* **279**: 1922–1925.

Yamada M, Mizuno Y, Mochizuki H (2005). Parkin gene therapy for alpha-synucleinopathy: A rat model of Parkinson's disease. *Human Gene Therapy* **16**: 262–270.

Yamada K, Mizuno M, Nabeshima T (2002). Role for brain-derived neurotrophic factor in learning and memory. *Life Sciences* **70**: 735–744.

Yamamoto Y, Ueta Y, Date Y, Nakazato M, Hara Y, Serino R, Nomura M, Shibuya I, Matsukura S, Yamashita H (1999). Down regulation of the prepro-orexin gene expression in genetically obese mice. *Molecular Brain Research* **65**: 14–22.

Yamada K, Wada E, Wada K (2000a). Bombesin-like peptides: Studies on food intake and social behaviour with receptor knock-out mice. *Annals of Medicine* **32**: 519–529.

Yamada K, Wada E, Wada K (2000b). Male mice lacking the gastrin-releasing peptide receptor (GRP-R) display elevated preference for conspecific odors and increased social investigatory behaviors. *Brain Research* **870**: 20–26.

Yamada K, Wada E, Wada K (2001). Female gastrin-releasing peptide receptor (GRP-R) deficient mice exhibit altered social preference for male conspecifics: Implications for GRP/GRP-R modulation of GABAergic function. *Brain Research* **894**: 281–287.

Yamanaka S, Johnson MD, Grinberg A, Westphal H, Crawley JN, Taniike M, Suzuki K, Proia RL (1994). Targeted disruption of the *Hexa* gene results in mice with biochemical and pathological features of Tay-Sachs disease. *Proceedings of the National Academy of Sciences USA* **91**: 9975–9979.

Yan Q, Zhang J, Liu H, Babu-Khan S, Vassar R, Biere AL, Citron M, Landreth G (2003). Anti-inflammatory drug therapy alters β-amyloid processing and deposition in an animal model of Alzheimer's disease. *Journal of Neuroscience* **23**: 7504–7509.

Yang K, Clifton GL, Hayes RL (1997). Gene therapy for central nervous system injury: The use of cationic liposomes. *Journal of Neurotrauma* **14**: 281–297.

Yazulla S, Studholme KM, Pinto LH (1997). Differences in the retinal GABA system among control, *spastic* mutant and *retinal degeneration* mutant mice. *Vision Research* **37**: 3471–3482.

Ye X, Rivera VM, Zoltick P, Cerasoli F, Schnell MA, Gao G, Hughes JV, Gilman M, Wilson JM (1999). Regulated delivery of therapeutic proteins after in vivo somatic cell gene transfer. *Science* **283**: 88–91.

Yekta S, Shih IH, Baratel DP (2004). MicroRNA-directed cleavage of HOXB8 mRNA. *Science* **304**: 594–596.

Yen TT, Steinmetz J, Wolff GL (1970). Lipolysis in genetically obese and diabetes-prone mice. *Hormone and Metabolic Research* **2**: 200–203.

Yeomans JS, Li L, Scott BW, Frankland PW (2002). Tactile, acoustic and vestibular systems sum to elicit the startle reflex. *Neuroscience and Biobehavioral Reviews* **26**: 1–11.

Yerkes RM (1973). The dancing mouse. In *Classics in Psychology*. Arno, New York.

Yilmazer-Hanke DM, Roskoden T, Zilles K, Schwegler H. Anxiety-related behavior and densities of glutamate, GABAA, acetylcholine and serotonin receptors in the amygdala of seven inbred mouse strains (2003). *Behavioural Brain Research* **145**: 145–159.

Yin JCP, Del Vecchio MD, Zhou H, Tully T (1995). CREB as a memory modulator: Induced expression of a dCREB2 activator isoform enhances long-term memory in Drosophila. *Cell* **81**: 107–115.

Yin JCP, Wallach JS, Del Vecchio M, Wilder EL, Zhou H, Quinn WG, Tully T (1994). Induction of a dominant negative CREB transgene specifically blocks long-term memory in *Drosophila*. *Cell* **79**: 49–58.

Yokel RA, Wise RA (1975). Increased lever pressing for amphetamine after pimozide in rats: implications for a dopamine theory of reward. *Science* **187**: 547–549.

Yokoi M, Kobayashi K, Manabe T, Takahashi T, Sakaguchi I, Katsuura G, Shigemoto R, Ohishi H, Nomura S, Nakamura K, Nakao K, Katsuki M, Nakanishi S (1996). Impairment of hippocampal mossy fiber LTD in mice lacking mGluR2. *Science* **273**: 645–649.

Yoo JH, Lee SY, Loh HH, Ho IK, Jang CG (2004). Loss of nicotine-induced behavioral sensitization in micro-opioid receptor knockout mice. *Synapse* **51**: 219–223.

Yoshimoto Y, Lin Q, Collier TJ, Frim DM, Breakfield XO, Bohn MC (1995). Astrocytes retrovirally transduced with BDNF elicit behavioural improvement in a rat model of Parkinson's disease. *Brain Research* **691**: 25–36.

Young JW, Finlayson K, Spratt C, Marston HM, Crawford N, Kelly JS, Sharkey J (2004). Nicotine improves sustained attention in mice: Evidence for involvement of the alpha7 nicotinic acetylcholine receptor. *Neuropsychopharmacology* **29**: 891–900.

Young RC, Gibbs J, Antin J, Holt J, Smith GP (1974). Absence of satiety during sham feeding in the rat. *Journal of Comparative and Physiological Psychology* **87**: 795–800.

Young WS, Shepard E, Amico J, Hennighausen L, Wagner KU, LaMarca ME, McKinney C, Ginns EI (1996). Deficiency in mouse oxytocin prevents milk ejection, but not fertility or parturition. *Journal of Neuroendocrinology* **8**: 847–853.

Youngentob SL, Pyrski MM, Margolis FL (2004). Adenoviral vector-mediated rescue of the OMP-null behavioral phenotype: Enhancement of odorant threshold sensitivity. *Behavioral Neuroscience* **118**: 636–642.

Yuferov V, Nielsen D, Butelman E, Kreek MJ (2005). Microarray studies of psychostimulant-induced changes in gene expression. *Addiction Biology* **10**: 101–118.

Zachariou V, Benoit-Marand M, Allen PB, Ingrassia P, Fienberg AA, Gonon F, Greengard P, Picciotto MR (2002). Reduction of cocaine place preference in mice lacking the protein phosphatase 1 inhibitors DARPP 32 or Inhibitor 1. *Biological Psychiatry* **51**: 612–620.

Zachariou V, Brunzell DH, Hawes J, Stedman DR, Bartfai T, Steiner RA, Wynick D, Langel U, Picciotto MR (2003). The neuropeptide galanin modulates behavioral and neurochemical signs of opiate withdrawal. *Proceedings of the National Academy of Sciences USA* **100**: 9028–9033.

Zahn TP, Kreusi MJ, Rapoport JL (1991). Reaction time indices of attention deficits in boys with disruptive behavior disorders. *Journal of Abnormal Child Psychology* **19**: 233–252.

Zamanillo D, Sprengel R, Hvalby O, Jensen V, Burnashev N, Rozov A, Kaiser KMM, Köster HJ, Borchardt T, Worley P, Lubke J, Frotscher M, Kelly PH, Sommer B, Andersen P, Seeburg PH, Sakmann B (1999). Importance of AMPA receptors for hippocampal synaptic plasticity but not for spatial learning. *Science* **284**: 1805–1811.

Zamore PD (2001). RNA interference: listening to the sound of silence. *Nature Structural Biology* **8**: 746–750.

Zamore PD, Haley B (2005). Ribo-gnome: The big world of small RNAs. *Science* **309**: 1519–1524.

Zatorre RJ, Perry DW, Beckett CA, Westbury CF, Evans AC (1998). Functional anatomy of musical processing in listeners with absolute pitch and relative pitch. *Proceedings of the National Academy of Sciences USA* **95**: 3172–3177.

Zebrowska-Lupina I, Stelmasiak M, Porowska A (1990). Stress-induced depression of basal motility: effects of antidepressant drugs. *Polish Journal of Pharmacology and Pharmaceutics* **42**: 97–104.

Zeineh MM, Engel SA, Thompson PM, Bookheimer SY (2003). Dynamics of the hippocampus during encoding and retrieval of face-name pairs. *Science* **299**: 577–580.

Zeitlin S, Liu JP, Chapman DL, Papaioannou VE, Efstratiadis A (1995). Increased apoptosis and early embryonic lethality in mice nullizygous for the Huntington's disease gene homologue. *Nature Genetics* **11**: 155–163.

Zeringue HC, Constantine-Paton M (2004). Post-transcriptional gene silencing in neurons. *Current Opinion in Neurobiology* **14**: 654–659.

Zetterstrom RH, Solomin L, Jansson L, Hoffer BJ, Olson L, Perlmann T (1997). Dopamine neuron agenesis in Nurr1-deficient mice. *Science* **276**: 248–250.

Zetterstrom T, Sharp T, Ungerstedt U (1983). In vivo measurement of dopamine and its metabolites by intracerebral dialysis: Changes after *d*-amphetamine. *Journal of Neurochemistry* **41**: 1769–1773.

Zhang J, Wu X, Qin C, Qi J, Ma S, Zhang H, Kong Q, Chen D, Ba D, He W (2003). A novel recombinant adeno-associated virus vaccine reduces behavioral impairment and beta-amyloid plaques in a mouse model of Alzheimer's disease. *Neurobiology of Disease*. **14**: 365–379.

Zhang M, Balmadrid C, Kelley AE (2003). Nucleus accumbens opioid, GABAergic, and dopaminergic modulation of palatable food motivation: Contrasting effects revealed by a progressive ratio studied in the rat. *Behavioral Neuroscience* **117**: 202–211.

Zhang W, Narayanan M, Friedlander RM (2003). Additive neuroprotective effects of minocycline with creatine in a mouse model of ALS. *Annals of Neurology* **53**: 267–270.

Zhang Y, Burk JA, Glode BM, Mair RG (1998). Effects of thalamic and olfactory cortical lesions on continuous olfactory delayed nonmatching-to-sample and olfactory discrimination in rats (*Rattus norvegicus*). *Behavioral Neuroscience* **112**: 39–53.

Zhang Y, Mantsch JR, Schlussman SD, Ho A, Kreek MJ (2002). Conditioned place preference after single doses or "binge" cocaine in C57BL/6J and 129/J mice. *Pharmacology Biochemistry and Behavior* **73**: 655–662.

Zhang Y, Proenca R, Maffel M, Barone M, Leopold L, Friedman JM (1994). Positional cloning of the mouse *obese* gene and its human homologue. *Nature* **372**: 425–432.

Zhang Y, Riesterer C, Ayrall AM, Sabilitzky F, Littlewood TD, Reth M (1996). Inducible site-directed recombination in mouse embryonic stem cells. *Nucleic Acids Research* **24**: 543–548.

Zhao GQ, Zhang Y, Hoon MA, Chandrashekar J, Erlenbach I, Ryba NJ, Zucker CS (2003). The receptors for mammalian sweet and umami taste. *Cell* **115**: 255–266.

Zhao Z, Davis M (2004). Fear-potentiated startle in rats is mediated by neurons in the deep layers of the superior colliculus/deep mesencephalic nucleus of the rostral midbrain through the glutamate non-NMDA receptors. *Journal of Neuroscience* **24**: 10326–10334.

Zhou QY, Palmiter RD (1995). Dopamine-deficient mice are severely hypoactive, adipsic, and aphagic. *Cell* **83**: 1197–1209.

Zhuang X, Gross C, Santarelli L, Compan V, Trillat AC, Hen R (1999). Altered emotional states in knockout mice lacking 5-HT1A or 5-HT1B receptors. *Neuropsychopharmacology* **21**(2 Suppl): 52S–60S.

Zimmer A (1996). Gene targeting and behaviour: A genetic problem requires a genetic solution. *Trends in Neurosciences* **19**: 470.

Zimmer A, Zimmer AM, Baffi J, Usdin T, Reynolds K, König M, Palkovits M, Mezey E (1998). Hypoalgesia in mice with a targeted deletion of the tachykinin 1 gene. *Proceedings of the National Academy of Sciences USA* **95**: 2630–2635.

Zimmer A, Zimmer AM, Hohmann AG, Herkenham M, Bonner TL (1999). Increased mortality, hypoactivity, and hypoalgesia in cannabinoid CB1 receptor knockout mice. *Proceedings of the National Academy of Sciences USA* **96**: 5780–5785.

Zippelius HM, Schleidt WM (1956). Ultraschall-aute bei jungen Mausen. *Naturwissenschaften* **43**: 502–503.

Zobeley E, Sufalko DK, Adkins S, Burmeister M (1998). Fine genetic and comparative mapping of the deafness mutation Ames waltzer on mouse chromosome 10. *Genomics* **50**: 260–266.

Zoghbi HY (2003). Postnatal neurodevelopmental disorders: Meeting at the synapse? *Science* **302**: 826–830.

Zorner B, Wolfer DP, Brandis D, Kretz O, Zacher C, Madani R, Grunwald I, Lipp HP, Klein R, Henn FA, Gass P (2003). Forebrain-specific trkB-receptor knockout mice: Behaviorally more hyperactive than "depressive." *Biological Psychiatry* **54**: 972–982.

Zorrilla EP, Taché Y, Koob GF (2003). Nibbling at CRF receptor control of feeding and gastrocolonic motility. *Trends in Pharmacological Sciences* **24**: 421–427.

Index

References followed by t indicate material in tables.